Management of Forest Pests and Diseases

Management of Forest Pests and Diseases

Guest Editors

Young-Seuk Park
Won Il Choi

MDPI

Basel • Beijing • Wuhan • Barcelona • Belgrade • Novi Sad • Cluj • Manchester

Guest Editors
Young-Seuk Park
Department of Biology
Kyung Hee University
Seoul
Republic of Korea

Won Il Choi
Division of Forest
Entomology and Pathology
National Institute of Forest Science
Seoul
Republic of Korea

Editorial Office
MDPI AG
Grosspeteranlage 5
4052 Basel, Switzerland

This is a reprint of the Special Issue, published open access by the journal *Forests* (ISSN 1999-4907), freely accessible at: https://www.mdpi.com/journal/forests/special_issues/forest_pests.

For citation purposes, cite each article independently as indicated on the article page online and as indicated below:

Lastname, A.A.; Lastname, B.B. Article Title. *Journal Name* **Year**, *Volume Number*, Page Range.

ISBN 978-3-7258-6023-4 (Hbk)
ISBN 978-3-7258-6024-1 (PDF)
https://doi.org/10.3390/books978-3-7258-6024-1

Contents

About the Editors

Young-Seuk Park

Young-Seuk Park is a professor at the Department of Biology, Kyung Hee University, Seoul, Republic of Korea. His laboratory studies the effects of environmental changes on biological systems at different hierarchical levels from molecules, individuals, populations, and communities through ecological modelling and ecological informatics approaches. In particular, his research is focused on the effects of global changes and alien species on ecosystems, and ecological monitoring and assessment for sustainable ecosystem management. He is interested in computational approaches such as machine learning techniques and advanced statistical methods. He served as the president of Korean Society for Mathematical Biology. He is an Editor-in-Chief of the Forest Health Section in Forests, and an associate editor of *Ecological Modelling and International Journal of Limnology*.

Won Il Choi

Won Il Choi is a senior researcher in the field of Forest Entomology and Pathology at the National Institute of Forest Science in Republic of Korea. He has been conducting research on forest insect pests since 2006. His primary research interests include the population dynamics of forest insect pests, invasive forest insects, and the impacts of climate change on forest pest outbreaks. He carries out these studies using ecological modeling and statistical analysis methods.

Preface

Forests are complex and dynamic ecosystems, sustained by intricate interactions among countless organisms. Among these, insects and microorganisms are indispensable components, functioning as critical regulators of nutrient cycling and energy flow. Despite their essential ecological roles, many of these organisms also pose significant challenges—acting as pests and pathogens that threaten the health and stability of forest ecosystems.

The nature of these organisms presents a paradox: while they contribute to ecological balance, their proliferation under certain conditions can disrupt ecosystem services, reduce biodiversity, and incur substantial economic losses in the forestry sector. In recent years, the prevalence and severity of forest pests and diseases have been exacerbated by two converging global forces: the introduction of alien species and the accelerating impacts of climate change.

Globalization, through increased trade and travel, has enabled the movement of species beyond their native ranges, often resulting in biological invasions that forests are ill-equipped to withstand. At the same time, rising global temperatures and shifting climate patterns have altered habitat conditions, expanding the range and reproductive capacity of many pest and pathogen species. These environmental changes have been closely linked to recent outbreaks, underscoring the urgent need for proactive and adaptive management approaches.

In this context, surveillance and monitoring of forest pests and diseases have become critical tools in sustainable forest management. Systematic data collection not only allows for early detection and rapid response but also supports the development of targeted control strategies and long-term resilience planning. Furthermore, continuous monitoring contributes to scientific understanding, enabling the evolution of innovative methods for impact assessment and ecological intervention.

This Special Issue brings together a diverse collection of studies that reflect the growing need for interdisciplinary approaches to forest health management. By addressing the complex interactions between pests, pathogens, climate, and human activity, the research presented here aims to inform policy, guide management practices, and enhance our understanding of forest ecosystem dynamics.

We hope that this compilation will serve as a valuable resource for researchers, practitioners, and policymakers alike, fostering collaboration and innovation in the pursuit of healthier, more resilient forests.

Young-Seuk Park and Won Il Choi
Guest Editors

forests

MDPI

Editorial

Management of Forest Pests and Diseases

Won Il Choi [1] and Young-Seuk Park [2,*]

[1] Division of Forest Ecology, National Forest Research Institute, Dongdaemun, Seoul 02445, Korea
[2] Department of Biology, Kyung Hee University, Dongdaemun, Seoul 02447, Korea
* Correspondence: parkys@khu.ac.kr; Tel.: +82-2-961-0946

Abstract: The occurrence patterns of forest insect pests and diseases have been altered by global events such as climate change. Recent developments in improved monitoring methods and tools for data analyses provide new opportunities to understand the causes and consequences of such changes. Using a variety of management tools, forest pest management programs can mitigate the influence of global changes on forest health. The goal of this Special Issue is to improve our understanding of the root causes of changes that have induced global changes. Fifteen papers are included in this Special Issue, covering several issues in forest pest management. One paper reviews the causes of Korean oak wilt, and another paper discusses fourteen invasive tree pests in Russia. The remaining thirteen papers cover issues related to the monitoring and management of forest pests. These studies provide a better understanding of the causes of change in the patterns of forest pests under the influence of global changes. These reviews also contribute to the development of forest-pest-management strategies to mitigate such impacts on forests due to global changes.

Keywords: climate change; forest pests; global change; invasive species; management; remote sensing

1. Introduction

Changes in the distribution of forest pests, increases in the numbers of and damage from invasive species, and outbreaks of new species have been observed in many countries, apparently due to climate change and the increase in international trade [1]. Responses of forest pests to global changes are complex, and their prediction is difficult [2]. Recent developments in remote sensing technology offer the ability to rapidly monitor large areas damaged by forest pests [3,4]. Additionally, recent developments in data analysis using machine learning offer tools better suited to understanding the causes and impacts of pests at forest scales. Various management options, including biological control, offer a chance to develop a management strategy to mitigate the influence of global changes on forest health.

The applicability of remote sensing to assess damage from forest pests has been reviewed recently [3,4]. Images acquired from satellites or unmanned aerial vehicles (UAVs) are potentially available to assess forest damage from forest insect pest and disease. Moreover, changes in forest damage at the spatial and temporal scales can be monitored quickly and with precision. Hall, Castilla, White, Cooke and Skakun [3] used this technology to evaluate forest damage from several forest pests, including gypsy moth (*Lymantria dispar* [L.]), mountain pine beetle (*Dendroctonus ponderosae* Hopkins), and forest tent caterpillar (*Malacosoma disstria* Hübner). Detecting the damage caused by forest pests using UAV has also been widely attempted [5–7].

Newly developed data analysis techniques also offer fresh insights into the influence of global changes on the occurrence patterns of forest pests. Through a better understanding of the causes and patterns of forest pests, more effective management strategies can be constructed. For example, the potential distribution of pine wilt disease (PWD), caused by pine wood nematode (*Bursaphelenchus xylophilus* (Steiner and Buhrer) Nickle), in Korea was predicted though machine learning methods using local first occurrence data [8].

The authors found that new occurrences of PWD were associated with human activities and vehicle movement. This finding was supported by independent studies using a path-finding algorithm and dispersal models in Korea [8,9]. These results showed that movement of PWD through human activities was a major factor in the dispersal of PWD in Korea, and the analysis provided basic information useful in building a management strategy for PWD in Korea.

A variety of forest pest management options have been developed to the level of practical use, following laboratory and field evaluations. The aim of this Special Issue "Management of Forest Pests and Diseases" is to discuss proposed or current management strategies for the control of forest pests whose effects are worsening due to global changes. The monitoring and analysis of pest occurrences form the basis for evaluation of new pest-control methods. Based on the understanding of the causes of new pests, pest outbreaks, or pest distributions and the role of global changes in each case, the authors of the papers in this issue discuss potential mitigation and prevention tools.

2. Papers in This Issue

Fifteen papers are included in this issue, focusing on the monitoring and management of forest pests: Choi, et al. [10] reviews Korean Oak wilt (KOW) (a new syndrome of oak decline in Korea) and describes its cause, distribution, history, disease cycle, and management. An increase in the number of stressed trees among the potential hosts of KOW has been the major cause of KOW in Korea due to an increase in the average age of trees in oak stands, as well as some effect of climate change. Musolin, et al. [11] reports 14 species of invasive tree pests in Russian forest and urban ecosystems, including *Leptoglossus occidentalis*, *Agrilus fleischeri*, and *Ips amitinus*. All of these species (except for *Ips amitinus* and *Acrocercops brongniardella*) were associated with human activity and are expected to expand geographically in Russia.

Six papers discuss the management of forest pests, while seven focus on monitoring. Interestingly, four papers are on pine wilt disease caused by pine wood nematode, showing that this nematode is a severe threat to forest health globally. The six papers on forest pest management discuss diverse approaches, including chemical and biological control, mass trapping, and overall management strategy. Sun, et al. [12] extracted two chemicals with nematocidal activities (cyclo-[Pro-Phe] and 2-Coumaranone) from a bacterium, *Lysinimonas* sp., collected in the rhizosphere of *Pinus thunbergii*. The nematocidal activity of these two compounds was confirmed through in vivo experiments using pine seedlings. Kim, et al. [13] found a parasitoid, *Spathius verustus* Chao, parasitizing *Monochamus* spp., vectors of pine wood nematode in Korea. The wasp is a gregarious ectoparasitoid that prefers *Monochamus alternatus* to *Monochamus saltuarius*. The possible use of this wasp as a biological control agent against vectors of PWD is discussed. Bălăcenoiu, et al. [14] discuss the use of pesticides to control populations of *Corythucha arcuata*, an invasive species in Romania. The contact pesticide alpha-cypermethrin and a systemic pesticide, acetamiprid, were applied to oak forests by helicopter. Chemical control reduced the nymphal population by up to 96%, but the application of the systemic pesticide was more effective in reducing the pest population. Fora and Balog [15] present information on the population density of two bark beetles (*Ips typographus* and *Pityogenes chalcographus*) in spruce (*Picea* spp.) in forests of the Apuseni Natural Park in Romania. They found that the ranges of both species were expanding and that their densities were high regardless of the level of forest management practiced. They attributed this bark beetle problem to climate change and concluded that new approaches, such as biological control using natural enemies, were needed. Olivieri, et al. [16] present results on the control efficiency of two *Bacillus thuringiensis kurstaki* (Btk) strains for the control of gypsy moth larvae in Mediterranean cork oak forests. Aerial applications using helicopters reduced the density of gypsy moth larvae by approximately 70%, showing the possibility of the use of Btk as a biological control agent for this pest and forest. Resnerová, et al. [17] compare the efficiency of three trapping methods for the bark beetle *Ips cembrae* in European larch forests. Trap trees, pheromone traps, and insecticide-baited traps were used for experiments, and the trap tree method proved to be the most efficient method in terms of the number of bark beetles captured. Therefore, although all traps tested were useful for the monitoring of this bark beetle, trap trees were most effective in reducing the density of this bark beetle.

Finally, seven other papers focus on monitoring forest pests. Davydenko, et al. [18] studied emerald ash borer (EAB), *Agrilus planipennis*, as an invasive species in Ukraine. This species had the potential to kill European ash, *Fraxinus excelsior*. The ash tree is also threatened by an invasive pathogen *Hymenoscyphus fraxineus*. Ash trees infected by this fungus are more vulnerable to infestation by EAB. Zhang, et al. [19] identified an olfactory gene from *Monochamus saltuarius*, a vector of pine wood nematode in China, and found that this gene in the male beetle was down-regulated when the beetle was infected with pine wood nematodes. Male beetles infected by pine wood nematode became insensitive to olfactory stimuli from female beetles or host plants. The resulting increased period of maturation feeding before mating provided the nematode with a longer time to transfer into pine trees. Gathercole et al. [20] studied the leaf microbiome of oaks in relation to acute oak decline (AOD) in *Quercus robur* and *Q. petraea* in Britain. Four bacteria including *Brenneria goodwinii* were collected from a lesion of AOD. Whole genomic DNA analysis collected from 421 trees nationwide showed that these bacteria were found regardless of AOD symptom, suggesting these four bacteria are part of the normal oak microbiome rather than causes of AOD. Gray et al. [21] used a Bayesian hierarchical model to predict the occurrence of eastern spruce dwarf mistletoe (ESDM), *Arceuthobium pusillum* and assessed the predictive performance of the model. The model's overall accuracy was only 52%. The model is useful for understanding the occurrence pattern of ESDM and aiding the monitoring and management of ESDM. de la Fuente and Saura [22] modeled the expansion of the distribution of pine wood nematode in the Iberian Peninsula using a process-based network model. Climate was not the most important factor for the dispersal of PWN because the nematode can disperse through areas that are already climatically suitable. Differences in the host-species-specific susceptibility to PWN were decisive for the expansion of PWN, showing the need for host-species-specific management strategies. Choi, et al. [23] quantified areas of gypsy moth defoliation in 2020 using Landsat satellite images in Wonju, Korea. About 13.0% of the forested area in Wonju was defoliated by gypsy moth in 2020. Moreover, in Wonju, the gypsy moth preferentially occurred in *Larix kaempferi* forests as compared to oak and other deciduous trees. Koreň, et al. [24] used machine learning algorithms (MLA) to model the spatial and temporal infestation patterns of the bark beetle *Ips typographus* in the Czech Republic. Among MLAs, the extra tree-classifier method was the most useful in assessing the spatial infestation of the beetle. This model is expected to be useful for designing new forest management strategies to reduce damage by this bark beetle.

3. Conclusions

New pest invasions or changes in the distributions of forest pests have been observed in many parts of the world. Impacts of invasive species on forest ecosystems are of major concern globally. PWD is an invasive species in Asia and Europe that has been a target of several research programs, focused variously on monitoring, modeling, and management. These studies suggest that PWD is one of the most severe threats to forest health globally. To mitigate the impacts of global changes on forest health, suitable forest pest management strategies need to be developed based on scientific monitoring and analysis. We hope that this Special Issue will lead to an improved understanding of the causes of changes in the occurrence patterns of forest pests and new ways to mitigate their damage.

Author Contributions: Conceptualization, W.I.C. and Y.-S.P.; writing—original draft preparation, W.I.C. and Y.-S.P.; writing—review and editing, W.I.C. and Y.-S.P. All authors have read and agreed to the published version of the manuscript.

Funding: This study was supported by the National Institute of Forest Science (FE-0100-2018-11-2022), Republic of Korea.

Acknowledgments: We would like to thank all contributors in this Special Issue and all reviewers who provided very constructive and helpful comments to evaluate and improve the manuscripts.

Conflicts of Interest: The authors declare no conflict of interest.

References

1. Ramsfield, T.D.; Bentz, B.J.; Faccoli, M.; Jactel, H.; Brockerhoff, E.G. Forest health in a changing world: Effects of globalization and climate change on forest insect and pathogen impacts. *Forestry* **2016**, *89*, 245–252. [CrossRef]
2. Jactel, H.; Koricheva, J.; Castagneyrol, B. Responses of forest insect pests to climate change: Not so simple. *Curr. Opin. Insect Sci.* **2019**, *35*, 103–108. [CrossRef] [PubMed]
3. Hall, R.J.; Castilla, G.; White, J.C.; Cooke, B.J.; Skakun, R.S. Remote sensing of forest pest damage: A review and lessons learned from a Canadian perspective. *Can. Entomol.* **2016**, *148* (Suppl. S1), S296–S356. [CrossRef]
4. Duarte, A.; Borralho, N.; Cabral, P.; Caetano, M. Recent advances in forest insect pests and diseases monitoring using UAV-based data: A systematic review. *Forests* **2022**, *13*, 911. [CrossRef]
5. Yu, R.; Luo, Y.; Zhou, Q.; Zhang, X.; Wu, D.; Ren, L. Early detection of pine wilt disease using deep learning algorithms and UAV-based multispectral imagery. *For. Ecol. Manag.* **2021**, *497*, 119493. [CrossRef]
6. Xia, L.; Zhang, R.; Chen, L.; Li, L.; Yi, T.; Wen, Y.; Ding, C.; Xie, C. Evaluation of deep learning segmentation models for detection of pine wilt disease in unmanned aerial vehicle images. *Remote Sens.* **2021**, *13*, 3594. [CrossRef]
7. Ecke, S.; Dempewolf, J.; Frey, J.; Schwaller, A.; Endres, E.; Klemmt, H.J.; Tiede, D.; Seifert, T. UAV-based forest health monitoring: A systematic review. *Remote Sens.* **2022**, *14*, 3205. [CrossRef]
8. Lee, D.-S.; Choi, W.I.; Nam, Y.; Park, Y.-S. Predicting potential occurrence of pine wilt disease based on environmental factors in South Korea using machine learning algorithms. *Ecol. Inform.* **2021**, *64*, 101378. [CrossRef]
9. Takahashi, D.; Park, Y.-S. Spatial heterogeneities of human-mediated dispersal vectors accelerate the range expansion of invaders with source–destination-mediated dispersal. *Sci. Rep.* **2020**, *10*, 21410. [CrossRef]
10. Choi, W.I.; Lee, D.-H.; Jung, J.B.; Park, Y.-S. Oak decline syndrome in Korean forests: History, biology, and prospects for Korean oak wilt. *Forests* **2022**, *13*, 964. [CrossRef]
11. Musolin, D.L.; Kirichenko, N.I.; Karpun, N.N.; Aksenenko, E.V.; Golub, V.B.; Kerchev, I.A.; Mandelshtam, M.Y.; Vasaitis, R.; Volkovitsh, M.G.; Zhuravleva, E.N.; et al. Invasive insect pests of forests and urban trees in Russia: Origin, pathways, damage, and management. *Forests* **2022**, *13*, 521. [CrossRef]
12. Sun, Y.; Wang, C.; Du, G.; Deng, W.; Yang, H.; Li, R.; Xu, Q.; Guo, Q. Two nematicidal compounds from *Lysinimonas* M4 against the pine wood nematode, *Bursaphelenchus xylophilus*. *Forests* **2022**, *13*, 1191. [CrossRef]
13. Kim, M.-S.; Jung, J.-K.; Hong, K.-J.; Kim, C.-J.; Lee, B.-W.; Kim, I.-K. Discovery and biology of *Spathius verustus* Chao (Hymenoptera: Braconidae), a potential biological agent on two *Monochamus* vectors of the pinewood nematode. *Forests* **2022**, *13*, 955. [CrossRef]
14. Bălăcenoiu, F.; Nețoiu, C.; Tomescu, R.; Simon, D.C.; Buzatu, A.; Toma, D.; Petrițan, I.C. Chemical control of *Corythucha arcuata* (Say, 1832), an invasive alien species, in oak forests. *Forests* **2021**, *12*, 770. [CrossRef]
15. Fora, C.G.; Balog, A. The effects of the management strategies on spruce bark beetles populations (*Ips typographus* and *Pityogenes chalcographus*), in Apuseni Natural Park, Romania. *Forests* **2021**, *12*, 760. [CrossRef]
16. Olivieri, M.; Mannu, R.; Ruiu, L.; Ruiu, P.A.; Lentini, A. Comparative efficacy trials with two different *Bacillus thuringiensis* Serovar *kurstaki* strains against gypsy moth in Mediterranean cork oak forests. *Forests* **2021**, *12*, 602. [CrossRef]
17. Resnerová, K.; Holuša, J.; Surový, P.; Trombik, J.; Kula, E. Comparison of *Ips cembrae* (Coleoptera: Curculionidae) capture methods: Small trap trees caught the most beetles. *Forests* **2020**, *11*, 1275. [CrossRef]
18. Davydenko, K.; Skrylnyk, Y.; Borysenko, O.; Menkis, A.; Vysotska, N.; Meshkova, V.; Olson, Å.; Elfstrand, M.; Vasaitis, R. Invasion of emerald ash borer *Agrilus planipennis* and ash dieback pathogen *Hymenoscyphus fraxineus* in Ukraine-A concerted action. *Forests* **2022**, *13*, 789. [CrossRef]
19. Zhang, S.; Wang, X.; Zhang, Y.; Zheng, Y.; Fan, Z.; Zhang, R. Identification of olfactory genes in *Monochamus saltuarius* and effects of *Bursaphelenchus xylophilus* infestation on their expression. *Forests* **2022**, *13*, 258. [CrossRef]
20. Gathercole, L.A.P.; Nocchi, G.; Brown, N.; Coker, T.L.R.; Plumb, W.J.; Stocks, J.J.; Nichols, R.A.; Denman, S.; Buggs, R.J.A. Evidence for the widespread occurrence of bacteria implicated in acute oak decline from incidental genetic sampling. *Forests* **2021**, *12*, 1683. [CrossRef]
21. Gray, E.R.; Russell, M.B.; Windmuller-Campione, M.A. The difficulty of predicting eastern spruce dwarf mistletoe in lowland black spruce: Model benchmarking in Northern Minnesota, USA. *Forests* **2021**, *12*, 843. [CrossRef]
22. de la Fuente, B.; Saura, S. Long-term projections of the natural expansion of the pine wood nematode in the Iberian Peninsula. *Forests* **2021**, *12*, 849. [CrossRef]
23. Choi, W.-I.; Kim, E.-S.; Yun, S.-J.; Lim, J.-H.; Kim, Y.-E. Quantification of one-year gypsy moth defoliation extent in Wonju, Korea, using landsat satellite images. *Forests* **2021**, *12*, 545. [CrossRef]
24. Koreň, M.; Jakuš, R.; Zápotocký, M.; Barka, I.; Holuša, J.; Ďuračiová, R.; Blaženec, M. Assessment of machine learning algorithms for modeling the spatial distribution of bark beetle infestation. *Forests* **2021**, *12*, 395. [CrossRef]

forests

MDPI

Review

Oak Decline Syndrome in Korean Forests: History, Biology, and Prospects for Korean Oak Wilt

Won Il Choi [1], Dong-Hyeon Lee [2], Jong Bin Jung [1] and Young-Seuk Park [3,*]

1 Forest Ecology Division, National Institute of Forest Science, Dongdaemun, Seoul 02445, Korea; wchoi71@korea.kr (W.I.C.); jbjung1373@korea.kr (J.B.J.)
2 Forest Entomology and Pathology Division, National Institute of Forest Science, Dongdaemun, Seoul 02445, Korea; leedh2009@korea.kr
3 Department of Biology, Kyung Hee University, Dongdaemun, Seoul 02447, Korea
* Correspondence: parkys@khu.ac.kr; Tel.: +82-2-961-0946

Abstract: Oak decline syndrome has been observed in South Korea and Japan and variously referred to as Korean oak wilt (KOW) and Japanese oak wilt (JOW). We reviewed aspects of the historical occurrence of KOW, disease cycle, and its potential causes. KOW has been seen principally in Mongolian oak (*Quercus mongolica* Fisch. ex Ledeb.). The first occurrence of KOW was in 2004 in Seongnam, South Korea. KOW is associated with the fungus *Raffaelea quercus-mongolicae*, which is vectored by the ambrosia beetle *Platypus koryoensis*. In addition, it has been suggested that yeasts have evolved a symbiotic relationship with the vector without antagonism for the pathogenic fungus. The number of trees in Korea killed by KOW increased to about 331,000 in 2011, but then decreased to about 157,000 in 2019. We hypothesized that trees infected by *R. quercus-mongolicae* accumulate phenolic compounds in the sapwood and the vector (*P. koryoensis*) beetles then avoid these trees. Therefore, the number of hosts available for the beetle after a KOW outbreak decrease due to accumulation of phenolic compounds in the wood, although most oak trees survive outbreaks. Therefore, *P. koryoensis* beetles move longer distances after outbreaks to find susceptible host trees. Novel occurrence of KOW in Korea may be due to either (1) an increase in tree age (size) in Korean oak stands, for which the beetle has a known preference or (2) climate change, which may be increasing the number of weakened or stressed trees, for which *P. koryoensis* also shows a preference.

Keywords: ambrosia beetle; climate change; long distance dispersal; mutualistic interaction; *Platypus*; *Raffaelea*

1. Introduction

Oak decline is global phenomenon and one of the severe concerns in forest ecosystem health [1,2]. Among 450 species in *Quercus* genus in the world [3], oak declines were reported for several species such as *Quercus rubra* L., *Quercus velutina* Lam., *Quercus coccinea* Muenchh. And *Quercus alba* L. in USA; *Q. rubra* and *Quercus petraea* (Matt.) Liebl. In Great Britain; *Quercus ilex* L. and *Quercus suber* L. in Spain [1]. Oak decline is related to stresses from various abiotic and biotic factors, alone or in combination [4] that facilitate attack by secondary pests and micro-organisms. Abiotic stressors such as drought, frost, excess moisture, and silvicultural practices can induce attack of secondary pests [2]. Drought is considered particularly important, and various agents such as wood-borers and symbiotic micro-organisms have attacked stressed oak trees in Europe and North America [1].

Declines of cork oak (*Q. suber*) and holm oak (*Q. ilex*) in the western Mediterranean region have been related to attacks of plant pathogenic fungi associated with the buprestid beetle *Coraebus florentinus* (Herbst) [5]. The decline of the oaks is also especially affected by *Phytophthora cinnamomi* Rands [6]. In Iran, acute oak decline of *Quercus castaneifolia* C.A. Mey was associated with presence of the bacteria *Brenneria* spp. and *Rahnella victoriana*

Brady et al. [7]. Among the biotic factors damaging forests, insect pests and diseases are expected to increase as they invade new areas or move to adapt to climate change [1].

The risk of outbreaks of ambrosia beetles and bark beetles is expected to increase as trees are weakened or stressed by climate change [8]. In the case of ambrosia beetles, their mutualistic interaction with fungi and their eusocial brood care increases their ecological fitness under the influence of environmental change [9,10]. In fact, outbreak and biological invasions of ambrosia beetles and their damage to forests are extensively reported [8,10,11]. Korea and Japan have both experienced similar oak decline problems: Korean oak wilt (KOW) [11] and Japanese oak wilt (JOW) [12]. Both diseases occurred in oak species, mainly *Q. mongolica* and *Quercus crispula* Blume for KOW and JOW, respectively [11,12]. However, the fungus and its insect vector are different for KOW and JOW. KOW was induced by *Raffaelea quercus-mongolicae* K.H. Kim, Y.J. Choi and H.D. Shin, sp. nov. vectored by *Platypus koryoensis* (Murayama), whereas JOW was caused by *Raffaelea quercivorus* Kubono et Shin. Ito vectored by *Platypus quercivorus* (Murayama) [13,14].

During the past five decades, the main forest pests in Korea have been species attacking conifers because of such trees' high dominance in Korea [15]. After the successful afforestation efforts made in the 1970s, the dominancy of coniferous species in Korea decreased, while that of oaks increased [15]. With increased oak dominance, the risk of oak tree mortality due to forest insect pests and diseases logically also increased [16]. Therefore, there is a need to better understand the current status of KOW and determine how to manage oak decline effectively.

Here, we reviewed aspects of the historical occurrence of KOW, disease cycle, and its potential causes. Based on the literature, we considered three possible explanations for oak decline in Korea: (1) the roles of pathogen in tree mortality, (2) an increase in size of oaks, and (3) climate change [11]. Finally, the future of KOW was discussed.

2. History of Korean Oak Wilt (KOW)

The first known occurrence of KOW was observed at Ibaejae in Seongnam, South Korea in August, 2004 (Figure 1a), infecting mainly dead Mongolian oak (*Q. mongolica*) and a few *Q. aliena* Blume and *Q. serrata* L., showing exit holes of wood-boring insects [13]. The main wood borer associated with KOW is the ambrosia beetle *P. koryoensis* [17]. Simultaneously, a fungus collected from the dead or infected trees and its associated ambrosia beetle were identified as a new fungal species, *R. quercus-mongolicae* [13]. Based on 18s rDNA and morphological analysis, this species is phylogenetically closest to *R. quercivora*, the causal agent of JOW, than to other *Raffaelea* species [13].

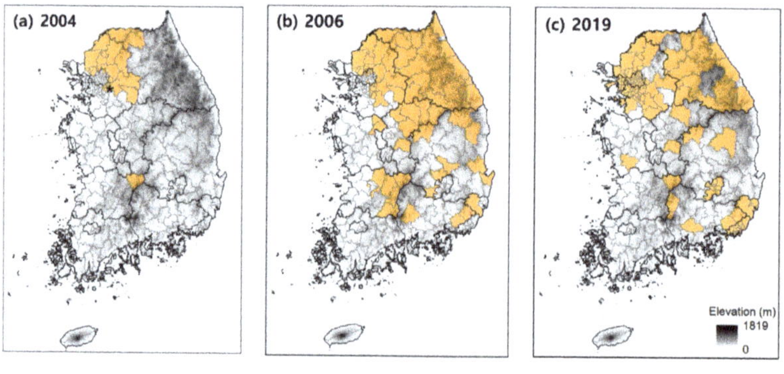

Figure 1. The occurrence areas (yellow color) of KOW in 2004 (**a**), 2006 (**b**) and 2019 (**c**) in South Korea. The asterisk indicates the first occurrence area (i.e., Ibaejae in Seongnam, South Korea) of KOW reported in 2004. The gray color indicates elevation. The graphs were drawn based on data from Annual Report of Monitoring for Forest Insect Pests and Diseases in Korea [18].

In a nationwide survey in 2004, KOW was detected in 18 cities or counties, mainly in the vicinity of Seongnam (Figure 1a). By 2006, KOW had been recorded in 61 cities or counties in the middle of the Korean peninsula (Figure 1a), and the occurrence area expanded slightly in 2019 compared to in 2006 (Figure 1c). The vector beetle of KOW was firstly reported in Korea in 1930 [19], and distribute throughout the country including Jeju island [20]. Therefore, the vector beetle of KOW is considered a native species in Korea. The occurrence pattern showing high occurrence in the northern part of South Korea and the limited occurrence in the southern part would be related with climate variables such as mean daily temperature difference and annual precipitation, elevation and distribution of host plants [21]. The occurrence area of KOW was 1944 ha in 2006 and 4087 ha in 2008. After several fluctuations through to 2013, the area declined to 1576 ha by 2019 (Figure 2a) (http://www.forest.go.kr (accessed on 1 March 2022)). A similar pattern occurred in numbers of trees killed by KOW (Figure 2b).

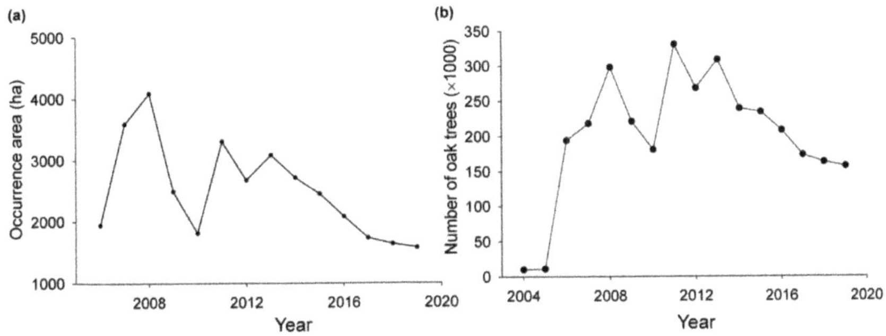

Figure 2. Annual changes in area (ha) damaged (**a**) and the number of dead oak trees (**b**) caused by KOW from 2004 to 2019. The graphs were drawn based on data from Korean Forest Service (www.forest.go.kr (accessed on 1 March 2022)).

Until 2014, the areas with the highest levels of KOW were in and around Seoul [22], although *P. koryoensis* is distributed throughout South Korea, including Jeju island [20]. Monitoring in the vicinity of Seoul from 2011 to 2014 showed that the main KOW areas shifted southward [23]. In 2011, an outbreak of KOW occurred in the Bukhan Mountain National Park in Seoul, and in 2012, 1,585,937 trees spread over 7943.4 ha were infected [24]. In this area, 58.5% of all oak trees were affected by KOW and the mortality of trees infected by KOW was 4.1%. After the mass outbreak of KOW in Bukhan Mountain National Park, the number of infected trees there decreased until 2014 and outbreak of KOW was not observed until 2020 [22].

The distribution of KOW at the regional level was analyzed using satellite images from 2009 to 2014 [22]. Damaged areas were classified based on the density of dead oak trees for the three mountains in Seoul or its vicinity (Bukhan, Cheonggye and Suri mountains). This mapping showed that the highest concentration of KOW was located in the broadleaf or mixed-stand forests with elevations of 200~400 m and slopes of 20~40 degrees, and that a spatial shift of the area with the highest KOW levels occurred from Cheonggye mountain to Suri mountain located 12.5 km to the southwest.

3. Disease Cycle of KOW

3.1. Host and Symptom

The typical symptom of KOW is systematic wilting of oak trees, which occurs when the number of entry holes by *P. koryoensis* is high (Figure 3) [11]. Attack density by the ambrosia beetle vector is the principal factor determining the level of oak mortality caused by KOW [11]. The beetle preferentially attacks the lower trunk, usually on the downslope side because the higher moisture content and thicker phloem on this face are beneficial to the survival of the vector's progeny in the lower trunk, even though more energy is required to penetrate thicker bark in this area [25]. The reason for this is thought to be the less dense tissue in that portion of the tree [11]. When the density of beetles increases, attack also occurs on the higher trunk, possibly to avoid intense intra-specific competition. The presence of *P. koryoensis* entrance holes above 1.5 m indicates high beetle density [11,25]. The spatial distribution of entrance holes was changed from random or contagious to random to reduce intra-specific competition [25].

Figure 3. Mongolian oak (*Q. mongolica*) attacked by *P. koryoensis* (**left**) and wilted by KOW (**right**) (Photos from Park, J.H.).

After attack by *P. koryoensis*, oak trees are also invaded by *R. quercus-mongolicae*, which affects the water-conducting tissues [13]. However, there are no studies on the mechanism by which the fungus causes wilting. In the case of *R. quercivorus*, the causal agent of JOW, cavitation, discoloration, desiccation, and dysfunction of xylem vessels leads to blockage of the upward movement of water after invasion by the fungus. Extensive development of these symptoms in the vessels is considered the cause of severe wilting of the foliage and the subsequent tree death [26]. The presence of extensive discoloration in the wood of trees affected by KOW suggests that trees heavily affected by KOW are killed by a similar mechanism to that of JOW.

The disease cycle of KOW is closely related to the life cycle of *P. koryoensis.* Male beetles make entry holes in oak trees in May and beetle frass is found in the trunk and soil around the trunk [20]. Heavily attacked oak trees die due to systematic wilting in late July to early September (Figure 4). The leaves of dead trees remain on the tree after autumn because oak trees die without the formation of an abscission layer [20] (Figure 3).

Although several different oak species (*Q. mongolica*, *Q. serrata*, *Q. aliena*, *Q. acutissima* Carruth. and *Carpinus laxiflora* (Siebold and Zucc.) Blume) are known to be hosts of KOW in South Korea, *Q. mongolica* is one of the main hosts and is a widely distributed species in South Korea [13]. Suitable habitat for *Q. mongolica* is mainly in the middle of the Korean peninsula, or at areas above 500 m elevation in the southern of Korea [27]. Due to climate change, it was predicted that the area of suitable habitat for this oak would decrease in South Korea in the future [27]. Interestingly, the density of the beetle's entrance holes increased with high dominance of the *Quercus* species [16], suggesting that the probability of KOW increases with available resources of the beetle vector. Similarly, the occurrence of JOW was concentrated in stands with high volumes of *Q. crispula*, a major host of JOW [28].

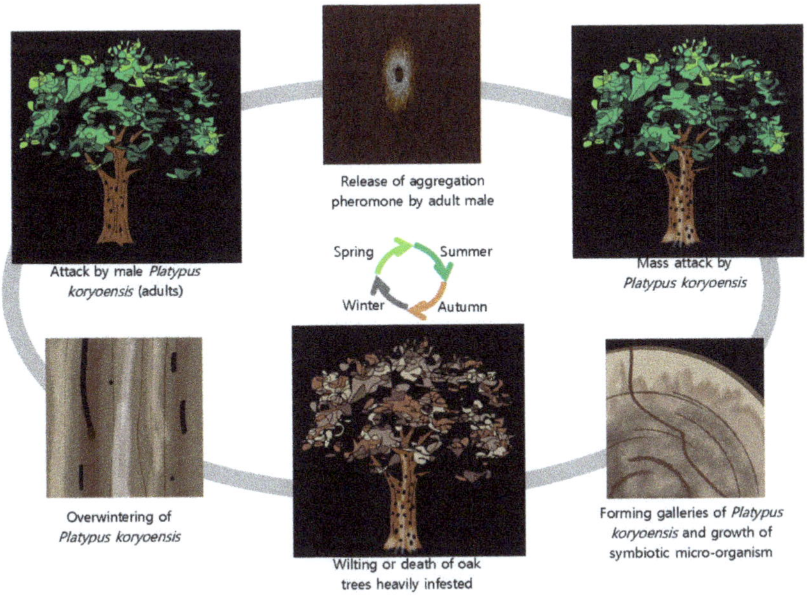

Figure 4. Life cycle of KOW in Korea. The graph was redrawn based on data from research on disease cycle and management of KOW [20].

3.2. Vector Beetle

Ambrosia beetles such as *P. koryoensis* (Platypodinae, Curculionidae) form obligate mutualisms with ophiostomatoid fungi [29]. Platypodinae is a major beetle group showing the fungus gardening behavior [29]. These fungus gardening beetles require a gallery-cleaning behavior by their parents or sisters to avoid contamination of galleries from unwanted fungi [30]. This behavior represents some degree of sociality, and this social behavior is favorable for adaptation to new environments [10]. New outbreaks or invasions to new areas by ambrosia beetles have likely increased due to their social behaviors [8,9].

Among the ambrosia beetles, periodic population outbreaks of the subgroup Platypodinae have been frequently reported throughout the world, and concerns for their economic and ecological impacts have increased [11]. Examples of tree-killing platypodines include *P. quercivorus* on *Quercus* spp. in Japan [31,32], *P. koryoensis* on *Quercus* in Korea [13,33], *Platypus cylindrus* Fabricius on *Castanea* spp., *Fagus* spp. and *Quercus* spp. in Europe [34–36], *Platypus oxyurus* Dufour on *Abies* spp. in Europe [37], and *Megaplatypus mutatis* Chapuis on *Populus* spp. in South America and Europe [38].

The basic biology of *P. koryoensis* was little studied until 2004 when KOW was first reported in South Korea. The beetle was initially identified as *Crossotarsus koryoensis* by Murayama in 1930 [19], and it was collected in 1987 [39]. Its biology was reported by Lee and Chung [40] but there was no report about tree damage until 2004 [33]. Its geographical distribution includes Korea, Taiwan, Japan, and the Russian Far East, although its distribution in Japan is not well known [33]. The presence of *P. koryoensis* in Kyushu, Japan was confirmed through specimen examination (personal communication with Dr. Goto).

P. koryoensis beetles mostly attack larger trees [11], seldom infesting trees less than 20 cm in DBH [25]. The galleries of *P. koryoensis* are horizontal to the ground and mainly in sapwood [11]. It is likely that differences in the chemical composition of sapwood and heartwood explain the beetle's avoidance of heartwood. In case of *Quercus faginea* Lam., phenolic levels were higher in heartwood than sapwood, although lignin and sugar composition were similar [41]. Interestingly, levels of the lignan (−)-lyoniresinol and various phenolic compounds [42] increased in the sapwood of *Q. crispula* in response to attack by *R. quercivora*. This increase in phenolic compounds led *P. koryoensis* to avoid infested wood for new ovipositions, presumably because the beetle interpreted infested sapwood as heartwood [42].

The spatial distribution of the beetle within oak stands, based on attack density in the lower trunks, showed a density-dependent pattern [43]. In low-density beetle populations, attack was concentrated on a few trees for mass attack to overcome host defense. As attack density in stands increases, attacks also occur on non-mass attacked trees. In stand with relatively high beetle density, beetles massively attack new hosts rather than already attacked hosts. Due to this pattern of attacks, the number of damaged trees increases rapidly when beetles are more abundant.

Most species of platypodid beetles exhibit male-initiated monogamy [44], in which male-produced pheromones attract con-specific females. In addition, aggregation pheromones are produced by males of *P. koryoensis*, which have been identified as a mixture of citronellol, nerol, neral, geraniol, and geranial [13]. Although ethanol can attract *P. koryoensis* (as well as other ambrosia beetles, such as *Xyleborus crassiusculus* (Motschulsky), *Xyleborus mutilates* (Blandford) and *Ambrosiodmus lewisi* (Blandford)), the highest catch of *P. koryoensis* was in traps baited with citral or a mixture of neral and geranial, not ethanol, under field conditions [45]. Similarly, quercivorol, (1S,4R)-*p*-Menth-2-en-1-ol, was identified as the aggregation pheromone of *P. quercivorus* [46].

Trap catches of *P. koryoensis* in four oak forests suggested that *P. koryoensis* is univoltine in central Korea [47]. Destructive sampling at monthly intervals of trees infested by *P. koryoensis* provided data on the phenology and age structure of the beetle's life stages [48]. Eggs, larvae, and adults were found throughout the year. However, pupae were abundant only in April and May, although a few were present in July and August, suggesting the beetle is univoltine. Most likely, pupae detected in July and August had developed from eggs laid by adults moving into the forest study site in June. Similarly, the first eggs laid by the ambrosia beetle *Xyleborinus saxeseni* (Ratzeburg) developed into females that either remained in galleries or dispersed late in the season. These females cooperated in care of other brood by tending the fungus gardens, extending the gallery, and feeding and cleaning offspring. The presence of pupae in July and August is also indirect evidence for sister care in *P. koryoensis* and some degree of eusociality.

Using a phenology model based on degree days and field observations, Nam et al. [47] determined that the lower threshold temperature for the development of the beetle is 5.8 °C. The model predicted the median flight date of the beetle within five days [47]. Historical data suggested that the date of the median flight day advanced nine days from 1970 to 2010, probably due to climate change [47].

The beetle's flight distance was estimated using mark–release–recapture methods, marking beetles with fluorescent powders and recovering marked beetles with sticky traps deployed in the oak forest studied [49]. At least 300 individuals were released in each trial at 11:00 a.m. in late June 2013 when beetle flight propensity was greatest, and recaptured individuals were counted after three hours. No marked beetle was recaptured the next day [50]. Recapture efficiency was 6.0 ± 1.6% and the mean dispersal distance was 18.0 ± 1.3 m. The annual movement distance in infested stands in Jinan, Jeollabuk-do was 24.1 m and 19.9 m from 2012 to 2013 and from 2013 to 2014, respectively, showing that the beetle's limited dispersal distance reasonably explained the slow movement of infested stands. A mechanism of long-distance dispersal was proposed to explain movement of beetles over several kilometers. The proposed mechanism postulates that when the beetle density in a tree becomes high, the beetle forms entrance holes in the upper trunk rather than the lower trunk as usual, to reduce intensive intra-specific competition [25]. The beetles emerging from galleries in the upper trunk probably have the potential to fly longer distances, possibly aided by wind (Figure 5). Satellite image analysis showed that KOW hot spots had jumped about 12.5 km per year [22].

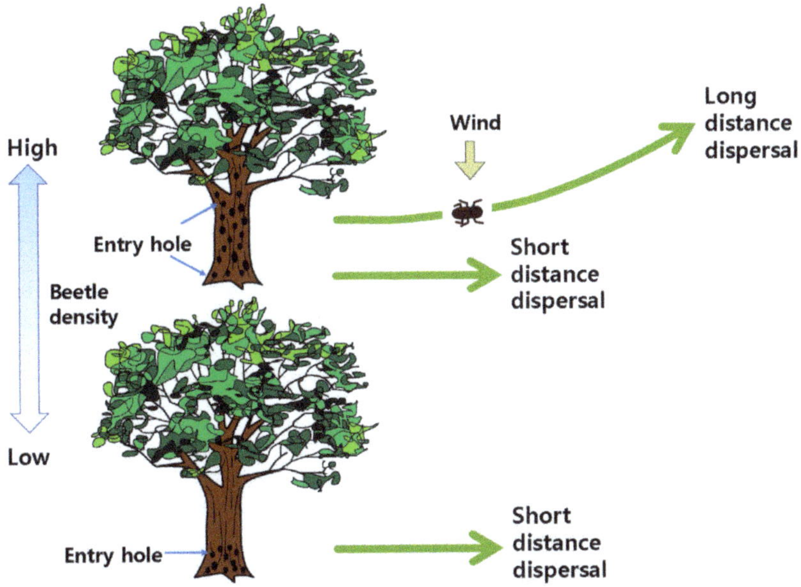

Figure 5. A diagram for hypothesis to explain long distance dispersal of *P. koryoensis* in density-dependent manner. The diagram was drawn based on a hypothesis suggested by Lee et al. [49].

Dispersal pattern of JOW was similar to that of KOW. For *P. quercivorus*, at least half of all trapped adults were found in traps installed <2 m above ground, and trap catch decreased with increasing trap height [51], showing that most beetles flew at less than 2 m height. Local dispersal distances of JOW were also relatively short and less than 100 m but its long-distance dispersal was over 10 km [14]. Flight mill experiments suggested that the flight capacity of *P. quercivorus* is at least 27 km [52], showing the beetle's capacity for long distance dispersal.

3.3. Fungus as the Causal Agent of KOW

Ophiostomatoid fungi have mutualistic interactions with bark or ambrosia beetles, and they have been recognized as potential causal agents for tree mortality in many countries [53–55]. Among these fungi, *Raffaelea* species potentially contributed several observed mass tree mortality events [56–59]. Laurel wilt disease due to *R. lauricola* has caused widespread mortality of several tree species in the family Lauraceae, including avocado in the southeastern United States [60,61]. The impacts of several *Raffaelea* species were considered black swan events (an unpredictable event that is beyond what is normally expected) in tree pathology given that pathogenicity of *Raffaelea* species was either shown to be weak or not proven by artificial inoculation experiments [62].

To test the pathogenicity of *R. quercus-mongolicae*, inoculation trials were performed on both saplings and twigs of mature trees of *Q. mongolica* [63]. The fungus had an ability to colonize sapwood, to contribute to sapwood discoloration and disrupt sap flows, but the pathogenicity of the fungus was not confirmed. The fungus also induced discoloration in the wood of chestnut oak (*Q. acutissima*) when inoculated by artificial injection using Chemjet tree injector, an injector with a steady low pressure [64]. When Japanese oak was artificially inoculated with the fungus, *R. quercivora*, the oak died rapidly with increase in gallery density, which resulted in many offspring. Furthermore, trees with high density of galleries have more discolored areas in wood, and the fungus was frequently re-isolated from the inoculated trees. These recoveries clearly showed that the mass mortality of Japanese oak was caused by mass attack of *P. quercivorus* and its associated fungus, *R. quercivora* [65]. Studies on the spatial distribution of *R. quercivora* in the xylem showed that the hyphae of the fungi were located in the nearby inoculation site and failure of the xylem vessels was associated with the hyphae distribution, suggesting that the oak mortality was caused by massive failure of the xylem vessels due to mass attack by the beetle and its associated fungus [66].

To understand the origin of *R. quercus-mongolicae* in South Korea, the population diversity of 54 isolates of the fungus, collected nationwide (except Jeju island), was determined using the restriction-site-associated DNA (RAD) marker [67]. Based on the lower genetic diversity and no apparent population structure of *R. quercus-mongolicae*, it was suggested that the fungus was introduced to the country, which led to the sudden emergence of KOW on native oaks. However, this view of *R. quercus-mongolicae* as an invasive fungal pathogen is based only on a lower level of genetic diversity and is not in line with other studies. In particular, Lee et al. [68] determined the genetic diversity of 81 isolates of *R. quercus-mongoicae* that were collected from inland areas covering 36 geographical locations in Korea, using microsatellite markers, and found higher levels of genetic indices, supported by other studies [69,70]. In addition to the genetic indices determined by Lee et al. [68], two different mating types of *R. quercus-mongolicae* occurred in equal proportions in the country [71]. Furthermore, the fungus has been isolated from a wide range of native hosts, including *C. laxiflora*, *Castanea* crenata Siebold and Zucc., *Q. acutissima*, *Q. aliena* and *Q. variabilis* Blume without no visible symptoms on the hosts being developed [68], leading to the conclusion that the fungus is a native fungal species rather than an exotic species.

3.4. Yeast

Bark beetles, including the ambrosia beetles, have evolved symbiotic relationships with microorganisms, especially the ophiostomatoid fungi. In addition to the ophiostomatoids, other microbes have symbiotic associations with ambrosia beetles that are potentially related to tree mortality. Yun et al. [72] examined the symbiotic relationships of various yeast species with *P. koryoensis* that were associated with KOW. Eight yeast species belonging to five genera were recovered from *P. koryoensis*. More yeast isolates were recovered from female beetles than males or larvae. Female *P. koryoensis* beetles have mycangia on their body surface where fungal associates are retained. The yeasts isolated from *P. koryoensis* had no antagonistic effects on *R. quercus-mongolicae* when the two were cocultivated in vitro. In addition, among the yeast species isolated from *P. koryoensis*, each produced different extracellular enzymes, suggesting that polymeric wood components would be more efficiently utilized not by a single yeast species, but by a diverse yeast assemblage. It has been suggested that various microorganisms co-occur in the wood of oak, which all have different functions in the formation and maintenance of the galleries of *P. koryoensis*. Similarly, two yeasts, *Candida* sp. and *Candida kashinagacola* Endoh et al., were identified as symbionts of *P. quercivorus* in Japan [73]. These yeasts were abundant in the interior of the gallery of *P. quercivorus*. Volatiles from yeast fermentation such as ethanol attracted beetles because the chemicals can be signal for presence food sources such sugar [74].

4. Management of KOW

The management of KOW has focused primarily on control of the beetle vector. The first control method is to cut heavily infested trees or ones killed by the beetle. The trunks of these trees are then cut into 1 m long bolts and fumigated with metham sodium or dimethylsulfate. As the entrance holes of the beetle are concentrated in lower trunk, the stumps are also fumigated. If the fumigation is not possible, the trunk is cut into 1 m bolts, and these are then split into half and left in the forest without any chemical treatments. During winter, the beetles inside split bolts are generally failed to form galleries, probably due to bolt dehydration. Some of the oak trees in stands heavily damaged by KOW are cut and used as charcoal, wood chips, or sawdust [75,76].

The second control method is to reduce the density of beetles with trapping methods. Mass-trapping of the vector using tree baits is one option. Tree baits are installed in infested stands in late April. Alternatively, to prevent further infestations by beetles from heavily infested beetles or to reduce density of beetle, a two-part mass-trapping method is used. First, sticky vinyl is used to cover the lower trunk up to 2 m from ground to catch emerging beetles and prevent infestation of nearby trees. This method also catches various non-target organisms such as collembola, insects and even small birds [76]. The second method is to replace the sticky vinyl with non-sticky vinyl sheets that lead to a collecting plastic bottle. The probability of catching non-target organisms was reduced [76].

The third control method is to spray the insecticide fenitrothion on the surface of the trunks of all oak trees in an infested stand. Three applications are applied at 10-day intervals in June. This method prevents the infestation of healthy trees. Injection of fungicides were also tested to directly suppress the fungus *R. quercus-mongolicae* [77]. This method has not yet been put to practical use.

5. Causes of KOW

Three hypotheses explaining the sudden outbreak of *P. koryoensis* in Korea have been proposed [11]. The first hypothesis is that *R. quercus-mongolicae* is a new invasive species recently reaching South Korea and novel association between fungus and oak is a potential cause of tree mortality. This hypothesis is partially supported by the low genetic diversity of *R. quercus-mongolicae* based on molecular techniques [67]. However, the existence of two different mating types of *R. quercus-mongolicae* and its higher genetic diversity estimated by microsatellite markers showed that genetic diversity of the fungus was higher than originally thought, and consequently, the fungus is now considered a native rather than an exotic species [68].

The second hypothesis is that warmer temperatures due to climate change have favored *P. koryoensis* over the cold-adapted Mongolian oak. The distribution of this oak ranges from Korea, China, and Siberia to Mongolia, and South Korea is on the southern edge of its distribution. Indeed, KOW is the most concentrated in toward the southern limit of *Q. mongolica*. Although the beetle can attack and successfully colonize other *Quercus* species [33], the beetles preferentially attack *Q. mongolica*. This host preference can be attributed at least partially to climate change because ambrosia beetles prefer to attack weakened or stressed trees [8]. This explanation is consistent with the pattern of JOW in Japan, for which tree mortality is high at lower elevations and extremely low at higher, colder sites (>1000 m) [28].

The last one is related to the age of the Mongolian oaks throughout Korea, given that *P. koryoensis* preferentially attacks larger oak trees and mortality from KOW was higher in large oak trees than in small ones [11,25]. The dominant age class of Mongolian oaks in Korean forests ranged from 30 to 50 years old in 2015 [78]. This tree age structure was attributed to the history of the country's afforestation policy and the restriction on cutting trees in the 1970s [15]. The outbreak of native red oak borer (*Enaphalodes rufulus* Haldeman) on oak species in Arkansas, United State of America in the 1990s and 2000s was observed in stand with over 50-year-old oak trees, suggesting that the increase in tree age is one of important causes for the outbreak [79,80].

Similarly, increase in JOW in Japan was attributed to climate change [81] or abandonment of the management of oak forests [82]. According to Kamata's hypothesis [81], recent climate changes have induced a northern range expansion of *P. quercivorus*, and a novel association between the beetle and *Q. crispula* was formed. This association induced mass mortality of the oak trees because the trees lacked co-evolutionary adaptations to co-exist with the beetle. In contrast, Kobayashi and Ueda's hypothesis [82] is that oak forests in Japan now contain older, larger trees due to cessation of oak forest management including the cutting of older trees. Older oaks in the forest stimulated the outbreak of *P. quercivorus* because such trees were more vulnerable to attack by ambrosia beetles. Shoda-Kagaya, Saito, Okada, Nozaki, Nunokawa and Tsuda [14] further examined both these hypotheses about the cause of JOW by determining the genetic structure of the vector, *P. quercivorus*, using DNA microsatellites. They found that the genetic structure of the beetle's populations was similar to that of the host *Q. crispula* as reported by Okaura et al. [83]. This finding suggests that the geographical genetic structure of the beetle was shaped by co-evolution with the host species. Therefore, Kobayashi and Ueda's hypothesis [82] was better supported by the molecular analysis of the genetic structure of the beetle populations [14].

6. Discussion

The ambrosia beetle *P. koryoensis* cultivates at least two types of microorganisms and engages in eusocial behaviors of kin care with the parental galleries [10]. To cultivate its fungus, *P. koryoensis* forms its gallery low on the trunk because of high moisture content in that part of the tree [25]. Oak wood is not consumed by the beetle, but rather by the symbiotic fungus, which in turn is consumed by the beetle. After entering the tree in newly formed galleries, adults remain there until their offspring emerge from the galleries the following year. Trees died from previous ambrosia beetle attacks are unfavorable for continued cultivation of the symbiotic fungi because of the dryness of the wood. Therefore, tree mortality is unfavorable to the ambrosia beetle.

In contrast, bark beetles attack trees and develop under the bark, feeding on the cambium as larvae. Bark beetles *Ips, Tomicus, Dendroctonus*, and *Scolytus* often kill their hosts when their densities are high because the extensive destruction of cambium from larval feeding destroys the xylem vessels [84]. Additionally, for bark beetles, wood is the basic food for larvae, although some bark beetles acquire nutrients from fungi that colonize infested wood [85].

Microorganisms play several different ecological roles in KOW. The first is as a food source for the beetle vector. Yeasts and their metabolites are potential food sources for ambrosia beetles because ethanol, a major metabolite of yeasts, attracts ambrosia beetles. The role of *R. quercus-mongolicae*, a weak pathogen, is to suppress the local defense of oak and provide nutrients for yeast growth by fostering the local destruction oak cells. It is likely that ambrosia beetles benefit mutualisms with weak pathogens because tree mortality is not favorable to the beetle. The roles of microorganisms in KOW need to be further studied to understand their ecological roles and contributions to tree mortality.

7. Conclusions

It is likely that the area in South Korea affected by KOW can be expanded, but that the degree of actual damage can be decreased. Most of the oak trees that survive in forests after an outbreak of KOW were infected by *R. quercus-mongolicae*, which probably makes them unsuitable as hosts for *P. koryoensis*. Therefore, the beetles must expand their distribution to find new suitable host trees, exposing beetles to potential mortality factors. Therefore, the probability of a re-occurrence of KOW in forests that have already been damaged by KOW would be low. The fate of trees infected by *R. quercus-mongolicae* is uncertain. Due to extensive discoloration, water conductivity decreases, but the influence of such decreased water conductivity is not yet clear. Physiological or drought-related mass mortality of oak trees already infected by the fungus was not reported during the last ten years. Further research or monitoring of physiological changes in oak trees damaged by KOW are needed. The recent outbreak of KOW suggests that the risk of outbreaks of new pests in deciduous forests is likely to increase because deciduous trees are increasingly dominating Korean forests.

Author Contributions: Conceptualization, W.I.C. and Y.-S.P.; writing—original draft preparation, W.I.C., D.-H.L., J.B.J. and Y.-S.P.; writing—review and editing, all authors. All authors have read and agreed to the published version of the manuscript.

Funding: This study was supported by the National Institute of Forest Science and the 'R&D Program for Forest Science Technology (FTIS 2017042B10-2223-CA01) provided by Korea Forest Service (Korea Forestry Promotion Institute).

Institutional Review Board Statement: Not applicable.

Informed Consent Statement: Not applicable.

Data Availability Statement: Not applicable.

Acknowledgments: The authors would like to thank Ye-Eun Kim for her help to draw figures.

Conflicts of Interest: The authors declare no conflict of interest.

References

1. Haavik, L.J.; Billings, S.A.; Guldin, J.M.; Stephen, F.M. Emergent insects, pathogens and drought shape changing patterns in oak decline in North America and Europe. *For. Ecol. Manag.* **2015**, *354*, 190–205. [CrossRef]
2. Gottschalk, K.W.; Wargo, P.M. Oak decline around the world. In *Proceedings, US Department of Agriculture Interagency Gypsy Moth Research Forum 1996, 16–19 January 1996*; Annapolis, MD. Gen. Tech. Rep. NE-230; Fosbroke, S.L.C., Gottschalk, K.W., Eds.; US Department of Agriculture, Forest Service, Northeastern Forest Experiment Station: Radnor, PA, USA, 1997; pp. 3–13.
3. Burlacu, E.; Nisca, A.; Tanase, C. A comprehensive review of phytochemistry and biological activities of *Quercus* species. *Forests* **2020**, *11*, 904. [CrossRef]
4. Thomas, F.M.; Blank, R.; Hartmann, G. Abiotic and biotic factors and their interactions as causes of oak decline in Central Europe. *For. Pathol.* **2002**, *32*, 277–307. [CrossRef]
5. Pinna, C.; Linaldeddu, B.T.; Deiana, V.; Maddau, L.; Montecchio, L.; Lentini, A. Plant pathogenic fungi associated with Coraebus florentinus (Coleoptera: Buprestidae) attacks in declining oak forests. *Forests* **2019**, *10*, 488. [CrossRef]
6. da Clara, M.I.E.; de Almeida Ribeiro, N.M.C. Decline of Mediterranean oak trees and its association with Phytophthora cinnamomi: A review. *Eur. J. For. Res.* **2013**, *132*, 411–432.
7. Moradi-Amirabad, Y.; Rahimian, H.; Babaeizad, V.; Denman, S. *Brenneria* spp. and *Rahnella victoriana* associated with acute oak decline symptoms on oak and hornbeam in Iran. *For. Path.* **2019**, *49*, e12535. [CrossRef]
8. Choi, W.I. Influence of global warming on forest coleopteran communities with special reference to ambrosia and bark beetles. *J. Asia Pac. Entomol.* **2011**, *14*, 227–231. [CrossRef]
9. Biedermann, P.H.W.; Vega, F.E. Ecology and evolution of insect–fungus mutualisms. *Annu. Rev. Entomol.* **2020**, *65*, 431–455. [CrossRef]
10. Hulcr, J.; Stelinski, L.L. The ambrosia symbiosis: From evolutionary ecology to practical management. *Ann. Rev. Entomol.* **2017**, *62*, 285–303. [CrossRef]
11. Lee, J.-S.; Haack, R.A.; Choi, W.I. Attack pattern of Platypus koryoensis (Coleoptera: Platypodidae) in relation to crown dieback of Mongolian oak in Korea. *Environ. Entomol.* **2011**, *40*, 1363–1369. [CrossRef]
12. Nakajima, H.; Ishida, M. Decline of Quercus crispula in abandoned coppice forests caused by secondary succession and Japanese oak wilt disease: Stand dynamics over twenty years. *For. Ecol. Manag.* **2014**, *334*, 18–27. [CrossRef]
13. Kim, K.-H.; Choi, Y.-J.; Seo, S.-T.; Shin, H.-D. *Raffaelea quercus-mongolicae* sp. nov. associated with *Platypus koryoensis* on oak in Korea. *Mycotaxon* **2009**, *110*, 189–197. [CrossRef]
14. Shoda-Kagaya, E.; Saito, S.; Okada, M.; Nozaki, A.; Nunokawa, K.; Tsuda, Y. Genetic structure of the oak wilt vector beetle *Platypus quercivorus: Inferences* toward the process of damaged area expansion. *MBC Ecol.* **2010**, *10*, 21. [CrossRef]
15. Choi, W.I.; Nam, Y.; Lee, C.Y.; Choi, B.K.; Shin, Y.J.; Lim, J.-H.; Koh, S.-H.; Park, Y.-S. Changes in major insect pests of pine forests in Korea over the last 50 years. *Forests* **2019**, *10*, 692. [CrossRef]
16. Um, T.-W.; Chun, J.-H.; Kim, K.H. Stand structure characteristics of oak wilt infected forest, Korea. *Korean J. Environ. Ecol.* **2009**, *23*, 220–232.
17. Park, S.; Lyu, D. Checklist of the family Platypodidae (Coleoptera) in Korea. *J. Asia-Pac. Entomol.* **2007**, *10*, 275–280. [CrossRef]
18. Korea Forest Research Institute. *Annual Report of Monitoring for Forest Insect Pests and Diseases in Korea*; SeongMunSa: Seoul, Korea, 2020.
19. Murayama, J. Révisions des Familiiies des Ipides et des Platypides de Corée. *J. Chosen Nat. Hist. Soc.* **1930**, *11*, 3–38.
20. Korea Forest Research Institute. *Researches on Disease Cycle and Management of Korean Oak Wilt*; Usgo Munhwasa: Seoul, Korea, 2010.
21. Park, J.H.; Nam, Y.; Lee, S.G.; Jung, J.K.; Lee, D.J.; Jung, B.N.; Jang, Y.; Eom, C.-D.; Lee, H.M.; Kim, G.H.; et al. *Development of Customized Management Methods for Korean Oak Wilt (KOW) and Research on Utilization of Woods Damaged by KOW*; National Institute of Forest Science: Seoul, Korea, 2021.
22. Cha, S.; Lee, W.-K.; Kim, M.; Lee, S.G.; Jo, H.-W.; Choi, W.I. Characterizing the spatial distribution of oak wilt disease using remote sensing data. *J. Korean For. Soc.* **2017**, *106*, 310–319.
23. Korea Forest Research Institute. *Annual Report of Monitoring for Forest Insect Pests and Diseases in Korea*; SeongMunSa: Seoul, Korea, 2014.
24. Yeum, J.H.; Han, B.H.; Choi, J.-W.; Jeong, H.-U. Mapping of the damaged forest by oak wilt disease in Bukhansan National Park. *Korean J. Environ. Ecol.* **2013**, *27*, 704–717.
25. Choi, W.I.; Lee, J.-S.; Choi, K.-S.; Kim, J.-K.; Shin, S.-C. Tree trunk level distribution of entry hole by *Platypus koryoensis* (Coleoptera: Platypodidae) and its implication to tree damage. *Korean J. Appl. Entomol.* **2008**, *47*, 127–131. [CrossRef]
26. Kuroda, K. Responses of *Quercus sapwood* to infection with the pathogenic fungus of a new wilt disease vectored by the ambrosia beetle *Platypus quercivorus*. *J. Wood Sci.* **2001**, *47*, 425–429. [CrossRef]

27. Lee, Y.G.; Sung, J.H.; Chun, J.H.; Shin, M.Y. Effect of climate changes on the distribution of productive areas for *Quercus mongolica* in Korea. *J. Korean For. Soc.* **2014**, *103*, 605–612. [CrossRef]
28. Nakajima, H. Region-wide mass mortality of Japanese oak due to ambrosia beetle infestation: Mortality factors and change in oak abundance. *For. Ecol. Manag.* **2019**, *449*, 117468. [CrossRef]
29. Farrell, B.D.; Sequeira, A.S.; O'Meara, B.C.; Normark, B.B.; Chung, J.H.; Jordal, B.H. The evolution of agriculture in beetles (Curculionidae: Scolytinae and Platypodinae). *Evolution* **2001**, *55*, 2011–2027. [CrossRef]
30. Peer, K.; Taborsky, M. Delayed dispersal as a potential route to cooperative breeding in ambrosia beetles. *Behav. Ecol. Sociobiol.* **2007**, *61*, 729–739. [CrossRef]
31. Kubono, T.; Ito, S.I. *Raffaelea quercivora* sp. nov. associated with mass mortality of Japanese oak, and the ambrosia beetle (*Platypus quercivorus*). *Mycoscience* **2002**, *43*, 0255–0260. [CrossRef]
32. Yamasaki, M.; Sakimoto, M. Predicting oak tree mortality caused by the ambrosia beetle *Platypus quercivorus* in a cool-temperate forest. *J. Appl. Entomol.* **2009**, *133*, 673–681. [CrossRef]
33. Hong, K.J.; Kwon, Y.D.; Park, S.W.; Lyu, D.P. *Platypus koryoensis* (Murayama) (Platypodidae: Coleoptera), the vector of oak wilt disease. *Korean J. Appl. Entomol.* **2006**, *45*, 113–117.
34. Baker, J.M. Ambrosia beetles and their fungi, with particular reference to *Platypus cylindrus* FAB. *Symp. Soc. Gen. Microbiol.* **1963**, *13*, 232–265.
35. Henriques, J.; de Lurdes Inácio, M.; Sousa, E. Ambrosia fungi in the insect-fungi symbiosis in relation to cork oak decline. *Rev. Iberoam. Micol.* **2006**, *23*, 185–188. [CrossRef]
36. Akbulut, S.; Keten, A.; Yüksel, B. Wood destroying insects in Düzce province. *Turk. J. Zool.* **2008**, *32*, 343–350.
37. Ponel, P.; Orgeas, J.; Samways, M.J.; Andrieu-Ponel, V.; de Beaulieu, J.-L.; Reille, M.; Roche, P.; Tatoni, T. 110,000 years of quaternary beetle diversity change. *Biodivers. Conserv.* **2003**, *12*, 2077–2089. [CrossRef]
38. Alfaro, R.I.; Humble, L.M.; Gonzalez, P.; Villaverde, R.; Allegro, G. The threat of the ambrosia beetle *Megaplatypus mutatus* (Chapuis) (=*Platypus mutatus* Chapuis) to world poplar resources. *Forestry* **2007**, *80*, 471–479. [CrossRef]
39. Choo, H.Y.; Woo, K.S. Supplementary notes on the bark and Ambrosia beetles of Korea. *Korean J. Appl. Entomol.* **1989**, *28*, 4–9.
40. Lee, B.Y.; Chung, Y.J. *Tree Insect Pests in Korea*; Sungandang: Seoul, Korea, 1997.
41. Miranda, I.; Sousa, V.; Ferreira, J.; Pereira, H. Chemical characterization and extractives composition of heartwood and sapwood from *Quercus faginea*. *PLoS ONE* **2017**, *12*, e0179268. [CrossRef]
42. Imai, K.; Yamauchi, K.; Mitsunaga, T. Extractives of *Quercus crispula* sapwood infected by the pathogenic fungus *Raffaelea quercivora* II: Isolation and identification of phenolic compounds from infected sapwood. *J. Wood Sci.* **2013**, *59*, 517–521. [CrossRef]
43. Nam, Y.; Choi, W.I.; Won, D.-S.; Kim, J.-K. Density related plasticity in stand-level spatial distribution of the ambrosia beetle, *Platypus koryoensis* (Coleoptera: Curculionidae). *Popul. Ecol.* **2013**, *55*, 3–10. [CrossRef]
44. Kobayashi, M.; Ueda, A. Preliminary study of mate choice in *Platypus quercivorus* (Murayama) (Coleoptera: Platypodidae). *Appl. Entomol. Zool.* **2002**, *37*, 451–457. [CrossRef]
45. Kim, H.K.; Seo, J.W.; Kang, W.J.; Lee, J.S.; Cho, W.S.; Seo, S.T.; Kwon, Y.-D.; Kwon, G.H.; Kim, G.H. Attractant effect of citral on *Platypus koryoensis* (Coleoptera: Curculionidae). *Entomol. Res.* **2018**, *48*, 27–31. [CrossRef]
46. Kashiwagi, T.; Nakashima, T.; Tebayashi, S.-i.; Kim, C.-S. Determination of the Absolute Configuration of Quercivorol, (1S,4R)-p-Menth-2-en-1-ol, an Aggregation Pheromone of the Ambrosia Beetle Platypus quercivorus (Coleoptera: Platypodidae). *Biosci. Biotechnol. Biochem.* **2006**, *70*, 2544–2546. [CrossRef]
47. Nam, Y.; Koh, S.-H.; Won, D.-S.; Kim, J.-K.; Choi, W.I. An empirical predictive model of the flight period of the ambrosia beetle *Platypus koryoensis* (Coleoptera: Platypodinae). *Appl. Entomol. Zool.* **2013**, *48*, 515–524. [CrossRef]
48. Park, G.H. *Ecological Characteristics and Chemical Control of Platypus kortoensis (Coleoptera: Platypodidae)*; Chungbuk National University: Cheong-Ju, Korea, 2008.
49. Lee, C.Y.; Nam, Y.; Seo, Y.O.; Bae, Y.J.; Choi, W.I. Characterization dispersal and flight behavior of *Platypus koryoensis* (Coleoptera: Curculionidae) by mark-release-recapture. *J. Econ. Entomol.* **2019**, *112*, 720–728. [CrossRef]
50. Nam, Y.; Choi, W.I. An empirical predictive model for the spring emergence of *Thecodiplosis japonensis* (Diptera: Cecidomyiidae): Model construction and validation on the basis of 25 years field observations data. *J. Econ. Entomol.* **2014**, *107*, 1136–1141. [CrossRef] [PubMed]
51. Igeta, Y.; Esaki, K.; Kato, K.; Kamata, N. Spatial distribution of a flying ambrosia beetle *Platypus quercivorus* (Coleoptera: Platypodidae) at the stand level. *Appl. Entomol. Zool.* **2004**, *39*, 583–589. [CrossRef]
52. Okada, R.; Pham, D.L.; Ito, Y.; Yamasaki, M.; Ikeno, H. Measuring the flight ability of the ambrosia beetle, *Platypus quercivorus* (Murayama), using a low-cost, small, and easily constructed flight mill. *J. Vis. Exp.* **2018**, *138*, e57468. [CrossRef]
53. Inácio, M.L.; Marcelino, J.; Lima, A.; Sousa, E.; Nóbrega, F. *Raffaelea quercina* sp. nov. associated with cork oak (*Quercus suber* L.) decline in Portugal. *Forests* **2021**, *12*, 513. [CrossRef]
54. Saragih, S.A.; Kusumoto, D.; Takemoto, S.; Torii, M.; Kamata, N. Virulence of fungi isolated from ambrosia beetles to *Acer amoenum* branches. *Plant Dis.* **2021**, *105*, 3087–3091. [CrossRef]
55. Trollip, C.; Carnegie, A.J.; Dinh, Q.; Kaur, J.; Smith, D.; Mann, R.; Rodoni, B.; Edwards, J. Ophiostomatoid fungi associated with pine bark beetles and infested pines in south-eastern Australia, including *Graphilbum ipis-grandicollis* sp. nov. *IMA Fungus* **2021**, *12*, 24. [CrossRef]

56. Li, Y.; Huang, Y.T.; Kasson, M.T.; Macias, A.M.; Skelton, J.; Carlson, P.S.; Yin, M.; Hulcr, J. Specific and promiscuous ophiostomatalean fungi associated with *Platypodinae ambrosia* beetles in the southeastern United States. *Fungal Ecol.* **2018**, *35*, 42–50. [CrossRef]
57. Olatinwo, R.; Hwang, J.; Johnson, W. First report of laurel wilt disease caused by *Raffaelea lauricola* on spicebush in Louisiana. *Plant Dis.* **2021**, *105*, 2250. [CrossRef]
58. Procter, M.; Nel, W.J.; Marincowitz, S.; Crous, P.W.; Wingfield, M.J. A new species of Raffaelea from beetle-infested *Leucaena leucocephala*. *Fungal Syst. Evol.* **2020**, *6*, 305–314. [CrossRef]
59. Ibarra-Juarez, L.A.; Burton, M.A.J.; Biedermann, P.H.W.; Cruz, L.; Desgarennes, D.; Ibarra-Laclette, E.; Latorre, A.; Alonso-Sánchez, A.; Villafan, E.; Hanako-Rosas, G.; et al. Evidence for succession and putative metabolic roles of fungi and bacteria in the farming mutualism of the Ambrosia beetle *Xyleborus affinis*. *mSystems* **2020**, *5*, e00541-20. [CrossRef]
60. Mayfield, A.E.; Smith, J.A.; Hughes, M.; Dreaden, T.J. First report of laurel wilt disease caused by a *Raffaelea* sp. on avocado in Florida. *Plant Dis.* **2008**, *92*, 976. [CrossRef]
61. Harrington, T.; Stephen, F.; Aghayeva, D. *Raffaelea lauricola*, a new ambrosia beetle symbiont and pathogen on the Lauraceae. Mycotaxon. *Mycotaxon* **2008**, *104*, 399–404.
62. Ploetz, R.C.; Hulcr, J.; Wingfield, M.J.; de Beer, Z.W. Destructive tree diseases associated with ambrosia and bark beetles: Black swan events in tree pathology? *Plant Dis.* **2013**, *97*, 856–872. [CrossRef]
63. Torii, M.; Matsuda, Y.; Seo, S.T.; Kim, K.H.; Ito, S.-i.; Moon, M.J.; Kim, S.H.; Yamada, T. The effect of *Raffaelea quercus-mongolicae* inoculations on the formation of non-conductive sapwood of *Quercus mongolica*. *Mycobiology* **2014**, *42*, 210–214. [CrossRef]
64. Yi, S.H.; Lee, J.H.; Seo, S.T.; Lee, J.K. In vivo pathogenicity test of oak wilt fungus (*Raffaelea quercus-mongolicae*) on oriental chestnut oak (*Quercus acutissima*). *J. For. Environ. Sci.* **2017**, *33*, 342–347.
65. Kinuura, H.; Kobayashi, M. Death of *Quercus crispula* by inoculation with adult *Platypus quercivorus* (Coleoptera: Platypodidae). *Appl. Entomol. Zool.* **2006**, *41*, 123–128. [CrossRef]
66. Takahashi, Y.; Matsushita, N.; Hogetsu, T. Spatial distribution of *Raffaelea quercivora* in xylem of naturally infested and inoculated oak trees. *Phytopathology* **2010**, *100*, 747–755. [CrossRef]
67. Kim, M.S.; Hohenlohe, P.A.; Kim, K.H.; Seo, S.T.; Klopfenstein, N.B. Genetic diversity and population structure of *Raffaelea quercus-mongolicae*, a fungus associated with oak mortality in South Korea. *For. Pathol.* **2016**, *46*, 164–167. [CrossRef]
68. Lee, D.H.; Jung, J.M.; Seo, S.T. Population genetic structure of *Raffaelea quercus-mongolicae* indicates a recent fungal introduction event to Jeju Island from inland areas of South Korea. *Plant Pathol.* **2021**, *70*, 1871–1882. [CrossRef]
69. Seo, S.T.; Kim, K.H.; Lee, S.H.; Kwon, Y.N.; Shin, C.H.; Kim, H.J.; Lee, S.Y. Genotypic characterization of oak wilt pathogen *Raffaelea quercus-mongolicae* and *R. quercivora* strains. *Res. Plant Dis.* **2010**, *16*, 219–223. [CrossRef]
70. Son, S.Y. Analysis of Pathogenicity, Genetic Diversity of *Raffaelea quercus-mongolicae* and Control of Oak Wilt Disease. Ph.D. Thesis, Chonbuk National University, Jeonju, Korea, 2017.
71. Lee, D.-H.; Son, S.-Y.; Seo, S.-T.; Lee, J.K. Investigation of the mating-type distribution of *Raffaelea quercus-mongolicae* in South Korea. *For. Pathol.* **2020**, *50*, e12590. [CrossRef]
72. Yun, Y.H.; Suh, D.Y.; Yoo, H.D.; Oh, M.H.; Kim, S.H. Yeast associated with the ambrosia beetle, *Platypus koryoensis*, the pest of oak trees in Korea. *Mycobiology* **2015**, *43*, 458–466. [CrossRef]
73. Endoh, R.; Suzuki, M.; Okada, G.; Takeuchi, Y.; Futai, K. Fungus symbionts colonizing the galleries of the ambrosia beetle *Platypus quercivorus*. *Microb. Ecol.* **2011**, *62*, 106–120. [CrossRef]
74. Madden, A.A.; Epps, M.J.; Fukami, T.; Irwin, R.E.; Sheppard, J.; Sorger, D.M.; Dunn, R.R. The ecology of insect-yeast relationships and its relevance to human industry. *Proc. R. Soc. B Biol. Sci.* **2018**, *285*, 20172733. [CrossRef]
75. Korea Forest Service. *Guideline for the Control of Forest Disease and Insect Pests*; Korea Forest Service: Daejeon, Korea, 2015.
76. Park, I.-K.; Nam, Y.; Seo, S.-T.; Kim, S.-W.; Jung, C.-S.; Han, H.-R. Development of a mass trapping device for the ambrosia beetle, Platypus koryoensis, an insect vector of oak wilt disease in Korea. *J. Asia Pac. Entomol.* **2016**, *19*, 39–43. [CrossRef]
77. Son, S.Y.; Seo, S.T.; Park, J.H. Control efficacy of fungicide injection on oak wilt in the field. *Res. Plant Dis.* **2014**, *20*, 295–298. [CrossRef]
78. Korea Forest Service. *Statistical Yearbook of Forestry*; Korea Forest Service: Daejeon, Korea, 2016.
79. Fan, Z.; Kabrick, J.M.; Spetich, M.A.; Shifley, S.R.; Jensen, R.G. Oak mortality associated with crown dieback and oak borer attack in the Ozark highlands. *For. Ecol. Manag.* **2008**, *255*, 2297–2305. [CrossRef]
80. Riggins, J.J.; Galligan, L.D.; Stephen, F.M. Rise and fall of red oakborer (Coleoptera: Cerambycidae) in the Ozark mountains of Arkansas, USA. *Fla. Entomol.* **2009**, *92*, 426–433, 428. [CrossRef]
81. Kamata, N.; Esaki, K.; Kato, K.; Igeta, Y.; Wada, K. Potential impact of global warming on deciduous oak dieback caused by ambrosia fungus *Raffaelea* sp. carried by ambrosia beetle *Platypus quercivorus* (Coleoptera: Platypodidae) in Japan. *Bull. Entomol. Res.* **2002**, *92*, 119–126. [CrossRef]
82. Kobayashi, M.; Ueda, A. Wilt disease of Fagaceae trees caused by *Platypus quercivorus* (Murayama) (Coleoptera: Platypodidae) and the associated fungus: Aim is to Clarify the Damage Factor. *J. Jpn. For. Soc.* **2005**, *87*, 435–450. [CrossRef]
83. Okaura, T.; Quang, N.D.; Ubukata, M.; Harada, K. Phylogeographic structure and late Quaternary population history of the Japanese oak *Quercus mongolica* var. *crispula* and related species revealed by chloroplast DNA variation. *Genes Genet. Syst.* **2007**, *82*, 465–477. [CrossRef]

84. Reeve, J.D.; Ayres, M.P.; Lorio, P.L., Jr. Host suitability, predation, and bark beetle population dynamics. In *Population Dynamics: New Approaches and Synthesis*; Cappuccino, N., Price, P.W., Eds.; Academic Press: San Diego, CA, USA, 1995; pp. 339–357.
85. Six, D.L. Ecological and evolutionary determinants of bark beetle—Fungus symbioses. *Insects* **2012**, *3*, 339–366. [CrossRef]

forests

MDPI

Review

Invasive Insect Pests of Forests and Urban Trees in Russia: Origin, Pathways, Damage, and Management

Dmitry L. Musolin [1,†], Natalia I. Kirichenko [1,2,3,†], Natalia N. Karpun [1,4], Evgeniy V. Aksenenko [5], Viktor B. Golub [5], Ivan A. Kerchev [6,7], Michail Y. Mandelshtam [1], Rimvys Vasaitis [8,*], Mark G. Volkovitsh [9], Elena N. Zhuravleva [4] and Andrey V. Selikhovkin [1]

1 Department of Forest Protection, Wood Science and Game Management, Saint Petersburg State Forest Technical University, Institutskiy Per. 5, 194021 Saint Petersburg, Russia; musolin@gmail.com (D.L.M.); nkirichenko@yahoo.com (N.I.K.); nkolem@mail.ru (N.N.K.); amitinus@mail.ru (M.Y.M.); a.selikhovkin@mail.ru (A.V.S.)
2 Department of Forest Zoology, Sukachev Institute of Forest, Siberian Branch of the Russian Academy of Sciences, Federal Research Center "Krasnoyarsk Science Center SB RAS", Akademgorodok 50/28, 660036 Krasnoyarsk, Russia
3 Institute of Ecology and Geography, Siberian Federal University, Svobodny Pr. 79, 660041 Krasnoyarsk, Russia
4 Federal Research Centre the Subtropical Scientific Centre of the Russian Academy of Sciences, Yana, Fabritsiusa Str., 2/28, 354002 Sochi, Russia; zhuravleva.cvet@mail.ru
5 Department of Zoology and Parasitology, Voronezh State University, Universitetskaya Pl. 1, 394018 Voronezh, Russia; entoma@mail.ru (E.V.A.); v.golub@inbox.ru (V.B.G.)
6 Institute of Monitoring of Climatic and Ecological Systems, Siberian Branch of the Russian Academy of Sciences, Akademicheskiy Pr. 10, 634055 Tomsk, Russia; ivankerchev@gmail.com
7 Department of Forestry and Landscape Construction, Biological Institute, National Research Tomsk State University, Lenina Pr. 36, 634050 Tomsk, Russia
8 Department of Forest Mycology and Plant Pathology, Swedish University of Agricultural Sciences, P.O. Box 7026, SE-75007 Uppsala, Sweden
9 Zoological Institute, Russian Academy of Sciences, Universitetskaya Emb., 1, 199034 Saint Petersburg, Russia; polycest@zin.ru
* Correspondence: rimvys.vasaitis@slu.se
† These authors contributed equally to this work.

Abstract: Invasive alien insects cause serious ecological and economical losses around the world. Here, we review the bionomics, modern ranges (and their dynamics), distribution pathways, monitoring, and control measures of 14 insect species known to be important invasive and emerging tree pests in forest and urban ecosystems of Russia: *Leptoglossus occidentalis* (Hemiptera: Heteroptera: Coreidae), *Halyomorpha halys* (Hemiptera: Heteroptera: Pentatomidae), *Corythucha arcuata* (Hemiptera: Heteroptera: Tingidae), *Agrilus fleischeri, A. mali, A. planipennis, Lamprodila* (*Palmar*) *festiva* (Coleoptera: Buprestidae), *Ips amitinus, Polygraphus proximus* (Coleoptera: Curculionidae: Scolytinae), *Cydalima perspectalis* (Lepidoptera: Crambidae), *Acrocercops brongniardella, Cameraria ohridella, Phyllonorycter issikii*, and *P. populifoliella* (Lepidoptera: Gracillariidae). We identified three major scenarios of tree pest invasions in the country and beyond: (1) a naturally conditioned range expansion, which results in the arrival of a pest to a new territory and its further naturalization in a recipient region; (2) a human-mediated, long-distance transfer of a pest to a new territory and its further naturalization; and (3) a widening of the pest's trophic niche and shift to new host plant(s) (commonly human-introduced) within the native pest's range frequently followed by invasion to new regions.

Keywords: forest entomology; forest health; biological invasions; outbreaks; insect pests; range expansion; urban trees; detection; controlling measures

1. Introduction

Insect invasions are a widely recognized problem. Indeed, many invasive insect species pose a serious threat to biodiversity and ecosystem functioning worldwide, affecting the well-being of modern human society [1–4]. With increasing intercontinental trade (including the trade of living plants and plant origin products) and global changes happening in the environment (e.g., climate change, deforestation, ecosystem degradations, etc.), the impact of invasive species will continue to grow [5–7].

Invasive insect species are often not pests in their natural range, where they are well-controlled by biotic and abiotic factors [8]. Furthermore, some of these species become known to science only after their invasions [9,10]. On one hand, alien pests that are often released from their native enemies can switch on non-coevolved hosts, resulting in successful invasion, with notable damage to the invaded ecosystem [11]. On the other hand, alien plants introduced into a new region can be a target to native pests that can benefit from the susceptibility of the novel hosts [12]. The ability of invaders to form novel trophic associations may present a threat to plants in their native range in cases of a pest accidental introduction. Our knowledge of the biology of novel pests, their invasive potential and impacts, as well as possible methods of monitoring and control is usually limited [13]. Often, such information is unavailable or scarcely published on a national language of a country from where a pest is originally known and is thus not accessible to an international audience.

Russia occupies a huge territory in Europe and Asia with different climatic zones, and the country is actively involved in international trade. Thus, it is not surprising that harmful organisms, arriving to the country via international trade (e.g., airfreight or shipping containers carrying crops, ornamental plants, wood, etc.) or through the movement of people (e.g., global tourism), are regularly detected here [14], and those that succeed in becoming established cause significant economic losses to Russia [15]. The opposite is also true. Pests are regularly found on plants and plant material in vehicles driving from or through the territory of Russia and, as a result, are recorded in new regions of the country or beyond, in foreign countries [8]. Information about the biology, monitoring, and control of alien invasive species that actively expand their ranges and have recently arrived in Russia or, in contrast, native species that are able to spread from Russia to other countries, is of a high importance for international specialists dealing with pest quarantine and forest protection in their homelands.

In this review, we analyzed the biology, distribution, pathways, monitoring, and control approaches of 14 tree insect species, which are considered highly important invasive or emerging insect pests of forests and urban woody plants in Russia [8,15]. It should be understood that they are not the only actual or potential invasive or emerging insect pests in Russia deserving attention. However, in our opinion, they can be considered as representative models of insect pest invasions at a national level.

2. Materials and Methods

In order to provide up-to-date knowledge on insect pests of forests and urban woody plants in Russia, we reviewed numerous studies, mostly published in a national language, and the references are given in each species' essay below. We have included the latest papers delivered at the All-Russia biennial conference "Dendrobiotic Invertebrates and Fungi and their Role in Forest Ecosystems (The Kataev Memorial Readings)", which serves as the main platform for discussing the most important issues on forest entomology, pathology, plant protection, and quarantine in the country [15]. We also examined "grey" and not easily accessible national literature (e.g., proceedings, conference notes, papers in regional editions, etc.). As a result, we compiled 14 brief essays on important invaders in Russia (mainly within its European part). Most of these pests (both alien invasive species which penetrated Russia and native national pests) can potentially further expand their ranges and pose a threat to woody plants in other countries.

The essays are arranged in taxonomic order. They briefly overview history of invasion, biology, basic monitoring and control measures, and are accompanied with photographs of pests and damage symptoms. A schematic map of the natural range (colored in yellow) and invasive range (colored in red) in Russia is also provided. In most cases, we outlined the entire administrative region/unit for published records, except for the cases when the area of the record is relatively small, whereas the region is large. The current ranges were mapped using ArcGIS 9.3 [16].

As there are discrepancies in the English naming of Russian administrative units and settlements, we applied the most commonly used names. To designate administrative units, the following terms were used: "Oblast" for Province and "Krai" for Administrative Territory.

3. Results

3.1. *The Western Coniferous Seed Bug, Leptoglossus occidentalis Heidemann, 1910 (Hemiptera: Heteroptera: Coreidae)*

Leptoglossus occidentalis (Figure 1) is a North American species that invaded Eurasia, North Africa, and Central America. The native range of the species covers the west of North America, from the south of Canada to Mexico [17]. In Europe, it was first recorded in 1999 in northern Italy [18], where it was accidently introduced from North America, probably by plane [19]. Over the next 10 years, the bug spread from Italy across Europe northward (to Norway) and eastward (to Russia and Turkey) (Supplimentary Table S1). At present, the invasive range of *L. occidentalis* in Eurasia covers significant parts of the Palaearctic. In Asia, it is known from Turkey, Israel, Armenia, Kazakhstan, Northeast China, Japan, and South Korea [17,20–25]. In Africa, it is recorded in Tunisia [26], Morocco [27], and South Africa [28]. It is also recorded in Chile [29], Brazil [28], Argentina [30], and Uruguay [31]. The first records in European Russia refer to Rostov-on-Don (but also to the neighboring Crimean Peninsula) [17]. Up to now, it has been recorded also in the Smolensk Oblast [32], Voronezh City [33], Volgograd City [34], Stavropol Krai [27], Krasnodar Krai, and North Ossetia (central Causasus) (Figure 1c) [19,27,35]. The species has spread along the entire Black Sea coast of the Caucasus (Krasnodar Krai) to the border with neighboring region of Abkhazia [19,35], from which it is also known [27]. In addition, it is documented from Transcaucasia: the neighboring region of South Ossetia [27], Georgia [36], and Armenia [25].

The routes of introduction into Russia are not known. Adults of *L. occidentalis* have good flying abilities and, thus, they can fly for long distances over regions having suitable habitats and host plants [17]. The bugs could have also been accidently introduced with coniferous plants, either with plants for planting or Christmas trees (whole or cut branches) [17,37]. The rate of spreading of *L. occidentalis* from the place of its first detection in Europe is approximately the same in all directions of its expansion and is estimated to be about 200 km per year [19]. Most likely, in addition to active distribution, the invader is moved passively with various types of transport [19].

Based on the record of the species in Voronezh City (the forest-steppe zone; approx. 500 km from Rostov-on-Don) and in the Smolensk Oblast (the forest zone; approx. 1000 km northward from Rostov-on-Don), a significant expansion of the invaded range northward in the European part of Russia should be expected. Successful establishment of the species in the forest-steppe zone is now evident at least in urban conditions, keeping in mind that average temperatures are higher in cities than in the surrounding environment [33]. The probability of spreading and establishment further eastward remains unknown.

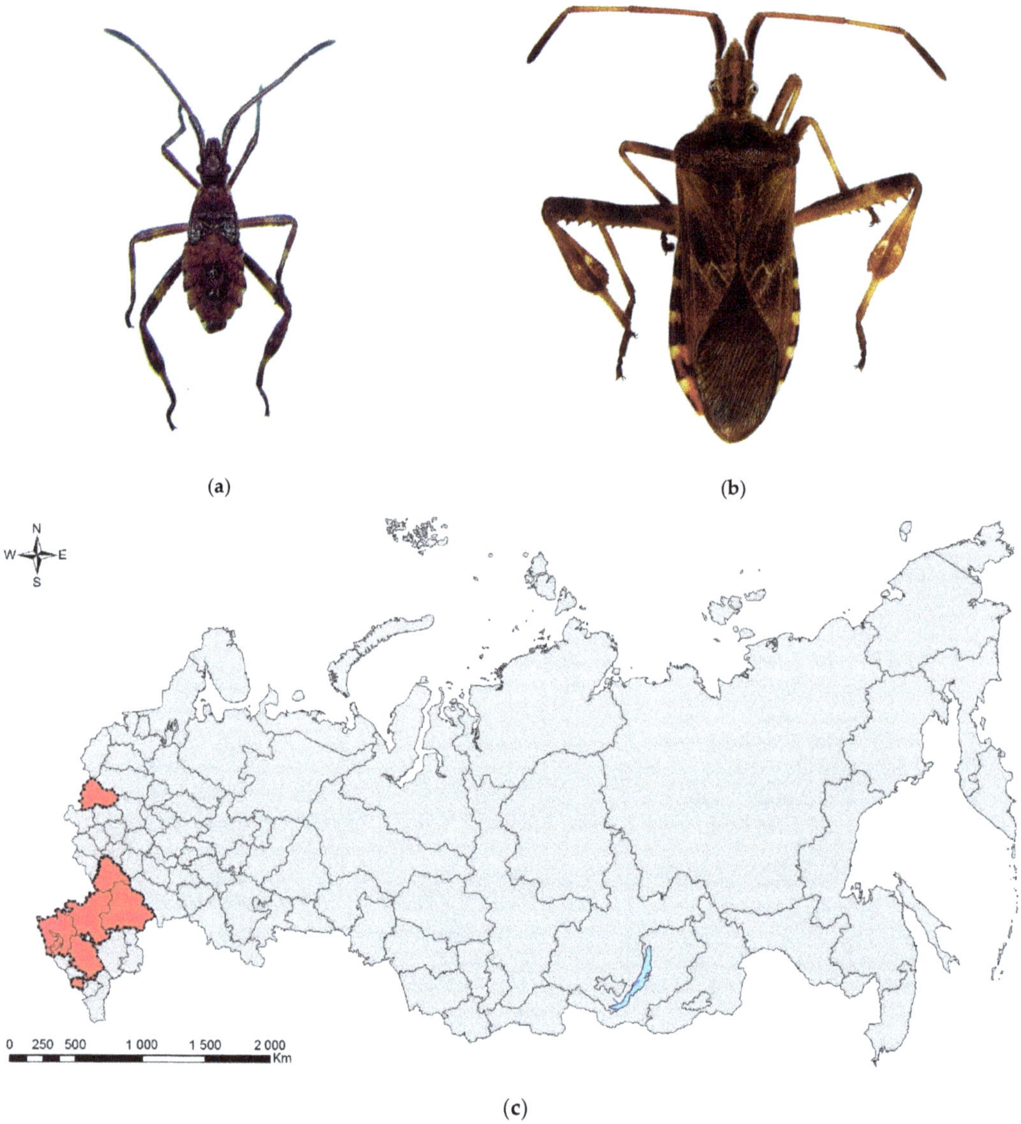

Figure 1. The Western coniferous seed bug, *Leptoglossus occidentalis*: (**a**) a nymph of IV instar; (**b**) an adult, Voronezh City, September 2020; (**c**) the current invasive range in Russia (colored in red). Photo by E.V. Aksenenko.

The species is trophically associated with a wide range of plants from Pinaceae (e.g., *Abies*, *Cedrus*, *Picea*, *Pinus*, *Pseudotsuga*, and *Tsuga*), Cupressaceae (e.g., *Calocedrus*, *Cupressus*, and *Juniperus*) [17,20,35,38], and exceptionally documented on Anacardiaceae (*Pistacia vera*) [39]. In Russia, feeding on spruce *Picea glauca*, *P. pungens*, and fir *Abies cocolor* has been recorded [19].

The bug overwinters as adults or nymphs of the last (V) instar. For overwintering, they aggregate in cracks in the bark and other similar microhabitats in forests and can also penetrate dwellings, cars, etc. They leave overwintering shelters in April to May and start feeding on young microstrobili and megastrobili, causing growth impairment, deformation, and reducing the production of microspores [17]. Then, the adults mate and the females lay eggs. Females lay about 30 eggs (up to 80 maximal) in clusters along the needles [40]. Hatched nymphs and adults feed on maturing and ripe seeds and also suck sap from the apical shoots throughout summer. Adults emerge close to the end of August.

Within its native range in northern California, *L. occidentalis* produces one generation per year [37,41]. In Italy, it has two or three generations on the plain, and only one generation in the highlands [42]. In the south of Krasnodar Krai, two generations per year were recorded [19]. In Dagestan (North Causasus), at least one generation per year is possible, and in the area between Anapa and Novorossiysk cities, two generations can occur [19]. Further northward, thermal conditions are not favorable for this species.

The damage caused by *L. occidentalis* has not yet been fully evaluated in Europe. In North America, it destroys up to 26% of the seeds of *Pinus monticola* and up to 41% of the seeds of different *Pseudotsuga* species [41]. In West Europe, the species has not caused any significant damage so far [43], probably because: (1) its abundance has not yet reached the critical level in the newly colonized territories, and (2) causing damage only to seeds, the bug practically does not harm the trees themselves. In Russia, a 3-year study of the influence of *L. occidentalis* on seed germination of the Crimean pine and Scots pine in the Rostov Oblast demonstrated the high harmfulness of the seed bug [19].

The bug carries spores of the pathogenic fungus *Sphaeropsis sapinea* (=*Diplodia pinea*), which causes diplodia tip blight: a necrosis of the needles and shoot bark, as well as the withering of seedlings and young plants [44,45]. Thus, the cumulative damage to pines can be highly significant. In its native range, *L. occidentalis* is parasitized by *Trichopoda pennipes* (Diptera: Tachinidae) [46] and hymenopteran egg parasitoids, *Gryon pennsylvanicum* (Scelionidae), *Anastatus pearsalli* (Eupelmidae), and *Ooencyrtus* spp. (Encyrtidae) [47]. In Italy, the hymenopterans, *Anastatus bifasciatus* (Eupelmidae), *Tetrastichus servadeii* (Eulophidae), and *Ooencyrtus pityocampae* (Encyrtidae) parasitize the eggs of *L. occidentalis*. In Russia, parasitoids that able to control the bug are not known, except *Ectophasia crassipennis* (Diptera: Tachinidae) [35].

Leptoglossus occidentalis can be effectively controlled using pyrethroid preparations in the years when the pine yield is low and the bug's population density is high; however, this and other chemical control measures against this pest, as well as biological control focusing on its natural enemies, should be developed in Russia [35].

3.2. The Brown Marmorated Stink Bug, Halyomorpha halys (Stål, 1855) (Hemiptera: Heteroptera: Pentatomidae)

The native range of *Halyomorpha halys* (Figure 2), covers Southeast Asia: China, the Korean Peninsula, Japan, Taiwan, and Vietnam [48,49]. The history of the pest invasion began in North America, where the species was first recorded in 1996, and since then, it has actively spread throughout the country. By 2015, the bug had been reported in 41 states [50–52]. In Canada, *H. halys* was documented in 2010 in Ontario and Quebec [53]. In Central and South America, *H. halys* has not yet been recorded, except for the only detection in the Chilean border zone in 2017 [54]. In Europe, *H. halys* was first detected in 2004 in Liechtenstein [55] and spread quickly to at least 30 countries (Supplimentary Table S2). The species expansion in Europe was recently analyzed based on the distribution of different haplotypes [56]. In 2010, live specimens of the bug were intercepted in England and New Zealand in the luggage of passengers traveling by air. In New Zealand, it was also found in used cars brought from Japan, but the species did not spread across the countries [57,58].

(a)

(b)

(c)

(d)

Figure 2. The brown marmorated stink bug, *Halyomorpha halys*: (**a**) a nymph of the final (V) instar, Sochi, August 2017; (**b**) an adult on grape, Sochi, September 2017; (**c**) an overwintering aggregation in warehouse, Sochi, January 2018; (**d**) the current invasive range in Russia (colored in red). Photo: (**a**) by V.Ye. Zakharchenko and (**b**) and (**c**) by N.N. Karpun.

In Russia (Figure 2), *H. halys* was detected for the first time in 2014 [59] in two parks in Sochi, and already in 2015, an outbreak was observed in the humid subtropics of Russia and neighboring region of Abkhazia [60]. The species occurs on the Black Sea coast of the Caucasus at altitudes up to 800 m a.s.l. [61]. In the fall of 2016, the bug was noted in Krasnodar City, and during 2017, it invaded the territory of Krasnodar Krai and Adygea (West Causasus) [62–64]. In 2018, *H. halys* was detected in the Stavropol Krai and in neighboring Crimean Peninsula (in Simferopol, Sevastopol as well as in the nearby villages of Inkerman, Balaklava, on Cape Fiolent) [65,66]. Since 2019, the species adults were regularly caught in the City of Rostov-on-Don and in the Rostov Oblast [67,68]. In 2020, nymphs and adults were intercepted in the Stavropol City [69] and in Dagestan (North Caucasus, Samur Forest) [70]. It is likely that goods of non-plant origin transported during the winter are the major vector of *H. halys* distribution in Russia [63].

The brown marmorated stink bug is a polyphagous pest damaging 300+ species of plants [50,71,72]. In southern Russia, nymphs and adults feed on 107 species of cultivated and wild plants, and during the year, the species migrates from one crop to another. It locally migrates to forest and forest plantations at the end of August or in September, concentrating in the crowns of beech, linden, hazel, and ash [73]. In 2016–2017, *H. halys* caused severe losses in the yield of fruit, subtropical, and vegetable crops [60,63,74]. In woody species, the pest destructs fruits. Thus, in hazel, kernels fail to properly form, while in Rosaceae, fruits are deformed and do not ripen [73]. On the Black Sea coast of Russia, two generations develop per year, and only one is likely to develop further north [74,75]. The monitoring of *H. halys* is possible by regular visual inspections of plantations during the summer and overwintering sites in the autumn and winter, as well as by pheromone trapping [62]. The traps are placed in crowns of woody plants or on poles in fields or other open areas. The composition of the aggregation pheromone of *H. halys* was identified as a mixture of two stereoisomers: (3S, 6S, 7R, 10S)-10,11-epoxy-1-bisabolene-3-ol and (3R, 6S, 7R, 10S)-10,11-epoxy-1-bisabolene-3-ol, in a ratio of 3.5: 1, respectively [76]. Among the pheromones produced in Russia, the most effective is a mixture consisting of an aggregation pheromone obtained from citronelal racemate and methyl (E, E, Z)-2,4,6-decatrienoate [77].

In the south of Russia, no effective predators and parasitoids of *H. halys* have been found so far. As biological control measures, the use of preparations based on entomopathogenic fungi (strains) of the following species is considered promising: *Beauveria bassiana*, *Metarhizium anisopliae*, *Isaria fumosorosea*, and *Ophiocordyceps nutans* [74,78–82]. Pheromone traps catch more than 100 individuals per week and can be effectively used in autumn [77,83]. To control nymphs and adults, neonicotinoids (based on imidacloprid, acetamiprid, thiacloprid, and thiamethoxam) and pyrethroids (based on bifenthrine, cypermethrin, lambda-cyhalothrin, alphacypermethrin, etc.) are effective [48,74,84–90].

3.3. The Oak Lace Bug, *Corythucha arcuata* (Say, 1832) (Hemiptera: Heteroptera: Tingidae)

Corythucha arcuata (Figure 3) is a North American species that is naturally distributed in the USA and the southern regions of Canada [91,92]. In Europe, the oak lace bug was first recorded in 2000 in Italy [93], and by now, it has quickly spread in almost 20 countries (Supplimentary Table S3). In Asia, it was found in 2002 (2003) in Northwest Turkey, and by 2008, it had spread about 600 km eastward [94,95]. In 2005, a specimen of the bug was caught in the north of Iran at the latitude of 1370 m a.s.l. in the vicinity of Urmia (Urmiech) [96]; however, no data on further acclimatization of the species in Iran are available.

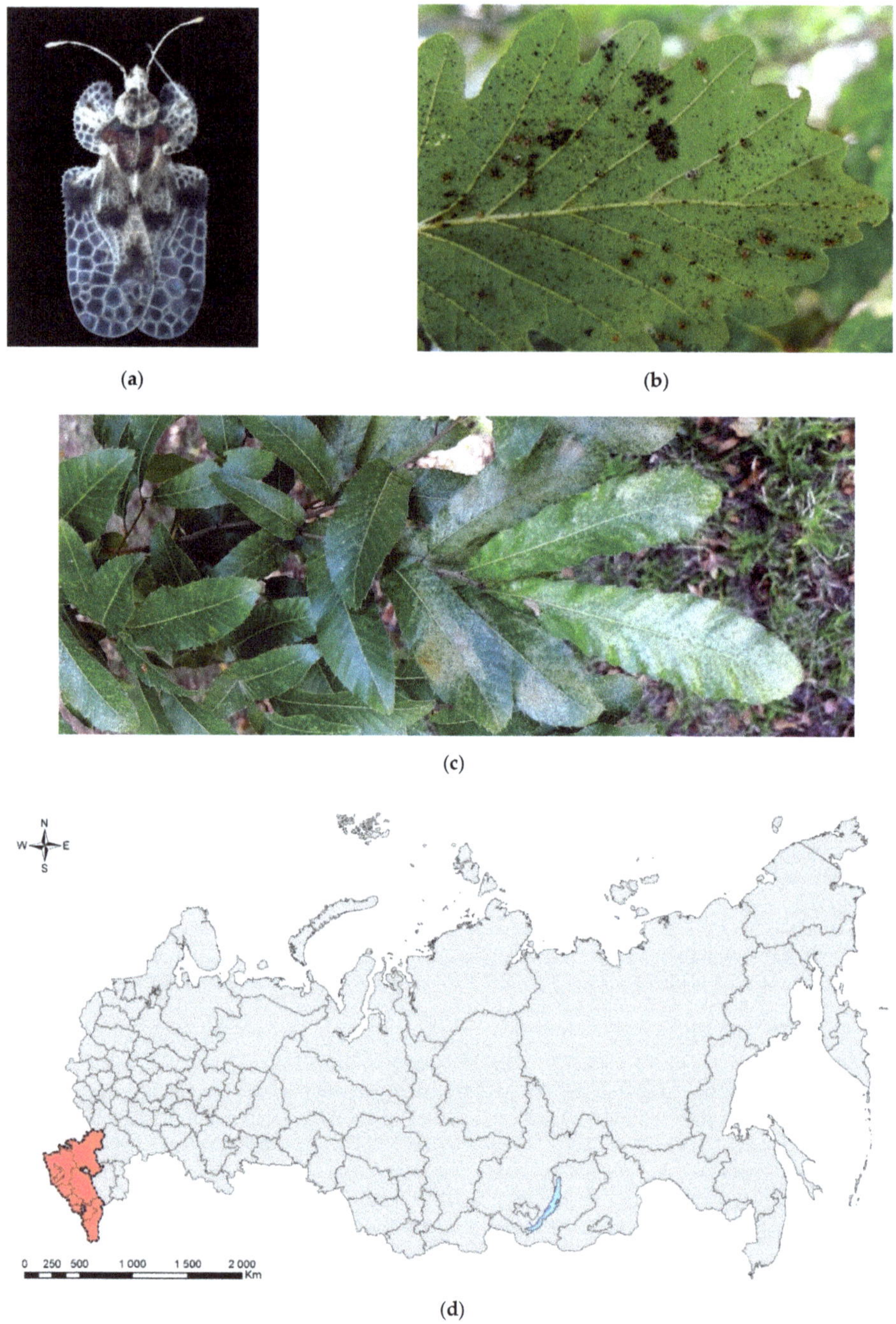

Figure 3. The oak lace bug, *Corythucha arcuata*: (**a**) an adult, Krasnodar Krai, June 2016; (**b**) eggs, nymphs, and adults on a leaf of *Quercus hartwissiana*, Sochi, September 2017; (**c**) heavy discoloration of *Q. variabilis*, Sochi, September 2017; (**d**) the current invasive range in Russia (colored in red). Photo: (**a**) by V.A. Soboleva, with permission, (**b**) and (**c**) by V.Ye. Zakharchenko.

In Russia (Figure 3d), the first local outbreak of this pest was recorded in the summer of 2015 in Krasnodar City. By the autumn 2016, the species had spread across the Krasnodar Krai and was found in Adygea [90,97,98]. In September 2017, *C. arcuata* was found in the subtropical zone of the Black Sea coast of the Caucasus [99] and in the neighboring Crimean Peninsula [100], where it was later repeatedly documented [101,102]. In 2018, the bug was recorded in the Rostov Oblast and many regions of the North Caucasus (Stavropol Krai, Karachay-Cherkessia, Kabardino-Balkaria, North Ossetia, Chechnya, and Dagestan) [103]. Further pest expansion outside Russia to natural forests of Central Asia and the countries of the Caucasus is likely.

Corythucha arcuata spreads over long distances mainly due to human-mediated transportation (land, air, and water vehicles); the distribution on air currents and transportation with seedlings of plants and other plant material are not ruled out [104]. Within its natural range, it damages many oaks (*Quercus* spp.; Fagaceae) by sucking juice from leaves. Additionally, it has been noted feeding on toothed chestnut *Castanea dentata* (Fagaceae), willows (*Salix* spp.; Salicaceae), and *Cercis canadensis* (Fabaceae) [92,105]. In southern Russia, it damages various species of oak, as well as a number of other woody plants (e.g., *Castanea sativa, Corylus avellana, Acer platanoides, Alnus glutinosa, A. incana*, etc.) [106,107].

In North America, the oak lace bug produces two or three generations per year [108]. Under favorable conditions of the North Caucasus, Transcaucasia, and the Crimean Peninsula, up to three generations can develop in April to November (sometimes a part of its population can produce the fourth generation) [109].

Nymphs and adults suck juices from leaves, resulting in a pronounced discoloration, which is first seen as a marbled diffuse yellowing of foliage, and by the middle of summer, as complete discoloration due to chlorophyll loss [106,108]. Despite growing population densities, the bug impact on oak forests in Russia has not been estimated yet. Significant destruction of foliage in tree crowns during several years can greatly weaken the oaks, causing their gradual decline [103,109]. In addition, the decorative quality of trees massively infested by the oak lace bugs is significantly reduced, which may be critically important in urbanized environments [63].

The oak lace bug should be monitored through regular visual ground surveys in forest and urban plantations [103,106], and remote sensing should be developed [110], since the damage caused to trees is clearly visible. Sweeping with classical entomological net in tree crowns and shrubs can also be used for surveys [102].

In Russia, predators and parasitoids of the oak lace bug have not yet been studied [111]. Nevertheless, it is believed that entomopathogenic fungi (e.g., *Beauveria* spp., *Metarhizium* spp., and *Cordyceps* spp.) can be potentially used for the biological control of *C. arcuata* [112,113]. Of the chemical preparations, avermectins, pyrethroids, and neonicotinoids are considered effective against this pest [114–116].

3.4. The Spotted Poplar Borer, Agrilus fleischeri Obenberger, 1925 (Coleoptera: Buprestidae)

Agrilus fleischeri (Figure 4), is a North Asian pest of poplars (*Populus* spp.) and, to a lesser extent, willows (*Salix* spp.), which produced outbreaks and caused tree mortality in poplar plantations in China (see below in this essay). Because of these circumstances, it is considered a potentially invasive pest of poplars in Europe and North America [117–119] and included in the European and Mediterranean Plant Protection Organization (EPPO) A2 List of pests recommended for regulation as quarantine pests [120].

The species was described from East Siberia (Transbaikalia: Berezovka settlement) [121]. In Russia, it is distributed across its Asian part: in West Siberia (Tyumen Oblast, Krasnoyarsk and Altai Krais, the Altai Republic), East Siberia (Tyva and Buryatia), and Far East (Primorskii and Khabarovsk Krais, Sakhalin, and Amur region) (Figure 4c).

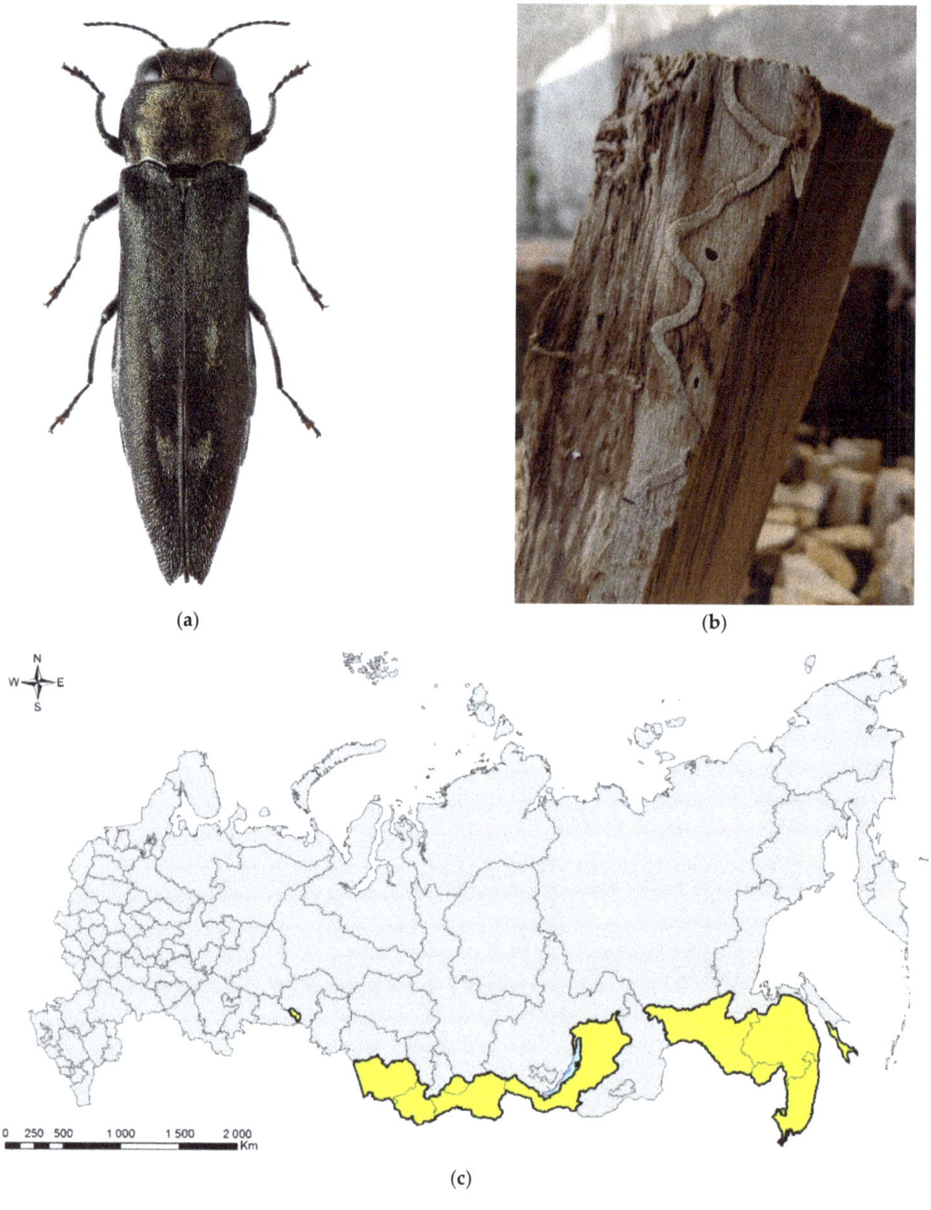

Figure 4. The spotted poplar borer, *Agrilus fleischeri*: (**a**) a female; (**b**) a fragment of a larval gallery and an exit hole on a piece of firewood (*Populus* sp.), Jilin Province, China; (**c**) the native range in Russia (colored in yellow). Photo: (**a**) by A.V. Kovalev, after [119]; (**b**) by E. Jendek, both with permission.

In Asia, it is also known from Kazakhstan (eastern and, probably, northern parts; report from Karaganda Oblast on the east needs confirmation), Mongolia (Töv Aimag), China (Northeast and Central China: Heilongjiang, Liaoning, Beijing, Sichuan, and Shaanxi Provinces), North Korea, South Korea, and Japan (Hokkaido, Honshu) [119,122]. In the westernmost part, the range of *A. fleischeri* reaches, but does not overlap with, the range of closely related European *Agrilus ater* [119].

The spotted poplar borer is considered a potentially invasive species. Although it has never been reported beyond its native range, it is known to outbreak on planted exotic poplars [117–119]. Outbreaks of *A. fleischeri* on introduced *Populus nigra* var. *italica* resulting in tree mortality were first observed in China (Liaoning Province) in 2013 to 2015 [123].

The species may potentially spread to Europe and North America, where poplars are native and where cultivated poplar species from China, Asian Russia, or East Kazakhstan are planted. In Canada, adults of *A. fleischeri* have been intercepted on two occasions being transported with wood packaging material and wood dunnage [120]. The pest can be accidently introduced with poplar plants for planting, round and sawn wood, wood residues, and wood packaging material. Natural spread (e.g., from Asian Russia via European Russia to EU) or hitchhiking with transport are considered unlikely but possible pathways [120].

In China (Liaoning Province), in addition to feeding on the indigenous species of poplars and willows, *A. fleischeri* can also severely damage introduced *Populus nigra* var. *italica* [123]. The current situation with *A. fleischeri* is reminiscent of the early stages of the invasion of the emerald ash borer, *A. planipennis,* which switched from the indigenous ash species to the cultivated American ash species, triggering its outbreaks in Eastern China. Then, the pest was accidentally imported to North America and European Russia [119] (see also Section 3.6 below).

Currently, no special control measures have been developed in Russia. In China, sanitary felling of infested, dead, and dying trees is performed [120]. The dimethoate is considered an effective insecticide against adults during the flight period of different *Agrilus* species in China [120,124]. Some hymenopteran parasitoids (i.e., *Oobius saimaensis, O. fleischeri* [Encyrtidae]; *Polystenus rugosus, Spathius* sp. [Braconidae]; *Paramblynotus* sp. [Liopteridae]; and *Euderus fleischeri* [Eulophidae]) and entomopathogenic fungi *Beauveria bassiana* (Ascomycota: Cordycipitaceae), which were found infesting *A. fleischeri* eggs and larvae in China, can be considered potential agents for the biological control of the pest [120].

3.5. *The Apple Buprestid, Agrilus mali Matsumura, 1924 (Coleoptera: Buprestidae)*

Agrilus mali (Figure 5), originally described as a destructive pest of apple plantations in Korea, is currently regarded a dangerous quarantine pest of cultivated and wild apple trees in Russian East Siberia and the Far East, East China, and both Koreas (see below in detail). Shortly before 1993, *A. mali* was unintentionally introduced in the Ili River valley (Xinjiang Province, China) from Shandong and later caused extensive tree mortality in the wild apple, *Malus sieversii,* forests in the Tien Shan Mountains [125–130]. Thereby, *A. mali* is considered a potentially invasive pest threatening cultivated apple trees outside its native range, wild apple trees and, in particular, *M. sieversii,* which is an important tree of mountain deciduous forests in Northwestern China, Kazakhstan, and adjacent countries of the Middle Asia [118,119] and a key ancestor of the domestic apple tree [127–130]. *Agrilus mali* is a quarantine species in the Russian Federation, the Eurasian Economic Union [131], and the EPPO region [132].

Currently, *A. mali* is distributed in Asian Russia—East Siberia (Chita Oblast, Zabaikalskii Krai) and the Far East (Amur Oblast, Khabarovsk and Primorskii Krais) (Figure 5c)—Mongolia (Dornod and Tuv Aimags), China (Gansu, Guangxi, Hebei, Heilongjiang, Henan, Hubei, Nei Mongol, Qinghai, Sichuan, Shandong, Xinjiang, and Xizang Provinces), North Korea (Pyongyang), and South Korea (Gyeonggi-do, Inchon, Seoul, Taegu) [119,122,133], but some authors consider the Koreas and the Xinjiang Province of China as invasive range [126–130,134].

(a)

(b)

(c)

Figure 5. The apple buprestid, *Agrilus mali*: (**a**) a male; (**b**) exit holes of *A. mali* on a dead *Malus* sp. tree, Beijing District, China; (**c**) the native range in Russia (colored in yellow). Photo (**a**) by A.V. Kovalev, after [119]; (**b**) by E. Jendek, both with permission.

It is supposed that *A. mali*, was introduced to Korea from East China (Liaoning Province) [134], while the population in Xinjiang Province presumably originated from Shandong [127,128]. Imported apple tree seedlings are univocally indicated as a vector in both regions. It is believed that the potential invasive range of *A. mali* can cover the entire apple orchards' cultivation area [135]. Based on the first findings in Korea in 1915 [134], the invasion of *A. mali* to that country could have happen in the early 1900s and the invasion to Xinjiang Province probably happened at the end of the 1980s to the early 1990s because the first outbreak took place there in 1993 [125–130]. Thus, the massive foci of *A. mali* on *M. sieversii* in Xinjiang Province may represent a newly forming invasive range. The proximity of these sites to the wild apple stands and cultivated apple plantations in the Kazakhstan's part of Tien-Shan Mountains is a direct threat to Kazakhstan and other Central Asian countries [118,119].

Agrilus mali was reported to damage apple trees (i.e., *Malus* spp., *M. asiatica* var. *rinki*, *M. baccata*, *M. domestica*, *M. prunifolia*, *M. pumila*, *M. sieversii*, and *M. spectabilis*; Figure 5b), and it was also recorded to feed on other Rosaceae (i.e., *Cydonia oblonga*, *Prunus* spp., and *Pyrus* spp.) [127,128,133,136–138]. The record on *Emmenopterys henryi* (Rubiaceae) needs verification, and references to *Juglans* (Juglandaceae) and *Salix* (Salicaceae) are erroneous [133], while a report on *Sorbus* (Rosaceae) can refer to *A. mendax* [119].

An instance of large-scale damage by *A. mali* to cultivated apple trees was documented in Korea [134]. In Russia, it was also reported as a dangerous pest of apple trees in East Siberia and the Far East [135,139,140]. In the latter, the infested area covered about 320 ha [135]. Since the establishment in Xinjiang Province in 1993, *A. mali* has killed millions of wild apple trees and has infested about 95% of the total area of wild apple forests in the Ili valley [130].

Russia has a ban on import of apple seedlings from the areas infested by *A. mali* [135]. Agro-technical measures, mainly pruning and the destruction of infested branches and trees, as well as chemical treatments, are suggested as the control measures [139,140]. A parasitoid of emerald ash borer, *Sclerodermus pupariae* (Hymenoptera: Bethylidae), has been experimentally released in Xinjiang (China) and showed some positive results against *A. mali* [130]. This parasitoid, together with braconids (i.e., *Atanycolus ivanowi*, *Doryctes undulatus*, *Pareucorystes varinervis*, *Polystenus rugosus*, *Spathius sinicus*, and *S. brevicaudis*; Hymenoptera: Braconidae) parasitizing on *A. mali* were suggested as potential agents of biological control [141]. Aerial insecticide sprays during the beetle flight period can also be effective [128,130]. The combination of biocontrol and spraying/pruning turned out to be the most effective management approaches [130].

3.6. The Emerald Ash Borer, Agrilus planipennis Fairmaire, 1888 (Coleoptera: Buprestidae)

Agrilus planipennis (Figure 6) is described from China (Beijing) [142]. Since the beginning of the XXI century, the emerald ash borer has turned from a little-known East Asian species into one of the most devastating pests of ash trees in the world [143]. It has killed millions of trees in the forests and urban plantings in North America, European Russia, and East Ukraine [144–148].

The species native range is confined to East Asia: Russian Far East (Khabarovsk and Primorskii Krais), China (Beijing, Hebei, Heilongjang, Jilin, Liaoning, Nei Mongol, Shandong, Sichuan, Taiwan (syn. *A. feretrius*), Tianjin, Xinjiang Provinces), Mongolia (syn. *A. marcopoli*), North Korea, South Korea, and Japan (Hokkaido, Honshu, Kyushu, Shikoku) (syn. *A. marcopoli ulmi*) [146]. There are some uncertainties regarding the native range and synonymy. As such, the records for Nei Mongol and Sichuan need confirmation, whereas the records for North Korea are still lacking, though *A. planipennis* occurs in all adjacent countries. In addition, the taxonomic state of *A. feretrius* and *A. marcopoli ulmi* needs verification, and the type locality of *A. marcopoli* (Mongolia or China) is still unclear.

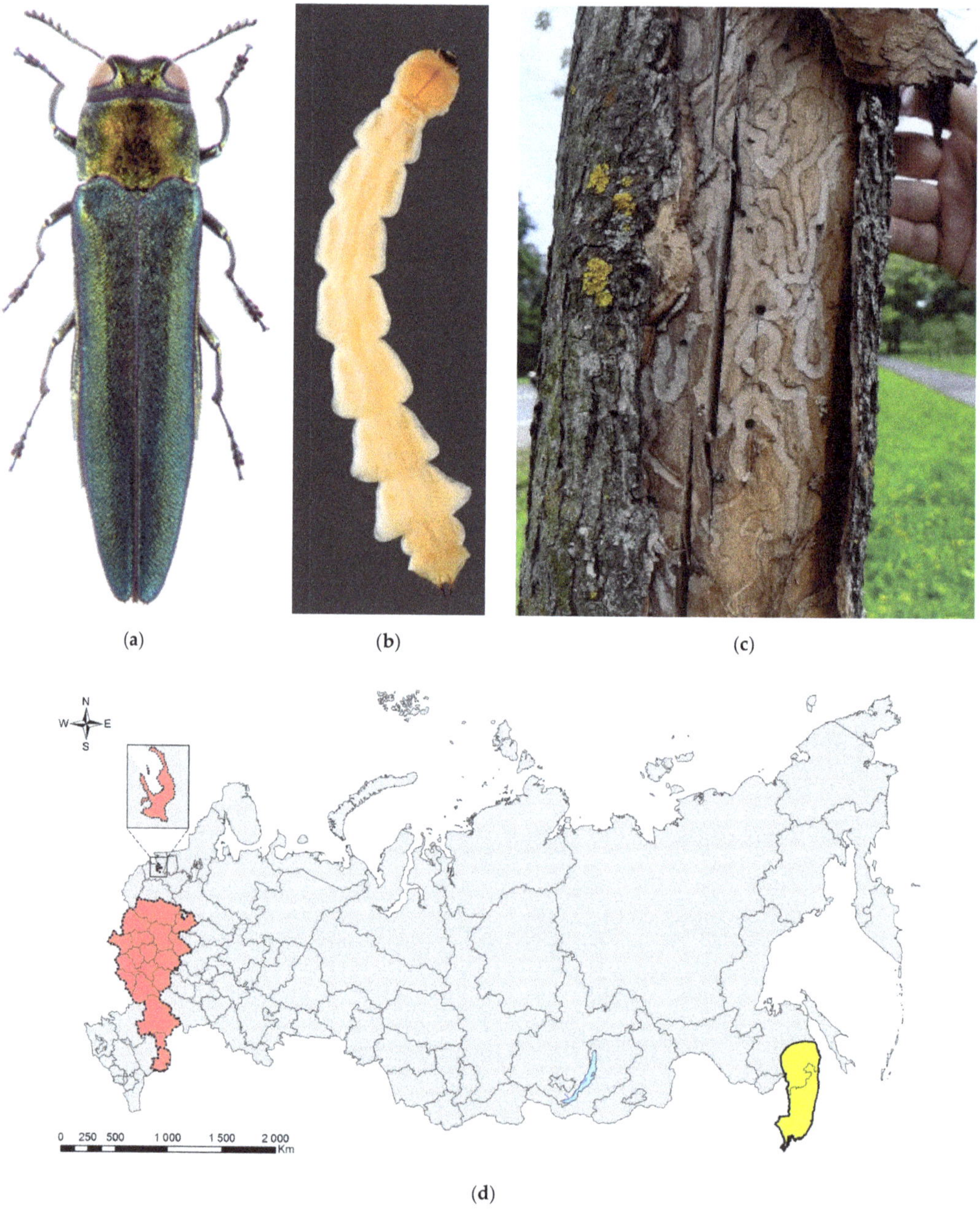

Figure 6. The emerald ash borer, *Agrilus planipennis*: (**a**) an adult; (**b**) a mature larva; (**c**) larval galleries and exit holes in young tree, Saint Petersburg, September 2020; (**d**) the current distribution in Russia: the native range (colored in yellow) and invasive range (colored in red). Photo: (**a**) by K.V. Makarov, (**b**) by A.V. Kovalev, after [149], both with permission, (**c**) by M.G. Volkovitsh.

Outside its native range, *A. planipennis* was first recorded in the USA (Michigan) in 2002 [150] and in the European part of Russia—in Moscow City in 2003 [144,145,151,152].

By early 2022, the emerald ash borer was recorded in 5 provinces of Canada and 35 states of the USA [153]; 18 provinces of European Russia, Moscow City, St. Petersburg City (Figure 6d); and two provinces of Ukraine (Luhansk and Kharkiv) [147,148,154–157].

The pest could be introduced to European Russia unintentionally in the late 1980s to early 1990s with wooden crafts, ash seedlings, or packaging material [158]. Taking into account that its outbreaks in North America and Moscow began almost simultaneously, an introduction from a common source (most probably from China) seems the most likely.

Under natural conditions, the emerald ash borer feeds on the East Asian ash tree species without causing significant damage to them due to the resistance of local ash species to the pest. However, in the areas of invasion, it completely destroys mature ash trees within a few years, depending largely on tree size and local *A. planipennis* population density [143]. Not all ash species are equally susceptible to *A. planipennis*. The examination of the ash collection in the N.V. Tsitsin Main Botanical Garden of the Russian Academy of Sciences (Moscow) revealed that only two Asian species (i.e., Chinese ash, *F. chinensis*, and Manchurian ash, *F. mandshurica*) were resistant to the emerald ash borer, whereas the pest killed the trees of both North American (i.e., *F. pennsylvanica* and *F. americana*) and European origins (i.e., *F. excelsior*, *F. angustifolia*, and *F. ornus*) [159].

The emerald ash borer can spread due to active flight, the human-mediated movement of infested wood (in North America, particularly, firewood), or the occasional transportation by vehicles and ships. Presumably, the beetles can also be transported over long distances by wind [143,158].

Since the first detection in Moscow in 2003, *A. planipennis* significantly expanded its invasive range in European Russia. In 2006, it was found in the Moscow Oblast, whereas by 2013, it was detected already in 11 oblasts of European Russia [160] and by early 2021, in 18 oblasts and in Saint Petersburg City [148,154,161]. The most recent detected localities are Saint Petersburg City and the Astrakhan Oblast [148,156,157].

The emerald ash borer has killed over tens of millions of healthy ash trees at a cost of billions of US dollars since its occurrence in North America [143]. In Europe, the potential losses can reach USD 1.81 billion. By this indicator, the species ranks fourth among the most "costly" invasive pests [162]. The actual losses (observed costs) from *A. planipennis* invasion in European Russia are estimated at as much as USD 258.9 million during 2011–2016 [163,164]. The impact might be even higher in the regions where the invasive range of *A. planipennis* overlaps with the invasive range of the devastating ash dieback fungus, *Hymenoscyphus fraxineus* (Ascomycota) [145].

Infestation usually begins in the canopy, with subsequent attacks lower along the trunk, while in small-diameter trees, infestation often starts along the main trunk. Canopy drying, epicormic branching along the lower trunk, bark crevices, and D-shaped exit holes on the bark are the most common symptoms of infestation [143,158,160].

Since its initial detections in Moscow City and the Moscow Oblast, monitoring in European Russia has been performed and, at the beginning, mostly by volunteers and enthusiasts. Initially, the monitoring was limited to the search for new foci and heavily infested ash trees in the urban environments. By now, seven quarantine phytosanitary zones in five oblasts of European Russia are declared by the National Plant Protection Organization (Rosselkhoznadzor) [148]. The pest is included in A Unified List of Quarantine Pests of the Eurasian Economic Union [131], and it is also included in the priority pest lists of the EPPO [120] and the European Food Safety Authority (EFSA) [165].

A recent analysis suggests that *A. planipennis* can potentially invade most European countries; however, it would probably not be able to establish in some regions of Norway, Sweden, Finland, Ireland, and Great Britain because of low annual heat accumulation in these regions [166].

Control measures include the cutting of infested trees (mostly detected by dry branches and typical D-shaped exit holes on the bark), with a clear-cut zone, replacement of North American and European ash trees with more resistant Asian ash species (first of all, *F. chinensis* and *F. mandshurica*) or possibly hybrids, and chemical and biological control [145,167]. Currently, the biological control agents are intensively looked for within the native and invasive ranges [145,168–170].

3.7. The Cypress Jewel Beetle, *Lamprodila (Palmar) festiva* (Linnaeus, 1767) (Coleoptera: Buprestidae)

Lamprodila festiva (Figure 7) is originally a relatively rare Mediterranean species, which, about 20 years ago, suddenly became a devastating pest of ornamental cypress trees (Cupressaceae) in urban plantings and nurseries all over central and East Europe, including south of European Russia and Eastern Ukraine [160,171] (Supplimentary Table S4). In addition to the Mediterranean, the native range of *L. festiva* covers North Africa, South Europe, the southern part of central Europe (*L. f. festiva*, and *L. f. cretica*) and Southwest Asia (*L. f. holzschuhi*) [160,171,172].

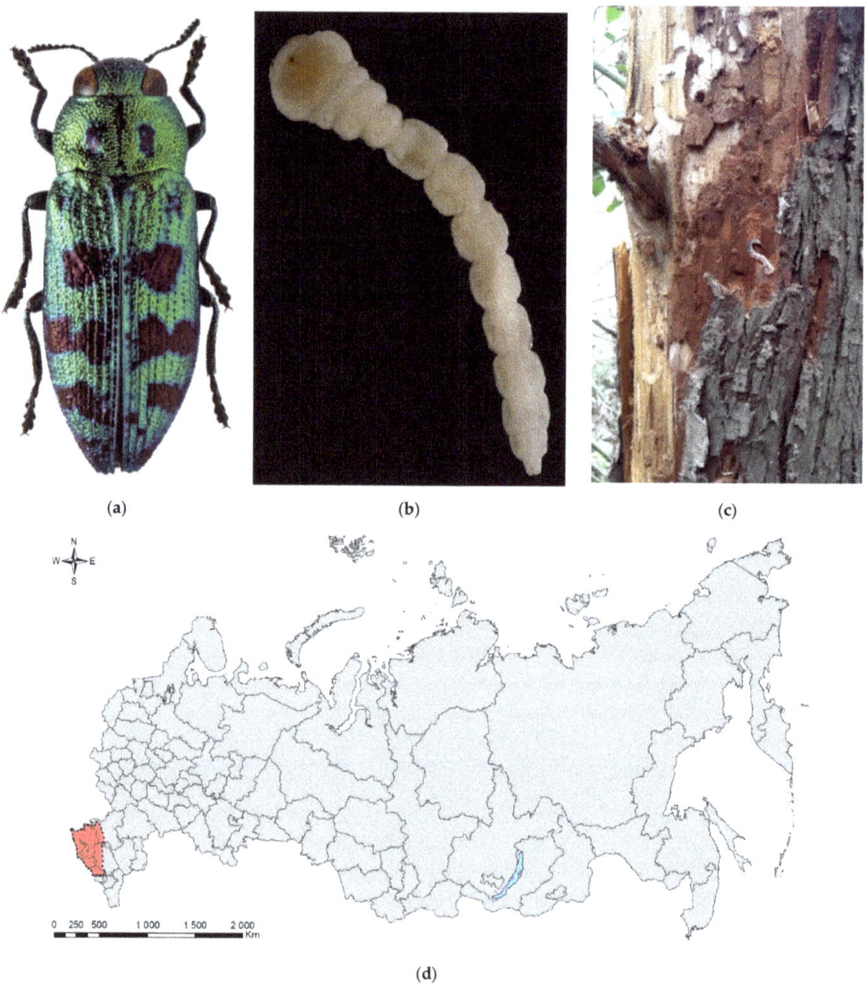

Figure 7. The cypress jewel beetle, *Lamprodila (Palmar) festiva*: (**a**) an adult; (**b**) a larva; (**c**) larval tunnels with larvae on the trunk of *Thuja plicata*, Sochi City, 2016; (**d**) the invasive range in Russia (colored in red). Photos: (**a**) and (**b**) by A.V. Kovalev, after [171], with permission; (**c**) by N.N. Karpun, after [172].

Until recently, *L. festiva* has never been recorded in Russia or the former USSR. The first symptoms of infestation were detected in the Black Sea coast (Greater Sochi) in 2013, and the first larvae and adults were collected in 2016 [172,173]. Most probably, the pest was unintentionally introduced together with cypress planting stock from the European nurseries (likely from Italy, Spain, or Montenegro) in 2011 to 2013, during intense landscaping in Sochi before the XXII Winter Olympic Games in 2014 [172,174]. Presently in Russia, the cypress jewel beetle is distributed along the Black Sea coast from Sochi to Anapa, in North Caucasus (Dombai, Krasnodar, Nalchik, Rostov, Stavropol, Nevinnomyssk, and Cherkessk Cities), and the south coast of the neighbouring Crimean Peninsula, from Sevastopol to Gurzuf (Figure 7d) [107,160,171,172,175–178].

During 2013–2021, *L. festiva* expanded its invasive range as much as approx. 500 km north (Donetsk, Ukraine), about 240 km northeast (Stavropol), and 310 km east (Nalchik). It is distributed along the Black Sea coast for about 360 km from Vityazevo settlement at the north to Sukhum City in the south, with the maximal distance from the initial record locality (Sochi) of 250 km (Vityazevo). Actually, the first reported findings outside Sochi were from Gelendzhik (170 km northwest from Sochi) in 2017 [177]. These records indicated a high rate of distribution of the pest, most likely associated with the spread of infested planting stock.

Until recently, the cypress jewel beetle was regarded as a secondary pest of wild and some introduced Cupressaceae (*Thuja* spp.), without significant damage [172]. Currently, the beetle is recorded on a wide range of cultivated cypress species from *Callitris, Chamaecyparis, Cupressus, Juniperus, Platycladus, Tetraclinis*, and *Thuja,* including their hybrids and cultivars [171]. Recently, *L. festiva* was reported to attack *Sequoia sempervirens*, weakened by fungal decease in the Nikitsky Botanical Garden (Crimean Peninsula) [176]. Thus, now, *L. festiva* is considered a dangerous pest of ornamental cypress trees in many European countries [160,172,179–184]. There are still no quantitative data available on the economic losses caused by this pest to urban plantings; however, the estimated damage to ornamental plants is considered to be significant [160,174,181,183].

The early symptoms of tree infestation with Cypress jewel beetle are brown drying needles, the death of damaged branches, and, in case of heavy infestation, drying and death of a whole plant. The secondary symptoms are fissures, resin leakage, swellings, bark detachment, presence of frass on the trunks and on the soil under the plants, exit holes, and larval galleries under the bark (Figure 7c) [172].

Monitoring of *L. festiva* in Russia is performed locally since the first discovery of damaged cypress trees in the Black Sea coast [172–174,176,177,185,186]. The monitoring is mainly restricted to the search for new foci. The pest control has not been elaborated so far.

The parasitoids of the cypress jewel beetle, such as *Pyemotes* sp. (Acarina: Pyemotidae) and *Metacolus unifasciatus* (Hymenoptera: Pteromalidae), are reported from Bulgaria [184]. On the Black Sea coast, chemical control is carried out during the *L. festiva* adult flight period (preventive treatments of cypress trees with insecticides from the pyrethroid and organophosphorus compounds to prevent colonization of new plants) [107].

3.8. The Small Spruce Bark Beetle, Ips amitinus (Eichhoff, 1872) (Coleoptera: Curculionidae: Scolytinae)

Ips amitinus (Figure 8) is a European species which has been considered a secondary pest associated with dying conifers, including different spruce species (e.g., *Picea abies* and *P. omorica*) and several species of pines (e.g., *Pinus cembra, P. mugo, P. peuce*, and *P. sylvestris*) [187]. Its native range spreads from The Netherlands to Turkey [188]. In the XX and early XXI centuries, the range of *I. amitinus* rapidly expanded into Northern Europe [189–192]. The first recorded specimens of *I. amitinus* from Estonia are dated by 1900 [193–195]. In the following decades, the species spread northward, and it had occupied the whole of Estonia by the end of the 1930s [194,196,197] and was recorded from southern Finland (Tuusula Administrative Region) in 1951 [198,199]. The species spread rapidly northward in Finland [200], so that in 1968, it was already recorded in Northeast Finland at the boundary with the modern Paanajärvi National Park [201], where it was most probably absent in 1960 [202]. The small spruce bark beetle invaded northern Sweden, apparently from Finland, in 2012 and continues to move southward [203].

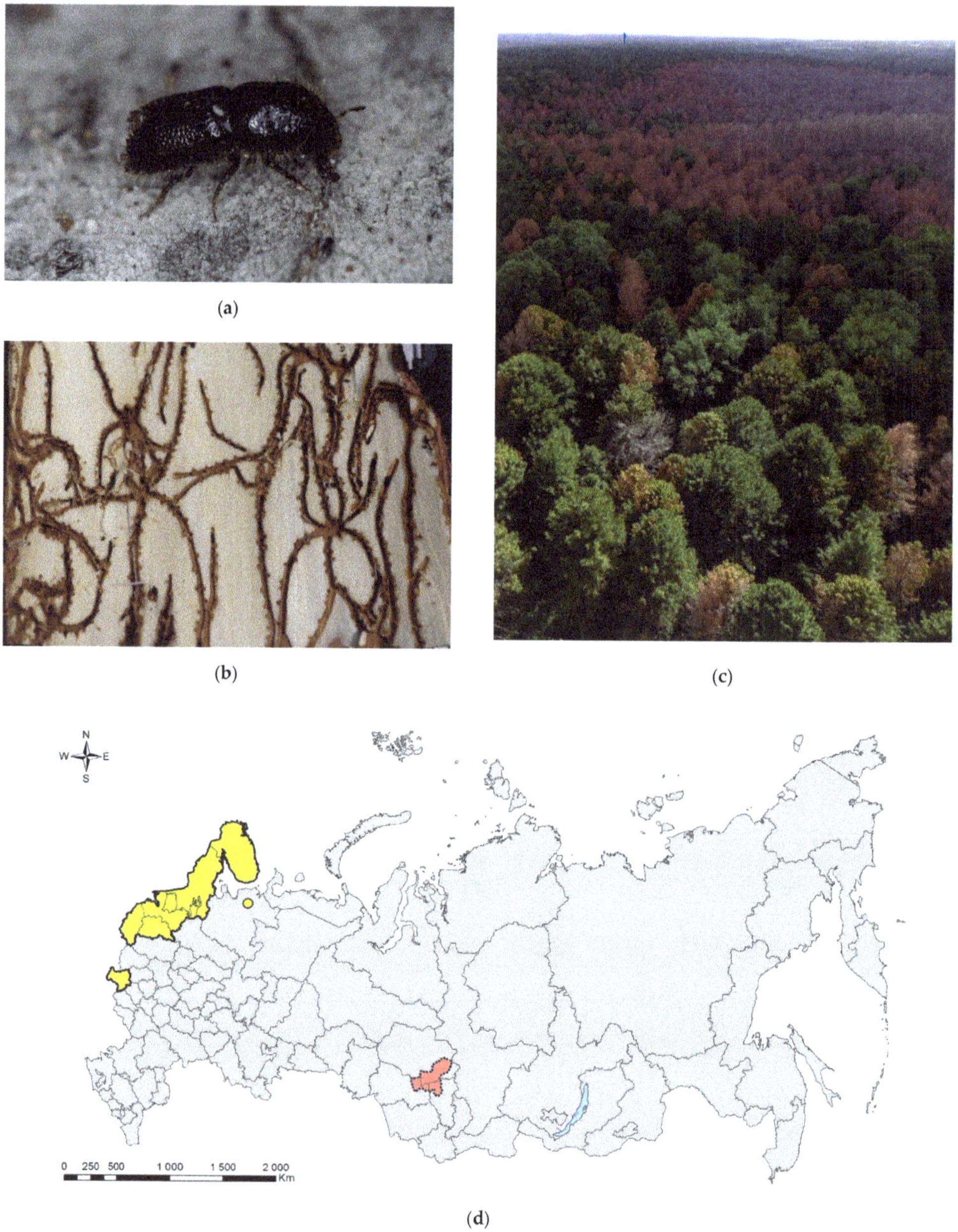

Figure 8. The small spruce bark beetle, *Ips amitinus*: (**a**) an adult on a branch of *Pinus sibirica*, Tomsk Oblast, 2021; (**b**) galleries on *P. sibirica*, Tomsk Oblast, 2021; (**c**) *I. amitinus* foci in the Siberian pine forest near settlement Luchanovo, Tomsk Oblast, June 2020; (**d**) the current distribution in Russia: the native range (colored in yellow) and invasive range (colored in red). Photo by I.A. Kerchev.

The first record in Russia refers to the Briansk Oblast and is dated back to 1920 [204, 205]. The "newly" inhabited territories of *I. amitinus* in Russia includes the Leningrad Oblast (Zelenogorsk in the Karelian Isthmus, 1976), though its appereance here was expected much earlier [205,206]. In the 1980s, the species was already widely distributed in Russian Karelia. Investigation of Titova [207] did not reveal *I. amitinus* in Karelia, but Yakovlev and coauthors [208] reported the species in southern Olonets and Loimola, as well as the central part of Karelia (Voloma). In 1989, it spread up to the Murmansk Oblast, Kandalaksha Nature Reserve [206,209]. In 2011, *I. amitinus* was recorded on pines nearby the northern range of a conifer forest in Pasvik Nature Reserve (Melkefoss; on the border between Russia and Norway) [210].

The easternmost record of *I. amitinus* in Europe known to date is the Pinega settlement in the Arkhangelsk Oblast, where the species was collected in 2013 on pine logs, but no infestations on spruces were recorded [189,211]. In 2000, the beetle was found in Pudozh, east to Onega Lake [206]. Thus, *I. amitinus'* range spreading continues not only northward but also eastward. Natural barriers (such as Ladoga Lake) provide limitations to the species spreading and only in 2007 was *I. amitinus* trapped on Valamo Island and only in a wood harbor on the logs imported from the mainland [205,212].

The species invaded Western Siberia relatively recently (presumably, in 2005–2010), attacking a new host tree, the Siberian pine (*Pinus sibirica*), and, at least in the Tomsk Oblast, it was absent until 2012 [213]. The first symptoms of tree damage caused by the small spruce bark beetle were noted in 2014 in the Kemerovo Oblast on the borderline with the Tomsk Oblast [213–215]. A rapid growth of the *I. amitinus* population in 2016–2017 was provoked by the outbreak of the Siberian moth, *Dendrolimus sibiricus,* and favorable weather conditions, which altogether resulted in a considerable number of heavily defoliated and snow-broken trees [213,216]. By the time the invasion was registered in 2019, *I. amitinus* had already switched to healthy trees and started to breed on the tree stands nearby undamaged by the Siberian moth [217]. By 2019, the area of the outbreak foci had reached 237 ha in the Tomsk Oblast and 1033 ha in the Kemerovo Oblast. By the end of 2020, it was already 1207.5 and 1232.4 ha, respectively. The invasive range currently covers Tomsk, Kemerovo, and the Novosibirsk Oblasts, having a total area of 31.2 mln ha [218].

The most probable pathway of *I. amitinus* invasion is natural range expansion in the European part of Russia and Fennoscandia, and transportation with wood by railroads from European part of Russia to Western Siberia. This is confirmed by a number of proofs, such as a long distance between its native and invasive ranges, the proximity of the first recorded outbreak foci in Siberia to the Trans-Siberian Railway, and the rates of invasive range expansion observed at the moment [217]. In Western Siberia, the median density of *I. amitinus'* nuptial chambers on *P. sibirica* was 2.7 (up to 20.0) per dm^2 [216], i.e., 10 times higher than on trap spruce trees in the Czech Republic [219].

The strategy to control this new pest has not been developed in Russia yet. The application of a complex of classical forest protection measures are hampered by a ban on sanitary felling in specially protected natural areas. The first attempts to control *I. amitinus* carried out in the Tomsk and Kemerovo Oblasts were inherently flawed because a species-specific attractant of *I. typographus* was used for the mass pheromone trapping. Moreover, this approach had a negative effect, as it attracted to traps a natural enemy, a checkered beetle *Thanasimus femoralis* (Coleoptera: Cleridae) [215].

The outbreak of *I. amitinus* can be slowed down by sanitary cutting of infested trees and by using trap trees, as well as by timely removal of wind-broken and snow-fallen trees; however, these measures have not been undertaken so far [215]. In the Tomsk Oblast, the death of trees has reached a catastrophic rate [220]. As the attacks mostly begin on the tree tops, drones carrying digital cameras can help to accurately assess the damage [220].

In 2020, *I. amitinus* caused the death of introduced mountain pine (*Pinus mugo*), Korean pine (*P. koraiensis*), and some individuals of native conifers (e.g., *P. sibirica*, *P. sylvestris*, and *Picea obovata*) in the Arboretum "Kedr" of the Institute of Monitoring of Climatic and Ecological Systems, Siberian Branch of the Russian Academy of Sciences (Tomsk Oblast).

The noted trophic preferences of the bark beetle indicate that it strongly prefers the five-needle pines, a trait that has likely developed in the course of coevolution with *P. cembrae* and *P. peuce* in its native range [217].

We believe that further expansion of *I. amitinus* range will continue eastward rather than northward, and the pest will be able to attack eastern pines, for instance, *P. koraiensis* in the Russian Far East [217]. In Western Russia, the species' arrival is also expected up to the Ural Mountains.

Keeping in mind that the species is highly aggressive to Siberian pine and has a potential to outbreak in spruce stands in Europe, its populations should be monitored both in European and Asian parts of Russia. Dramatic rates of Siberian pine stands decline caused by the small spruce bark beetle in West Siberia has shown that the species invasion potential had been underestimated.

3.9. *The Four-Eyed Fir Bark Beetle, Polygraphus proximus Blandford, 1894 (Coleoptera: Curculionidae: Scolytinae)*

Polygraphus proximus is a conifer pest species of East Asian origin, which, in the last few decades spread in Russia westwards [221,222]. It naturally occurs in Japan, North and South Koreas, Northwest China (Heilongjiang and Jilin Provinces) [188], and in the south of the Russian Far East (Khabarovsk and Primorskii Krais, Sakhalin Island, and Southern Kurils) (Figure 9). In its native range, it develops on firs (i.e., *Abies nephrolepis, A. sachalinensis, A. holophylla*, and *A. mayriana*; Pinaceae) [223,224]. According to the list of the fir bark beetle's host plants from the native range [225] and analysis of the trees colonized by the beetle in artificial plantations [226,227], it was concluded that firs from the sections Balsamea and Grandis are most preferred, while those from the section Abies are least preferred by this bark beetle [225]. Less common *P. proximus* can be found on other genera of Pinaceae: *Picea* (*P. jezoensis*), *Pinus* (including Korean pine, *P. koraiensis*), *Larix*, and *Tsuga* [224,225].

The first record outside of the species' native range is dated back to 1999, to the Leningrad Oblast [228]. As the species was never documented in this region again, we consider this record to be doubtful [190]. The species was unintentionally introduced to Moscow City [226], where it was recorded in 2006 on artificial stands of *Abies sibirica* and *A. balsamea* along Kurkino highway in the Khimki District. It was also discovered in 2006–2007 on Norway spruce (*P. abies*) and *A. sibirica* in Pushkin, Odintsovo, and Podolsk Districts of the Moscow Oblast [226]. In the East European Plain, fir bark beetle's outbreaks have been registered in the Agryz region of Tatarstan, in the Sarapulsky, Kiyasovsky, Zavyalovsky, and Malopurginsky regions of the Udmurt Republic [229,230].

The species invaded Siberia, from where it was recorded for the first time in 2008, and the beetles were found in pheromone traps for *Ips sexdentatus* in *P. sibrica* stands nearby Tomsk City [231]. In 2009, the presence of *P. proximus* was confirmed in Krasnoyarsk Krai, and it then became clear that this species (and not *Xylechinus pilosus* as it had been though before) was the real cause of Siberian fir stands decline [221–234]. The most probable pathway of *P. proximus* invasion is the transportation of unbarked timber or packaging materials from the Far East. Presently in Siberia, the four-eyed fir bark beetle is found in Tomsk (except the north part), the Kemerovo, Novosibirsk, and Irkutsk Oblasts, Altai, and Khakassia (Figure 9d) [221–236]. It was recorded at an altitude of 100–200 m a.s.l., and in mountainous areas, it inhabits the belt of 300 to 1490 m a.s.l. [225].

The dendrochronological analysis revealed that the beetle killed the first trees in Krasnoyarsk Krai in about 1976 [237] and in the Tomsk Oblast before 2020 [238]. It was suggested that the pest could have been introduced to the Siberian fir range much earlier, e.g., in the 1960s when fir lumber from the Russian Far East was brought to the Kemerovo Oblast for construction of mines [229,237].

Figure 9. The four-eyed fir bark beetle, *Polygraphus proximus*: (**a**) an adult on *Abies sibirica*, Tomsk Oblast, 2021; (**b**) maternal and larval galleries with pupal chambers on a stem of *A. sibirica*, Tomsk Oblast, 2021; (**c**) *P. proximus* foci in the dark coniferous forest near settlement Barbig, Tomsk Oblast, July 2017; (**d**) the current distribution in Russia: the native range (colored in yellow) and invasive range (colored in red). Photo by I.A. Kerchev.

The four-eyed fir bark beetle was introduced into the Tomsk Oblast and the Udmurtia at approximately the same time, in the mid-1990s, since the beetles taken from both localities belong to the same haplogroup, as suggested by the cytochrome oxidase subunit 1 (COI) mitDNA analysis [239], whereas the specimens from the Krasnoyarsk Krai and Khakassia belong to a divergent group [239,240].

Since the first record in Siberia, mass infestations of *P. proximus* were recorded only on the Siberian fir, but not on other tree species, whereas laboratory tests demonstrated the pest's ability to develop on other Pinaceae species (e.g., *Pinus sibirica*, *Picea obovata*, and *Larix sibirica*) [241]. Later, in the outbreak foci, infestation of wind-broken spruces, Scots pine (*Pinus sylvestris*) and Siberian pine (*P. sibirica*), by *P. proximus* was documented [225].

The beetle can vector an obligate symbitic fungus *Grosmannia aoshimae* (Ascomycota: Ophiostomataceae), and development of the fungus leads to rapid weakening of the host trees [242]. This fungus arrived at Siberia together with *P. proximus* [242,243]. Laboratory tests demonstrated that the fungus could colonize and develop on the same unusual host tree species as *P. proximus* [244]. This work revealed the possibility of a hidden existence of invasive complexes even on the hosts that are not favorable for reproduction of the pest.

Polygraphus proximus kills fir stands in Siberia causing significant changes in the structure of the ecosystem [245]. In declining forests, the microclimate changes, creating preconditions for future replacement of taiga with deciduous trees. Falling down of dead trees leads to a significant imbalance between the intensity of production and destruction processes in the ecosystems [246]. Significant changes are also observed in the vegetation cover and structure of the local insect fauna [233,247,248]. Natural enemies of *P. proximus* [225,249,250] and entomopathogens [251] are not able to control the pest in its invasive range [225]. One of the control methods that has shown high efficiency against the fir bark beetle is the traditional practice of keeping cut trees submerged in water [252]. An integrated approach including treatment with systemic insecticides, timely sanitary cutting and removal of colonized trees used in the N.V. Tsitsin Main Botanical Garden of the Russian Academy of Sciences (Moscow), showed high efficiency in suppression of an outbreak on a relatively small territory [253].

As mentioned above, firs from the sections Balsamea and Grandis are most preferred by the beetle [225]. Thus, possible accidental introduction of *P. proximus* to North America, where firs from these sections are widely distributed, would result in enormous damage. In Europe, firs are represented by *Abies alba* and *A. nordmanniana* from the section Abies [254], which is less preferred as a host by the beetle than representatives of sections Balsamea and Grandis [225]. Nevertheless, phytosanitary and monitoring measures aimed at preventing the invasion of this species and early detection should be prioritized. In the regions where *P. proximus* spreads widely, efforts should be put to minimize damage by conducting timely forest sanitary measures.

3.10. *The Box Tree Moth, Cydalima perspectalis (Walker, 1859) (Lepidoptera: Crambidae)*

Cydalima perspectalis (Figure 10) is an East Asian species naturally distributed in China, Japan, the Korean Peninsula, the Russian Far East, and India [255,256]. Outside of its native range, the pest was detected in Germany in 2006, where it was presumably imported with boxwood planting material from China [257]. Further invasion of the species in Europe proceeded rapidly (Supplimentary Table S5). In 2007, it was included in the EPPO Alert List, but already in 2011, it was excluded as a species widespread in the region [120]. In 2021, the boxwood moth was found in North America (USA: Connecticut, Michigan, and South Carolina) [258].

(**a**)

(**b**)

(**c**)

(**d**)

Figure 10. Box tree moth, *Cydalima perspectalis*: (**a**) a larva of VI instar, Sochi, August 2013; (**b**) an adult moth, Sochi, September 2014; (**c**) severely damaged box tree, *Buxus sempervirens*, Sukhum District, Abkhazia, July 2015; (**d**) the current distribution in Russia: the native range (colored in yellow) and invasive range (colored in red). Photo: (**a**) and (**c**) by N.N. Karpun, (**b**) by E.N. Zhuravleva.

In Russia (Figure 10d), the larvae of *C. perspectalis* were first recorded on the planting stock of *Buxus sempervirens* in 2012, in the nursery with planting material imported for landscaping the territory of the Main Olympic Village in Sochi [259]. In 2013, the species gave an outbreak in the ornamental plantations in the city of Sochi [260] and was recorded in the plantations in the city of Grozny (Chechnya) [261], and by 2014, it had spread across the Krasnodar Krai and the neighboring region of Abkhazia [262]. In 2015, the species was recorded in the Crimean Peninsula, in Adygea (West Caucasus), in the Mineralnye Vody region (Stavropol Krai), and North Ossetia (North Caucasus) [97,263–265], while in 2016 and in 2017, it was recorded in Rostov-on-Don [266] and in the neighboring region of South Ossetia [267], respectively. In 2019, the boxwood moth was found in Dagestan (North Caucasus) [268] and in 2020 in the Kaliningrad Oblast (southeast of the Baltic Sea) [269]. The main pathway of *C. perspectalis* invasion was the introduction of infested boxwood planting material [63,257,259,270].

The pest causes dramatic defoliation of boxwood in both natural and ornamental plantations (Figure 10c). In 2015–2017, it had almost completely destroyed the natural plantations of boxwood in the North Caucasus and the Black Sea coast of Russia and further eastwards situated region of Abkhazia [63]. Caterpillars of the early (I–III) instars gnaw the lower part of the leaf without touching the upper epidermis, and matured caterpillars consume the whole leaf. At high population densities, they also eat the bark of branches and trunks.

Within the native range, *C. perspectalis* damages *Buxus* spp. (Buxaceae), *Euonymus japonicus* (Celastraceae), and *Ilex purpurea* (Aquifoliaceae). In Europe, it feeds on all native and cultivated species of boxwood, while in Russia, it feeds on 9 boxwood species and 13 cultivars [271].

In the humid subtropics of the Black Sea coast of Russia, the pest produces three generations per year and, under favourable conditions, four generations per year can develop. The species overwinter at different stages: egg, caterpillar, and pupa. As a result, different generations overlap [63].

The boxwood moth is monitored by regular visual inspections of plantations and by pheromone trapping. Standard glue deltoid traps are used and they are placed on the branches of boxwood or nearby. The pheromone combines unsaturated aldehydes: (Z)-11-hexadecenal (Z11-16:Ald), (E)-11-hexadecenal (E11-16:Ald), and (Z)-11-hexadecenol (Z11-16:OH) [272–274].

Within the invasive range in Southern Russia, only a few birds and parasitoid wasps (Hymenoptera) attack the alien pest. Despite the fact that in Europe, parasitoids, entomopathogenic fungi, and nematodes are considered ineffective in the control of this pest under natural conditions [275,276], in the humid subtropics of Russia, the early results of the application of entomopathogenic fungi are encouraging [277]. *Bacillus thuringiensis*-based preparations are effective against early-instar caterpillars [276,278,279], whereas avermectins, organophosphorus compounds, pyrethroids, and neonicotinoids are considered effective against older-instar caterpillars [279–282]. Further range expansion of *C. perspectalis* from Russia is likely into natural forests across the Caucasus (Transcaucasia) and to countries located further south.

3.11. The Leaf Blotch Miner Moth, Acrocercops brongniardella (Fabricius, 1798) (Lepidoptera: Gracillariidae)

Acrocercops brongniardella (Figure 11) is a micromoth species widely distributed in Europe and European Russia (except northern and southern regions) [283,284]. The presence of *A. brongniardella* in the Russian Far East is questionable. A single record of the species in Khabarovsk Krai, in fact, refers to *Caloptilia infuscatus* [285], a taxon which might represent a senior synonymy of the East Asian sister species, *Acrocercops amurensis* [286].

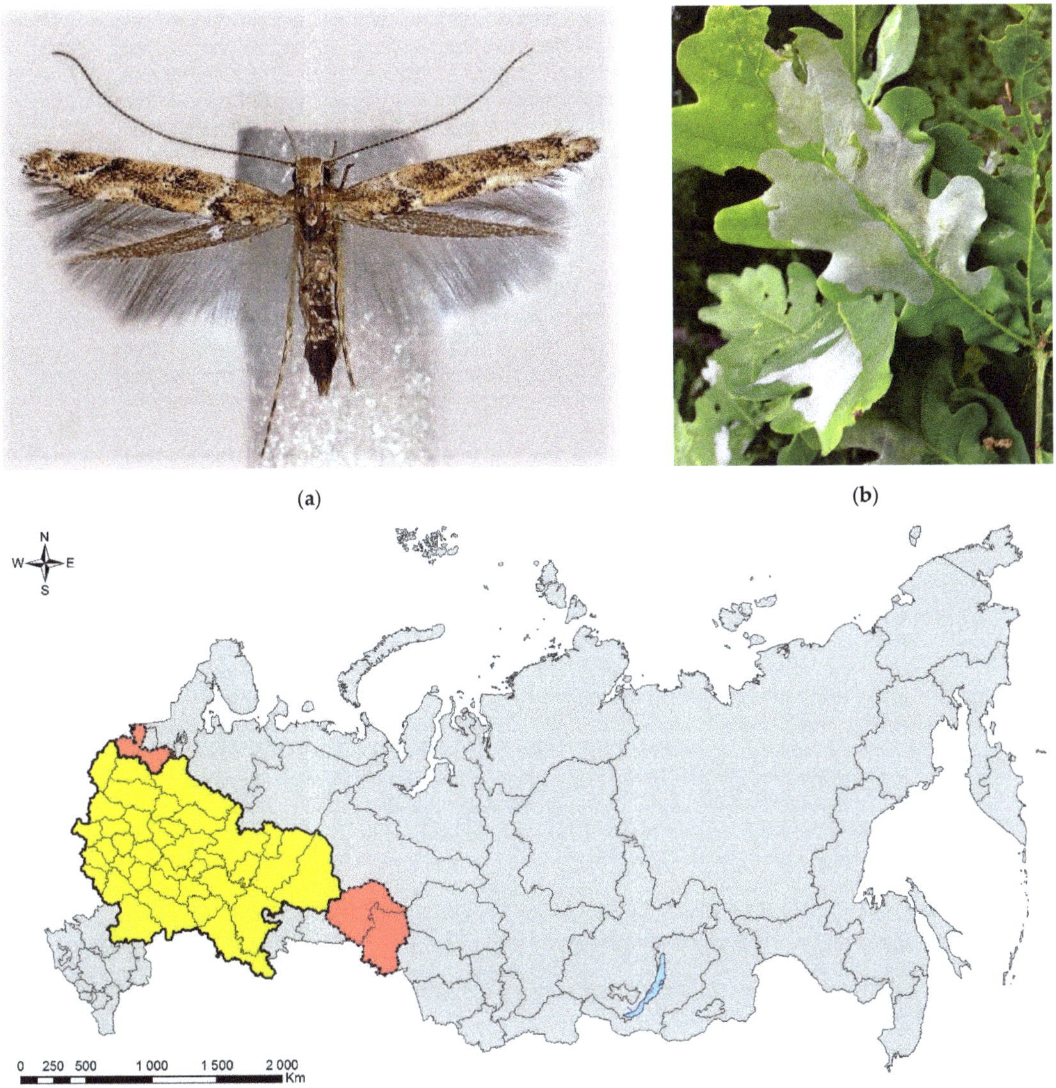

(a) (b) (c)

Figure 11. The leaf blotch miner moth, *Acrocercops brongniardella*: (**a**) an adult moth; (**b**) leaves of *Quercus robur* with mines; (**c**) the current distribution in Russia: the native range (colored in yellow) and invasive range (colored in red). Photo: (**a**) by D. Lees; (**b**) by I.A. Utkina, both with permission.

In European Russia, *A. brongniardella* has been found to expand its native range to northern territories, in particular to the Leningrad Oblast and Saint Petersburg City, where it was detected for the first time in 2018 [287,288]. In the following years, the species was recorded south and north of Saint Petersburg, in Gatchina and Sestroretsk, respectively [289,290]. In addition, *A. brongniardella* was also documented in a few localities in Western Siberia, Omsk, and Tyumen Oblasts (Figure 11c) [291–293], where it is considered an alien species. In the city plantations of Omsk, it produces noticeable outbreaks [291].

The pest mines the leaves of oaks (*Quercus* spp.), including evergreen species (e.g., *Q. ilex*), whereas the most common host is the deciduous oak, *Q. robur* [284]; occasionally, the mines can be found on leaves of *Castanea sativa* (Fagaceae) [294]. The mines are big blotches on the upper sides of leaves, often covering a significant area of the leaf lamina (Figure 11b). Fully developed mines are easy to spot by the whitish color. Occasionally, up to 16 young mines can be found on a leaf, later on merging into a big blotch [291,295]. Leaf mining starts in May, then the larvae vacate the mines in July and pupate externally. Adult emergence is prolonged. Moths overwinter in bark crevices and dwellings [291,295]. In Ukraine, the species produces two generations per year [295,296], whereas in Russia, one generation per year was documented [291,297–299].

It is one of the most important folivore insect pests of oaks in Russia. The intense and prolonged outbreaks have been recorded in central parts of European Russia [299,300]. In a foci, often more than 75% of leaves carry mines and increased population density can be observed for several successive years [297,298]. Long-lasting outbreaks can lead to decrease of radial growth, drying of crowns, and death of oaks [301]. However, the impact of this species remains insufficiently well studied [297–299].

In the Leningrad Oblast, where *Q. robur* is a native tree and where the northern limit of its range in Russia is known [302], so far, the moth has been found in low density [287,289,290]. In contrast, in Western Siberia (in particular the Omsk Oblast), the moth shows a tendency to outbreak and harm *Q. robur* [291], an introduced tree in Western Siberia. The first planting of *Q. robur* in the Omsk Oblast is dated to 1898, whereas, in the city of Omsk, it has been planted since 1948. The moth could have arrived to the region with plants for planting, with vehicles or by itself, naturally expanding its range.

In the northeast of European Russia (Udmurtia), no outbreaks have been recorded in the last 35 years [303,304]. We hypothesize that insufficiently warm summers in the north of Russia could limit the moth population density in Saint Petersburg area. The observed climate warming in Saint Petersburg [288] and availability of the main host plant, *Q. robur*, in the city plantations may favor the pest outbreaks in the following years.

In cities, the control of *A. brongniardella* populations is complicated. Removal of leaf litter in order to reduce the population density will unlikely lead to desirable results as with the pest, its native enemies (in particular parasitoids) are killed. The use of insecticides is often impossible in urban areas. Injections of systemic insecticides into trees in high-value plantings and protected areas might be a perspective to keep in mind, given their effectiveness against other pestiferous invasive gracillariids [305–307]. However, so far, this approach has not been tested against this pest in Russia.

3.12. The Horse-Chestnut Leaf Miner, Cameraria ohridella Deschka et Dimić, 1986 (Lepidoptera: Gracillariidae)

Cameraria ohridella (Figure 12) is a tiny moth that became widely known due to its fast distribution across Europe and scenic damage to the horse chestnut, *Aesculus hippocastanum* (Sapindaceae) [308]. It is the only species of the genus *Cameraria* occurring in Europe, while the majority of the genus representatives (83 species) are distributed in North America, with a few species found in Asia and Africa [284]. Curiously, *C. ohridella* was not even known to science before it started invading Europe (Supplimentary Table S6). In the 1980s, it was discovered in high densities in North Macedonia near the Ohrid Lake, from where it was formally described [9]. In the following years, the moth was recorded in many European countries and, in approx. three decades, it occupied most of Europe [308,309].

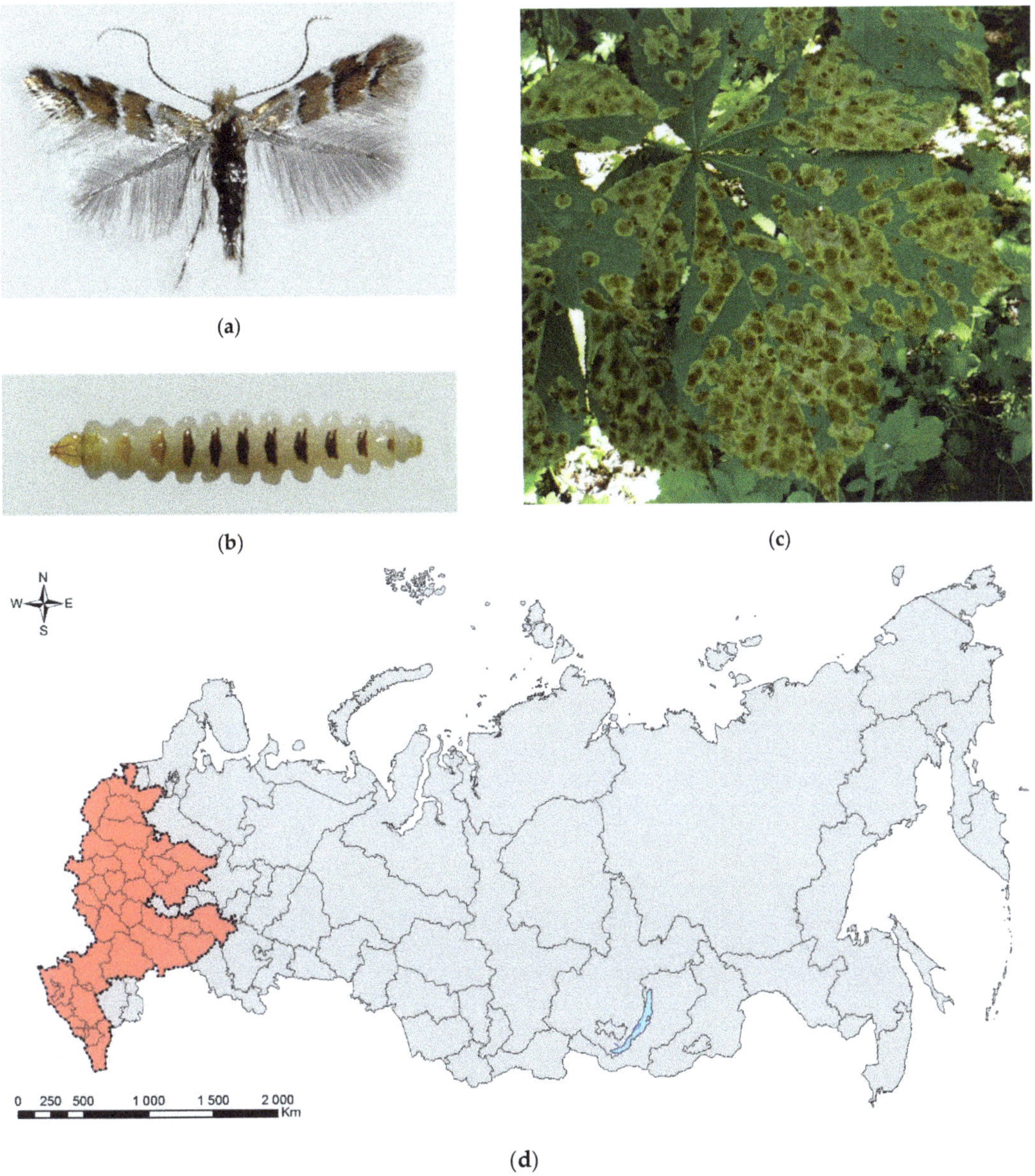

Figure 12. The horse-chestnut leaf miner, *Cameraria ohridella*: (**a**) an adult moth; (**b**) a mature larva; (**c**) damaged leaves of *Aesculus hippocastanum*, Anapa City, 2021; (**d**) the invasive range in Russia (colored in red). Photo by N.I. Kirichenko.

For more than 20 years, the origin of the species was unclear. Some authors hypothesized it to be a relict species that, together with its host, survived the glaciation of the Tertiary Period [9]. Hellrigl [310] suggested that the species originated from southeastern Europe, where it could have shifted from maple (*Acer* spp.) to *Ae. hippocastanum*. Finally, others suggested the species to be alien for Europe and probably originating from North America or Asia [311]. A phylogeographic study showed that the most probable motherland of the species is the Balkans [312,313]. Furthermore, a survey of historical herbaria confirmed the long-term presence of the species in the Balkans, since the mines were found in pressed leaves sampled more than a century ago, that is, long before the moth was recorded in other parts of Europe [313].

Progressively spreading on the European continent, the moth arrived in European Russia [314,315]. In Russia, its first record is dated to 2003, and refers to Kaliningrad City (southeastern coast of the Baltic Sea) [315]. In 2005, the moth was documented in Moscow [316], about 1000 km east from Kaliningrad and about 450 km from the border with Ukraine and Belarus, where the pest was already present by that time [315,317]. In the following years, the species was found in several regions of European Russia, where the horse chestnut, an exotic plant for the country, is grown in ornamental plantings. By 2015, *C. ohridella* was recorded in 13 central regions of European Russia [318]. Shortly after that, the species occupied the south of European Russia, in particular the Black Sea coast, where it gives spectacular outbreaks in resort areas [319,320]. It also invaded the neighboring Crimean Peninsula (including Sevastopol City), Ciscaucasia [321,322], including North Caucasus [323]. On the north, *C. ohridella* was documented in Saint Petersburg in 2013 [287,324]. The easternmost region where the pest has been found in European Russia is the Volga region [325,326].

In European Russia, *C. ohridella* develops from early May to early October and produces two or three generations per year [306,318,321]. In Moscow and Pskov, it develops in two generations and the third generation is usually incomplete [306,327]. In the south, in particular in Stavropol City, it is able to complete four generations per year [321].

The up-to-date distribution map, compiled based on the literature and modern records, indicates that the species has already occupied a significant part of European Russia (Figure 12d). By now, *C. ohridella* was confirmed in 40 out of 58 (i.e., 69%) administrative units of European Russia, mostly in its central and southern parts. Since the horse chestnut is used ornamentally in nearly all regions of European Russia, further distribution of *C. ohridella* eastward up to the Urals, in our opinion, can be expected. Further expansion east of the Urals will not be possible, as *Ae. hippocastanum* is hardly used in plantings in Western Siberia due to the harsh climate [328].

The female lays eggs individually on the upper sides of leaves [306], and the average fecundity is about 40 eggs per female [329]. Larvae develop in blotches of reddish-brown mines between secondary veins (Figure 12c). Up to 250 mines per leaf have been reported from outbreaking populations [329]. Larvae live solitary in mines, but in heavily infested trees, individual mines merge into big ones [321]. Pupation occurs in the mine. The species hibernates as pupae in the mines of the fallen leaves [321], whereas, in Moscow, overwintering at the adult stage was also documented [306]. Multivoltinism, high fecundity, and easy distribution (that is largely associated with human activities) in short-to-long distances can explain the successful invasion and high population densities of *C. ohridella* in European Russia [306,329].

The main host plant is the horse chestnut, *Ae. hippocastanum*, a native species to the Balkans that started being widely planted in Europe in the early XVII century [330]. In European Russia, this tree has been actively used as an ornamental since the XIX century [328]. Other horse chestnut species, such as Japanese *Ae. turbinate*, American *Ae. octandra* (=*flava*), and *Ae. Glabra*, are also suitable hosts for the moth [331,332]. According to the recent observations in Kaliningrad City, *Ae. carnea*, *Ae. glabra*, and *Ae. pavia* are resistant to the pest [318], while in Moscow, the two latter species are the least damaged [333]. In Europe, the moth can also occasionally attack maples, *Acer pseudoplatanus* and *A. platanoides* (Sapindaceae, the same family as of the horse chestnut), when neighboring horse chestnut trees are infested [310,334]. Mines of *C. ohridella* were also found on *A. platanoides* in the N.V. Tsitsin Main Botanical Garden of the Russian Academy of Sciences (Moscow) in 2006–2009; however, the larvae did not succeed to pupate [306]. In Moscow, it can also successfully develop on *Ae. glabra* [306].

The moth spreads fast through natural dispersion and anthropogenic transportation [306]. In European countries, local dispersal involves both adult flight and the dissemination of infested leaves by the wind [335]. Over long distance, the dispersal of *C. ohridella* is possible by wind (as a part of the aerial plankton), but mostly it happens via passive transportation of infested leaves and adult moths as stowaway in/on cars, trucks, and other vehicles, as well as via the movement of infested seedlings [335,336]. Within Russia, the transportation of live plants of *Ae. hippocastanum* (with leaves that may already contain mines of the pest) undoubtedly facilitates the insect distribution. Furthermore, adults from outbreaking populations can be easily moved over significant distances with any plants and vehicles [306].

Similar to the European countries where *C. ohridella* causes severe aesthetic damage to horse chestnut trees [337], it is a highly important pest of *Ae. hippocastanum* in European Russia [306,329]. This species is listed among the Top 100 most dangerous invasive species in Russia [338]. Heavily infested trees start losing foliage already in July, resulting in pronounced aesthetic (i.e., social) impact in urban areas [318,327]. In addition to *C. ohridella*, since 2012–2013, in the Krasnodar Krai and Adygea (West Caucasus), the horse chestnut trees were affected by an invasive bacterium, *Pseudomonas syringae* pv. *aesculi*, which causes bleeding canker disease [339]. No link between the damage caused by the moth and the infestation by the bacteria was found, rejecting the hypothesis that *C. ohridella* can facilitate the distribution of the disease or increase its impact [340,341].

In European countries, at least 60 parasitoid species are known to attack *C. ohridella* [284]. However, parasitism rates remain generally low, usually within 10–25% in southeast Europe and generally lower than 10% in recently invaded countries of Europe [342–345]. The parasitoid *Pediobius saulius* (Walker) (Hymenoptera: Eulophidae), the most abundant species in the Balkans, gradually followed the invasion of its host, *C. ohridella* [346,347]. Exceptionally, on the south of European Russia, in particular in Krasnodar City, where *C. ohridella* was for the first time recorded in 2010, the parasitism level reached 33.6% already by 2013 [348]. The parasitoid complex in Krasnodar includes 44 parasitoid wasps from Braconidae, Eupelmidae, Ichneumonidae, Pteromalidae, and Eulophidae (Hymenoptera), with most of them having diverse trophic relations with local insects [348]. This parasitoid complex seemed to be effective in controlling *C. ohridella* in Krasnodar, dropping the population density significantly by 2014 [348]. In contrast, in another southern region (Stavropol Krai), no entomophages were recorded in a rather dense pest population, despite the following predators being found: three ladybirds (*Adalia bipunctata*, *Adonia variegata*, *Coccinella septempunctata*), the common earwig (*Forficula auricularia*), common orb-weaving spider (*Araneus* sp.), and tit (*Parus* sp.) [321]. However, this predator complex did not appear to be efficient in controlling the pest and stopping its outbreak [321].

Bearing in mind that the aesthetic damage caused to *Ae. hippocastanum* by the invasive moth can be severe, the strategies to control *C. ohridella* have been extensively explored in European Russia. Since the larvae of *C. ohridella* are endophagous, the use of chemicals with contact-intestinal action cannot provide reliable protection against the pest [306]. A high mortality rate of larvae can be reached by the application of systemic insecticides (e.g., new BI-58, Danadim, etc.); however, their use in settlements is impossible because of their toxicity [306]. Chemical control by aerial spraying is efficient but expensive and not adapted to the urban area [349], while in some European countries, such a measure has raised public concern [350]. Other pesticides with fewer non-target effects can be feasible [351], and in particular, stem injections can provide satisfactory results [352]. The 2-year experiment carried on the horse chestnut in Moscow, demonstrated the high efficiency of the stem injection with Abasol, already the next year after application [352].

In European Russia, the use of pheromone trapping utilizing the synthetic pheromone of *C. ohridella* developed in the Czech Republic [353] was intensively explored. Unitrap pheromone traps and Delta adhesive traps showed high efficiency in *C. ohridella* monitoring and could be also effective in its control [329,354]. Biocontrol approaches, such as the release of specialist parasitoids attacking the leafminer in its native range, have been studied but are still far from practical applications in European Russia [306].

The complete removal of leaf litter in which pupae hibernate can be an effective control measure [321]. However, bearing in mind that damaged leaves start shading early, foliage should be removed repeatedly starting already in mid-summer onward [306]. The majority of adults can be prevented from emerging when leaves are properly composted (by mulching damaged leaves with soil) [321]. However, such approach will likely adversely impact parasitoids, which might stay in *C. ohridella* mines [355].

Replacing horse chestnut with other species is one of the discussed measures [321], but evidently it will be economically very expensive for big cities. Furthermore, the feasibility of tree species chosen for replacement should be carefully assessed.

3.13. The Lime Leaf Miner, *Phyllonorycter issikii* (Kumata, 1963) (Lepidoptera: Gracillariidae)

Phyllonorycter issikii (Figure 13) is a tiny moth of East Asian origin that in the mid-1980s invaded the western part of the Palearctic, becoming a notable pest of limes *Tilia* spp. (Supplimentary Table S7; [288,356–358]). The moth is known to naturally occur in Japan [356], the Russian Far East (Primorskii Krai) [357], and South Korea [358]. A recent study based on the survey of historical herbaria and sequencing of the larvae and pupae found in the mines revealed a wide distribution of *P. issikii* in China [359,360].

The earliest record of *P. issikii* outside its native range is dated to 1985 and refers to the ornamental plantations of Moscow [361]. Two years later, the species was documented in Voronezh City, i.e., 500 km southwest of Moscow, already at a high population density [362]. During the next 20 years, it spread over the territory of Europe [363,364]. In Russia, the pest occupied most of the European part, except the southernmost and northernmost regions (Figure 13). In 2021, our surveys did not reveal the pest on limes in Sochi City and in settlements of Krasnodar and the Stavropol Krais. Remarkably, the species invaded Western Siberia, with the first record dated to 2006 in Tyumen City [365]. Presently, *P. issikii* is found in Siberia up to the river Yenisei, despite the fact that limes are rare elements of Siberian flora and are mostly used as ornamental plants in cities and smaller settlements [309]. Further expansion to Eastern Siberia is limited due to the absence of its host plants on the territory from the Krasnoyarsk Krai to the Amur Oblast [360].

A phylogeographic study of its currently known range discovered a high genetic diversity of *P. issikii* in Europe vs. East Asia, suggesting multiple pest introductions and its further dispersal [366]. Extensive surveys of 250-year-old herbaria confirmed the East Asian origin and the invasive status of the species in the western Palearctic, as well as highlighted the contribution of China to the species' invasion westward [360]. Furthermore, the historical herbaria clarified the absence of the species from North America [360].

The pest larvae feed exclusively on limes, *Tilia* spp. (Malvaceae), and reference to *Betula* spp. as the host [356] should be regarded as an error. In its native range, the species develops on East Asian limes, among which *T. amurensis* and *T. mandshurica* are most regularly documented hosts [356,358,366]. In the invaded regions, *P. issikii* shifted to novel hosts, for example, *T. cordata* (a main host), *T. platyphyllos*, *T. sibirica*, and others [363,366,367]. Notably, *P. issikii* willingly attacks American lime (*T. americana*) planted in Russian botanical gardens, in particular in Moscow City [368]. It is a common plant in eastern USA [369], and in case of accidental *P. issikii* introduction, *T. americana* will likely fit as a favorable host plant for the pest.

In the invaded regions of Russia, the species develops in one or two generations per year, from May to September, depending on the region [364,368]. Larvae live in the lower side tentiform mines (Figure 13b,c), and occasionally, upper side mines occur in dense populations [360]. In exceptional cases, up to 29 mines per leaf were recorded in Novosibirsk on *T. cordata* [357]. The species hibernates at the adult stage.

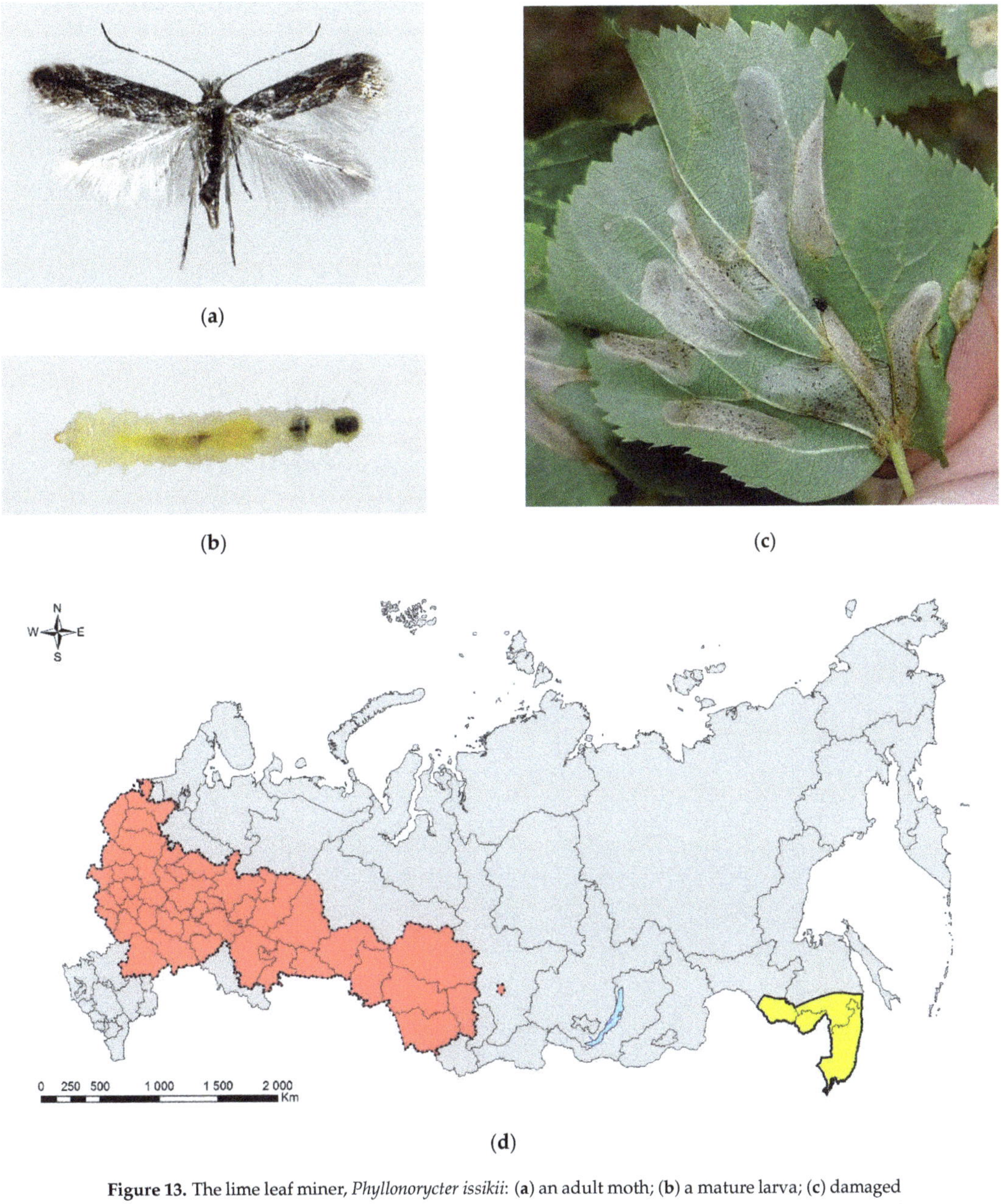

Figure 13. The lime leaf miner, *Phyllonorycter issikii*: (**a**) an adult moth; (**b**) a mature larva; (**c**) damaged *Tilia cordata* in Central Siberian Botanical Garden of the Siberian Branch of the Russian Academy of Science, Novosibirsk. (**d**) the current distribution in Russia: the native range (colored in yellow) and invasive range (colored in red). Photo by N.I. Kirichenko.

Adults of *P. issikii* fly only over short distances [363]. Long-distance dispersal on air currents is possible, as well as the distribution of *P. issikii* by hitchhiking with various land and air transports, bearing in mind that adult moths hide themselves in crevices and other shelters and thus can be passively moved to remote territories [308,363,370].

The species is generally known as an ornamental pest causing aesthetic damage to limes in parks and gardens [363]. In addition, *P. issikii* was also reported as affecting natural stands in European Russia [371] and Western Siberia (Kemerovo Oblast) [366]. In high densities, the lime leaf miner negatively impacts lime growth; furthermore, the pest may adversely affect the sugar content in nectar, subsequently decreasing honey production, as documented in the Udmurtia (east of Volga River) [371]. In Western Siberia, the pest threatens vulnerable Tertiary relic limes groves [366]. Notably, so far, the species has been known by its exceptionally outbreaking character in the invaded regions of Russia [364,366]. Exploration of the lime herbarium specimens provided evidence of a population density increase in *P. issikii* in Primorskii Krai (Far East) in some years in the period of 1914 to 1958 [372].

The control methods have not been effectively developed against this pest. No data on pheromone monitoring and sex disruption are available in Russia. In European Russia, the parasitoid complex of *P. issikii* accounts for at least 43 species (Braconidae, Eulophidae, and Pteromalidae; Hymenoptera) vs. 13 parasitoid species known from its native range (East Asia), with only 6 species in common [373,374]. In European Russia and Siberia, the parasitism level varies within 1.4–37.0% but still remains low [368,375]; the highest value (up to 37.0%) was documented in the Volga Basin [376]. In the invaded regions of Russia, the commonly recorded hymenopteran parasitoids are *Sympiesis gordius*, *Pnigalio soemius* (both Eulophidae), and *Minotetrastichus frontalis* (Chalcidoidea) [373]. The latter species is also present in East Asia [374]. These are generalist parasitoids, with a wide host range among Gracillariidae [374]. Thus, their potential application against the invasive lime leaf miner is questionable, keeping in mind that they may also attack other native gracillariids, including rare species. Nevertheless, *M. frontalis* has been proposed as a biocontrol of *P. issikii* in Russia [377]. In East Asia, little data are available about parasitism rate in *P. issikii* populations [374], whilst these data would be of a high importance for the biocontrol of the pest.

3.14. The Poplar Leafminer, *Phyllonorycter populifoliella* (Treitschke, 1833) (Lepidoptera: Gracillariidae)

Phyllonorycter populifoliella (Figure 14), is a micromoth species widely distributed across the Palearctic [284]. In Russia, this native species occupies large territory from Kaliningrad City (southeastern coast of the Baltic Sea) to the Far East, except northern regions in the Asian part of the country (Figure 14d) [283,378,379]. Recently, the pest was detected beyond its known range, in northern India, in high densities on introduced poplars fueling the hypothesis about its accidental introduction [380].

The moth develops on poplars, *Populus* spp., preferring black and balsam poplars (i.e., Aigeiros and Tacamahaca sections, respectively), to other poplars [379,381–383]. The larvae make distinctive blotch mines on the lower side of leaves, while in dense populations, upper side mines can also occur [384].

The poplar leafminer is known as a severe pest of balsam poplars and their hybrids with black poplars widely planted as ornamental in European Russia, in the Ural region and in Siberia [382–388]. In European Russia, the first outbreak of the pest was documented in the 1930s in Moscow [389]. Further east, in Siberia, the moth foci are known since the middle of the XX century in Irkutsk [385], and later, they were documented in other cities [382,386]. Interestingly, in some regions of European Russia where the poplar leaf miner was never known as an outbreaking pest, nowadays it provides noticeable damage to poplars. For instance, in Saint Petersburg, *P. populifoliella* was known as a rare species with the first record of the moth dating to 1936 [387,390], and the next finding was done only in 1974 [391]. In 1991, a sharp increase in population density was documented, with an outbreak in Saint Petersburg lasting from 1992 to 1999 [392,393]. For almost the next 20 years, the population density of this species was very low again, but since 2017, the new outbreak began in Saint Petersburg [387], where the pest showed the tendency to develop foci on new territories [289–390,394].

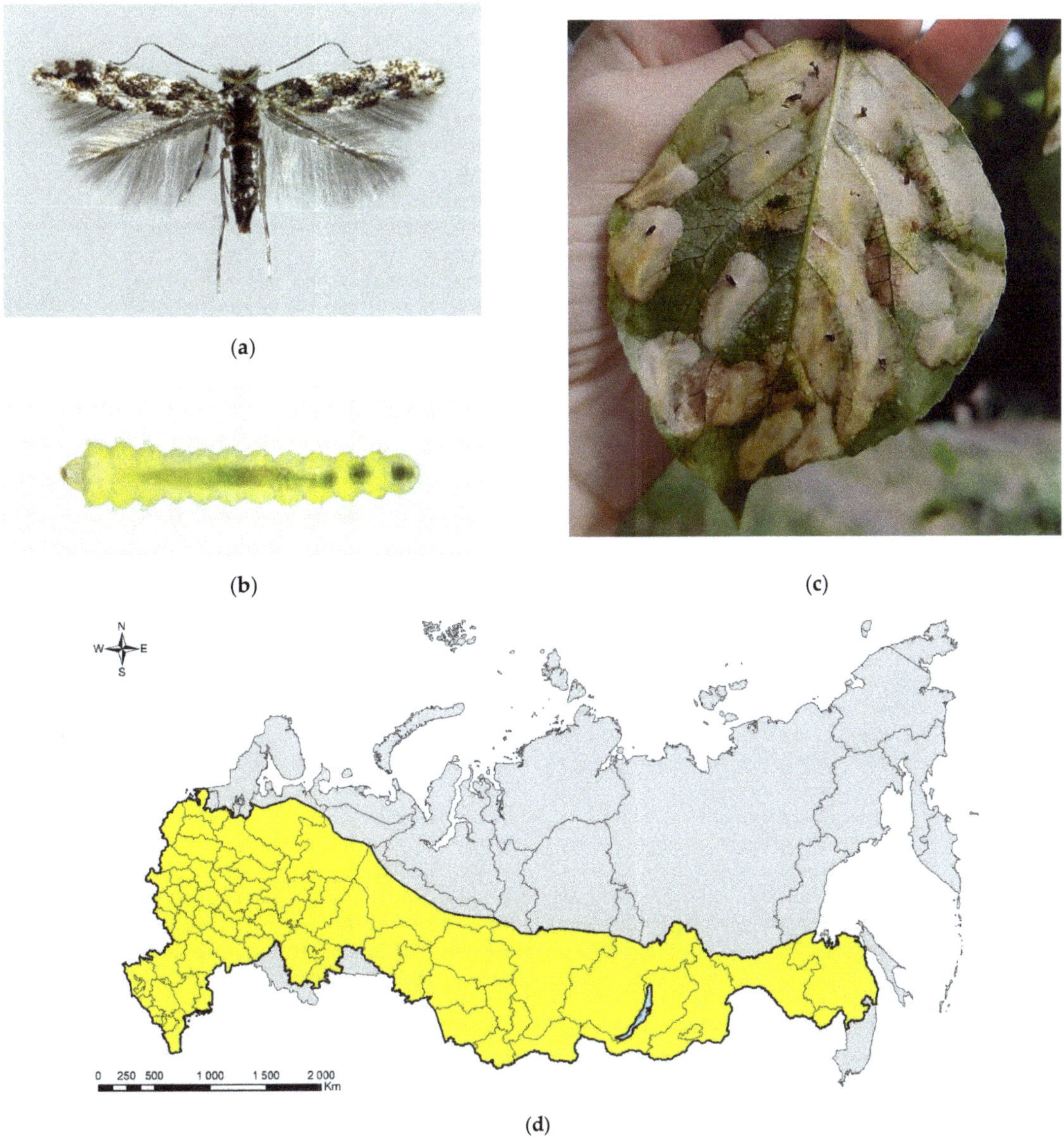

Figure 14. The poplar leafminer, *Phyllonorycter populifoliella*: (**a**) an adult moth; (**b**) a mature larva; (**c**) a down side of a leaf of *Populus* × *berolinensis* heavily damaged by *P. populifoliella*, Saint Petersburg, August 2020; (**d**) the native range in Russia (colored in yellow). Photo: (**a**) and (**b**) by N.I. Kirichenko and (**c**) by N.A. Mamaev, with permission.

Severely damage poplars start losing their leaves already in late July to early August, which significantly reduces the aesthetic and ecological functioning of poplars in settlements and big cities. The pest can affect tree growth [395]. The mass death of poplars in Saint Petersburg in the early 2000s might have a link to the preceding pest outbreak [387]. Presently in Saint Petersburg, the *P. populifoliella* gives only one generation per year, since the larvae of the second generation are not able to complete their development and die not reaching the pupal stage. Climate warming will favor the second generation, thus increasing the pest population density [289,290]. Currently, there is no regular monitoring of the poplar leaf miner in Russia. Injection of poplars with systemic insecticides is a possible way to reduce the pest population densities, but this approach is not used due to its high economic and labor cost. The removal of fallen leaves (carrying mines with larvae and pupae of the pest) is not an effective approach as it can also affect survival of native enemies (parasitoids) that can be helpful in pest control [396]. The complexes of parasitoids attacking *P. populfoliella* in European Russia, includes overall 68 species [383], which in some years, can kill up to 90% of the second generation of the moth as found in Moscow [383,396].

The gradual replacement of poplars with other tree species could help decrease the pest's impact. Black and balsam poplars commonly used as ornamental in Russian cities should be planted rather in small groups in order to avoid the formation of large foci of *P. populifoliella* [383,395].

Bearing in mind the *P. populifoliella* outbreaks on balsam poplars and their hybrids with black poplars in Russia, the accidental introduction of the moth to North America, where balsam and black poplars (e.g., *Populus balsamifera, P. trichocarpa, P. deltoides,* and *Populus × canadensis*) have a wide range, may potentially result in significant tree damage.

4. Discussion

4.1. Taxonomy

This review analyzes the most recent data on 14 species of insects from 3 orders: Hemiptera-Heteroptera (3 species from 3 families), Coleoptera (6 species from 2 families), and Lepidoptera (5 species from 2 families) that are known as highly invasive in Russia or native to the country but also pose a potential danger to woody plants elsewhere (Table 1). The very approximate estimations of the insect fauna of Russia suggest that it includes 80–100 thousand species [397], which is roughly close to the earlier assessments of the insect fauna of the former USSR (approximately the same, 81 and 119 thousand species, respectively [398]). Overall, 192 species of phytophagous alien insects from 48 families and 8 orders had been documented in the European part of Russia by 2011 [8]. Undoubtedly, their number has increased since that time, both because of the escalating invasions and the increased detection of alien species due to growing survey efforts. Looking at all these estimations, it is unjustified to conclude that the listed three orders and seven families of insects include all the most dangerous pests, and it is impossible to say that in other taxa there are no species which are already identified or potentially can be considered pestiferous for woody plants. Out of 14 species, 4 species belong to Gracillariidae, including *Cameraria ohridella*, which intensively invades urban environments providing scenic outbreaks. Four species, actual or potential invaders, represent the family Buprestidae. The question remains open whether or not there are specific family-level traits that drive the family representative to invade new territories. Taxonomic estimates are still very rough and cannot be used to estimate the invasive potential of any particular taxon. In the coming years, in our opinion, in addition Coleoptera and Lepidoptera, economically important invaders can also be well expected in Hemiptera-Homoptera, Hymenoptera, and Diptera.

Table 1. Invasive and emerging insect pests of forest and urban woody plants in Russia: possibility of further invasions, damage, monitoring, and control (Summary).

No.	Species	Native Range	Invasive Range	Availability of Host Plants Outside Russia	The Regions and/or Countries to Where the Invasion Is Possible from Russia	Damage Level (in Forests/ Urban Areas)	Monitoring Measures *	Control Measures *	
								General and in Forests	Specific in Urban Areas
1	Conifer seed bug, *Leptoglossus occidentalis* (Hemiptera: Heteroptera: Coreidae)	West of North America: from South of Canada to Mexico	Most of Europe: from South of Norway to Mediterranean Sea, from British Isles to European Russia; Asia: Kazakhstan, Northeast China, Japan, South Korea; Africa: North and South; South America: Chile, Argentina, Uruguay	All conifer, native and introduced Pinaceae	Mongolia, Kyrgyzstan, Asian Turkey	Moderate to high level of damage in forests, coniferous tree nurseries, and urban areas	Visual inspection; analysis of seed damage and germination of seeds from infested trees	Biological and chemical control	Biological control
2	Brown marmorated stink bug, *Halyomorpha halys* (Hemiptera: Heteroptera: Pentatomidae)	Southeast Asia: China, Taiwan, Korea, Japan, Vietnam	North America; Europe (including South of European Russia); Asia: Kazakhstan	300+ species of vascular plants	Transcaucasia, Middle Asia	Major damage in fruit, berry, and nut plantations	Inspections of overwintering quarters; visual inspection of host plants; pheromone traps	Quarantine measures; pheromone traps; biological and chemical control	Pheromone traps
3	Oak lace bug, *Corythucha arcuata* (Hemiptera: Heteroptera: Tingidae)	North America	Europe (including South of European Russia); Asia Minor	*Quercus* spp. and other Fagaceae, some species of Salicaceae, Rosaceae, Fabaceae	Middle Asia, Transcaucasia, Ukraine, Moldova, Belarus	Major in urban areas; moderate in forests	Inspections of overwintering quarters; visual inspection of host plants	Biological and chemical control	Biological control
4	Spotted poplar borer, *Agrilus fleischeri* (Coleoptera: Buprestidae)	North Asia: Russia (West and East Siberia, Far East), East Kazakhstan, Mongolia, Northeast and central China, North Korea, South Korea, Japan	unclear	Native and introduced *Populus* spp. and *Salix* spp.	West Kazakhstan, Europe (including east of European Russia), North America	In urban areas and forests (minor; during outbreaks—major)	Visual, color, and pheromone traps	Quarantine measures; sanitation felling; chemical control should be developed; biological control (egg and larval parasitoids) should be developed	Elimination of individual infested trees; chemical control should be developed (stem injections with insecticides); biological control (egg and larval parasitoids) should be developed

Table 1. *Cont.*

No.	Species	Native Range	Invasive Range	Availability of Host Plants Outside Russia	The Regions and/or Countries to Where the Invasion Is Possible from Russia	Damage Level (in Forests/ Urban Areas)	Monitoring Measures *	Control Measures *	
								General and in Forests	Specific in Urban Areas
5	Apple buprestid, *Agrilus mali* (Coleoptera: Buprestidae)	East Asia: Russia (East Siberia, Far East), Mongolia, Northeast and central China, North Korea (**), South Korea (**)	Asia: China (**): Xinjiang, North Korea (**), South Korea (**)	Cultivated and wild *Malus* spp., *Cydonia* spp., *Prunus* spp., *Pyrus* spp.	Kazakhstan and other Central Asian countries; entire apple orchard cultivation area	In agricultural areas and forests (major)	Visual, color, and pheromone traps	Quarantine measures; biological control (egg and larval parasitoids) and chemical control should be developed	Pruning; elimination of individual infested trees; chemical control (stem injections with insecticides) and biological control (egg and larval parasitoids) should be developed
6	Emerald ash borer, *Agrilus planipennis* (Coleoptera: Buprestidae)	East Asia: Russian Far East, Northeast China, Japan, Mongolia, North Korea, South Korea	Europe: Central, Northwest, South and Southeast of European Russia, Eastern Ukraine; North America	Native and introduced *Fraxinus* spp., *Chionanthus* ssp., potentially other Oleaceae	Ukraine, Belarus, Poland, Baltic countries, Finland, and other European countries	In urban areas and forests (major)	Visual, color and pheromone traps; checking nests of *Cerceris* spp. (Hymenoptera)	Quarantine measures; sanitation felling; chemical control should be developed; biological control (egg and larval parasitoids)	Elimination of individual infested trees; chemical control (stem injections with insecticides); biological control (egg and larval parasitoids)
7	Cypress jewel beetle, *Lamprodila (Palmar) festiva* (Coleoptera: Buprestidae)	Europe: Mediterranean countries, South Europe, South of central Europe; North Africa; Southwestern Asia	Europe: Central and East Europe, South of European Russia, Eastern Ukraine	Native and introduced Cupressaceae	Transcaucasia, entire Ukraine, Moldova, Belorussia, Poland, Baltic countries, North America	In urban areas (major) and forests	Visual, color, and pheromone traps	Quarantine measures; sanitation felling; chemical control (insecticides during adult flight period); biological control (egg and larval parasitoids) should be developed	Elimination of infested trees; chemical control (pyrethroid insecticides during adult flight period, if allowed and stem injections with insecticides); biological control (egg and larval parasitoids) should be developed

Table 1. *Cont.*

No.	Species	Native Range	Invasive Range	Availability of Host Plants Outside Russia	The Regions and/or Countries to Where the Invasion Is Possible from Russia	Damage Level (in Forests/ Urban Areas)	Monitoring Measures *	Control Measures *	
								General and in Forests	Specific in Urban Areas
8	Small spruce bark beetle, *Ips amitinus* (Coleoptera: Curculionidae: Scolytinae)	Europe: Central, South, and North Europe; Baltic countries; Northwest Russia	Asia: West Siberia	Different conifers, especially *Pinus* spp. and *Picea* spp.	China, North America	Major pest of all conifer forests; possible economic effect is unpredictable	Surveys; pheromone traps	Pheromone traps; sanitation felling	Pheromone traps; elimination of individual infested trees
9	Four-eyed fir bark beetle, *Polygraphus proximus* (Coleoptera: Curculionidae: Scolytinae)	Asia: Northeast China, Japan, Korea, Russian Far East	Europe: central Russia; Asia: West and East Siberia	Different conifers, especially *Abies* spp.	North and Central Europe, North America	Major pest of firs in forests and urban areas	Phytosanitary quarantine and surveys	Pheromone traps; sanitation felling	Pheromone traps; elimination of individual infested trees
10	Box tree moth, *Cydalima perspectalis* (Lepidoptera: Crambidae)	Asia: China, Japan, Korea, Russian Far East, India	Europe: South of European Russia, Georgia, Turkey	*Buxus* spp.	Natural forests across the Caucasus (Transcaucasia) and to countries located further south	Major in urban areas and forests	Visual, UV-light, or pheromone traps	Pheromone traps; biological and chemical control; sanitation felling	Pheromone traps; elimination of individual infested trees; biological and chemical control
11	Leaf blotch miner moth, *Acrocercops brongniardella* (Lepidoptera: Gracillariidae)	Europe	Europe: St. Petersburg and the Karelian Isthmus	*Quercus* spp.	Siberia, Far East, North America	Mainly in urban areas	Visual (mines); pheromone traps	Biological control (egg and larval parasitoids) should be developed	Stem injections with insecticides should be developed
12	Horse-chestnut leaf miner, *Cameraria ohridella* (Lepidoptera: Gracillariidae)	Europe: The Balkans	Europe: East and West Europe, European Russia (except some northern and eastern regions)	Native and introduced *Aesculus* spp., and possibly *Acer* spp.	East Asia, East and West of North America	Mainly in urban areas	Visual; pheromone traps (might be not efficient al low density)	Biological control (parasitoids, disease agents)	Stem injections with insecticides; removal of leaf litter (populations with overwintering of pupae)

Table 1. *Cont.*

No.	Species	Native Range	Invasive Range	Availability of Host Plants Outside Russia	The Regions and/or Countries to Where the Invasion Is Possible from Russia	Damage Level (in Forests/ Urban Areas)	Monitoring Measures *	Control Measures *	
								General and in Forests	Specific in Urban Areas
13	Lime leaf miner, *Phyllonorycter issikii* (Lepidoptera: Gracillariidae)	East Asia: Japan, South Korea, China, Russian Far East	Europe: East and West Europe, European Russia (except some northern and southern regions); Asia: Western Siberia	Native and introduced *Tilia* spp.	North America	Mainly in urban areas, occasionally in forests	Visual; pheromone traps (might be not efficient al low density)	Biological control (parasitoids, disease agents)	Protection of natural enemies
14	Poplar leafminer, *Phyllonorycter populifoliella* (Lepidoptera: Gracillariidae)	Eurasia	Europe: North of European Russia; Asia: Siberia	*Populus* spp.	Far East, and North America	In urban areas	Visual (mines); pheromone traps	Biological control (egg and larval parasitoids) should be developed	Stem injections with insecticides should be developed

* Already used in Russia or elsewhere. ** Not clear, native or invasive populations.

4.2. Directions of Invasions

In terms of the directions of invasions, the pests included in this review can be clustered into four groups:

(1) Invasions from Asia to West or central Europe and then (or directly) to European Russia: This is the most numerous group, which consists of five species—*Cydalima perspectalis*, *Phyllonorycter issikii*, *Halyomorpha halys* (likely first invaded North America and only then Europe and European Russia; see above), and two species, which actually skipped West or central Europe and arrived directly to European Russia, namely, *Agrilus planipennis* and *Polygraphus proximus*;

(2) Invasions from North America to West or central Europe and then to European Russia: *Leptoglossus occidentalis* and *Corythucha arcuata*;

(3) Invasions from Europe to Asia: the case of *Ips amitinus*;

(4) Range expansions within Europe and invasions to European Russia: *Lamprodila festiva* and *Cameraria ohridella*.

Additionally, there is a group of potential invaders currently changing their ranges within European and/or Asian parts of Russia: As described above, two beetle species (*Agrilus fleischeri* and *A. mali*) have not yet noticeably expanded their native ranges in Asian Russia but demonstrate a potential of range expansion and/or host plant shift; somewhat similar could be said about *Acrocercops brongniardella* and *Phyllonorycter populifoliella*.

Thus, there is no common geographic pattern of invasion, and different invaders spread in very different directions.

4.3. Causes and Pathways of Invasions

Invasions of almost all discussed species were associated with human activity, except *Ips amitinus* (in North Europe) and *Acrocercops brongniardella*, which are believed to expand their ranges due to natural causes. Some species (e.g., *Leptoglossus occidentalis*, *Halyomorpha halys*, and *Corythucha arcuata*) at the beginning of their invasions were most likely transported by airplanes (from one continent to another) with goods of plant or non-plant origin. *Agrilus planipennis* was likely moved with wood products or packaging materials; *Polygraphus proximus* and *Ips amitinus* (in Siberia) with timber by railway roads; and *Lamprodila festiva*, *Cydalima perspectalis*, *Cameraria ohridella*, and *Phyllonorycter issikii* with plant materials and/or plants for planting. Very small insects such as *Corythucha arcuata* and *Cameraria ohridella* could use air flow for mass spreading. In other cases, the invasive processes (actively developing (e.g., in *Agrilus planipennis* and *Lamprodila festiva*) or just beginning at the early stages (e.g., in *Agrilus fleischeri*, *A. mali*, *Acrocercops brongniardella*, and *Phyllonorycter populifoliella*)) are probably more complicated and involve host plant shifts as a result of planting or introduction of susceptible host plant species or hybrids (see corresponding essays above).

4.4. Probability of Further Invasions to Neighboring and Distant Countries

All species reviewed are expected to spread further to (in) Europe, including countries of the Caucasus region (i.e., nine species: *Leptoglossus occidentalis*, *Halyomorpha halys*, *Corythucha arcuata*, *Agrilus fleischeri*, *A. mali*, *A. planipennis*, *Lamprodila festiva*, *Polygraphus proximus*, and *Cydalima perspectalis*) or to (in) Asia (i.e., nine species: *Leptoglossus occidentalis*, *Halyomorpha halys*, *Corythucha arcuata*, *Agrilus fleischeri*, *A. mali*, *Ips amitinus*, *A. brongniardella*, *Cameraria ohridella*, and *Phyllonorycter populifoliella*). Moreover, it is expected that nine species can potentially invade North America and/or other continents (i.e., *Agrilus fleischeri*, *A. mali*, *Lamprodila festiva*, *Ips amitinus*, *Polygraphus proximus*, *Acrocercops brongniardella*, *Cameraria ohridella*, *Phyllonorycter issikii*, and *P. populifoliella*) (Table 1).

4.5. Role of the Trophic Factor in Insect Pest Range Expansion

Analysis of invasion histories performed in the framework of this review, suggests that in some cases, invasions start with the widening of the pest's trophic niche and shifts to new host plant(s) (commonly human-introduced) within the native pest's range, frequently followed by invasion to new regions (e.g., *Agrilus fleischeri*, *A. mali*, *Acrocercops planipennis*, *Lamprodila festiva*, and *Phyllonorycter populifoliella*). The data reviewed above suggest that a few steps can be usually recognized within those five scenarios:

(1) A shift in an insect species (often not even a pest) within its native range from its usual host plant(s) to introduced and cultivated host plant(s) (usually from the same or close genus of woody plants and often non-resistant because of lack of co-evolution) (e.g., a shift in *Agrilus planipennis* in Russian Far East and China, from local Asian ash species to introduced North American ash species; a shift in *Phyllonorycter populifoliella* in European Russia and Siberia, from local poplars to widely cultivated introduced North American balsam poplar and hybrids; a shift in *Lamprodila festiva* from wild Cupressaceae to introduced *Thuja* and other cultivated representatives of this family, including their hybrids and cultivars in the Mediterranean region; a shift in *Agrilus fleischeri* to introduced poplars in China);
(2) A local niche expansion due to exploration of cultivated, recently introduced host plants in anthropogenic (urban or agricultural) landscapes; local population build-up and outbreaks (e.g., recorded earlier in *Agrilus planipennis*, *A. mali*, *Phyllonorycter populifoliella*, and *Lamprodila festiva*, and currently seen in *A. fleischeri*);
(3) A range expansion outside the limits of the native range through anthropogenic (urban or agricultural) landscapes, i.e., beginning of invasion (e.g., *Agrilus mali* and *Lamprodila festiva*);
(4) A distant invasion (e.g., invasions of *Agrilus planipennis* to European Russia or North America or *Ips amitinus* to Siberia);
(5) A secondary host plant shift to the native (for the invaded region) food plant(s) (e.g., shift in *Agrilus planipennis* to *Chionanthus virginicus* in North America [399] and to *Fraxinus excelsior* in European Russia; shift in *A. mali* from cultivated apples to the wild apple, *Malus sieversii* in China).

Separately, we should mention *Acrocercops brongniardella*. This species does not shift to new host plants but simply follows range expansion of its usual host plant (*Quercus robur*) [360].

4.6. Monitoring and Control Measures

The monitoring methods recommend for the reviewed species are mainly limited by visual inspections and application of pheromone and color traps (Table 1). These methods might be quite effective when used appropriately and systematically. However, as this review demonstrates, in many cases, the monitoring system does not work properly, and many invasive pests rapidly increase their secondary ranges in Russia (e.g., *Leptoglossus occidentalis*, *Halyomorpha halys*, *Agrilus planipennis*, *Cydalima perspectalis*, *Lamprodila festiva*, and *Cameraria ohridella*). The lack of effective integration between organizations responsible for the monitoring and management of forests, urban, and suburban woody plantations in Russia (on both the national and local levels) creates serious difficulties for the practical and systematic use of these methods, especially keeping in mind the vast territory of the country. To increase the effectiveness of the monitoring, it is advisable to create a fast and flexible system of information exchange between the public and the organizations responsible for the monitoring and management of forests, urban, and suburban woody plantations, as well as to increase the level of effective involvement of scientists, university professionals, and citizens in monitoring of invasive insect pests of woody plants.

The control measures understandably differ between forests and urban woody plantations. To control the spread of the reviewed invasive pests in forests, we mostly suggest developing and use biological and chemical methods (Table 1). Special attention should be paid to a braconid *Spathius polonicus* Niezabitowski (Hymenoptera: Braconidae: Doryctinae), a parasitoid, which seems to have potential to effectively control *Agrilus planipennis* in European Russia, although it is likely that a few years are needed to build up a sufficient parasitoid population [145,169,400]. The effective use of sanitation fellings in the forests of the Russian Federation is currently much overcomplicated by the existing regulatory framework, which does not allow timely felling of trees infested by insect pests. The use of pheromone traps proposed as a measure to control some invaders (e.g., *Halyomorpha halys*, *Corythucha arcuata*, *Ips amitinus*, *Polygraphus proximus*, and *Cydalima perspectalis*; Table 1) is possible and potentially effective only when we deal with a limited, relatively small forest area.

The situation is different with the invasive pest control in urban and suburban woody tree plantations. The use of chemical control is significantly legally limited in such environments, whereas the application of stem injections is not always possible (taking into account its cost) but advisable, especially against insect pests that live inside the plant tissues (e.g., mining insects, gallers, wood-borers, etc.) in individual trees in urban areas. Within a relatively small woody tree plantation, it is possible to use biological control methods, pheromone traps, elimination of individual infested trees and other specific methods. It should be kept in mind that overall such control methods might be very expensive.

5. Conclusions

The majority of the reviewed insect pests have demonstrated their invasive behaviour in Russia during the last 10–30 years. In most cases, first reports of these invaders in Russia were unexpected for the stakeholders, who were supposed to provide forest and urban woody plants' monitoring and management. These invasions led to significant ecological and economic losses and caused negative social consequences [15,63,338,360].

We suggest that there are three major scenarios of invasions of woody plant's pests: (1) a naturally conditioned range expansion, which results in the arrival of a pest to a new territory and its further naturalization; (2) a human-mediated long-distance transfer of a pest to a new territory and its further naturalization; and (3) a widening of the pest's trophic niche and shift to new host plant(s) (commonly human-introduced) within the native pest's range, frequently followed by invasion to new regions.

Bearing in mind these and many other examples of devastating invasions of tree pests as well as unpredictable emergence of novel invaders, it is essential to stress the importance of insect pest monitoring (including the episodes of host plant shift) and more effective use of early detection programs (e.g., sentinel plantings) [7,401,402] and application of new and developing species identification techniques (e.g., DNA-barcoding) [403].

Supplementary Materials: The following supporting information can be downloaded at: https://www.mdpi.com/article/10.3390/f13040521/s1; Supplimentary Table S1: Timeline of the first records of the Western coniferous seed bug, *Leptoglossus occidentalis* in its invasive range in Europe; Supplimentary Table S2: Timeline of the first records of *Halyomorpha halys* in its invasive range in Europe; Supplimentary Table S3: Timeline of the first records of *Corythucha arcuata* in its invasive range in Europe; Supplimentary Table S4: Timeline of the first records of *Lamprodila (Palmar) festiva* in its invasive range in Europe; Supplimentary Table S5: Timeline of the first records of *Cydalima perspectalis* in its invasive range in Europe; Supplimentary Table S6: Timeline of the first records of *Cameraria ohridella* in its invasive range in Europe; Supplimentary Table S7: Timeline of the first records of *Phyllonorycter issikii* in its invasive range in Europe.

Author Contributions: Conceptualization, D.L.M., N.I.K., A.V.S. and R.V.; methodology, D.L.M., N.I.K. and A.V.S.; formal analysis, D.L.M., N.I.K., N.N.K., E.V.A., V.B.G., I.A.K., M.Y.M., R.V., M.G.V., E.N.Z. and A.V.S.; writing—original draft preparation, D.L.M., N.I.K., N.N.K., E.V.A., V.B.G., I.A.K., M.Y.M., M.G.V., E.N.Z. and A.V.S.; writing—review and editing, D.L.M., N.I.K., N.N.K., E.V.A., V.B.G., I.A.K., M.Y.M., R.V., M.G.V., E.N.Z. and A.V.S.; visualization, D.L.M., N.I.K., N.N.K., E.V.A., I.A.K., M.G.V. and E.N.Z.; supervision, D.L.M., R.V. and A.V.S.; project administration, D.L.M.; funding acquisition, R.V., A.V.S. and D.L.M. All authors have read and agreed to the published version of the manuscript.

Funding: The research was funded by the Russian Science Foundation (data collection and analysis for Hemiptera-Heteroptera and Lepidoptera species in general, preparation of maps: project No. 21-16-00050, D.L.M.; data collection and analysis for *Ips amitinus* and Coleoptera species in general: project No. 21-16-00065, A.V.S.). N.I.K. was partially funded by the basic project of Sukachev Institute of Forest SB RAS, Russia (Project No. 0287-2021-0011) [review of *Phyllonorycter issikii*]. I.A.K. was partially supported by the Russian Foundation for Basic Research (project No. 20-04-00587) [review of *Polygraphus proximus*]. N.N.K. and E.N.Z. were partially supported by the State assignment of the Federal Research Centre of the Subtropical Scientific Centre RAS, project No. FGRW-2022-0006 [review of *Cydalima perspectalis*]. The study by M.G.V. [review of *Lamprodila* (*Palmar*) *festiva*, *Agrilus fleischeri*, *A. planipennis*, and *A. mali*] was performed in the frame of state research project No. 122031100272-3. The study by R.V. was funded by a grant from the Forest Damage Centre, Swedish University of Agricultural Sciences.

Institutional Review Board Statement: Not applicable.

Informed Consent Statement: Not applicable.

Data Availability Statement: Data are available upon email request to the authors.

Acknowledgments: We sincerely acknowledge V.A. Soboleva (Voronezh State University, Voronezh, Russia), E. Jendek (Czech University of Life Sciences, Prague, Czech Republic), K.V. Makarov (Moscow Pedagogical State University, Moscow, Russia), A.V. Kovalev (All-Russian Plant Protection Institute, Saint Petersburg, Russia), D. Lees (Museum of Natural History, London, UK), I.A. Utkina (Institute of Forest Science, Moscowskya Oblast, Russia), and N.A. Mamaev (Saint Petersburg State Forest Technical University, Saint Petersburg, Russia) for permissions to use their photographs of insects or damage symptoms and I.A. Mikhailova (Sukachev Institute of Forest SB RAS, Krasnoyarsk, Russia) for help with mapping. We thank Eddy Velez for language improvement and two annonimous reviwers for their useful comments and helpful criticism.

Conflicts of Interest: The authors declare no conflict of interest. The funders had no role in the design of the study; in the collection, analyses, or interpretation of data; in the writing of the manuscript, or in the decision to publish the results.

References

1. Pimentel, D.; McNair, S.; Janecka, J.; Wightman, J.; Simmonds, C.; O'Connell, C.; Won, E.; Russel, L.; Zern, J.; Aquino, T.; et al. Economic and environmental threats of alien plant, animal, and microbe invasions. *Agric. Ecosys. Environ.* **2001**, *84*, 1–20. [CrossRef]
2. Fei, S.; Phillips, J.; Shouse, M. Biogeomorphic impacts of invasive species. *Annu. Rev. Ecol. Evol. Syst.* **2014**, *45*, 69–87. [CrossRef]
3. Liebhold, A.M.; Brockerhoff, E.G.; Nunez, M.A. Biological invasions in forest ecosystems: A global problem requiring international and multidisciplinary integration. *Biol. Invasions* **2017**, *19*, 3073–3077. [CrossRef]
4. Seebens, H.; Blackburn, T.M.; Dyer, E.E.; Genovesi, P.; Hulme, P.E.; Jeschke, J.M.; Pagad, S.; Pyšek, P.; van Kleunen, M.; Winter, M.; et al. The global rise in emerging alien species results from increased accessibility of new source pools. *Proc. Natl. Acad. Sci. USA* **2018**, *115*, E2264–E2273. [CrossRef]
5. Hulme, P.E. Trade, transport and trouble: Managing invasive species pathways in an era of globalization. *J. Appl. Ecol.* **2009**, *46*, 10–18. [CrossRef]
6. Paap, T.; Burgess, T.I.; Wingfield, M.J. Urban trees: Bridge-heads for forest pest invasions and sentinels for early detection. *Biol. Invasions* **2017**, *19*, 3515–3526. [CrossRef]
7. Panzavolta, T.; Bracalini, M.; Benigno, A.; Moricca, S. Alien invasive pathogens and pests harming trees, forests, and plantations: Pathways, global consequences and management. *Forests* **2021**, *12*, 1364. [CrossRef]
8. Maslyakov, V.Y.; Izhevsky, S.S. *Alien Phytophagous Insects Invasions in the European Part of Russia*; Institute of Geography ot Russian Academy of Sciences: Moscow, Russia, 2011; pp. 1–272. (In Russian)
9. Deschka, G.; Dimić, N. *Cameraria ohridella* n. sp. (Lepidoptera, Lithocolletidae) aus Mazedonien, Jugoslawien. *Acta Entomol. Jugosl.* **1986**, *22*, 11–23.
10. van Nieukerken, E.; Wagner, D.; Baldessari, M.; Mazzon, L.; Angeli, G.; Girolami, V.; Duso, C.; Doorenweerdet, C. *Antispila oinophylla* new species (Lepidoptera, Heliozelidae), a new North American grapevine leafminer invading Italian vineyards: Taxonomy, DNA barcodes and life cycle. *ZooKeys* **2012**, *170*, 29–77. [CrossRef]
11. Mlynarek, J.J. Testing the enemy release hypothesis in a native insect species with an expanding range. *PeerJ* **2015**, *3*, e1415. [CrossRef] [PubMed]
12. Dang, Y.; Wei, K.; Wang, X.; Duan, J.J.; Jennings, D.E.; Poland, T.M. Introduced plants induce outbreaks of a native pest and facilitate invasion in the plants' native range: Evidence from the emerald ash borer. *J. Ecol.* 2022, *in press*. [CrossRef]

13. Venette, R.C.; Hutchison, W.D. Invasive insect species: Global challenges, strategies & opportunities. *Front. Insect Sci.* **2021**, *1*, 1. [CrossRef]
14. Akulov, E.N.; Ponomarenko, M.G.; Kirichenko, N.I. Exploring fauna of Microlepidoptera in South Siberia: Novel regional records and interception of quarantine species. *J. Asia-Pac. Biodivers.* **2019**, *12*, 597–612. [CrossRef]
15. Dendrobiotic Invertebrates and Fungi and their Role in Forest Ecosystems. In *The Kataev Memorial Readings—XI, Proceedings of the All-Russia Conference with International Participation, St. Petersburg, Russia, 24–27 November 2020*; Musolin, D.L., Kirichenko, N.I., Selikhovkin, A.V., Eds.; Saint Petersburg State Forest Technical University: St. Petersburg, Russia, 2020; p. 452. Available online: http://spbftu.ru/wp-content/uploads/2020/10/Kataev-Readings-XI-2020.pdf (accessed on 20 January 2022).
16. ESRI. *ArcGIS Desktop: Release 9.3*; Environmental Systems Research Institute: Redlands, CA, USA, 2008. Available online: http://www.esri.com/software/arcgis/eval-help/arcgis-93 (accessed on 20 January 2022).
17. Gapon, D.A. First records of the western conifer seed bug *Leptoglossus occidentalis* Heid. (Heteroptera, Coreidae) from Russia and Ukraine, regularities in its distribution and possibilities of its range expansion in the Palaearctic region. *Entomol. Rev.* **2013**, *93*, 174–181. [CrossRef]
18. Tescari, G. *Leptoglossus occidentalis* coreide nearctico rinveuto in Italia (Heteroptera, Coreidae). *Lavor. Societa Veneziana Scie. Nat.* **2001**, *26*, 3–5.
19. Gninenko, Y.I.; Cheplyansky, I.Y.; Chernova, U.A.; Rakov, A.G.; Khegai, I.V.; Latyshova, N.S.; Gimranov, R.I. *Methodical Recommendations for Protection against Pine Seed Bug (for Industrial Trial)*; VNIILM: Pushkino, Russia, 2019; p. 28.
20. Fent, M.; Kment, P. First record of the invasive western conifer seed bug *Leptoglossus occidentalis* (Heteroptera: Coreidae) in Turkey. *North-West. J. Zool.* **2011**, *7*, 72–80. Available online: https://www.researchgate.net/publication/228493133_First_record_of_the_invasive_western_conifer_seed_bug_Leptoglossus_occidentalis_Heteroptera_Coreidae_in_Turkey (accessed on 20 January 2022).
21. Ahn, S.J.; Son, D.; Choo, P.Y.; Parka, C.G. The first record on *Leptoglossus occidentalis* (Hemiptera: Coreidae) in Korea, a potential pest of the pinaceous tree species. *J. Asia-Pac. Entomol.* **2013**, *16*, 281–284. [CrossRef]
22. Aukema, B.; Rieger, C.; Rabitsch, W. (Eds.) *Catalogue of the Heteroptera of the Palaearctic Region. Vol. 6. Supplement*; The Netherlands Entomological Society: Wageningen, The Netherlands, 2013; 629p.
23. Barclay, M.; Nikolaeva, S. Arrival in Kazakhstan of *Leptoglossus occidentalis* (Hemiptera: Heteroptera: Coreidae); a North American invasive species expands 2500 kilometres to the east. *Klapalekiana* **2018**, *54*, 1–3.
24. van der Heyden, T. *Leptoglossus occidentalis* Heidemann (Heteroptera: Coreidae: Coreinae: Anisoscelini) in Israel. *Revista Chilena Entomol.* **2019**, *45*, 435–437. [CrossRef]
25. Kalashian, M.Y.; Ghrejyan, T.L.; Karagyan, G.H. First finding of Western conifer seed bug *Leptoglossus occidentalis* Heid. (Heteroptera, Coreidae) in Armenia. *Russ. J. Biol. Invasions* **2021**, *12*, 274–276. [CrossRef]
26. Ben Jamâa, M.L.; Mejri, M.; Naves, P.; Sousa, E. Detection of *Leptoglossus occidentalis* Heidemann, 1910 (Heteroptera: Coreidae) in Tunisia. *Afr. Entomol.* **2013**, *21*, 165–167. [CrossRef]
27. Gapon, D.A.; Busarova, N.V.; Komarov, Y.E. New records of the western conifer seed bug *Leptoglossus occidentalis* (Heteroptera: Coreidae) in Russia and in adjacent territories. *Cauc. Entomol. Bull.* **2016**, *12*, 221–222. (In Russian) [CrossRef]
28. van der Heyden, T.; Faúndez, E.I. First records of *Leptoglossus occidentalis* Heidemann, 1910 (Hemiptera: Heteroptera: Coreidae) in Brazil and South Africa. *Boletín Mus. Nac. Hist. Nat. Parag.* **2020**, *24*, 28–30. Available online: https://www.researchgate.net/publication/341832223_First_records_of_Leptoglossus_occidentalis_Heidemann_1910_Hemiptera_Heteroptera_Coreidae_in_Brazil_and_South_Africa (accessed on 20 January 2022).
29. Faúndez, E.I.; Rocca, J.; Villablanca, J. Detection of the invasive western conifer seed bug *Leptoglossus occidentalis* Heidemann, 1910 (Heteroptera: Coreidae: Coreinae) in Chile. *Arq. Entomol.* **2017**, *17*, 317–320. Available online: https://www.researchgate.net/publication/316277401_Detection_of_the_invasive_western_conifer_seed_bug_Leptoglossus_occidentalis_Heidemann_1910_Heteroptera_Coreidae_Coreinae_in_Chile (accessed on 20 January 2022).
30. Carpintero, D.L.; Farina, J.L.; De Biase, S. Reporte para la provincia de Buenos Aires de tres especies de Heteroptera (Hemiptera) introducidas en Argentina. *Hist. Nat. Terc. Ser.* **2019**, *9*, 63–70.
31. Faúndez, E.I.; Silvera, M. Sobre la presencia de la chinche de las coníferas occidental *Leptoglossus occidentalis* Heidemann (Heteroptera: Coreidae) en Uruguay. *Rev. Chil. Entomol.* **2019**, *45*, 549–551. [CrossRef]
32. Gildenkov, M.Y. Invasive bug species *Leptoglossus occidentalis* Heidemann, 1910 (Heteroptera: Coreidae) in ecosystems of the Smolensk area. In *Modern Scientific Research: Trends and Prospects, Proceedings of the 2nd All-Russia Conference with International Participation, Chistopol, Tatarstan, Russia, 26 February 2021*; Nazarov, E.A., Ed.; Astor i Ya Publisher: Kazan, Russia, 2021; pp. 5–8. (In Russian)
33. Golub, V.B.; Aksenenko, E.V.; Soboleva, V.A.; Kornev, I.I. New data on the distribution of the tropical bed bug *Cimex hemipterus* and the western conifer seed bug *Leptoglossus occidentalis* (Heteroptera: Cimicidae, Coreidae) in the European part of Russia. *Russ. J. Biol. Invasions* **2020**, *11*, 97–100. [CrossRef]
34. Golub, V.B. *Personal Communication*; Voronezh State University: Voronezh, Russia, 2021.
35. Gninenko, Y.I.; Gapon, D.A.; Shchurov, V.I.; Bondarenko, A.S. A western conifer seed bug *Leptoglossus occidentalis* (Heteroptera, Coreidae) emerged in Russia. *Plant Prot. Quar.* **2014**, *6*, 38–40. Available online: https://www.elibrary.ru/download/elibrary_21561963_71479165.pdf (accessed on 20 January 2022). (In Russian)
36. van der Heyden, T. New data on the distribution of *Leptoglossus occidentalis* Heidemann (Heteroptera: Coreidae: Coreinae: Anisoscelini), including the first record of the species in Georgia. *Rev. Chil. Entomol.* **2018**, *44*, 433–435.

37. Koerber, T.W. *Leptoglossus occidentalis* (Hemiptera, Coreidae), a newly discovered pest of coniferous seed. *Ann. Entomol. Soc. Am.* **1963**, *56*, 229–234. [CrossRef]
38. Werner, D.J. Die amerikanische Koniferen-Samen-Wanze *Leptoglossus occidentalis* (Heteroptera: Coreidae) als Neozoon in Europa und in Deutschland: Ausbreitung und Biologie. *Entomol. Heute* **2011**, *23*, 31–68. Available online: https://www.zobodat.at/pdf/Entomologie-heute_23_0031-0068.pdf (accessed on 20 January 2022).
39. Rice, R.E.; Uyemoto, J.K.; Ogawa, J.M.; Pemberton, W.M. New findings on pistachio problems. *Calif. Agric.* **1985**, *39*, 15–18.
40. Barta, M. Biology and temperature requirements of the invasive seed bug *Leptoglossus occidentalis* (Heteroptera: Coreidae) in Europe. *J. Pest. Sci.* **2016**, *89*, 31–44. [CrossRef]
41. Hedlin, A.F.; Yates, H.O., III; Tovar, D.C.; Ebel, B.H.; Koerber, T.W.; Merkel, E.P. *Cone and Seed Insects of North American Conifers*; Environment Canada, Forestry Service, Pacific Forestry Centre: Victoria, BC, Canada; The United States Forest Service: Wahington, DC, USA; Secretaría de Agricultura y Recursos Hidráulicos: Tulla de Allende, Mexico, 1981; pp. 1–122. Available online: https://cfs.nrcan.gc.ca/publications?id=2026 (accessed on 2 January 2022).
42. Bernardinelli, I.; Rovato, M.; Zandigiacomo, P. Life history and laboratory rearing of *Leptoglossus occidentalis*. In *IUFRO Working Party 7.III.10, Proceedings of the Workshop, Gmunden, Austria, 11–14 September 2006*; Federal Research and Training Centre for Forests, Natural Hazards and Landscape Forest Training Centre Ort: Gmunden, Austria, 2006; p. 225.
43. Rabitsch, W. Alien true bugs of Europe (Insecta: Hemiptera: Heteroptera). *Zootaxa* **2008**, *1827*, 1–44. [CrossRef]
44. Luchi, N.; Mancini, V.; Feducci, M.; Santini, A.; Capretti, P. *Leptoglossus occidentalis* and *Diplodia pinea*: A new insect-fungus association in Mediterranean forests. *For. Pathol.* **2012**, *42*, 246–251. [CrossRef]
45. Tamburini, M.; Maresi, G.; Salvadori, C.; Battisti, A.; Zottele, F.; Pedrazzoli, F. Adaptation of the invasive western conifer seed bug *Leptoglossus occidentalis* to Trentino, an alpine region (Italy). *Bull. Insectol.* **2012**, *65*, 161–170. Available online: https://www.researchgate.net/publication/233100158_Adaptation_of_the_invasive_western_conifer_seed_bug_Leptoglossus_occidentalis_to_Trentino_an_alpine_region_Italy (accessed on 20 January 2022).
46. Cargnus, E.; Buian, F.M.; Zandigiacomo, P. Presenza di *Trichopoda pennipes* (Diptera, Tachinidae) nell'Italia nord-orientale. *Boll. Soc. Nat. Silvia Zenari* **2011**, *35*, 123–130.
47. Maltese, M.; Caleca, V.; Guerrieri, E.; Strong, W.B. Parasitoids of *Leptoglossus occidentalis* Heidemann (Heteroptera: Coreidae) recovered in western North America and first record of its egg parasitoid *Gryon pennsylvanicum* (Ashmead) (Hymenoptera: Platygastridae) in California. *Pan-Pac. Entomol.* **2012**, *88*, 347–355. [CrossRef]
48. Lee, D.-H.; Short, B.D.; Joseph, S.V.; Bergh, J.C.; Leskey, T.C. Review of the biology, ecology, and management of *Halyomorpha halys* (Hemiptera: Pentatomidae) in China, Japan, and the Republic of Korea. *Environ. Entomol.* **2013**, *42*, 627–641. [CrossRef]
49. Haye, T.; Gariepy, T.; Hoelmer, K.; Rossi, J.-P.; Streito, J.-C.; Tassus, X.; Desneux, N. Range expansion of the invasive brown marmorated stink bug, *Halyomorpha halys*: An increasing threat to field, fruit and vegetable crops worldwide. *J. Pest Sci.* **2015**, *88*, 665–673. [CrossRef]
50. Hoebeke, E.R.; Carter, M.E. *Halyomorpha halys* (Stål) (Heteroptera: Pentatomidae): A polyphagous plant pest from Asia newly detected in North America. *Proc. Entomol. Soc. Wash.* **2003**, *105*, 225–237. Available online: https://www.researchgate.net/publication/279897494_Halyomorpha_halys_Stal_Heteroptera_Pentatomidae_A_polyphagous_plant_pest_from_Asia_newly_detected_in_North_America (accessed on 20 January 2022).
51. Lee, D.-H. Current status of research progress on the biology and management of *Halyomorpha halys* (Hemiptera: Pentatomidae) as an invasive species. *Appl. Entomol. Zool.* **2015**, *50*, 277–290. [CrossRef]
52. Hamilton, G.C.; Ahn, J.J.; Bu, W.; Leskey, T.C.; Nielsen, A.L.; Park, Y.-L.; Rabitsch, W.; Hoelmer, K.A. Halyomorpha halys (Stål). In *Invasive Stink Bugs and Related Species (Pentatomoidea): Biology, Higher Systematics, Semiochemistry and Management*; McPherson, J.E., Ed.; CRC Press: Boca Raton, FL, USA, 2018; pp. 243–292. [CrossRef]
53. Fogain, R.; Graff, S. First records of the invasive pest, *Halyomorpha halys* (Hemiptera: Pentatomidae), in Ontario and Quebec. *J. Entomol. Soc. Ont.* **2011**, *142*, 45–48.
54. Faúndez, E.I.; Rider, D.A. The brown marmorated stink bug *Halyomorpha halys* (Stål, 1855) (Heteroptera: Pentatomidae) in Chile. *Arquivos Entomol.* **2017**, *17*, 305–307. Available online: https://www.researchgate.net/publication/316277383_The_brown_marmorated_stink_bug_Halyomorpha_halys_Stal_1855_Heteroptera_Pentatomidae_in_Chile (accessed on 12 December 2021).
55. Arnold, K. *Halyomorpha halys* (Stål, 1855), eine für die europäische Fauna neu nachgewiesene Wanzenart (Insecta: Heteroptera: Pentatomidae: Cappaeini). *Mitt. Des. Thüring. Entomol.* **2009**, *16*, 19.
56. Gariepy, T.D.; Musolin, D.L.; Konjević, A.; Karpun, N.N.; Zakharchenko, V.Y.; Zhuravleva, E.N.; Tavella, L.; Bruin, A.; Haye, T. Diversity and distribution of cytochrome oxidase I (COI) haplotypes of the brown marmorated stink bug, *Halyomorpha halys* Stål (Hemiptera, Pentatomidae), along the eastern front of its invasive range in Eurasia. *NeoBiota* **2021**, *68*, 53–77. [CrossRef]
57. Harris, A.C. *Halyomorpha halys* (Hemiptera: Pentatomidae) and *Protaetia brevitaris* (Coleoptera: Scarabaeidae: Cetoniinae) intercepted in Dunedin. *Weta* **2010**, *40*, 42–44. Available online: https://citeseerx.ist.psu.edu/viewdoc/download?doi=10.1.1.694.5502&rep=rep1&type=pdf (accessed on 20 January 2022).
58. Malumphy, C. Second interception of *Halyomorpha halys* (Stål) (Hemiptera: Pentatomidae) in Britain. *Het News* **2014**, *21*, 4–5.
59. Mityushev, I.M. First record of *Halyomorpha halys* in Russia. *Plant Prot. Quar.* **2016**, *3*, 48. Available online: https://www.elibrary.ru/download/elibrary_25606587_18888986.pdf (accessed on 2 December 2021). (In Russian)

60. Musolin, D.L.; Konjević, A.; Karpun, N.N.; Protsenko, V.Y.; Ayba, L.Y.; Saulich, A.K. Invasive brown marmorated stink bug *Halyomorpha halys* (Stål) (Heteroptera: Pentatomidae) in Russia, Abkhazia, and Serbia: Range expansion, early stages of establishment and first records of damage to local crops. *Arthropod-Plant Interact.* **2018**, *12*, 517–529. [CrossRef]
61. Borisov, B.A.; Karpun, N.N.; Protsenko, V.Y. New data on trophic relationships of the invasive brown marmorated stink bug *Halyomorpha halys* Stål (Heteroptera: Pentatomidae) in the subtropical zone of the Black Sea coast of the Caucasus. In *Monitoring and Biological Control Methods of Woody Plant Pests and Pathogens: From Theory to Practice, Proceedings of the 2nd International Conference, Moscow, Russia, 22–26 April 2019*; Baranchikov, Y.N., Ed.; SIF SN RASc.: Krasnoyarsk, Russia, 2019; pp. 33–35. (In Russian)
62. Karpun, N.N.; Grebennikov, K.A.; Protsenko, V.E.; Ayba, L.Y.; Borisov, B.A.; Mityushev, I.M.; Zhimerkin, V.N.; Ponomaryov, V.L.; Chekmaryov, P.A.; Dolzhenko, V.I.; et al. Brown marmorated stink bug monitoring and identification methods *Halyomorpha halys* Stål, 1855. *Plant Quar. Sci. Pract.* **2018**, *2*, 7–11. Available online: https://elibrary.ru/download/elibrary_35209732_42561776.pdf (accessed on 18 January 2022). (In Russian)
63. Karpun, N.N. The Structure of Complexes of Harmful Organisms of Woody Plants in the Humid Subtropics of Russia and the Biological Basis of Protection Measures. Ph.D. Thesis, Russian State Agricultural University, Moscow, Russia, 2018. Available online: http://www.old.timacad.ru/catalog/disser/dd/karpun/disser.pdf (accessed on 20 January 2022). (In Russian)
64. Zamotajlov, A.S.; Belyi, A.I.; Esipenko, L.P. Harmfulness of herbivorous bugs of the family Pentatomidae (Insecta, Heteroptera) on tomatoes. In *Results of Research Work for 2017, Proceedings of the 73th Research Conference of the Kuban' State Agricultural University, Krasnodar, Russia, 14 March 2018*; Koshaev, A.G., Ed.; Kuban State Agricultural University: Krasnodar, Russia, 2018; pp. 43–44. (In Russian)
65. Stryukova, N.M.; Stryukov, A.A. The first detection of the brown marmorated stink bug in Crimea. In *Sustainable Neospheric Development, Proceedings of the Scientific Inter-university Conference Devoted to the 156th Anniversary of V.I. Vernadsky, Simferopol, Russia, 15 March 2019*; Bashta, A.I., Ed.; IP Zueva, T.V. Publisher: Simferopol, Russia, 2019; pp. 68–70. (In Russian)
66. Zhuravleva, E.N.; Karpun, N.N. On the discovery of the brown marmorated stink bug *Halyomorpha halys* Stål (Heteroptera: Pentatomidae) in Sevastopol. In *Monitoring and Biological Control Methods of Woody Plant Pests and Pathogens: From Theory to Practice, Proceedings of the 2nd International Conference, Moscow, Russia, 22–26 April 2019*; Baranchikov, Y.N., Ed.; SIF SN RASc.: Moscow, Russia; Krasnoyarsk, Russia, 2019; pp. 74–75. (In Russian)
67. Gapon, D.A. The brown marmorated stink bug *Halyomorpha halys* (Stål, 1855) (Heteroptera: Pentatomidae): Expansion of its range in the European part of Russia, description of the imago and larvae, and the diagnostics of the species. *Cauc. Entomol. Bull.* **2019**, *15*, 241–247. [CrossRef]
68. Bulgakov, T.S. (Federal Research Centre the Subtropical Scientific Centre of the Russian Academy of Sciences, Sochi, Russia). Personal communication, 2020.
69. Chenikalova, E.V. Marmorated stink bug *Halyomorpha halys* Stål in Stavropol. *Proc. Stavropol Branch Russ. Entomol. Soc.* **2020**, *16*, 46–49.
70. Petrov, A.V. (Institute of Forest Science of Russian Academy of Science, Moscow, Russia). Personal communication, 2020.
71. Watanabe, K. Specific characteristics of damage to cherry caused by *Lycogoris* (*Apolygus*) *lucorum* (Meyer-Dür) (Heteroptera: Miridae) and *Halyomorpha halys* (Stål) (Heteroptera: Pentatomidae). *Annu. Rep. Plant Prot. Soc. N. Jpn.* **1996**, *1996*, 143–144.
72. Bergmann, E.; Bernhard, K.M.; Bernon, G.; Bickerton, M.; Gill, S.; Gonzales, C.; Hamilton, G.C.; Hedstrom, C.; Kamminga, K.; Koplinka-Loehr, C.; et al. Host Plants of the Brown Marmorated Stink Bug in the U.S. Available online: https://www.stopbmsb.org/where-is-bmsb/host-plants (accessed on 2 January 2022).
73. Zakharchenko, V.; Karpun, N.; Borisov, B. Trophic connections of the brown marmorated stink bug *Halyomorpha halys* Stål in the conditions of the invasive area on the Black Sea coast of the Caucasus. *BIO Web. Conf.* **2020**, *21*, 7. [CrossRef]
74. Zakharchenko, V.Y. Bioecological Features of the Brown Marmorated Stink Bug (*Halyomorpha halys* Stål) in the Humid Subtropics of Russia and Control Measures. Ph.D. Thesis, Russian State Agricultural University, Moscow, Russia, 2021. Available online: http://www.old.timacad.ru/catalog/disser/kd/zaharchenko/zaharchenko_disser.pdf (accessed on 10 January 2022). (In Russian)
75. Musolin, D.L.; Dolgovskaya, M.Y.; Protsenko, V.Y.; Karpun, N.N.; Reznik, S.Y.; Saulich, A.K. Photoperiodic and temperature control of nymphal growth and adult diapause induction in the invasive Caucasian population of the Brown marmorated stink bug, *Halyomorpha halys*. *J. Pest Sci.* **2019**, *92*, 621–631. [CrossRef]
76. Khrimian, A.; Zhang, A.; Weber, D.C.; Ho, H.-Y.; Aldrich, J.R.; Vermillion, K.E.; Siegler, M.A.; Shirali, S.; Guzma, F.; Leskey, T.C. Discovery of the aggregation pheromone of the Brown marmorated stink bug (*Halyomorpha halys*) through the creation of stereoisomeric libraries of 1-bisabolen-3-ols. *J. Nat. Prod.* **2014**, *77*, 1708–1717. [CrossRef] [PubMed]
77. Sinitsyna, E.V.; Protsenko, V.E.; Karpun, N.N.; Mityushev, I.M.; Lobur, A.Y.; Todorov, N.G. First field trials of Russia-produced pheromone preparations for monitoring and control of the Brown marmorated stink bug, *Halyomorpha halys* Stål. *Bull. Timiryazev Agricult. Acad.* **2019**, *3*, 60–79. [CrossRef]
78. Gouli, V.; Gouli, S.; Skinner, M.; Hamilton, G.; Kim, J.S.; Parker, B.L. Virulence of select entomopathogenic fungi to the brown marmorated stink bug, *Halyomorpha halys* (Stål) (Heteroptera: Pentatomidae). *Pest Manag. Sci.* **2012**, *68*, 155–157. [CrossRef] [PubMed]
79. Parker, B.L.; Skinner, M.; Gouli, S.; Gouli, V.; Kim, J.S. Virulence of BotaniGard® to second instar brown marmorated stink bug, *Halyomorpha halys* (Stål) (Heteroptera: Pentatomidae). *Insects* **2015**, *6*, 319–324. [CrossRef] [PubMed]

80. Gorgadze, O.; Bakhtadze, G.; Kereselidze, M.; Lortkhipanidze, M. The efficacy of entomopathogenic agents against *Halyomorpha halys* (Hemiptera: Pentatomidae). *Int. J. Current Res.* **2017**, *9*, 62177–62180. Available online: https://www.journalcra.com/sites/default/files/issue-pdf/27484.pdf (accessed on 10 January 2022).
81. Borisov, B.A.; Protsenko, V.Y.; Karpun, N.N. Potential use of entomotrophic bacteria and fungi to contain the brown marmorated stink bug *Halyomorpha halys* Stål (Heteroptera: Pentatomidae). In *Biological Plant Protection as a Basis for Sustainable Agroecosystems, Proceedings of the X International Scientific and Practical Conference, Krasnodar, Russia, 11–13 September 2018*; All-Russia Research Institute of Biological Plant Protection: Krasnodar, Russia, 2018; pp. 165–168. (In Russian)
82. Protsenko, V.Y.; Borisov, B.A.; Karpun, N.N. The results of evaluating the insecticidal action of some species and strains of entomoparasitic fungi on the marmorated bug (*Halyomorpha halys*). In *New and Unconventional Plants and Prospects of their Application, Proceedings of the XIII International Conference; Sochi, Russia, 4–8 June 2018*; RUDN University Publishing: Moscow, Russia, 2018; pp. 591–594. (In Russian)
83. Kulava, L.D.; Karpun, N.N.; Zhuravleva, E.N.; Shoshina, E.I.; Ayba, L.Y. The effectiveness of pheromones of a brown marmorated stink bug and traps of various designs in mandarin agrocenoses in Abkhazia. *Subtrop. Ornam. Hortic.* **2021**, *77*, 161–169. [CrossRef]
84. Fujisawa, T. Damage and control of the brown marmorated stink bug in apple orchards. *Jpn. Agric. Technol.* **2001**, *45*, 42–47.
85. Willrich, M.M.; Leonard, B.R.; Cook, D.R. Arthropod management laboratory and field evaluations of insecticide toxicity to stink bugs (Heteroptera: Pentatomidae). *J. Cotton Sci.* **2003**, *7*, 156–163. Available online: https://www.cotton.org/journal/2003-07/4/upload/jcs07-156.pdf (accessed on 10 January 2022).
86. Nielsen, A.L.; Shearer, P.W.; Hamilton, G.C. Toxicity of insecticides to *Halyomorpha halys* (Hemiptera: Pentatomidae) using glass vial bioassays. *J. Econ. Entomol.* **2008**, *101*, 1439–1442. [CrossRef]
87. Leskey, T.C.; Lee, D.-H.; Short, B.D.; Wright, S.E. Impact of insecticides on the invasive *Halyomorpha halys* (Hemiptera: Pentatomidae): Analysis of insecticide lethality. *J. Econ. Entomol.* **2012**, *105*, 1726–1735. [CrossRef] [PubMed]
88. Funayama, K. Control effect on the brown-marmorated stink bug, *Halyomorpha halys* (Hemiptera: Pentatomidae), by combined spraying of pyrethroid and neonicotinoid insecticides in apple orchards in northern Japan. *Appl. Entomol. Zool.* **2012**, *47*, 75–78. [CrossRef]
89. Mooneyham, K.L.; Aigner, J.D., Jr.; Kuhar, T.P. Control of brown marmorated stink bug with insecticide-treated window screens. *Arthropod Manag. Tests* **2016**, *41*, tsw021. [CrossRef]
90. Kuhar, T.P.; Kamminga, K.L. Review of the chemical control research on *Halyomorpha halys* in the USA. *J Pest. Sci.* **2017**, *90*, 1021–1031. [CrossRef]
91. Osborn, H.; Drake, C.J. Notes on American Tingidae with descriptions of new species. *Ohio J. Sci.* **1917**, *17*, 295–307.
92. Torres-Miller, L. Additions to the West Virginia tingid fauna (Hemiptera: Heteroptera: Tingidae). *Insecta Mundi* **1995**, *9*, 281–282.
93. Bernardinelli, I.; Zandigiacomo, P. First record of the oak lace bug *Corythucha arcuata* (Say) (Heteroptera, Tingidae) in Europa. *Inf. Fitopatol.* **2000**, *50*, 47–49.
94. Mutun, S. First report of the oak lace bug, *Corythucha arcuata* (Say, 1832) (Heteroptera: Tingidae) from Bolu, Turkey. *Isr. J. Zool.* **2003**, *49*, 323–324.
95. Mutun, S.; Ceyhan, Z.; Sözen, C. Invasion by the oak lace bug, *Corythucha arcuata* (Say) (Heteroptera: Tingidae), in Turkey. *Turk. J. Zool.* **2009**, *33*, 263–268. [CrossRef]
96. Samin, N.; Linnavuori, R.E. A contribution to the Tingidae (Heteroptera) from north and northwestern Iran. *Entomofauna. Z. Entomol.* **2011**, *32*, 373–380.
97. Shchurov, V.I.; Bondarenko, A.S.; Skvortsov, M.M.; Shchurova, A.V. Alien forest insect pests revealed in the Northwest Caucasus in 2010–2016 and consequences of their uncontrolled dispersal. *Izv. St.-Peterbg. Lesoteh. Akad.* **2017**, *220*, 212–228. (In Russian) [CrossRef]
98. Neimorovets, V.V.; Shchurov, V.I.; Bondarenko, A.S.; Skvortsov, M.M.; Konstantinov, F.V. First documented outbreak and new data on the distribution of *Corythucha arcuata* (Say, 1832) (Hemiptera: Tingidae) in Russia. *Acta Zool. Bulg.* **2017**, *9*, 139–142. Available online: https://www.researchgate.net/publication/321496929_First_Documented_Outbreak_and_New_Data_on_the_Distribution_of_Corythucha_arcuata_Say_1832_Hemiptera_Tingidae_in_Russia (accessed on 10 January 2022).
99. Karpun, N.N.; Protsenko, V.Y.; Borisov, B.A.; Shiryaeva, N.V. Discovery of oak lace bug *Corythucha arcuata* (Say) (Heteroptera: Tingidae) in the subtropical zone of the Black Sea coast of the Caucasus. *Eurasian Entomol. J.* **2018**, *17*, 113–119. [CrossRef]
100. Stryukova, N.M.; Omel'yanenko, T.Z.; Golub, V.B. Oak lace bug in the Republic of Crimea. *Plant Prot. Quar.* **2019**, *9*, 43–44. Available online: https://www.elibrary.ru/download/elibrary_39322104_97303903.pdf (accessed on 10 January 2022). (In Russian).
101. Martynov, V.V.; Nikulina, T.V. Oak lace bug *Corythucha arcuata* (Say, 1832) (Hemiptera: Tingidae)—A new invasive pest in the forests of the southwestern part of mountain Crimea. *Subtrop. Ornam. Hortic.* **2020**, *72*, 124–138. (In Russian) [CrossRef]
102. Golub, V.B.; Golub, N.V.; Soboleva, V.A. Distribution and trophic relations of the oak lace bug *Corythucha arcuata* (Say) (Heteroptera: Tingidae) in Crimea. *Field Biol. J.* **2020**, *2*, 179–184. Available online: https://www.elibrary.ru/download/elibrary_44184062_33929269.pdf (accessed on 10 January 2022). (In Russian)
103. Gninenko, Y.I.; Chernova, U.A.; Gimranov, R.I.; Rakov, A.G.; Khegay, I.V. Oak lace bug expanses its area on the territory of Russia. *Plant Prot. Quar.* **2020**, *10*, 37–38. Available online: https://www.elibrary.ru/download/elibrary_43992021_73556954.pdf (accessed on 10 January 2022). (In Russian)

104. Blyummer, A.G. Invasive species of not-Arctic lace bugs of the genus *Corythucha* (Heteroptera, Tingidae) in Eurasia: Features of distribution and injuriousness. In *Ecological and Economic Consequences of Invasions of Dendrophilous Insects, Proceedings of the All-Russian Conference with International Participation, Krasnoyarsk, Russia, 25–27 September 2012*; Baranchikov, Y.N., Ed.; Forest Institute SB RAS: Krasnoyarsk, Russia, 2012; pp. 139–143. (In Russian)
105. Drake, C.J.; Ruhoff, F.A. Lacebugs of the World: A Catalog (Hemiptera: Tingidae). *Bull. U.S. Natl. Mus.* **1965**, *243*, 1–634. [CrossRef]
106. Borisov, B.A.; Karpun, N.N.; Bibin, A.R.; Grabenko, Y.A.; Shiryaeva, N.V.; Lyanguzov, M.Y. New data on trophic relations of the invasive oak lace bug *Corythucha arcuata* (Heteroptera: Tingidae) in the Krasnodar region and in the Republic of Adygea based on the research findings for the year 2018. *Subtrop. Ornam. Hortic.* **2018**, *67*, 188–203. (In Russian) [CrossRef]
107. Karpun, N.N.; Russian Research Institute of Floriculture and Subtropical Crops, Sochi, Russia. Personal communication, 2021.
108. Paulin, M.; Hirka, A.; Eötvös, C.B.; Gáspár, C.; Fürjes-Mikó, Á.; Csóka, G. Known and predicted impacts of the invasive oak lace bug (*Corythucha arcuata*) in European oak ecosystems—A review. *Folia Oecologica* **2020**, *47*, 131–139. [CrossRef]
109. Shchurov, V.I.; Zamotajlov, A.S.; Shchurova, A.V. Assessment of climatic conditions for the expansion of *Corythucha arcuata* (Say, 1832) (Heteroptera: Tingidae) in the European part of Russia. In *Dendrobiotic Invertebrates and Fungi and their Role in Forest Ecosystems. The Kataev Memorial Readings—XI, Proceedings of the All-Russia Conference with International Participation, St. Petersburg, Russia, 24–27 November 2020*; Musolin, D.L., Kirichenko, N.I., Selikhovkin, A.V., Eds.; SPbFTU: St. Petersburg, Russia, 2020; pp. 379–380. (In Russian)
110. Shchurov, V.I.; Zamotajlov, A.S.; Bondarenko, A.S.; Shchurova, A.V.; Skvortsov, M.M.; Glushchenko, L.S. The oak lace bug *Corythucha arcuata* (Say, 1832) (Heteroptera: Tingidae) in the Northwestern Caucasus: Phenology, biology, monitoring of the territorial expansion, and harmfulness. *Izv. St.-Peterbg. Lesoteh. Akad.* **2019**, *228*, 58–87. (In Russian) [CrossRef]
111. Gninenko, Y.I.; Chernova, U.A.; Nalepin, V.P. Oak lace bug: Stages of formation of the secondary range in Russia. In *Innovations in the Conservation and Sustainable Development of Forest Ecosystems, Proceedings of International and Practical Conference Dedicated to the 20th Anniversary of State National Park "Burabay" Creation, Burabay, Kazakhstan, 2–5 September 2020*; Bykov, S.V., Ed.; State National Park "Burabay": Burabay, Kazakhstan, 2020; pp. 66–68.
112. Kovač, M.; Gorczak, M.; Wrzosek, M.; Tkaczuk, C.; Pernek, M. Identification of entomopathogenic fungi as naturally occurring enemies of the invasive oak lace bug, *Corythucha arcuata* (Say) (Hemiptera: Tingidae). *Insects* **2020**, *11*, 679. [CrossRef] [PubMed]
113. Borisov, B.A. (AgroBioTechnology Ltd., Moscow, Russia). Personal communication, 2021.
114. Chernova, U.A.; Rakov, A.G.; Khegay, I.V. Mixed insecticide against oak lace bug. *Plant Prot. Quar.* **2019**, *11*, 51. Available online: https://www.elibrary.ru/download/elibrary_41239385_98490426.pdf (accessed on 10 January 2022). (In Russian)
115. Bălăcenoiu, F.; Nețoiu, C.; Tomescu, R.; Simon, D.C.; Buzatu, A.; Toma, D.; Petrițan, I.C. Chemical control of *Corythucha arcuata* (Say, 1832), an invasive alien species, in oak forests. *Forests* **2021**, *12*, 770. [CrossRef]
116. Besedina, E.N.; Ismailov, V.Y.; Nastasiy, A.S. Field evaluation of the effectiveness of biological and biorational insecticides against oak lace bug *Corythucha arcuata* Say (Hemiptera, Tingidae). *Agrochemistry* **2021**, *3*, 45–50. (In Russian) [CrossRef]
117. EPPO. Pest Risk Analysis for *Agrilus fleischeri*. 2019. Available online: https://gd.eppo.int/taxon/AGRLFL/documents (accessed on 2 January 2022).
118. Volkovitsh, M.G. On the invasive potential of buprestid beetles (Coleoptera: Buprestidae) damaging woody plants. In *Dendrobiotic Invertebrates and Fungi and their Role in Forest Ecosystems. The Kataev Memorial Readings—XI, Proceedings of the All-Russia Conference with International Participation, St. Petersburg, Russia, 24–27 November 2020*; Musolin, D.L., Kirichenko, N.I., Selikhovkin, A.V., Eds.; SPbFTU: St. Petersburg, Russia, 2020; pp. 113–116, (In Russian and English)
119. Volkovitsh, M.G.; Kovalev, A.V.; Orlova-Bienkowskaja, M.J. Current distribution and diagnostic features of two potentially invasive Asian buprestid species: *Agrilus mali* Matsumura and *A. fleischeri* Obenberger (Coleoptera: Buprestidae). *Insects* **2020**, *11*, 493. [CrossRef] [PubMed]
120. EPPO. A2 List of Pests Recommended for Regulation as Quarantine Pests (Version 2019-09). Available online: https://www.eppo.int/ACTIVITIES/plant_quarantine/A2_list (accessed on 15 August 2021).
121. Obenberger, J. De novis Buprestidarum regionis palaearcticae speciebus V. *Cas. Ceskoslovenske Spoiecnosti Entomol.* **1925**, *22*, 30–34.
122. Kubáň, V.; Jendek, E.; Kalashian, M.Y.; Volkovitsh, M.G. Superfamily Buprestoidea Leach, 1815. In *Catalogue of Palaearctic Coleoptera. Scarabaeoidea, Scirtoidea, Dascilloidea, Buprestoidea and Byrrhoidea*; Löbl, I., Löbl, D., Eds.; BRILL: Leiden, The Netherlands, 2016; Volume 3, pp. 19–32, 432–574.
123. Zang, K.; Wang, X.-Y.; Yang, Z.-Q.; Wei, K.; Duan, J.J. Biology and natural enemies of *Agrilus fleischeri* (Coleoptera: Buprestidae), a newly emerging destructive buprestid pest in Northeast China. *J. Asia-Pac. Entomol.* **2017**, *20*, 47–52. [CrossRef]
124. Wang, J.S.; Shang, P.H.; Li, F. Preliminary observations on the life history of *Agrilus* sp. *Shaanxi For. Sci. Technol.* **1995**, *3*, 40–42.
125. Ji, Y.; Ji, R.; Huang, R.-X. Invasive species—*Agrilus mali* Matsumura and damage in Xinjiang. *Xinjiang Agric. Sci.* **2004**, *41*, 31–33.
126. Cui, X.N.; Liu, D.G.; Li, A.H. Research progress in integrated management of *Agrilus mali*. *Plant Prot.* **2015**, *41*, 16–23.
127. Bozorov, T.A.; Luo, Z.; Li, X.; Zhang, D. *Agrilus mali* Matsumura (Coleoptera: Buprestidae), a new invasive pest of wild apple in western China: DNA barcoding and life cycle. *Ecol. Evol.* **2018**, *9*, 1160–1172. [CrossRef]
128. Cui, Z.-J.; Zhang, Y.-L.; Zhang, X.; Luo, Z.-H.; Zhang, P.; Golec, J.; Poland, T.M.; Zalucki, M.P.; Han, P.; Lu, Z.-Z. Life history and mortality factors of *Agrilus mali* Matsumura (Coleoptera: Buprestidae) in wild apples in Northwestern China. *Agric. For. Entomol.* **2019**, *21*, 309–317. [CrossRef]

129. Zhang, P.; Cui, Z.; Xu, H.; Ali, A.; Zhang, X.; Liu, X.; Zhang, Y.; Zhou, X.; Lu, Z. Thirst or malnutrition: The impacts of invasive insect *Agrilus mali* on the physiological status of wild apple trees. *Insects* **2020**, *11*, 440. [CrossRef]
130. Zhang, X.; Zhang, Y.-L.; Zhang, P.; Cui, Z.-J.; Han, P.; Gao, G.-Z.; Poland, T.M.; Zalucki, M.P.; Lu, Z.-Z. *Agrilus mali* Matsumura (Coleoptera: Buprestidae) density and damage in wild apple *Malus sieversii* (Rosales: Rosaceae) forests in Central Eurasia under four different management strategies. *Entomol. Gen.* **2021**, *41*, 257–266. [CrossRef]
131. A Unified List of Quarantine Pests of the Eurasian Economic Union. (As Amended by 8 August 2019.) (Decision of the Council of the Eurasian Economic Commission of 8 August 2019 No. 74). Available online: https://vniikr.ru/edinyij-perechen-karantinnyix-obektov-evrazijskogo-ekonomicheskogo-soyuza (accessed on 22 August 2021).
132. *Agrilus mali*. EPPO Global Database. 2021. Available online: https://gd.eppo.int/taxon/AGRLMA (accessed on 22 August 2021).
133. Jendek, E.; Grebennikov, V. *Agrilus (Coleoptera, Buprestidae) of East Asia*; Jan Farkač: Prague, Czech Republic, 2011; p. 362.
134. Matsumura, S. Life history of *Agrilus mali* Mats. (Buprestidae). In *Bionomics on Apple Trees in China and Korea*; Muramatsu, S., Ed.; Korea Agricultural Experiment Station: Seoul, Korea, 1924; pp. 1–21. (In Japanese)
135. Sokolov, E.A.; Atanov, N.M.; Zhimerikin, V.N.; Gura, N.A.; Komarova, G.F.; Nikritin, L.M.; Shakhramanov, I.K. Quarantine pests with limited distribution on the territory of the Russian Federation. *Plant Prot.* **1995**, *5*, 37–41. (In Russian)
136. Jendek, E.; Poláková, J. *Host Plants of World Agrilus (Coleoptera, Buprestidae). A Critical Review*; Springer: New York, NY, USA, 2014; pp. 1–706.
137. Liu, Z.Q.; Chen, W.M.; Xu, Z.; Liang, Q.L. *Malus sieversii* forest distribution and *Agrilus mali* Matsumura status of damage in the west part of Tianshan Mountains. *North. Hortic.* **2014**, *17*, 121–124.
138. Li, M.L.; Zhang, Z.Q. Discussion on biology and life history associated with *Agrilus mali* Matsumura. *J. Northwest For. Univ.* **2017**, *32*, 139–146.
139. Nikritin, L.M.; Shutova, N.N. Apple buprestid. *Plant Prot.* **1985**, *1*, 44. (In Russian)
140. Nikritin, L.M. Apple buprestid. *Plant Prot.* **1994**, *3*, 46. (In Russian)
141. Cao, L.M.; Zhang, Y.L.; van Achterberg, C.; Wang, Z.Y.; Wang, X.Y.; Zhao, W.X.; Yang, Z.Q. Notes on braconid wasps (Hymenoptera, Braconidae) parasitising on *Agrilus mali* Matsumura (Coleoptera, Buprestidae) in China. *ZooKeys* **2019**, *867*, 97–121. [CrossRef]
142. Fairmaire, L. Notes sur les Coléoptères des environs de Pékin (2e Partie). *Revue Ent. Caen* **1888**, *7*, 111–160.
143. Chamorro, M.L.; Jendek, E.; Haack, R.A.; Petrice, T.R.; Woodley, N.E.; Konstantinov, A.S.; Volkovitsh, M.G.; Yang, X.K.; Grebennikov, V.V. *Illustrated Guide to the Emerald Ash Borer, Agrilus planipennis Fairmaire and Related Species (Coleoptera, Buprestidae)*; Pensoft Publishers: Sofia, Bulgaria; Moscow, Russia, 2015; pp. 1–198.
144. Baranchikov, Y.; Mozolevskaya, E.; Yurchenko, G.; Kenis, M. Occurrence of the emerald ash borer, *Agrilus planipennis* in Russia and its potential impact on European forestry. *EPPO Bull.* **2008**, *38*, 233–238. [CrossRef]
145. Musolin, D.L.; Selikhovkin, A.V.; Shabunin, D.A.; Zviagintsev, V.B.; Baranchikov, Y.N. Between ash dieback and emerald ash borer: Two Asian invaders in Russia and the future of ash in Europe. *Balt. For.* **2017**, *23*, 316–333.
146. Orlova-Bienkowskaja, M.J.; Volkovitsh, M.G. Are native ranges of the most destructive invasive pests well known? A case study of the native range of the emerald ash borer, *Agrilus planipennis* (Coleoptera: Buprestidae). *Biol. Invasions* **2018**, *20*, 1275–1286. [CrossRef]
147. Meshkova, V.L.; Kucheryavenko, T.V.; Zinchenko, O.V.; Borysenko, A.I. Beginning of the spread of *Agrilus planipennis* Fairmaire (Coleoptera: Buprestidae) on the territory of Ukraine. *Izv. St.-Peterbg. Lesoteh. Akad.* **2021**, *236*, 163–184. (In Russian) [CrossRef]
148. Volkovitsh, M.G.; Bieńkowski, A.O.; Orlova-Bienkowskaja, M.J. Emerald ash borer approaches the borders of the European Union and Kazakhstan and is confirmed to infest European ash. *Forests* **2021**, *12*, 691. [CrossRef]
149. Volkovitsh, M.G.; Orlova-Bienkowskaja, M.J.; Kovalev, A.V.; Bieńkowski, A.O. An illustrated guide to distinguish emerald ash borer (*Agrilus planipennis*) from its congeners in Europe. *Forestry* **2020**, *93*, 316–325. [CrossRef]
150. Haack, R.A.; Jendek, E.; Liu, H.; Marchant, K.R.; Petrice, T.R.; Poland, T.M.; Ye, H. The emerald ash borer: A new exotic pest in North America. *Newsl. Mich. Entomol. Soc.* **2002**, *47*, 1–5. Available online: https://www.researchgate.net/publication/228544207_The_Emerald_Ash_Borer_A_New_Exotic_Pest_in_North_America (accessed on 10 January 2022).
151. Izhevskii, S.S. Threatening Findings of the Emerald Ash Borer *Agrilus planipennis* in the Moscow Region. 2007. Available online: http://www.zin.ru/Animalia/Coleoptera/rus/agrplaiz.htm (accessed on 6 July 2021). (In Russian)
152. Volkovitsh, M.G. Emerald Ash Borer *Agrilus planipennis*—New Extremely Dangerous Pest of Ash in the European Part of Russia. 2007. Available online: http://www.zin.ru/Animalia/Coleoptera/rus/eab_2007.htm (accessed on 7 July 2021).
153. Emerald Ash Borer Informative Network. Available online: http://www.emeraldashborer.info/ (accessed on 6 July 2021).
154. Orlova-Bienkowskaja, M.J.; Drogvalenko, A.N.; Zabaluev, I.A.; Sazhnev, A.S.; Peregudova, E.Y.; Mazurov, S.G.; Komarov, E.V.; Struchaev, V.V.; Martynov, V.V.; Nikulina, T.V.; et al. Current range of *Agrilus planipennis* Fairmaire, an alien pest of ash trees, in European Russia and Ukraine. *Ann. For. Sci.* **2020**, *77*, 1–14. [CrossRef]
155. Meshkova, V.L.; Skrylnik, Y.Y.; Terekhova, V.V.; Kucheryavenko, T.V. Emerald ash borer (*Agrilus planipennis*) in Kharkiv region. In *Modern Problems of Forestry and Ecology and Ways of Their Solution, Proceedings of the International Scientific and Practical Conference "Faculty of Forestry and Ecology—20 years", Zhytomyr, Ukraine, 7–8 October 2021*; Skidan, O.V., Romanchuk, L.D., Vishnevskiy, A.V., Siruk, Y.V., Kratiuk, O.L., Zhitova, O.P., Andreeva, O.Y., Shvets, M.V., Ishuk, O.V., Eds.; Poliskiy National University: Zhytomyr, Ukraine, 2021; pp. 125–126.

156. Musolin, D.L.; Selikhovkin, A.V.; Peregudova, E.Y.; Popovichev, B.G.; Mandelshtam, M.Y.; Baranchikov, Y.N.; Vasaitis, R. North-westward expansion of the invasive range of Emerald ash borer, *Agrilus planipennis* Fairmaire (Coleoptera: Buprestidae) towards the EU: From Moscow to Saint Petersburg. *Forests* **2021**, *12*, 502. [CrossRef]
157. Selikhovkin, A.V.; Musolin, D.L.; Popovichev, B.G.; Merkuryev, S.A.; Volkovitsh, M.G.; Vasaitis, R. Invasive populations of the Emerald ash borer *Agrilus planipennis* Fairmaire, 1888 (Coleoptera: Buprestidae) in Saint Petersburg, Russia: A hitchhiker? *Insects* **2022**, *13*, 191. [CrossRef] [PubMed]
158. Volkovitsh, M.G.; Mozolevskaya, E.G. The tenth «anniversary» of the invasion of emerald ash borer *Agrilus planipennis* Fairm. (Coleoptera: Buprestidae) in Russia: Results and prospects. *Izv. St.-Peterbg. Lesoteh. Akad.* **2014**, *207*, 8–19. (In Russian)
159. Baranchikov, Y.N.; Seraya, L.G.; Grinash, M.N. All European ash species are susceptible to emerald ash borer *Agrilus planipennis* Fairmaire (Coleoptera: Buprestidae)—A Far Eastern invader. *Siberian Forest J.* **2014**, *6*, 80–85. Available online: https://www.researchgate.net/publication/272822249_Baranchikov_Yu_N_Seraya_L_G_Grinash_M_N_All_European_ash_species_are_susceptible_to_emerald_ash_borer_Agrilus_planipennis_Fairmaire_Coleoptera_Buprestidae_-_a_Far_Eastern_invader_Sibirskiy_Lesnoy_Zurn (accessed on 10 January 2022). (In Russian)
160. Volkovitsh, M.G. Buprestidae. In *Inventory of Alien Beetles of European Russia*; Orlova-Bienkowskaja, M.J., Ed.; Mukhametov, G.V. Publisher: Livny, Russia, 2019; pp. 84–95. (In Russian)
161. Volkovitsh, M.G.; Suslov, D.V. The first record of the emerald ash borer, *Agrilus planipennis* Fairmaire (Coleoptera: Buprestidae), in St. Petersburg signals a real threat to the palace and park ensembles of Peterhof and Oranienbaum. In *Dendrobiotic Invertebrates and Fungi and their Role in Forest Ecosystems. The Kataev Memorial Readings—XI, Proceedings of the All-Russia Conference with International Participation, St. Petersburg, Russia, 24–27 November 2020*; Musolin, D.L., Kirichenko, N.I., Selikhovkin, A.V., Eds.; SPbFTU: St. Petersburg, Russia, 2020; pp. 119–122, (In Russian and English).
162. Haubrock, P.J.; Turbelin, A.J.; Cuthbert, R.N.; Novoa, A.; Taylor, N.G.; Angulo, E.; Ballesteros-Mejia, L.; Bodey, T.W.; Capinha, C.; Diagne, C.; et al. Economic costs of invasive alien species across Europe. *NeoBiota* **2021**, *67*, 153–190. [CrossRef]
163. Gninenko, Y.I.; Kliukin, M.S.; Khegai, I.V. Emerald ash borer: Catastrophe postponed? *Plant Health Res. Pract.* **2016**, *3*, 38–45. Available online: https://elibrary.ru/download/elibrary_28379702_35791451.pdf (accessed on 10 January 2022). (In Russian)
164. Kirichenko, N.; Haubrock, P.J.; Cuthbert, R.N.; Akulov, E.; Karimova, E.; Shneider, Y.; Liu, C.; Angulo, E.; Diagne, C.; Courchamp, F. Economic costs of biological invasions in terrestrial ecosystems in Russia. *NeoBiota* **2021**, *67*, 103–130. [CrossRef]
165. Commission Delegated Regulation (EU) 2019/1702 of 1 August 2019 Supplementing Regulation (EU) 2016/2031 of the European Parliament and of the Council by Establishing the List of Priority Pests. C/2019/5637. Available online: https://eur-lex.europa.eu/legal-content/EN/TXT/?qid=1570788497660&uri=CELEX:32019R1702 (accessed on 16 January 2022).
166. Orlova-Bienkowskaja, M.J.; Bienkowski, A.O. Low heat availability could limit the potential spread of the emerald ash borer to Northern Europe (Prognosis based on growing degree days per year). *Insects* **2022**, *13*, 52. [CrossRef] [PubMed]
167. EPPO. PM 9/14 (1) *Agrilus planipennis*: Procedures for official control. *Bull. OEPP/EPPO Bull.* **2013**, *43*, 499–509. [CrossRef]
168. Yang, Z.Q.; Strazanac, J.S.; Marsh, P.M.; Van Achterberg, C.; Choi, W.Y. First recorded parasitoid from China of *Agrilus planipennis*: A new species of *Spathius* (Hymenoptera: Braconidae: Doryctinae). *Ann. Entomol. Soc. Am.* **2005**, *98*, 636–642. [CrossRef]
169. Orlova-Bienkowskaja, M.J.; Belokobylskij, S.A. Discovery of the first European parasitoid of the emerald ash borer *Agrilus planipennis* (Coleoptera: Buprestidae). *Eur. J. Entomol.* **2014**, *111*, 594–596. [CrossRef]
170. Bauer, L.S.; Duan, J.J.; Gould, J.R. Emerald ash borer (Agrilus planipennis Fairmaire) (Coleoptera: Buprestidae). In *The Use of Classical Biological Control to Preserve Forests in North America*; Van Driesche, R., Reardon, R., Eds.; FHTET-2013-02; U.S. Department of Agriculture, Forest Service, Forest Health Technology Enterprise Team: Morgantown, WV, USA, 2014; pp. 189–209. Available online: https://www.fs.usda.gov/treesearch/pubs/48051 (accessed on 16 January 2022).
171. Volkovitsh, M.G.; Zykov, I.E.; Karpun, N.N.; Zakharchenko, V.Y.; and Kovalev, A.V. A description of the larva of the cypress jewel beetle, *Lamprodila (Palmar) festiva* (L.), with notes on the larval characters of Poecilonotini and Dicercini (Coleoptera, Buprestidae). *Entomol. Rev.* **2020**, *99*, 1304–1317. [CrossRef]
172. Volkovitsh, M.G.; Karpun, N.N. A new invasive species of Buprestid beetles in the Russian fauna: *Lamprodila (Palmar) festiva* (L.) (Coleoptera, Buprestidae), a pest of Cupressaceae. *Entomol. Rev.* **2017**, *97*, 425–437. [CrossRef]
173. Karpun, N.N.; Gnezdilov, A.A. Why do Arborvitae and False Cypresses Dry Up on the Black Sea Coast? 2016. Available online: http://www.vniisubtrop.ru/novosti/731-pochemu-usykhayut-tui-i-kiparisoviki-na-cheromorskompoberezhe.html (accessed on 3 August 2021). (In Russian)
174. Karpun, N.N.; Zhuravleva, Y.N.; Volkovitsh, M.G.; Procenko, V.Y.; Musolin, D.L. To the fauna and biology of new alien insect pest species of woody plants in humid subtropics of Russia. *Izv. St.-Peterbg. Lesoteh. Akad.* **2017**, *220*, 169–185. (In Russian) [CrossRef]
175. Gubin, A.I.; Martynov, V.V.; Nikulina, T.V. The first record of Cypress jewel beetle *Lamprodila (Palmar) festiva* (Linnaeus, 1767) (Coleoptera, Buprestidae) in the Donbass. *Subtrop. Ornam. Hortic.* **2020**, *75*, 96–107. [CrossRef]
176. Sinelnikov, K.Y. Cypress jewel beetle *Lamprodila festiva* (L.) on the Southern Coast of the Crimea (2019). Available online: https://vitusltd.ru/blog/lesozaschita/17147 (accessed on 3 August 2021). (In Russian)
177. Skvortsov, M.M. The Cypress Jewel Beetle is a New Object of State Forest Pathology Monitoring in Krasnodar Territory. 2017. Available online: http://www.czl23.ru/print.php?news.221 (accessed on 3 August 2021). (In Russian)
178. Zhuravleva, Y.N. (Research Centre the Subtropical Scientific Centre of RAS, Sochi, Russia). Personal communication, 2021.
179. Wermelinger, B. Der Grüne Wacholder-Prachtkäfer. *G'plus Gart.-Fachz.* **2011**, *3*, 30.

180. Razinger, J.; Zerjav, M.; Modic, S. *Thuja occidentalis* L. is commonly a host for Cypress jewel beetle (*Ovalisia festiva* L.) in Slovenia. In *Zbornik Predavanj in Referatov 11. Slovenskega Posvetovanja o Varstvu Rastlin z Mednarodno Udeleibo, Proceedings of Conference, Bled, Ljubljana, Slovenia, 5–6 March 2013;* Društvo za varstvo rastlin Slovenije: Ljubljana, Slovenia, 2013; pp. 359–365. Available online: http://www.dvrs.bf.uni-lj.si/spvr/2013/61Razinger.pdf (accessed on 18 January 2022).
181. Schmidt, G.; Diószegi, M.S.; Szabó, V.; Hrotkó, K. Cypress borer (*Lamprodila festiva*), a new urban pest in Hungary. In *Plants in Urban Areas and Landscape, Proceedings of International Symposium, Nitra, Slovakia, 14–15 May 2014;* Slovak University of Agriculture: Nitra, Slovakia, 2014; pp. 32–34.
182. Thoma, J.; Eickermann, M. Erstauftreten des Wacholderprachtkafers *Ovalisia festiva* (Linnaeus, 1767) in Luxemburg. *Bull. Soc. Nat. Luxemb.* **2014**, *115*, 227–229.
183. Nitzu, E.; Dobrin, I.; Dumbravă, M.; Gutue, M. The range expansion of *Ovalisia festiva* (Linnaeus, 1767) (Coleoptera: Buprestidae) in Eastern Europe and its damaging potential for Cupressaceae. *Trav. Muséum Natl. Hist. Nat.* **2016**, *58*, 51–57. [CrossRef]
184. Ruseva, S.; Todorov, I.; Pencheva, A. New data on *Ovalisia (Palmar) festiva* (Linnaeus) (Coleoptera: Buprestidae) and its natural enemies reported from Bulgaria. *Ecol. Montenegrina* **2020**, *28*, 53–60. Available online: https://www.researchgate.net/publication/339539961_New_data_on_Ovalisia_Palmar_festiva_Linnaeus_Coleoptera_Buprestidae_and_its_natural_enemies_reported_from_Bulgaria (accessed on 18 January 2022). [CrossRef]
185. Karpun, N.N.; Volkovitsh, M.G. Cypress jewel-beetle *Lamprodila (Palmar) festiva* (L.) (Coleoptera: Buprestidae)—A new invasive pest on the Black Sea coast of the Caucasus. In *The Kataev Memorial Readings–IX. Dendrobiotic Invertebrates and Fungi and their Role in Forest Ecosystems, Proceedings of the International Conference, St. Petersburg, Russia, 23–25 November 2016;* Musolin, D.L., Selikhovkin, A.V., Eds.; SPbFTU: St. Petersburg, Russia, 2016; pp. 45–46. Available online: https://www.researchgate.net/publication/310832354_Cypress_jewel-beetle_Lamprodila_Palmar_festiva_L_Coleoptera_Buprestidae_-_a_new_invasive_pest_on_the_Black_Sea_Coast_of_the_Caucasus_in_Russian (accessed on 18 January 2022). (In Russian)
186. Volkovitsh, M.G.; Karpun, N.N. Lamprodila (Palmar) festiva (L.), a New Invasive Pest of Cupressaceae in the Russian Fauna (Coleoptera: Buprestidae: Poecilonotini). 2016. Available online: http://www.zin.ru/animalia/coleoptera/pdf/Volkovitsh_Karpun_2016_Lamprodila_festiva.pdf (accessed on 5 August 2021). (In Russian)
187. Pfeffer, A. Zentral- und Westpaläarktishe Borken- und Kernkäfer (Coleoptera: Scolytidae, Platypodidae). *Entomol. Basiliensia* **1994**, *17*, 5–310.
188. Knížek, M. *Curculionidae: Scolytinae, In Catalogue of Palaearctic Coleoptera;* Löbl, I., Smetana, A., Eds.; Apollo Books: Stenstrup, Denmark, 2011; Volume 7, Curculionoidea I; pp. 86–88, 204–251.
189. Økland, B.; Flø, D.; Schroeder, M.; Zach, P.; Cocos, D.; Martikainen, P.; Siitonen, J.; Mandelshtam, M.Y.; Musolin, D.L.; Neuvonen, S.; et al. Range expansion of the small spruce bark beetle *Ips amitinus*: A newcomer in northern Europe. *Agric. For. Entomol.* **2019**, *21*, 286–298. [CrossRef]
190. Mandelshtam, M.Y.; Selikhovkin, A.V. Bark and ambrosia beetles (Coleoptera, Curculionidae: Scolytinae) of Northwest Russia: History of the study, composition and genesis of the fauna. *Entomol. Rev.* **2020**, *100*, 800–826. [CrossRef]
191. Mandelshtam, M.Y.; Selikhovkin, A.V. Comparative characteristics of species richness of the bark beetles (Coleoptera: Curculionidae, Scolytinae) fauna in north-west subregions of Russia. In *Dendrobiotic Invertebrates and Fungi and their Role in Forest Ecosystems. The Kataev Memorial Readings—XI, Proceedings of the All-Russia Conference with International Participation, St. Petersburg, Russia, 24–27 November 2020;* Musolin, D.L., Kirichenko, N.I., Selikhovkin, A.V., Eds.; SPbFTU: St. Petersburg, Russia, 2020; pp. 214–215. (In Russian)
192. Mandelshtam, M.Y.; Musolin, D.L.; Økland, B.; Flø, D.; Schroeder, M.; Zach, P.; Cocos, D.; Martikainen, P.; Siitonen, J.; Neuvonen, S.; et al. Range expansion of the small spruce bark beetle *Ips amitinus* (Coleoptera, Curculionidae: Scolytinae) in Northern Europe and Western Siberia. In *Dendrobiotic Invertebrates and Fungi and their Role in Forest Ecosystems. The Kataev Memorial Readings—XI, Proceedings of the All-Russia Conference with International Participation, St. Petersburg, Russia, 24–27 November 2020;* Musolin, D.L., Kirichenko, N.I., Selikhovkin, A.V., Eds.; SPbFTU: St. Petersburg, Russia, 2020; pp. 211–212. (In Russian)
193. Mikutowicz, J.M. Zur Koleopterenfauna der Ostseeprovinzen Russlands. *Korresp.-Bl. Des Nat.-Ver. Zu Riga* **1905**, *48*, 73–92.
194. Leius, K. Taiendavaid andmeid kodumaa urasklaste (Ipidae) fauna kohta. Erganzende Angaben uber die Borkenkafer-Fauna (Ipidae) in Estland (Zusammenfassung). *Eest. Metsanduse Aastaraam. (Estländisches Forstwirtsch. Jahrb. Tartu)* **1939**, *9*, 317–328. (In Estonian)
195. Voolma, K. (Estonian University of Life Sciences, Tartu, Estonia). Personal communication, 2021.
196. Zolk, K. Kodumaa ürasklased (Ipidae) ühes lühikese ülevaatega nende bionoomiast ja levimisest Eestis [Die Borkenkäfer (Ipidae) Estlands mit kurzer Berücksichtigung ihrer Bionomie und Verbreitung]. *Eest. Metsanduse Aastaraam. (Estländisches Forstwirtsch. Jahrb. Tartu)* **1932**, *6*, 127–176. (In Estonian)
197. Zolk, K. Metsakahjurite esinemine Eestis 1934 [Das Vorkommen der Forstschädlinge in Estland im Jahre 1934]. *Eest. Metsanduse Aastaraam. (Estländisches Forstwirtsch. Jahrb. Tartu)* **1935**, *7*, 614–638. (In Estonian)
198. Nuorteva, M. Fennoskandialle uusi kaarnakuoriaislaji *Ips amitinus* Eichh. tavattu Suomesta [*Ips amitinus* Eichh. (Col., Scolytidae) neu für Fennoscandien]. *Ann. Entomol. Fenn.* **1955**, *21*, 30–32.
199. Koponen, M. Distribution of *Ips amitinus* Eichh. (Coleoptera, Scolytidae) in Finland in 1950–1973. *Ann. Entomol. Fenn.* **1975**, *41*, 65–69.
200. Koponen, M. Distribution of *Ips amitinus* Eichh. (Coleoptera, Scolytidae) in Finland in 1974–1979. *Not. Entomol.* **1980**, *60*, 223–225.

201. Muona, J.; Viramo, J. The Coleoptera of the Koillismaa area (Ks), North-East Finland. *Oulanka Rep.* **1986**, *6*, 3–50.
202. Hansen, V. *Catalogus Coleopterorum Fennoscandiae et Daniae*; Lindroth, C.H., Ed.; Entomologiska Sällskapet i Lund: Lund, Sweden, 1960; pp. 1–476.
203. Lindelöw, Å. Väntad barkborre funnen i Sverige-fynd av *Ips amitinus* (Coleoptera, Scolytinae) [*Ips amitinus* (Coleoptera, Scolytinae) expected and found in Sweden]. *Entomol. Tidskr.* **2013**, *134*, 203–206.
204. Stark, V.N. Materials to knowledge of bark beetles of Briansk Gouvernement. *Zashchita Rasteniy* [Plant Protection]. **1926**, *III*, 330–339. (In Russian)
205. Mandelshtam, M. Notes on the current status of *Ips amitinus* Eichh. (Coleoptera, Scolytidae) in North-West Russia. *Entomol. Fenn.* **1999**, *10*, 29–34. [CrossRef]
206. Voolma, K.; Mandelshtam, M.Y.; Shcherbakov, A.N.; Yakovlev, E.B.; Õunap, H.; Süda, I.; Popovichev, B.G.; Sharapa, T.V.; Galasjeva, T.V.; Khairetdinov, R.R.; et al. Distribution and spread of bark beetles (Coleoptera: Scolytidae) around the Gulf of Finland: A comparative study with notes on rare species of Estonia, Finland and North-Western Russia. *Entomol. Fenn.* **2004**, *15*, 198–210. [CrossRef]
207. Titova, E.V. Bark beetles associated with coniferous coppice in forest clearings of Karelia. In *Studies of Forest Regeneration in Karelia, Proceedings of the Karelian Branch of the USSR Academy of Sciences, Petrozavodsk, Russia, 1959*; Shiperovich, V.Y., Kishenko, T.I., Lisenkov, A.F., Eds.; Karelian Branch of the USSR Academy of Sciences: Petrozavodsk, Russia, 1959; Volume 16, pp. 110–126. (In Russian)
208. Yakovlev, E.B.; Shorokhov, V.V.; Gorbunova, V.N. Materials on the fauna of xylophagous beetles of Karelia. In *The Fauna and Ecology of Arthropods in Karelia*; Yakovlev, E.B., Uzenbaev, S.D., Eds.; Forest Institute: Petrozavodsk, Russia, 1986; pp. 40–43. (In Russian)
209. Mozolevskaya, E.G.; Sharapa, T.V. Species composition of xylophagous insects of Murmansk Province. *Entomol. Obozr.* **1996**, *75*, 558–566. (In Russian)
210. Shcherbakov, A.N.; Nikitsky, N.B.; Polevoi, A.V.; Humala, A.E. On the fauna of beetles (Insecta, Coleoptera) of Pasvik Nature Reserve. *Lesn. Vestn.* **2013**, *6*, 16–21. (In Russian)
211. Mandelshtam, M.Y.; Musolin, D.L. Bark beetle *Ips amitinus* (Eichhoff, 1872) (Coleoptera: Curculionidae: Scolytinae) continues to expand its range in north-western and northern Russia. In *Monitoring and Biological Control Methods of Woody Plant Pests and Pathogens: From Theory to Practice, Proceedings of International Conference, Moscow, Russia, 18–22 April 2016*; Baranchikov, Y.N., Ed.; SIF SN RASc.: Moscow, Russia; Krasnoyarsk, Russia, 2016; pp. 129–130. (In Russian)
212. Mandelshtam, M.Y.; Khairetdinov, R.R. Additions to the check-list of bark beetles (Coleoptera, Curculionidae: Scolytinae) from Leningrad province, Russia. *Entomol. Rev.* **2017**, *97*, 893–899. [CrossRef]
213. Kerchev, I.A.; Mandelshtam, M.Y.; Krivets, S.A.; Ilinsky, Y.Y. Small spruce bark beetle *Ips amitinus* (Eichhoff, 1872) (Coleoptera, Curculionidae: Scolytinae): A new alien species in Western Siberia. *Entomol. Rev.* **2019**, *99*, 639–644. [CrossRef]
214. Skorokhodov, S.N. Save the neighboring stone pine woods. *Yashkinsky Vestn.* **2017**, *37*, 8564. (In Russian)
215. Krivets, S.A. Problems of protection of Siberian stone pine forests from the small spruce bark beetle *Ips amitinus* (Coleoptera: Curculionidae, Scolytinae). In *Dendrobiotic Invertebrates and Fungi and Their Role in Forest Ecosystems. The Kataev Memorial Readings—XI, Proceedings of the All-Russia Conference with International Participation, St. Petersburg, Russia, 24–27 November 2020*; Musolin, D.L., Kirichenko, N.I., Selikhovkin, A.V., Eds.; SPbFTU: St. Petersburg, Russia, 2020; pp. 183–184. (In Russian)
216. Kerchev, I.A.; Krivets, S.A.; Smirnov, N.A. Analysis of the population characteristics of the small spruce bark beetle *Ips amitinus* (Coleoptera: Curculionidae: Scolytinae) in the areas of invasion in Western Siberia. In *Dendrobiotic Invertebrates and Fungi and Their Role in Forest Ecosystems. The Kataev Memorial Readings—XI, Proceedings of the All-Russia Conference with International Participation, St. Petersburg, Russia, 24–27 November 2020*; Musolin, D.L., Kirichenko, N.I., Selikhovkin, A.V., Eds.; SPbFTU: St. Petersburg, Russia, 2020; pp. 170–171. (In Russian)
217. Kerchev, I.A.; Krivets, S.A. An attack of *Ips amitinus* (Coleoptera: Curculionidae: Scolytinae) on arboretum in West Siberia: New host of invasive bark beetle among exotic conifers. *J. Asia-Pac. Entomol.* **2021**, *24*, 148–152. [CrossRef]
218. Kerchev, I.A.; Krivets, S.A.; Bisirova, E.M.; Smirnov, N.A. Distribution of the small spruce bark beetle *Ips amitinus* (Eichhoff, 1872) in the Western Siberia. *Russ. J. Biol. Invasions* **2021**, *4*, 77–84. (In Russian) [CrossRef]
219. Holuša, J.; Lukasova, K.; Grodzki, W.; Kula, E.; Matousek, P. Is *Ips amitinus* (Coleoptera: Curculionidae) abundant in wide range of altitudes? *Acta Zool. Bulg.* **2012**, *64*, 219–228. Available online: https://www.researchgate.net/publication/236260984_Is_Ips_amitinus_Coleoptera_Curculionidae_Abundant_in_Wide_Range_of_Altitudes (accessed on 18 January 2022).
220. Bisirova, E.M.; Kerchev, I.A. Assessment of the state of the Siberian stone pine *Pinus sibirica* in the outbreak foci of the small spruce bark beetle *Ips amitinus* (Coleoptera: Curculionidae, Scolytinae) a new invasive pest in Siberia. In *Dendrobiotic Invertebrates and Fungi and Their Role in Forest Ecosystems. The Kataev Memorial Readings—XI, Proceedings of the All-Russia Conference with International Participation, St. Petersburg, Russia, 24–27 November 2020*; Musolin, D.L., Kirichenko, N.I., Selikhovkin, A.V., Eds.; SPbFTU: St. Petersburg, Russia, 2020; pp. 82–83. (In Russian)
221. Baranchikov, Y.N.; Krivets, S.A. About professional skills in determination of insects: How the occurrence of new aggressive invasive fir pest in Siberia was missed. *Ecol. South. Sib. Adjac. Territ.* **2010**, *1*, 50–52. (In Russian)
222. Baranchikov, Y.N.; Pet'ko, V.M.; Astapenko, S.A.; Akulov, E.N.; Krivets, S.A. Four-eyed fir bark beetle as a new aggressive pest of fir in Siberia. *Lesn. Vestn.* **2011**, *80*, 78–81. (In Russian)

223. Kurentsov, A.I. *The Bark Beetles of the USSR Far East*; Academy of Sciences of the USSR: Moscow, Russia; St. Petersburg, Russia, 1941; pp. 1–234. (In Russian)
224. Krivolutskaya, G.O. Fam. Scolytidae—The bark beetles. In *Key for Identification of Insects of the Russian Far East*; Ler, P.A., Ed.; Dal'nauka: Vladivostok, Russia, 1996; Volume 3, Part 3. Coleoptera, pp. 312–373. (In Russian)
225. Kerchev, I.A. Ecology of four-eyed fir bark beetle *Polygraphus proximus* Blandford (Coleoptera: Curculionidae, Scolytinae) in the West Siberian region of invasion. *Russ. J. Biol. Invasion* **2014**, *5*, 176–185. [CrossRef]
226. Chilakhsaeva, E.A. First record of *Polygraphus proximus* (Coleoptera: Scolytidae) in Moscow Province. *Bull. Mosc. Soc. Nat. Biol.* **2008**, *113*, 39–42. (In Russian)
227. Baranchikov, Y.N.; Seraya, L.G. For whom the bell tolls: Evaluation of potential invasive organisms and their role in regional biota. In *Monitoring and Biological Control Methods of Woody Plant Pests and Pathogens: From Theory to Practice. Proceedings of International Conference, Moscow, Russia, 18–22 April 2016*; Baranchikov, Y.N., Ed.; SIF SN RASc.: Moscow, Russia; Krasnoyarsk, Russia, 2016; pp. 25–26. (In Russian)
228. Mandelshtam, M.Y.; Popovichev, B.G. Annotated list of bark beetles (Coleoptera, Scolytidae) of Leningrad Province. *Entomol. Rev.* **2000**, *80*, 887–903.
229. Baranchikov, Y.N.; Efremenko, A.A.; Demidko, D.A.; Titova, V.V. Four-eyed fir bark beetle *Polygraphus proximus* Blandford (Coleoptera: Curculionidae, Scolytinae) in the western piedmont of the Ural Mountains: Where, whence and when? In *Dendrobiotic Invertebrates and Fungi and their Role in Forest Ecosystems. The Kataev Memorial Readings—XI, Proceedings of the All-Russia Conference with International Participation, St. Petersburg, Russia, 24–27 November 2020*; Musolin, D.L., Kirichenko, N.I., Selikhovkin, A.V., Eds.; SPbFTU: St. Petersburg, Russia, 2020; pp. 74–75. (In Russian)
230. Dedyukhin, S.V.; Titova, V.V. Finding of the bark beetle *Polygraphus proximus* Blandford, 1894 (Coleoptera, Curculionidae: Scolytinae) in Udmurtia. *Russ. J. Biol. Invasions* **2021**, *12*, 258–263. [CrossRef]
231. Krivets, S.A.; Kerchev, I.A.; Kizeev, Y.M.; Kozhurin, M.A.; Kozak, R.G.; Filimonov, M.N.; Chemodanov, A.V.; Chugin, B.S. Four-eyed fir bark beetle *Polygraphus proximus* Blandf. (Coleoptera, Scolytidae) in fir forests of Tomsk Oblast. In *Diseases and Pests in the Forests of Russia: XXI Century, Proceedings of the All-Russian Conference and V Annual Kataev Memorial Reading, Yekaterinburg, Russia, 20–25 September 2011*; Selikhovkin, A.V., Shavnin, S.A., Zalesov, S.V., Onuchin, A.A., Isaev, A.S., Mozolevskaya, E.G., Ponomarev, V.I., Baranchikov, Y.N., Petrova, I.V., Meshkova, V.L., et al., Eds.; Institute of Forest, SB RAS: Krasnoyarsk, Russia, 2011; pp. 53–55. (In Russian)
232. Krivets, S.A.; Bisirova, E.M.; Kerchev, I.A.; Pashenova, N.V.; Demidko, D.A.; Petko, V.M.; Baranchikov, Y.N. *Four-eyed Fir Bark Beetle in Siberian Forests (Distribution, Biology, Ecology, Detection and Survey of Damaged Stands). Manual*; Umium Publisher: Tomsk, Russia; Krasnoyarsk, Russia, 2015; pp. 1–48. (In Russian)
233. Krivets, S.A.; Bisirova, E.M.; Kerchev, I.A.; Pats, E.N.; Chernova, N.A. Transformation of taiga ecosystems in the Western Siberian invasion focus of four-eyed fir bark beetle *Polygraphus proximus* Blandford (Coleoptera: Curculionidae, Scolytinae). *Russ. J. Biol. Invasions* **2015**, *6*, 94–108. [CrossRef]
234. Krivets, S.A.; Kerchev, I.A.; Bisirova, E.M.; Demidko, D.A.; Pet'ko, V.M.; Baranchikov, Y.N. Distribution of the four-eyed fir bark beetle *Polygraphus proximus* Blandf. (Coleoptera, Curculionidae: Scolytinae) in Siberia. *Izv. St.-Peterbg. Lesoteh. Akad.* **2015**, *211*, 33–45. (In Russian)
235. Krivets, S.A.; Kerchev, I.A.; Bisirova, E.M.; Debkov, N.M. Current distribution and forecasted invasive area expansion of the four-eyed fir bark beetle *Polygraphus proximus* Blandford, 1894 in Tomsakaya Oblast' (Western Siberia). *Eurasian Entomol. J.* **2018**, *17*, 53–60. (In Russian)
236. Bystrov, S.O.; Antonov, I.A. First record of the four-eyed beetle *Polygraphus proximus* Blandford, 1894 (Coleoptera: Curculionidae, Scolytinae) from Irkutsk Province, Russia. *Entomol. Rev.* **2019**, *98*, 54–55. [CrossRef]
237. Baranchikov, Y.N.; Demidko, D.A.; Laptev, A.B.; Pet'ko, V.M. The dynamics of Siberian fir death in the outbreak focus of four-eyed fir bark beetle. *Lesn. Vestn.* **2014**, *18*, 78–81. (In Russian)
238. Demidko, D.A. The dates of invasion of four-eyed fir bark beetle *Polygraphus proximus* Blandford (Coleoptera: Curculionidae, Scolytinae) in Tomsk Oblast. *Izv. St.-Peterbg. Lesoteh. Akad.* **2014**, *207*, 225–234. Available online: https://www.researchgate.net/publication/273321170_Dating_of_Four-Eyed_Fir_Bark_Beetle_Polygraphus_proximus_Blandford_Coleoptera_Curculionidae_Scolytinae_Invasion_into_Tomsk_Region (accessed on 18 January 2022). (In Russian)
239. Bykov, R.; Kerchev, I.; Demenkova, M.; Ryabinin, A.; Ilinsky, Y. Sex-specific *Woldbachia* infection patterns in populations of *Polygraphus proximus* Blandford (Coleoptera: Curculionidae: Scolytinae). *Insects* **2020**, *11*, 547. [CrossRef]
240. Kononov, A.; Ustyantsev, K.; Blinov, A.; Fet, V.; Baranchikov, Y.N. Genetic diversity of aboriginal and invasive populations of four-eyed fir bark beetle *Polygraphus proximus* Blandford (Coleoptera, Curculionidae, Scolytinae). *Agric. For. Entomol.* **2016**, *18*, 294–301. [CrossRef]
241. Kerchev, I.A. Experimental study of the possibility of the emergence of new trophic connections of four-eyed fir bark beetle *Polygraphus proximus* Blandf. (Coleoptera: Curculionidae, Scolytinae) in the West Siberian region. *Vestn. Tomsk. Gos. Univ. Biol. (Tomsk State Univ. J. Biol.)* **2012**, *3*, 169–177. (In Russian)
242. Pashenova, N.V.; Kononov, A.V.; Ustyantsev, K.V.; Blinov, A.G.; Pertsovaya, A.A.; Baranchikov, Y.N. Ophiostomatoid fungi associated with the four-eyed fir bark beetle on the territory of Russia. *Russ. J. Biol. Invasion* **2018**, *9*, 63–74. [CrossRef]

243. Pashenova, N.V.; Petko, V.M.; Babichev, N.S.; Kerchev, I.A. Transfer of ophiostomatoid fungi by a four-eyed fir bark beetle *Polygraphus proximus* Blandf. (Coleoptera, Scolytidae) in Siberia. *Izv. St.-Peterbg. Lesoteh. Akad.* **2012**, *210*, 114–120. Available online: https://www.researchgate.net/publication/286452586_Transfer_of_Ophiostomatales_fungi_by_Polygraphus_proximus_Blandford_Coleoptera_Sco_lytidae_in_Sibera (accessed on 20 January 2022). (In Russian)
244. Pashenova, N.V.; Demidko, D.A.; Pertsovaya, A.A.; Baranchikov, Y.N. Phloem sensibility of Siberian conifers towards *Grosmannia aoshimae* (Ohtaka, Masuya & Yamaoka) Masuya & Yamaoka, a symbiotic fungus of the four-eyed fir bark beetle. In *Dendrobiotic Invertebrates and Fungi and their Role in Forest Ecosystems. The Kataev Memorial Readings—XI, Proceedings of the All-Russia Conference with International Participation, St. Petersburg, Russia, 24–27 November 2020*; Musolin, D.L., Kirichenko, N.I., Selikhovkin, A.V., Eds.; SPbFTU: St. Petersburg, Russia, 2020; pp. 247–248. (In Russian)
245. Bisirova, E.M.; Krivets, S.A. Dynamics of the state of Siberian fir tree stands damaged by the four-eyed fir bark beetle Polygraphus proximus Blandf. in Tomsk Oblast. *Vestn. Tomsk. Gos. Univ. Biol. (Tomsk State Univ. J. Biol.)* **2018**, *44*, 118–140. (In Russian) [CrossRef] [PubMed]
246. Mukhortova, L.V.; Sergeeva, O.V.; Demidko, D.A.; Krivobokov, L.V.; Baranchikov, Y.N. Dynamics of coarse woody debris stocks in the fir forests damaged by the bark beetle *Polygraphus proximus* Blandf. (Coleoptera: Curculionidae, Scolytinae). In *Dendrobiotic Invertebrates and Fungi and their Role in Forest Ecosystems. The Kataev Memorial Readings—XI, Proceedings of the All-Russia Conference with International Participation, St. Petersburg, Russia, 24–27 November 2020*; Musolin, D.L., Kirichenko, N.I., Selikhovkin, A.V., Eds.; SPbFTU: St. Petersburg, Russia, 2020; pp. 231–232. (In Russian)
247. Kerchev, I.A.; Krivets, S.A. Composition and number of xylophagous consort's members of Siberian fir associated with four-eyed fit bark beetle in Tomsk Oblast. In *Environmental and Economic Consequences of Invasions of Dendrophilous Insects, Proceedings of All-Russian Conference with International Participation, Krasnoyarsk, Russia, 25–27 September 2012*; Baranchikov, Y.N., Ed.; Institute of Forest, SB RAS: Krasnoyarsk, Russia, 2012; pp. 57–59. (In Russian)
248. Bisirova, E.M.; Kerchev, I.A.; Krivets, S.A.; Pats, E.N. The state of fir forests of Tashtagol District of Kemerovo Province, Russia, damaged by the four-eyed fir bark beetle *Polygraphus proximus* Blandf. (Coleoptera: Curculionidae, Scolytinae). In *Dendrobiotic Invertebrates and Fungi and their Role in Forest Ecosystems. The Kataev Memorial Readings—XI, Proceedings of the All-Russia Conference with International Participation, St. Petersburg, Russia, 24–27 November 2020*; Musolin, D.L., Kirichenko, N.I., Selikhovkin, A.V., Eds.; SPbFTU: St. Petersburg, Russia, 2020; pp. 84–85. (In Russian)
249. Krivets, S.A.; Kerchev, I.A. Insects inhabiting the galleries of the four-eyed fir bark beetle *Polygraphus proximus* Blandf. (Coleoptera, Curculionidae: Scolytinae) in Siberia. *Entomol. Rev.* **2016**, *96*, 545–558. [CrossRef]
250. Krivosheina, M.G.; Krivosheina, N.P.; Kerchev, I.A. Flies (Diptera) associated with *Polygraphus proximus* Blandford, 1894 (Coleoptera, Curculionidae) in Siberia and the Russian Far East. *Entomol. Rev.* **2018**, *98*, 156–164. [CrossRef]
251. Kerchev, I.A.; Kryukov, V.Y.; Yaroslavtseva, O.N.; Polovinko, G.P.; Tokarev, Y.S.; Glupov, V.V. The first data on fungal pathogens (Ascomycota, Hypocreales) in the invasive populations of four-eyed fir bark beetle *Polygraphus proximus* Blandf. *Russ. J. Biol. Invasions* **2017**, *8*, 34–40. [CrossRef]
252. Efremenko, A.A.; Demidko, D.A.; Baranchikov, Y.N. An old remedy for a new problem: Submersion of logs to prevent adult emergence of the invasive bark beetle *Polygraphus proximus* Blandford (Coleoptera: Curculionidae, Scolytinae). In *Dendrobiotic Invertebrates and Fungi and their Role in Forest Ecosystems. The Kataev Memorial Readings—XI, Proceedings of the All-Russia Conference with International Participation, St. Petersburg, Russia, 24–27 November 2020*; Musolin, D.L., Kirichenko, N.I., Selikhovkin, A.V., Eds.; SPbFTU: St. Petersburg, Russia, 2020; p. 159. (In Russian)
253. Seraya, L.G.; Pashenova, N.V.; Demidko, D.A.; Kozhenkova, A.A.; Efremenko, A.A.; Gninenko, Y.I.; Baranchikov, Y.N. Attempts of chemical control of invasive populations of the bark beetle *Polygraphus proximus* (Coleoptera: Curculionidae). In *The Kataev Memorial Readings—X, Proceedings of the International Conference, St. Petersburg, Russia, 22–25 October 2018*; Musolin, D.L., Selikhovkin, A.V., Eds.; SPbFTU: St. Petersburg, Russia, 2018; Volume 1, pp. 97–98. (In Russian)
254. Xiang, Q.P.; Wei, R.; Shao, Y.Z.; Yang, Z.Y.; Wang, X.Q.; Zhang, X.C. Phylogenetic relationships, possible ancient hybridization, and biogeographic history of Abies (Pinaceae) based on data from nuclear, plastid, and mitochondrial genomes. *Mol. Phylogenet. Evol.* **2015**, *82*, 1–14. [CrossRef]
255. Kirpichnikova, V.A. Pyralidae. In *Key for Identification of Insects of the Russian Far East*; Ler, P.A., Ed.; Dal'nauka: Vladivostok, Russia, 2005; Volume 5, Part 2, pp. 526–539. (In Russian)
256. Mally, R.; Nuss, M. Phylogeny and nomenclature of the box tree moth, *Cydalima perspectalis* (Walker, 1859) comb. n., which was recently introduced into Europe (Lepidoptera: Pyraloidea: Crambidae: Spilomelinae). *Eur. J. Entomol.* **2010**, *107*, 393–400. [CrossRef]
257. Kruger, E.O. *Glyphodes perspectalis* (Walker, 1859)—Neu für die Fauna Europas (Lepidoptera: Crambidae). *Entomol. Z.* **2008**, *118*, 81–83.
258. USDA Confirms Box Tree Moth and Takes Action to Contain and Eradicate the Pest. 2021. Available online: https://www.aphis.usda.gov/aphis/newsroom/stakeholder-info/sa_by_date/sa-2021/sa-05/box-tree-moth (accessed on 2 January 2022).
259. Gninenko, Y.I.; Shiryaeva, N.V.; Shchurov, V.I. Box tree moth—A new invasive organism in the forests of the Russian Caucasus. *Plant Quar. Sci. Prac.* **2014**, *7*, 32–36. Available online: https://elibrary.ru/download/elibrary_25051637_65094735.pdf (accessed on 18 January 2022).

260. Karpun, N.N.; Ignatova, Y.A. The first report about *Cydalima perspectalis* Walker on Black Sea Coast of Russia. In *Zprávy Vědecké Ideje—2013 (19), Proceedings of the IX International Scientific and Practical Conference, Prague, Czech Republic, 27 October—5 November 2013*; Publishing House "Education and Science" s.r.o.: Prague, Czech Republic, 2013; pp. 29–32. Available online: http://www.rusnauka.com/31_NNM_2013/Biologia/7_146134.doc.htm (accessed on 18 January 2022).
261. Proklov, V.V.; Karayeva, S.Z. New and interesting Lepidoptera records from Chechen Republic (Russia). *Cauc. Entomol. Bull.* **2013**, *9*, 281–282. Available online: https://www.ssc-ras.ru/files/files/11_%20Proklov.pdf (accessed on 18 January 2022). [CrossRef]
262. Karpun, N.N.; Ignatova, Y.A.; Zhuravleva, E.N. New species of harmful entomofauna on ornamental woody plants in humid subtropics of Krasnodar region. *Izv. St.-Peterbg. Lesoteh. Akad.* **2015**, *211*, 187–203. Available online: https://elibrary.ru/download/elibrary_23815865_37414681.pdf (accessed on 18 January 2022). (In Russian)
263. Shchurov, V.I.; Kuchmistaya, E.V.; Vibe, E.N.; Bondarenko, A.S.; Skvortsov, M.M. The box-tree moth *Cydalima perspectalis* (Walker, 1859)—The real threat to biological diversity of natural forests of the Northwest Caucasus. *Proc. Kuban State Agrar. Univ.* **2015**, *53*, 178–198. (In Russian)
264. Balykina, E.B.; Trikoz, N.N. Changes in the taxonomic structure of the ornamental plants phytophages complex in Crimean parks. *Veles* **2017**, *43*, 59–63.
265. Dobronosov, V.V. New Data on the Box Tree Moth *Cydalima perspectalis* (Walker, 1859) in the Central Caucasus. 2017. Available online: http://aeconomy.ru/news/agro/novye-dannye-o-samshitovoy-ognevke-.html (accessed on 2 January 2022). (In Russian)
266. Poltavskiy, A.N. Annotated catalog of insect pests of introduced plants in the Botanical Garden of the Southern Federal University. *Proc. Bot. Gard. South. Fed. Univ.* **2017**, *2*, 120–138. (In Russian)
267. Tuniyev, B.S.; Aliev, K.U.; Timukhin, I.N. Colchis boxwood and the new pest *Cydalima perspectalis* Walker, 1859 (Lepidoptera, Crambidae) in South Ossetia. *Bot. Her. North Cauc.* **2018**, *1*, 30–36. Available online: http://gorbotsad.ru/files/Tuniev_2018_1.pdf (accessed on 18 January 2022). (In Russian)
268. Il'ina, E.V.; Gasanova, N.M.-S. New Information about Invasive Insects in the Republic of Dagestan. *Plant Prot. Quar.* **2020**, *4*, 39–40. Available online: https://elibrary.ru/download/elibrary_42673425_76973702.pdf (accessed on 18 January 2022). (In Russian)
269. Rozhina, V.I.; Volodina, L.S.; Drotikova, A.V. First detection of *Cydalima perspectalis* (Walker, 1859) (Lepidoptera: Crambidae: Pyraustinae) in Kaliningrad region. In *ChemBioSeasons, Proceedings of the Abstracts of the Forum of Young Researchers, Kaliningrad, Russia, 23 April 2021*; Baltic Federal University Named after I. Kant: Kaliningrad, Russia, 2021; pp. 67–68. Available online: https://elibrary.ru/item.asp?id=46359179 (accessed on 4 January 2022).
270. *Cydalima perspectalis*. EPPO Global Database. 2021. Available online: https://gd.eppo.int/taxon/DPHNPE (accessed on 1 July 2021).
271. Karpun, N.N.; Trochov, E.S.; Ignatova, Y.A.; Zhuravleva, E.N.; Kaurova, Z.G. Analysis of the food specialization of the box tree moth (*Cydalima perspectalis* Walker). *Regul. Issues Vet. Med.* **2015**, *4*, 173–176. Available online: https://elibrary.ru/item.asp?id=24825731 (accessed on 18 January 2022).
272. Kawazu, K.; Honda, H.; Nakamura, S.; Adati, T. Identification of sex pheromone components of the box tree pyralid, *Glyphodes perspectalis*. *J. Chem. Ecol.* **2007**, *33*, 1978–1985. [CrossRef] [PubMed]
273. Kim, J.; Park, I.-K. Female sex pheromone components of the box tree pyralid, *Glyphodes perspectalis*, in Korea: Field test and development of film-type lure. *J. Asia-Pacific Entomol.* **2013**, *16*, 473–477. [CrossRef]
274. Nesterenkova, A.E.; Ponomarev, V.L.; Karpun, N.N.; Protsenko, V.E.; Glebov, V.E.; Danilenko, E.A.; Rastegaeva, V.M. Field testing of biological activity of pheromone box tree moth *Cydalima perspectalis* Walker. *Lesn. Vestn.* **2017**, *21*, 70–77. (In Russian) [CrossRef]
275. Göttig, S.; Herz, A. The box tree pyralid *Cydalima perspectalis*: New results of the use of biological control agents and pheromone traps in the field. *J. Plant Dis. Prot.* **2014**, *121*, 98–99.
276. Göttig, S.; Herz, A. Susceptibility of the box tree pyralid *Cydalima perspectalis* Walker (Lepidoptera: Crambidae) to potential biological control agents Neem (NeemAzal®-T/S) and entomopathogenic nematodes (Nemastar®) assessed in laboratory bioassays and field trials. *J. Plant Dis. Prot.* **2018**, *125*, 365–375. [CrossRef]
277. Borisov, B.A.; Karpun, N.N.; Zhuravleva, E.N.; Borisova, I.P. Evaluation of the possibility of biological control of box tree moth (*Cydalima perspectalis*) by entomoparasitic fungi. In *Monitoring and Biological Control Methods of Woody Plant Pests and Pathogens: From Theory to Practice, Proceedings of International Conference, Moscow, Russia, 18–22 April 2016*; Baranchikov, Y.N., Ed.; SIF SN RASc.: Moscow, Russia; Krasnoyarsk, Russia, 2016; pp. 41–43. (In Russian)
278. Kenis, M.; Nacambo, S.; Leuthardt, F.L.G.; Di Domenico, F.; Haye, T. The box tree moth, *Cydalima perspectalis*, in Europe: Horticultural pest or environmental disaster? *Aliens* **2013**, *33*, 38–41. Available online: https://www.researchgate.net/publication/271483672_The_box_tree_moth_Cydalima_perspectalis_in_Europe_horticultural_pest_or_environmental_disaster (accessed on 18 January 2022).
279. Agasieva, I.S.; Ismailov, V.Y.; Fedorenko, E.V.; Nefedova, M.V. Biological control of box tree moth. *Plant Prot. Quar.* **2017**, *8*, 21–23. Available online: https://elibrary.ru/download/elibrary_29879952_31932504.pdf (accessed on 18 January 2022).
280. Fora, C.G.; Sasu, L.; Poşta, D.; Berar, C. Chemical possibilities of *Cydalima perspectalis* Walk. (Lepidoptera: Crambidae) control. *J. Hortic. For. Biotechnol.* **2016**, *20*, 31–34. Available online: https://journal-hfb.usab-tm.ro/romana/2016/Lucrari%20PDF/Vol%2020(3)%20PDF/7Fora%20George.pdf (accessed on 18 January 2022).

281. Somsai, A.P.; Lukacs, L.; Florian, T.; Sestraş, A.F.; Bunea, C.I.; Hoble, A.; Vlaşin, H. Chemical control of box tree moth *Cydalima perspectalis* (Walker, 1859) (Lepidoptera: Crambidae) in North-Western Romania. *Agricultura* **2019**, *111*, 210–216. [CrossRef]
282. Plugutar, Y.V.; Sharmagy, A.K.; Shishkin, V.A. Effectiveness of bioinsecticides against the caterpillar of *Cydalima perspectalis* (Walker, 1859) on the Southern Coast of the Crimea. *Bull. State Nikitsk. Bot. Gard.* **2020**, *137*, 7–15. (In Russian) [CrossRef]
283. Baryshnikova, S.V. Gracillariidae. In *Catalogue of the Lepidoptera of Russia*, 2nd ed.; Sinev, S.Y., Ed.; Zoological Institute RAS: St. Petersburg, Russia, 2019; pp. 36–43. (In Russian)
284. De Prins, J.; De Prins, W. Global Taxonomic Database of Gracillariidae (Lepidoptera). World Wide Web Electronic Publication. 2006–2021. Available online: http://www.gracillariidae.net) (accessed on 10 November 2021).
285. Caradja, A. Beitrag zur Kenntnis der geographischen Verbreitungder Mikrolepidopteren des palaearktischen Faunengebietes nebst Beschreibung neuer Formen. *Dt. Ent. Z. Ires* **1920**, *34*, 75–179.
286. Baryshnikova, S.V.; Dubatolov, V.V. To the knowledge of small moths (Microlepidoptera) of the Bolshekhekhtsirskii Nature Reserve (Khabarovsk District). 2nd report. Bucculatricidae, Gracillariidae, Lyonetiidae]. *Anim. World Far East* **2007**, *6*, 47–50. Available online: https://elibrary.ru/download/elibrary_20453720_27411630.pdf (accessed on 18 January 2022). (In Russian)
287. Selikhovkin, A.V.; Drenkhan, R.; Mandelstam, M.Y.; Musolin, D.L. Invasions of insect pests and fungal pathogens of woody plants into the North-Western part of the European Russia. *Vestn. St. Petersburg Univ. Earth Sci.* **2020**, *65*, 263–283. (In Russian) [CrossRef]
288. Selikhovkin, A.; Merkuriev, S.; Khodachek, A. Native and alien tree insect pests: Climate change impact and economic losses in Northwestern Russia. In Proceedings of the 1st International Electronic Conference on Entomology, Basel, Switzerland, 1–15 July 2021; Available online: https://sciforum.net/paper/view/10412 (accessed on 4 January 2021).
289. Buy, D.D.; Baryshnikova, S.V.; Denisova, N.V.; Mamaev, N.A.; Musolin, D.L.; Selikhovkin, A.V. Changes in the complex of dendrobiotic phyllophagous insects in Saint Petersburg. In *Dendrobiotic Invertebrates and Fungi and their Role in Forest Ecosystems. In The Kataev Memorial Readings—XI, Proceedings of the All-Russia Conference with International Participation, St. Petersburg, Russia, 24–27 November 2020*; Musolin, D.L., Kirichenko, N.I., Selikhovkin, A.V., Eds.; SPbFTU: St. Petersburg, Russia, 2020; pp. 98–99. (In Russian)
290. Buy, D.D.; Baryshnikova, S.V.; Denisova, N.V.; Shevchenko, S.V.; Selikhovkin, A.V. Actual changes in the species composition and population density of phyllophagous insects in St. Petersburg. *Izv. St.-Peterbg. Lesoteh. Akad.* **2020**, *230*, 73–99. (In Russian) [CrossRef]
291. Chursina, V.A.; Vokhtantseva, K.V.; Gayvas, A.A. The main pest of English oak on the territory of the city of Omsk is the leaf blotch miner moth. In *Innovative Technologies in Agriculture, Proceedings of the II International Scientific Conference. St. Petersburg, Russia, 20–23 July 2012*; Akhmetov, I.G., Akhmetova, M.N., Ivanova, Y.V., Kalenskiy, A.V., Kutashov, V.A., Laktionova, K.S., Saraeva, N.M., Abdrasilov, T.K., Avdejuk, O.A., Aidarov, O.T., et al., Eds.; Svoye Izdatel'stvo Publisher: St. Petersburg, Russia, 2016; pp. 21–25. (In Russian)
292. Baryshnikova, S.V. (Zoological Institute RAS, Saint Petersburg, Russia). Personal communication, 2021.
293. Galich, D.E. (Siberian Forest Experiment Station, Tyumen, Russia). Personal communication, 2021.
294. Heckford, R.J. Previously unrecorded foodplants of three species of Microlepidoptera. *Entomol. Rec. J. Var.* **1993**, *105*, 93.
295. Nikitenko, H.M.; Fursov, V.M.; Hershenzon, Z.S.; Svyrydov, S.V. Oak broad-leaved moths and other mining insects on the oak. 2. Morphobiological and ecological characteristics of oak widespread moth and other mining pests of oak. *Vestn. Zool.* **2004**, *38*, 53–61. (In Ukrainian)
296. Apostolov, L.G. *Harmful Entomofauna of Forest Biogeocenoses in the Vicinity of the Central Dnieper*; Vishcha Shkola Publisher: Kiyev, Ukraine, 1981; pp. 1–232. (In Russian)
297. Utkina, I.A.; Rubtsov, V.V. The leaf blotch miner moth *Acrocercops brongniardella* F. (Lepidoptera: Gracillariidae) in Tellerman oakwood. In *The Kataev Memorial Readings—X, Proceedings of the International Conference, St. Petersburg, Russia, 22–25 October 2018*; Musolin, D.L., Selikhovkin, A.V., Eds.; SPbFTU: St. Petersburg, Russia, 2018; Volume 1. Insects and Other Invertebrates, pp. 104–105. Available online: https://elibrary.ru/download/elibrary_36373591_75292794.pdf (accessed on 18 January 2022). (In Russian)
298. Utkina, I.A.; Rubtsov, V.V. The leaf blotch miner moth—Long known, but still little studied species. *Izv. St.-Peterbg. Lesoteh. Akad.* **2019**, *228*, 42–56. (In Russian)
299. Golub, V.B.; Soboleva, V.A.; Aksenenko, E.V. Damage of oak by the leaf blotch miner moth *Acrocercops brongniardella* F. (Lepidoptera: Gracillariidae) in the Usman' Forest (Voronezh Province, Russia) 10 years after the fire. In *Dendrobiotic Invertebrates and Fungi and their Role in Forest Ecosystems. The Kataev Memorial Readings—XI, Proceedings of the All-Russia Conference with International Participation, St. Petersburg, Russia, 24–27 November 2020*; Musolin, D.L., Kirichenko, N.I., Selikhovkin, A.V., Eds.; SPbFTU: St. Petersburg, Russia, 2020; pp. 136–137. (In Russian)
300. Utkina, I.A.; Rubtsov, V.V. Relationship of different species of oak and phyllophages as an object of biogeocenotic research. *Lesovedenie* **2021**, *5*, 547–554. (In Russian) [CrossRef]
301. Golub, V.B.; Berezhnova, O.N.; Kornev, I.I. Outbreak of the leaf blotch miner moth (*Acrocercops brongniardella* F., Lepidoptera, Gracillariidae) in the Voronezh Region. *Izv. St.-Peterbg. Lesoteh. Akad.* **2009**, *187*, 96–102. Available online: https://elibrary.ru/download/elibrary_13020131_77590310.pdf (accessed on 18 January 2022). (In Russian)

302. Eaton, E.; Caudullo, G.; Oliveira, S.; de Rigo, D. Quercus robur and *Quercus petraea* in Europe: Distribution, habitat, usage and threats. In *European Atlas of Forest Tree Species*; San-Miguel-Ayanz, J., de Rigo, D., Caudullo, G., Houston Durrant, T., Mauri, A., Eds.; Publisher Off. EU: Luxembourg, Belgium, 2016; pp. 160–163.
303. Ermolaev, I.V.; Vasil'ev, A.A. Phytophagous insects associated with the pedunculate oak (*Quercus robur* L.) in the Siva River valley. In *Dendrobiotic Invertebrates and Fungi and their Role in Forest Ecosystems. The Kataev Memorial Readings—XI, Proceedings of the All-Russia Conference with International Participation, St. Petersburg, Russia, 24–27 November 2020*; Musolin, D.L., Kirichenko, N.I., Selikhovkin, A.V., Eds.; SPbFTU: St. Petersburg, Russia, 2020; pp. 155–156. (In Russian)
304. Ermolaev, I.V.; Ponomarev, V.I.; Vasil'ev, A.A.; Kumaeva, M.S. Phytophagous insects of the pedunculate oak (*Quercus robur*) in the northeast of its distribution area. *Zool. J.* **2021**, *100*, 640–651. (In Russian) [CrossRef]
305. Marion, D.F.; Larew, H.G.; Knodel, J.J.; Natoli., W. Systemic activity of neem extract against birch leafminer. *J. Arboric.* **1990**, *16*, 12–16.
306. Golosova, M.A.; Gninenko, Y.I.; Golosova, E.I. *Ohrid Leaf-Miner Cameraria ohridella is a Dangerous Quarantine Pest on Objects of Urban Ggreening*; VNIILM: Moscow, Russia, 2008; pp. 1–26. (In Russian)
307. Gous, S.; Richardson, B. Stem injection of insecticides to control herbivorous insects on *Eucalyptus niten*. *New Zealand Plant Prot.* **2008**, *61*, 174–178. [CrossRef]
308. Šefrová, H. Invasions of Lithocolletinae species in Europe—Causes, kinds, limits and ecological impact (Lepidoptera, Gracillariidae). *Ekológia* **2003**, *22*, 132–142. Available online: https://www.researchgate.net/publication/297685678_Invasions_of_Lithocolletinae_species_in_Europe_-_Causes_kinds_limits_and_ecological_impact_Lepidoptera_Gracillariidae (accessed on 18 January 2022).
309. Kirichenko, N.I.; Triberti, P.; Akulov, E.N.; Ponomarenko, M.G.; Lopez-Vaamonde, C. Novel data on the taxonomic diversity, distribution, and host plants of leafmining moths of the family Gracillariidae (Lepidoptera) in Siberia, based on DNA barcoding. *Entomol. Rev.* **2019**, *99*, 796–819. [CrossRef]
310. Hellrigl, K. Neue Erkenntnisse und Untersuchungen über die Roßkastanien-Miniermotte *Cameraria ohridella* Deschka & Dimić, 1986 (Lepidoptera, Gracillariidae). *Gredleriana* **2001**, *1*, 9–81.
311. Grabenweger, G.; Avtzis, N.; Girardoz, S.; Hrasovec, B.; Tomov, R.; Kenis, M. Parasitism of *Cameraria ohridella* (Lepidoptera, Gracillariidae) in natural and artificial horse-chestnut stands in the Balkans. *Agric. For. Entomol.* **2005**, *7*, 291–296. [CrossRef]
312. Valade, R.; Kenis, M.; Hernandez-Lopez, A.; Augustin, S.; Mari Mena, N.; Magnoux, E.; Rougerie, R.; Lakatos, F.; Roques, A.; Lopez-Vaamonde, C. Mitochondrial and microsatellite DNA markers reveal a Balkan origin for the highly invasive horse-chestnut leaf miner *Cameraria ohridella* (Lepidoptera, Gracillariidae). *Mol. Ecol.* **2009**, *18*, 3458–3470. [CrossRef] [PubMed]
313. Lees, D.C.; Lack, H.W.; Rougerie, R.; Hernandez-Lopez, A.; Raus, T.; Avtzis, N.; Augustin, S.; Lopez-Vaamonde, C. Tracking origins of invasive herbivores using herbaria and archival DNA: The case of the horse-chestnut leafminer. *Front. Ecol. Environ.* **2011**, *9*, 322–328. [CrossRef]
314. Gninenko, Y.I.; Shepelev, S.V. New phytophages and diseases of tree species. *Lesn. Khozyaistvo* **2004**, *3*, 48. (In Russian)
315. Gninenko, Y.I.; Orlinski, A.D. New insect pests of forest plantations. *Plant Prot. Quar.* **2004**, *4*, 33. (In Russian)
316. Golosova, M.A.; Gninenko, Y.I. The appearance of the horse-chestnut leafminer on horse chestnut in Moscow. *Lesn. Vestn.* **2006**, *2*, 43–46. Available online: https://elibrary.ru/download/elibrary_9273118_72290780.pdf (accessed on 18 January 2022). (In Russian)
317. Akimov, I.A.; Zerova, M.D.; Gershenson, Z.S.; Narolsky, N.B.; Kochanez, O.M.; Sviridov, S.V. First record of the horse-chestnut leafminer *Cameraria ohridella* (Lepidoptera, Gracillariidae) on *Aesculus hippocastanum* (Hippocastanaceae) in Ukraine. *Vestn. Zool.* **2003**, *37*, 3–12. (In Russian)
318. Rakov, A.G. The Horse-chestnut Leafminer and other Invasive Phyllophagous Species under the Conditions of Formation of their Ranges in the European Part of Russia. Ph.D. Thesis, Russian State Agricultural University, Moscow, Russia, 2015. (In Russian)
319. Zhuravleva, E.N. The first appearance of the Ohrid leaf miner *Cameraria ohridella* (Lepidoptera: Gracillariidae) on the horse chestnut in the Greater Sochi. In *The Kataev Memorial Readings—VIII, Proceedings of the International Conference, St. Petersburg, Russia, 18–20 November 2014*; Musolin, D.L., Selikhovkin, A.V., Eds.; SPbFTU: St. Petersburg, Russia, 2014; p. 32. (In Russian)
320. Karpun, N.N. Features of formation of dendrophagous invasive pest fauna in the humid subtropics of Russia at the beginning of the XXI century. *Izv. St.-Peterbg. Lesoteh. Akad.* **2019**, *228*, 104–119. (In Russian) [CrossRef]
321. Chenikalova, E.V. Balkan leafminer moth—A threat to the horse chestnut in the Stavropol Region. *Plant Prot. Quar.* **2018**, *11*, 38–39. Available online: https://elibrary.ru/download/elibrary_36498157_20089242.pdf (accessed on 18 January 2022). (In Russian)
322. Chenikalova, E.V. The chestnut moth *Cameraria ohridella* (Deschka et Dimić, 1986) and boxwood moth *Cydalima perspectalis* (Walker, 1859) in the tree stands of Stavropol and its environs. In *Dendrobiotic Invertebrates and Fungi and their Role in Forest Ecosystems. The Kataev Memorial Readings—XI, Proceedings of the All-Russia Conference with International Participation, St. Petersburg, Russia, 24–27 November 2020*; Musolin, D.L., Kirichenko, N.I., Selikhovkin, A.V., Eds.; SPbFTU: St. Petersburg, Russia, 2020; pp. 347–348. (In Russian)
323. Martynov, V.V.; Nikulina, T.V.; Shokhin, I.V.; Terskov, E.N. Contributions to the fauna of invasive insects of Ciscaucasia. *Field Biol. J.* **2020**, *2*, 99–122. (In Russian) [CrossRef]

324. Selikhovkin, A.V.; Popovichev, B.G.; Denisova, N.V.; Timofeeva, Y.A. Invasive mining moths (Lepidoptera: Gracillariidae) of woody plantations in St. Petersburg. In *XV Congress of the Russian Entomological Society, Proceedings of the Congress, Novosibirsk, Russia, 31 July—7 August 2017*; Azarkina, G.N., Baranchikov, Y.N., Barkalov, A.V., Belokobilskiy, S.A., Glupov, V.V., Grichanov, I.Y., Danilov, Y.N., Dubatolov, V.V., Dudko, R.Y., Kireychuk, A.G., et al., Eds.; Garamond Publisher: Novosibirsk, Russia, 2017; pp. 446–447. (In Russian)
325. Anikin, V.V. Present day bio-invasions in the Volga-Ural Region: From the South to the North or from the East to the West? *Cameraria ohridella* (Lepidoptera: Gracillariidae) in the Lower and Middle Volga. *Zootaxa* **2019**, *4624*, 583–588. [CrossRef]
326. Anikin, V.V.; Zolotuhin, V.V.; Polumordvinov, O.A. Mass damage of horse chestnut's leaves (*Aesculus hippocastanum*) by Ohrid leafminer (*Cameraria ohridella*) on the territory of Penza in 2019. *Bull. Botanic Garden Saratov State Univ.* **2019**, *17*, 235–241. (In Russian) [CrossRef]
327. Kryukova, A.V.; Nikolaeva, Z.V. Harmfulness of chestnut mining moth in Pskov region. *Bull. Velikoluk. State Agric. Acad.* **2018**, *1*, 1–7. Available online: https://elibrary.ru/download/elibrary_35034180_72824990.pdf (accessed on 18 January 2022). (In Russian)
328. Grigoriev, M.A.; Grigoriev, A.I. Ecological and biological characteristic of horse chestnut (*Aesculus hippocastanum* L.) in the conditions of the south of Western Siberia. *Bull. Omsk State Agric. Univ.* **2017**, *3*, 61–68. Available online: https://elibrary.ru/download/elibrary_30467968_44356667.pdf (accessed on 18 January 2022). (In Russian)
329. Triebel, S.A.; Gamanova, O.N. Monitoring chestnut miner moth. *Plant Prot. Quar.* **2009**, *2*, 45–47.
330. Lack, H.W. The discovery and rediscovery of the horse chestnut. *Arnoldia* **2002**, *61*, 15–19. [CrossRef]
331. Kenis, M.; Tomov, R.; Svatos, A.; Schlinsog, P.; Lopez-Vaamonde, C.; Heitland, W.; Grabenweger, G.; Girardoz, S.; Freise, J.; Avtzis, N. The horse-chestnut leaf miner in Europe—Prospects and constraints for biological control. In Proceedings of the Second International Symposium on Biological Control of Arthropods, Davos, Switzerland, 12–16 September 2005; Hoddle, M., Ed.; Forest Health Technology Enterprise Team: Morgantown, WV, USA, 2005; pp. 77–90.
332. D'Costa, L.; Koricheva, J.; Straw, N.; Simmonds, M.S.J. Oviposition patterns and larval damage by the invasive horse-chestnut leaf miner *Cameraria ohridella* on different species of Aesculus. *Ecol. Entomol.* **2013**, *38*, 456–462. [CrossRef]
333. Davydova, I.A.; Gradusov, V.M.; Rybinceva, A.L. Cameraria ohridella Deschka et Dimić (Lepidoptera: Gracillariidae) on different species of horse chestnut in Moscow landscaping. In *Dendrobiotic Invertebrates and Fungi and their Role in Forest Ecosystems. The Kataev Memorial Readings—XI, Proceedings of the All-Russia Conference with International Participation, St. Petersburg, Russia, 24–27 November, 2020*; Musolin, D.L., Kirichenko, N.I., Selikhovkin, A.V., Eds.; SPbFTU: St. Petersburg, Russia, 2020; pp. 141–142. (In Russian)
334. Pschorn-Walcher, H. Freiland-Biologie der eingeschleppten Roßkastanien-Miniermotte *Cameraria ohridella* Deschka and Dimić (Lep., Gracillariidae) im Wienerwald. *Linz. Biol. Beiträge* **1994**, *26*, 633–642.
335. Gilbert, M.; Grégoire, J.-C.; Freise, J.F.; Heitland, W. Long-distance dispersal and human population density allow the prediction of invasive patterns in the horse chestnut leafminer *Cameraria ohridella*. *J. Anim. Ecol.* **2004**, *73*, 459–468. [CrossRef]
336. Augustin, S.; Reynaud, P. Un nouveau ravageur pour le marronnier: *Cameraria ohridella*. *PHM Rev. Hortic.* **2000**, *418*, 44–45.
337. Kirichenko, N.; Augustin, S.; Kenis, M. Invasive leafminers on woody plants: A global review of pathways, impact and management. *J. Pest Sci.* **2019**, *92*, 93–106. [CrossRef]
338. *The Most Dangerous Invasive Species in Russia (TOP-100)*; Dgebuadze, Y.Y.; Petrosyan, V.G.; Khlyap, L.A. (Eds.) KMK Scientific Press Ltd.: Moscow, Russia, 2018; pp. 1–688. (In Russian)
339. Cherpakov, V.V. About bacterial damage caused by *Pseudomonas syringae* pv. *aesculi* on horse chestnut in the Russian Federation. *Actual Probl. For. Complex* **2013**, *36*, 72–80. Available online: https://elibrary.ru/item.asp?id=21267711 (accessed on 20 January 2022). (In Russian)
340. Straw, N.A.; Williams, D.T. Impact of the leaf miner *Cameraria ohridella* (Lepidoptera: Gracillariidae) and bleeding canker disease on horse-chestnut: Direct effects and interaction. *Agric. For. Entomol.* **2013**, *15*, 321–333. [CrossRef]
341. Koskella, B.; Meaden, S.; Crowther, W.J.; Leimu, R.; Metcalf, C.J.E. A signature of tree health? Shifts in the microbiome and the ecological drivers of horse chestnut bleeding canker disease. *New Phytol.* **2017**, *215*, 737–746. [CrossRef]
342. Freise, J.; Heitland, W.; Tosevski, I. Parasitism of the horse chestnut leafminer, *Cameraria ohridella* Deschka and Dimić (Lep., Gracillariidae), in Serbia and Macedonia. *Anz. Schädlingskunde* **2002**, *75*, 152–157. [CrossRef]
343. Bhatti, I.; Shaw, O.; Shaw, P. Parasitoids and parasitism rates of the horse chestnut leaf miner *Cameraria ohridella* Deschka and Dimić [Lepidoptera: Gracillariidae] across four sites in south-west London. *Arboricultural J.* **2013**, *35*, 147–159. [CrossRef]
344. Volter, L.; Kenis, M. Parasitoid complex and parasitism rates of the horse chestnut leafminer, *Cameraria ohridella* (Lepidoptera: Gracillariidae) in the Czech Republic, Slovakia and Slovenia. *Eur. J. Entomol.* **2006**, *103*, 365–370. [CrossRef]
345. Girardoz, S.; Volter, L.; Tomov, R.; Quicke, D.L.J.; Kenis, M. Variations in parasitism in sympatric populations of three invasive leaf miners. *J. Appl. Entomol.* **2007**, *131*, 603–612. [CrossRef]
346. Grabenweger, G.; Kehrli, P.; Zweimuller, I.; Augustin, S.; Avtzis, N.; Bacher, S.; Freise, J.; Girardoz, S.; Guichard, S.; Heitland, W.; et al. Temporal and spatial variations in the parasitoid complex of the horse chestnut leafminer during its invasion of Europe. *Biol. Invasions* **2010**, *12*, 2797–2813. [CrossRef]
347. Hernandez-Lopez, A.; Rougerie, R.; Augustin, S.; Lees, D.C.; Tomov, R.; Kenis, M.; Cota, E.; Kullaj, E.; Hansson, C.; Grabenweger, G.; et al. Host tracking or cryptic adaptation? Phylogeography of *Pediobius saulius* (Hymenoptera, Eulophidae), a parasitoid of the highly invasive horse-chestnut leafminer. *Evol. Appl.* **2012**, *5*, 256–269. [CrossRef]

348. Kostyukov, V.V.; Kosheleva, O.V.; Nakonechnaya, I.V.; Gunasheva, Z.M. First report on parasitoids of the horse chestnut moth in Russia. *Plant Prot. Quar.* **2014**, *9*, 41–42. Available online: https://elibrary.ru/download/elibrary_21841455_76822640.pdf (accessed on 18 January 2022). (In Russian)
349. Měsić, A.; Barčić, J.; Barčić, J.I.; Miličević, T.; Duralija, B.; Čuljak, T.G. A low environmental impact method to control horse chestnut leaf miner *Cameraria ohridella* (Deschka & Dimic). *J. Food Agric. Environ.* **2008**, *6*, 421–427. Available online: https://www.researchgate.net/publication/231167581_A_low_environmental_impact_method_to_control_horse_chestnut_leaf_miner_Cameraria_ohridella_Deschka_Dimic (accessed on 18 January 2022).
350. De Prins, J.; De Prins, W. The occurrence of *Cameraria ohridella* in Belgium (Lepidoptera: Gracillariidae). *Phegea* **2001**, *29*, 81–88. Available online: https://www.researchgate.net/publication/250928533_The_occurrence_of_Cameraria_ohridella_in_Belgium_Lepidoptera_Gracillariidae (accessed on 18 January 2022).
351. Jagiełło, R.; Walczak, U.; Iszkuło, G.; Karolewski, P.; Baraniak, E.; Giertych, M.J. Impact of *Cameraria ohridella* on *Aesculus hippocastanum* growth and long-term effects of trunk injection with pesticides. *Int. J. Pest Manag.* **2019**, *65*, 33–43. [CrossRef]
352. Larina, G.E.; Seraya, L.G.; Kashtanova, O.A.; Dymovich, A.M.; Baranchikov, Y.N. The effectiveness of endotherapy of trunks in protection of horse-chestnut *Aesculus hippocastanum* L. against mining insects. In *Dendrobiotic Invertebrates and Fungi and their Role in Forest Ecosystems. The Kataev Memorial Readings—XI, Proceedings of the All-Russia Conference with International Participation, St. Petersburg, Russia, 24–27 November 2020*; Musolin, D.L., Kirichenko, N.I., Selikhovkin, A.V., Eds.; SPbFTU: St. Petersburg, Russia, 2020; pp. 187–188. (In Russian)
353. Svatos, A.; Kalinová, B.; Hoskovec, M.; Kindl, J.; Hrdy, I. Chemical communication in horse-chestnut leafminer *Cameraria ohridella* Deschka & Dimic. *Plant. Prot. Sci.* **1999**, *35*, 10–13.
354. Atanov, N.M.; Kuzina, N.P.; Kuzin, A.A.; Chirskaya, M.V. Impact of modifiers on the attractiveness of the horse-chestnut leaf-miner's pheromone *Cameraria ohridella* Deschka et Dimič, 1986. *Plant Health Res. Pract.* **2016**, *4*, 47–49. Available online: http://www.rsn-msk.ru/files/karantin-18-2016.pdf (accessed on 18 January 2022).
355. Kehrli, P.; Lehmann, M.; Bacher, S. Mass-emergence devices: A biocontrol technique for conservation and augmentation of parasitoids. *Biol. Control* **2005**, *32*, 191–199. [CrossRef]
356. Kumata, T. Taxonomic studies on the Lithocolletinae of Japan. *Insecta Matsumurana* **1963**, *25*, 53–90.
357. Kirichenko, N.; Ryazanova, M.; Akulov, E.; Ponomarenko, M. Can native parasitoids control the invasive lime leaf-miner Phyllonorycter issikii (Lepidoptera: Gracillariidae) in Western Siberia? In Proceedings of the 1st International Electronic Conference on Entomology, Basel, Switzerland, 1–15 July 2021; Available online: https://sciforum.net/paper/view/10537 (accessed on 4 January 2021).
358. Kumata, T.; Kuroko, H.; Park, K. Some Korean species of the subfamily Lithocolletinae Gracillariidae, Lepidoptera. *Korean J. Plant Prot.* **1983**, *22*, 213–227.
359. Kirichenko, N.I.; Akulov, E.N.; Babichev, N.S.; Mikhailova, I.A.; Ponomarenko, M.G.; Lopez-Vaamonde, C. Past distribution of *Tilia*-feeding *Phyllonorycter* micromoth (Lepidoptera: Gracillariidae) in the Russian Far East based on survey of historical herbarium. *Far Eastern Entomol.* **2019**, *390*, 19–32. [CrossRef]
360. Kirichenko, N.I. Trophic Associations and Invasion Patterns of the Dendrophilous Gracillariid Moths (Lepidoptera: Gracillariidae) in the Asian Part of Russia. Ph.D. Thesis, Siberian Federal University, Krasnoyarsk, Russia, 2020. Available online: https://research.sfu-kras.ru/sites/research.sfu-kras.ru/files/Dissertaciya_Kirichenko.pdf (accessed on 18 January 2022). (In Russian)
361. Bednova, O.V.; Belov, D.A. Lime leaf miner (Lepidoptera, Gracillariidae) in green spaces of Moscow and Moscow Region. *Lesn. Vestn.* **1999**, *2*, 172–177. (In Russian)
362. Kozlov, M.V. Leaf-mining moth is a pest of limes. *Plant Prot. Quar.* **1991**, *4*, 46. (In Russian)
363. Šefrová, H. *Phyllonorycter issikii* (Kumata, 1963)—Bionomics, ecological impact and spread in Europe (Lepidoptera, Gracillariidae). *Acta Univ. Agric. Silvic. Mendel. Brun.* **2002**, *50*, 99–104.
364. Ermolaev, I.V.; Rubleva, E.A. History, rate and factors of invasion of lime leafminer *Phyllonorycter issikii* (Kumata, 1963) (Lepidoptera, Gracillariidae) in Eurasia. *Russ. J. Biol. Invasions* **2017**, *1*, 2–19. (In Russian) [CrossRef]
365. Gninenko, Y.I.; Kozlova, E.I. Lime leaf miner in Russia and problems of biological control. In Proceedings of the International Organization for Biological Control of Harmful Animals and Plants, East Palaearctic Regional Section, Abstracts, Poznan, Poland, 15–19 May 2006; IOBC: Poznan, Poland, 2006; p. 16.
366. Kirichenko, N.; Triberti, P.; Ohshima, I.; Haran, J.; Byun, B.-K.; Li, H.; Augustin, S.; Roques, A.; Lopez-Vaamonde, C. From east to west across the Palearctic: Phylogeography of the invasive lime leaf miner *Phyllonorycter issikii* (Lepidoptera: Gracillariidae) and discovery of a putative new cryptic species in East Asia. *PLoS ONE* **2017**, *12*, e0171104. [CrossRef]
367. Ermolaev, I.V.; Rubleva, E.A.; Rysin, S.L.; Ermolaeva, M.V. Forage plants of lime leafminer *Phyllonorycter issikii* (Kumata, 1963) (Lepidoptera, Gracillariidae). *Russ. J. Biol. Invasions* **2018**, *2*, 2–13. Available online: https://elibrary.ru/download/elibrary_35166252_58174722.pdf (accessed on 18 January 2022). (In Russian)
368. Kirichenko, N. The lime leafminer *Phyllonorycter issikii* in Western Siberia: Some ecological characteristics of the population of the recent invader. *Contemp. Probl. Ecol.* **2014**, *7*, 114–121. [CrossRef]
369. Pigott, C.D. *Lime-trees and Basswoods. A Biological Monograph of the Genus Tilia*; Cambridge University Press: New York, NY, USA, 2012; pp. 1–395.
370. Zolotukhin, V.V. About some invasive arthropods on the territory of the Ulyanovsk region. *Nat. Simbirsk Volga Reg.* **2002**, *2*, 200–203.

371. Ermolaev, I.V.; Zorin, D.A. Ecological consequences of invasion of *Phyllonorycter issikii* (Lepidoptera, Gracillariidae) in lime forests in Udmurtia. *Entomol. Rev.* **2011**, *91*, 592–598. [CrossRef]
372. Kirichenko, N.I.; Zakharov, E.V.; Lopez-Vaamonde, C. Revealing the invasion history of the lime leafminer *Phyllonorycter issikii* (Lepidoptera: Gracillariidae) using historical herbaria and next generation sequencing. In *Dendrobiotic Invertebrates and Fungi and their Role in Forest Ecosystems. The Kataev Memorial Readings—XI, Proceedings of the All-Russia Conference with International Participation, St. Petersburg, Russia, 24–27 November 2020*; Musolin, D.L., Kirichenko, N.I., Selikhovkin, A.V., Eds.; SPbFTU: St. Petersburg, Russia, 2020; pp. 172–173, 175. (In Russian)
373. Ermolaev, I.V.; Yefremova, Z.A.; Izhboldina, N.V. Parasitoids as a mortality factor for the lime leafminer (*Phyllonorycter issikii*, Lepidoptera, Gracillariidae). *Entomol. Rev.* **2011**, *91*, 326–334. [CrossRef]
374. Szőcs, L.; George, M.; Thuróczy, C.; Csóka, G. Parasitoids of the lime leaf miner *Phyllonorycter issikii* (Lepidoptera: Gracillariidae) recorded throughout the area it recently colonized. *Eur. J. Entomol.* **2015**, *112*, 591–598. [CrossRef]
375. Ermolaev, I.V.; Yefremova, Z.A.; Gerasimova, N.A.; Koryoleva, E.A.; Lushnikov, N.N.; Petrov, A.I.; Pchel'nikov, A.A. Parasitoids (Hymenoptera) of the lime leafminer (*Phyllonorycter issikii*, Lepidoptera, Gracillariidae) in different cities of the Russian Federation and their role in the mortality of the invasive pest. *Entomol. Rev.* **2019**, *99*, 485–493. [CrossRef]
376. Yefremova, Z.A.; Mishchenko, A.V. The parasitoid complex (Hymenoptera, Eulophidae) of the leafminer *Phyllonorycter issikii* (Kumata) (Lepidoptera, Gracillariidae) from the Middle Volga Basin. *Entomol. Rev.* **2008**, *88*, 178–185. [CrossRef]
377. Mishchenko, A.V. Method of mass rearing ichneumons Minotetrastichus frontalis (Nees). *Bull. Russ. Fed. Fed. Serv. Intellect. Property* **2015**, *10*, 1–8.
378. Baryshnikova, S.V.; Bolshakov, L.V. Microlepidoptera of Tula Province. 15. Families Bucculatricidae, Gracillariidae, and Lyonetiidae (Hexapoda: Lepidoptera). In *Biodiversity of Tula Region at the Turn of Centuries*; Bolshakov, L.V., Ed.; Grif & K: Tula, Russia, 2004; Volume 4, pp. 31–37. (In Russian)
379. Ermolaev, I.V.; Rubleva, E.A.; Rysin, S.L.; Kozhenkova, A.A.; Ermolaeva, M.V. Trophic specialization of the poplar leafminer *Phyllonorycter populifoliella* (Treitschke, 1833) (Lepidoptera, Gracillariidae). *Entomol. Rev.* **2020**, *100*, 287–300. [CrossRef]
380. Shashank, P.R.; Singh, N.; Harshana, A.; Sinha, T.; Kirichenko, N. First report of the poplar leaf miner, *Phyllonorycter populifoliella* (Treitschke) (Lepidoptera: Gracillariidae) from India. *Zootaxa* **2021**, *4915*, 435–450. [CrossRef] [PubMed]
381. Kozhanchikov, I.V.; Danilevskiy, A.S.; Djakonov, A.M. Order Lepidoptera. In *Pests of Forest (Reference Book)*; Shtakelberg, A.A., Ed.; Publishing House of the Academy of Sciences of the USSR: Moscow, Russia; St. Petersburg, Russia, 1955; Volume 1, pp. 35–287. (In Russian)
382. Tarasova, O.V.; Kovalev, A.V.; Sukhovolsky, V.G.; Khlebopros, R.G. *Phylophagous Insects of Green Plantations of Cities: Species Compositions and Population Dynamics*; Nauka: Novosibirsk, Russia, 2004; pp. 1–180. (In Russian)
383. Ermolaev, I.V. Ecological mechanisms of nonperiodical population wave: A case study of the poplar leafminer—*Phyllonorycter populifoliella* (Lepidoptera, Gracillariidae). *J. General Biol.* **2019**, *80*, 451–476.
384. Polezhaev, V.G. Struggle for existence in the poplar leafminer *Lithocolletis populifoliella* Tr. *Zool. Zhurnal.* **1934**, *13*, 67–75. (In Russian)
385. Frolov, D.N. The poplar leafminer as a landscaping pest in Irkutsk. *Proceedings Irkutsk State Univ. Ser. Biol.* **1948**, *3*, 19. (In Russian)
386. Selikhovkin, A.V. Population dynamics of Microlepidoptera under the conditions of industrial air pollution: A follow-up. *Izv. St.-Peterbg. Lesoteh. Akad.* **1996**, *162*, 26–38. (In Russian)
387. Selikhovkin, A.V.; Baryshnikova, S.V.; Denisova, N.V.; Timofeeva, Y.A. Species composition and population dynamics of dominant dendrophagous moths (Lepidoptera) in St. Petersburg and its environs. *Entomol. Rev.* **2018**, *98*, 963–978. [CrossRef]
388. Selikhovkin, A.V.; Egorov, A.A.; Sitnikova, D.D.; Mamaev, N.A. Occurrence of leaf miners *Phyllonorycter populifoliella* (Treitschke) and *Ph. pastorella* (Zeller) (Lepidoptera, Gracillariidae) on different species of poplars. *Entomol. Rev.* **2020**, *100*, 301–306. [CrossRef]
389. Rumyantsev, P.D. Biology of the poplar leaf miner *Lithocolletis populifoliella* in the conditions of Moscow. *Zool. Zhurnal.* **1934**, *13*, 257–279. (In Russian)
390. Mamaev, N.A.; Buy, D.D.; Selikhovkin, A.V. The second outbreak of the poplar leafminer *Phyllonorycter populifoliella* in St. Petersburg. *Izv. St.-Peterbg. Lesoteh. Akad.* **2020**, *233*, 81–94. (In Russian) [CrossRef]
391. Lvovsky, A.L. Lepidoptera (Insecta) within the city limits of St. Petersburg. *Proceedings Kharkiv Entomol. Soc.* **1994**, *2*, 4–48. (In Russian)
392. Bondarenko, E.A. An outbreak of the poplar leafminer *Phyllonorycter populifoliella* Tr. (Lepidoptera, Gracillariidae) in St. Petersburg. *Izv. St.-Peterbg. Lesoteh. Akad.* **2008**, *182*, 45–55. (In Russian)
393. Selikhovkin, A.V. Peculiarities of the population dynamics of the poplar mining moth *Phyllonorycter populifoliella* Tr. (Gracillariidae). *Izv. St.-Peterbg. Lesoteh. Akad.* **2010**, *192*, 220–235. Available online: https://elibrary.ru/download/elibrary_17047133_73905770.pdf (accessed on 18 January 2022). (In Russian)
394. Ermolaev, V.P. Ecologo-faunistic review of leafmining moths (Lepidoptera, Gracillariidae) of southern Primorye. *Tr. Zool. Inst. Akad. Nauk SSSR* **1977**, *70*, 98–116. (In Russian)
395. Đức, B.Đ.; Leont'ev, L.L.; Baryshnikova, S.V.; Selikhovkin, A.V. Consequences of the outbreaks of the poplar leaf miner and other mining Microlepidoptera in St. Petersburg. *Contemp. Probl. Ecol.* **2021**, *14*, 822–826. [CrossRef]
396. Sulkhanov, A.V. Ecology of urban populations of the poplar moth *Lithocolletis populifoliella* Tr. In *Dendrobiont Insects of Trees and Shrubs of Moscow*; Krivosheina, N.P., Striganova, B.R., Eds.; Nauka Publisher House: Moscow, Russia, 1992; pp. 70–97. (In Russian)

397. Chaika, S.Y. Insects. In *Great Russian Encyclopedia*; Electronic Version; 2017. Available online: https://bigenc.ru/biology/text/2250439 (accessed on 2 November 2021).
398. Kerzhner, I.M. An approach and perspective of study of Insecta in the territory of the former USSR. In *Biological Diversity: The Level of Taxonomic Study*; Sokolov, V.E., Reshetnikov, Y.S., Eds.; Nauka Publisher House: Moscow, Russia, 1994; pp. 65–69. (In Russian)
399. Cippolini, D. White fringetree as a novel larval host for emerald ash borer. *J. Econ. Entomol.* **2015**, *108*, 370–375. [CrossRef]
400. Orlova-Bienkowskaja, M.J. Cascading ecological effects caused by the establishment of the emerald ash borer *Agrilus planipennis* (Coleoptera: Buprestidae) in European Russia. *Eur. J. Entomol.* **2015**, *112*, 778–789. [CrossRef]
401. Eschen, R.; O'Hanlon, R.; Santini, A.; Vannini, A.; Roques, A.; Kirichenko, N.; Kenis, M. Safeguarding global plant health: The rise of sentinels. *J. Pest Sci.* **2018**, *62*, 29–36. [CrossRef]
402. Morales-Rodríguez, C.; Anslan, S.; Auger-Rozenberg, M.-A.; Augustin, S.; Baranchikov, Y.; Bellahirech, A.; Burokienė, D.; Čepukoit, D.; Çota, E.; Davydenko, K.; et al. Forewarned is forearmed: Preventive detection of potentially invasive pests and pathogens in sentinel plantings. *NeoBiota* **2019**, *47*, 95–123. [CrossRef]
403. Hebert, P.D.N.; Cywinska, A.; Ball, S.L.; de Waard, J.R. Biological identifications through DNA barcodes. *Proc. R. Soc. Lond.* **2003**, *270*, 313–321. [CrossRef] [PubMed]

forests

MDPI

Article

Two Nematicidal Compounds from *Lysinimonas* M4 against the Pine Wood Nematode, *Bursaphelenchus xylophilus*

Yixiu Sun [1], Chao Wang [1], Guicai Du [1], Wenjun Deng [1], Hong Yang [1], Ronggui Li [1], Qian Xu [2,*] and Qunqun Guo [1,*]

[1] College of Life Sciences, Qingdao University, Qingdao 266071, China; sun1786218@163.com (Y.S.); wangchao6903199@163.com (C.W.); duguicai@qdu.edu.cn (G.D.); dengwenjun@qdu.edu.cn (W.D.); 13963939639@139.com (H.Y.); lrg@qdu.edu.cn (R.L.)

[2] College of Agriculture and Biotechnology, Heze University, Heze 274000, China

* Correspondence: xq710301@163.com (Q.X.); gqunqun@163.com (Q.G.)

Abstract: A rich source of bioactive secondary metabolites from microorgannisms are widely used to control plant diseases in an eco-friendly way. To explore ideal candidates for prevention of pine wilt disease (PWD), a bacterial strain from rhizosphere of *Pinus thunbergii*, *Lysinimonas* M4, with nematicidal activity against pine wood nematode (PWN), *Bursaphelenchus xylophilus*, was isolated. Two nematicidal compounds were obtained from the culture of *Lysinimonas* M4 by silica gel chromatography based on bioactivity-guided fractionation and were subsequently identified as 2-coumaranone and cyclo-(Phe-Pro) by nuclear magnetic resonance (NMR) and mass spectrometry (MS). The 2-coumaranone and cyclo-(Phe-Pro) showed significant nematicidal activity against PWN, with LC_{50} values at 24 h of 0.196 mM and 0.425 mM, respectively. Both compounds had significant inhibitory effects on egg hatching, feeding, and reproduction. The study on nematicidal mechanisms revealed that 2-coumaranone and cyclo-(Phe-Pro) caused the accumulation of reactive oxygen species (ROS) in nematodes, along with a notable decrease in CAT and POS activity and an increase in SOD activity in nematodes, which might contribute to the death of pine wood nematodes. Bioassay tests demonstrated that the two compounds could reduce the incidence of wilting in Japanese black pine seedlings. This research offers a new bacterial strain and two metabolites for biocontrol against PWN.

Keywords: *Bursaphelenchus xylophilus*; 2-coumaranone; cyclo-(Phe-Pro); reactive oxygen species (ROS); nematicidal activity

1. Introduction

Pine wilt disease (PWD) is an extremely destructive forest disease caused by the pine wood nematode (PWN), *Bursaphelenchus xylophilus* [1]. The disease was discovered in Jiangsu Province in 1982 and subsequently spread rapidly in China [2]. The State Forestry and Grassland Administration of China announced that PWD was found in 721 county-level administrative regions across the country in 2021. It caused a serious economic loss and a negative influence on the ecological environment [3].

Owing to the rapid spread of PWD, effective and reliable prevention and control measures for the disease became more and more urgent. Physical controls, such as cutting and burning the infected pine trees, cannot fundamentally prevent the spread of PWN [4]. Spraying synthetic insecticides against PWN vectors, such as *Monochamus alternatus* and trunk injection of nematicides, to directly control PWN as the common measures for PWD management might cause environmental pollution and disruptions of the host or other beneficial organisms [5,6]. Biocontrol based on microorganisms is considered as a promising strategy in the green management of plant diseases. Fungi and actinomycetes producing mycotoxins or antibiotics are reported to be effective in PWN control [7–10]. However, bacteria are far more commonly reported for biocontrol of PWN than fungi and actinomycetes due to their high diversity and rapid growth. Yu et al. [11] reported the nematicidal effect of volatiles produced by *Pseudotalteromonas marina* strain H-42 and *Vibrio atlanticus* strain S-16 that were isolated from the seawater and bay scallops, respectively. Ponpandian et al. [12] screened two nematicidal bacteria from the endophytes of four Pinus species, and found that their metabolites had a significant inhibitory effect on egg hatching and development of PWN.

The plant growth-promoting rhizobacteria, a group of root-colonizing microorganisms beneficial to plants, provide promising and sustainable sources for the screening of biological control microbes against plant pathogens [13]. Plant rhizoshperic microbes can indirectly promote plant growth by production of toxins, biosurfactants, and lytic enzymes to inhibit plant pathogens and induce systemic resistance in host plants [14]. *Pseudomonas fluorescens* CHA0 and *Bacillus amyloliquefaciens* FZB42 were identified as antagonists of root-knot nematodes [15,16]. Moreover, Deng et al. [17] reported that the microbial diversity of the pine root system changed significantly when PWNs infected trees at early and late stages, and these changes were associated with the suppression effect of microbial community against PWNs.

In the present study, we screened a strain of rhizosphere bacteria from healthy Japanese black pine trees with nematicidal activity against PWN, and characterized the effective compounds in its culture, which provided reference to exploration of nematicidal agents against PWN.

2. Materials and Methods

2.1. Sample Preparation and Nematode Cultivation

Soil samples were collected from the roots of *Pinus thunbergii* on the campus of Qingdao University. PWNs were isolated from infected *P. thunbergii* by the Baermann funnel and morphologically confirmed [18]. The collected PWNs were washed 3 times with sterile water and cultured with *Botrytis cinerea* on potato dextrose agar (PDA) medium in dark at 25 °C for 1 week.

2.2. Isolation and Screening of Nematicidal Bacteria against PWN

An approximately 0.1 g soil sample was transferred to 49.9 mL sterile water and incubated at 30 °C for 30 min. The soil suspension was diluted serially with sterile water to obtain 10^{-1} to 10^{-6} dilutions [19]. 100 μL soil suspension from each dilution was plated on nutrient agar (NA) culture medium, and incubated at 30 °C for 48 h. Single colonies were picked from plates and sub-cultured until a pure culture with nematicidal activity against PWN was obtained.

The purified bacteria were fermented in a nutrient broth (NB) medium at 30 °C for 4 d. The culture was centrifuged at 10,000 rpm for 10 min and the supernatant was collected for subsequent experiments. The impregnation method was applied to screen the bacteria with nematicidal activity [20]. Next, 400 μL of the supernatant was mixed with 100 μL PWN suspension (containing approximately 100 nematodes) in sterile water and added into a well of 24-well plate, whereas control groups were carried out by mixing NB medium and PWN suspension. The plates were incubated at 25 °C for 72 h under dark conditions. Dead nematodes were observed and counted under a stereomicroscope (Motic BA200, Xiamen, China). Nematodes were considered dead if they were stiff and had no response to physical stimulus. Corrected mortality was calculated according to Schneider Orelli's formula [21]:

$$\text{corrected mortality (\%)} = [(\text{mortality in treatment} - \text{mortality in control})/(100 - \text{mortality in control})] \times 100 \quad (1)$$

The experiment was carried out with three biological replicates.

Three strains with obvious nematicidal activity against PWN were further studied for their nematicidal activity by diluting their supernatants to 2.5, 5, and 10-fold, respectively [22]. The nematicidal activity of diluted suspensions was tested following the same procedures as described above.

2.3. Identification of Bacteria with Nematicidal Activity

The bacterial strain M4 with the highest nematicidal activity was identified. After culturing the M4 strain on a NA plate at 30 °C for 3 days. The strain was characterized by gram staining, spore staining, capsular staining, and morphological observation under a microscope (Motic SW350B, Xiamen, China). In addition, biochemical tests including starch hydrolysis test and oil hydrolysis were also used as an indicator for the identification of the M4 strain [23].

The M4 strain was cultivated in NA liquid medium at 30 °C for 18 h, then the genomic DNA was extracted using the TIANGEN Genomic DNA Kit (TIANGEN Biotech, Beijing, China). The 16S rDNA was amplified by PCR using the genomic DNA and primer pair 27F (5′-AGAGTTTGATCCTGGCTCAG-3′) and 1492R (5′-TACCTTGTTACGACTT-3′) and sequenced by Sangon Biotech Company (Qingdao, China). The 16S rDNA sequence underwent a BLAST search in the GenBank database of the National Center for Biotechnology Information (NCBI). The phylogenetic tree was constructed in MEGA7 using the neighbor-joining (N-J) method [24,25].

2.4. Isolation and Structural Determination of the Nematicidal Compounds

The culture of strain M4 (5 L) was extracted with an equal volume of ethyl acetate three times [12]. The ethyl acetate phase was combined and concentrated to dryness by rotary evaporation at 25 °C under vacuum, and 0.5 g of crude extracts was obtained. The nematicidal bioactivity of the crude extracts and the left aqueous phase was then bioassayed. Next, 5 μL of ethyl acetate extract (10 mg/mL, dissolved in DMSO) was mixed with 95 μL nematode suspension (containing approximately 100 nematodes) in a 96-well plate at 25 °C for 72 h under dark conditions. Additionally, 5 μL of ethyl acetate extract of pure NB medium (10 mg/mL, dissolved in DMSO), mixed with 95 μL of the same nematode suspension, was served as control. Dead nematodes were observed and counted under a stereomicroscope (Motic BA200, Xiamen, China).

The nematicidal components in the crude extracts were isolated by silica gel column chromatography (200 to 300 mesh) with petroleum/ethyl acetate (2:1, 2:3, 1:3, and 1:5 *v/v*), ethyl acetate/methanol (20:1 *v/v*), and five fractions (Fr1, Fr2, Fr3, Fr4, and Fr5) were collected [26]. Fr4 and Fr5, which showed higher nematicidal activity, were further purified by silica gel chromatography based on bioactivity-guided fractionation. Compound **1** in Fr.4 was recrystallized with petroleum/ethyl acetate (1:5, *v/v*) and compound **2** in Fr.5 was recrystallized with ethyl acetate.

The NMR spectra were obtained using NMR spectrometer (JNM-ECZ600R, JOEL, Tokyo, Japan) operating at 600 MHz for ^{1}H NMR and 100 MHz for ^{13}C NMR in chloroform-*d* ($CDCl_3$) with tetramethylsilane (TMS) as the internal standard. The electron impact mass spectra (EIMS) were determined using gas chromatography mass spectrometry (7890A-5975C, Agilent, Santa Clara, CA, USA).

2.5. Nematicidal Activity of the Nematicidal Compounds against PWN In Vitro

Nematicidal activities of the isolated compounds were assessed by determination of the LC_{50} value for 24 h following the method as previously described [27]. Corrected mortalities were determined for each compound according to Equation (1), and the LC_{50} values were calculated according to the Probi model, respectively.

2.6. Effect of the Nematicidal Compounds on Egg Hatching

Eggs were obtained according to the method previously described by Liu et al. [6]. Specifically, approximately 100 nematode eggs were transferred to a 48-well culture plate containing 2-coumaranone or cyclo-(Phe-Pro) of different concentrations (0.1 mM, 0.2 mM, and 0.5 mM in 5% DMSO), while 5% DMSO was used as a control group. Plates were incubated at 25 °C and nematodes at the J2 stage were counted at 48 h. Three replicates were performed in this experiment. The hatching rate was calculated as follows:

$$\text{hatching rate (\%)} = [\text{juveniles}/(\text{eggs} + \text{juveniles}) \times 100] \quad (2)$$

2.7. Influence of the Nematicidal Compounds on Nematodes Feeding and Population

Approximately 100 nematodes were treated with 0.1 mM 2-coumaranone or cyclo-(Pro-Phe) solution in 5% DMSO for 24 h, and 5% DMSO and sterile water were used as controls. PWNs were transferred to a PDA plate covered with *B. cinerea* and incubated at 25 °C for 7 d. The feeding of PWNs to *B. cinerea* was observed and photographed daily. Nematodes in each group were then collected, washed with distilled water, and quantified under a stereomicroscope (Motic BA200, Xiamen, China) [28].

2.8. Detection of Reactive Oxygen Species (ROS) and Antioxidant Enzymes in Nematicide-Treated PWD

ROS production was determined by the fluorescein 2,7-Dichlorodihydrofluorescein -diacetate (DCFH-DA) probe method [29]. Approximately 200 adult nematodes were pretreated with 0.2 mM 2-coumaranone and 0.2 mM cyclo-(Phe-Pro), respectively. They were incubated in 96-well plates for 24 h, and 5% DMSO was used in control wells. At the end of incubation, the PWNs were washed three times with 0.01 M PBS and treated with a probe from ROS assay kit (Nanjing Jiancheng Bioengineering Institute, Nanjing, China). Reactive oxygen hydrogen donor-induced PWNs served as positive control. Nematodes were collected, centrifuged at 10,000 rpm for 5 min, and washed three times with 0.01 M PBS. Then, the fluorescence intensity was detected at an excitation wavelength of 485 nm and a blocking wavelength of 528 nm under a fluorescence microscope (Olympus IX73, Tokyo, Japan).

The treated nematodes were thoroughly ground on ice, the extracts were centrifuged at 4 °C, 10,000 rpm for 10 min, and the supernatants were collected. The protein content was determined according to the Bradford method. The activities of superoxide dismutase (SOD), peroxidase (POD), and catalase (CAT) were measured using azalantetrazole reduction, guaiacol, and hydrogen peroxide, respectively [30,31]. The absorbance of each reaction was measured at wavelengths of 560 nm, 470 nm, and 240 nm, using a UV spectrophotometer (GENESY-50, Thermo Fisher Scientific, Waltham, MA, USA), respectively.

2.9. Nematicidal Activity of the Nematicidal Compounds against PWN In Vivo

Healthy Japanese black pine seeds were soaked with $HgCl_2$ for 10 min and rinsed with sterile water three times. Seeds were placed in Petri dishes containing sterilized filter paper, which was kept moist and incubated in the dark at 25 °C for 5 d. When the seeds germinated with roots about 1.5 cm long, they were transferred to nutrient soil on daylight exposure for 16 h/d cultivation for 30 d. Next, 5 μL of nematode suspension (50 PWNs/μL) was injected into pine seedlings by micro syringe [32]. On post-inoculation day 2 (PID 2), each seedling in the experimental groups was injected with 10 μL of 2-coumaranone (50 mM) and cyclo-(Phe-Pro) (50 mM), respectively. The same volume of 5% DMSO was injected into pine seedlings for the control group. In addition, the pine seedlings without inoculation of PWN were injected with 10 μL of nematicidal compounds (50 mM) to observe whether they would have an unexpected effect on the seedlings, and the same dose of sterile water was injected into another group of seedlings as the control.

2.10. Statistical Analysis

The means and standard deviations (SD) of the corrected mortality data were calculated in Microsoft Excel. One-way analyses of variance (ANOVA) with least significant differences (LSD) test ($p = 0.05$) and Student's t-tests ($p = 0.05$ and 0.01) were performed to compare among treatment levels, and the lethal concentration (LC_{50}) values were calculated by probit analysis in SPSS version 25.0 (SPSS Inc., Chicago, IL, USA).

3. Results

3.1. Screening and Identification of Nematicidal Strains

A total of 40 bacterial strains were isolated from soil samples. The corrected mortality of PWNs was determined by exposing PWNs to a 1.25 times diluted cell-free culture for 72 h (Table 1). The strains with higher than 80% corrected mortality were considered as nematicidal candidates. The corrected mortalities of the cultures from BG-1, S-1-3, and M4 strains were 86.63%, 88.63%, and 91.30%, respectively, which were significantly higher than that of other strains. After further dilution for 2.5, 5, and 10 times, the culture of the M4 strain showed the strongest nematicidal effect among the three forementioned strains (Figure 1). Thus, the M4 strain was selected for subsequent characterization studies.

Table 1. Nematicidal activities of cell-free culture of different strains.

Stains	Corrected Mortality (%)	Stains	Corrected Mortality (%)	Stains	Corrected Mortality (%)	Stains	Corrected Mortality (%)
NA-1	5.36 ± 1.52 c	BG-1	86.63 ± 1.53 o	S-1-1	19.4 ± 1.53 g	M1	8.70 ± 1.73 e
NA-2	49.83 ± 3.00 n	BG-2	7.02 ± 2.08 d	S-1-2	37.79 ± 3.00 l	M2	1.00 ± 1.52 a
NA-3	2.01 ± 2.08 a	BG-3	29.77 ± 3.61 i	S-1-3	88.63 ± 1.15 o	M3	4.01 ± 2.08 a
NA-4	5.02 ± 1.53 b	BG-4	16.38 ± 1.53 g	S-1-4	11.71 ± 1.73 f	M4	91.3 ± 1.53 o
NA-5	9.03 ± 1.53 e	BG-5	7.02 ± 0.58 d	S-1-5	5.69 ± 1.73 c	M5	6.69 ± 1.73 d
NA-6	5.36 ± 1.52 c	BG-6	16.38 ± 1.53 g	S-2-1	16.34 ± 2.08 g	M6	11.71 ± 1.73 f
NA-7	27.09 ± 2.52 i	BG-7	20.07 ± 1.15 g	S-2-2	5.35 ± 1.52 c	M7	19.4 ± 2.08 g
NA-8	0.34 ± 0.58 a	BG-8	1.00 ± 1.53 a	S-2-3	36.11 ± 1.15 k	M8	1.00 ± 1.52 a
NA-9	50.17 ± 2.52 n	BG-9	22.07 ± 3.06 h	S-2-4	40.80 ± 2.64 m	M9	32.78 ± 2.00 j
NA-10	30.77 ± 3.00 i	BG-10	37.45 ± 2.52 l	S-2-5	18.06 ± 2.31 g	M10	37.12 ± 2.51 l
						Control	0.33 ± 0.75 a

The mean corrected mortality of nematodes was determined after exposure to 1.25-fold dilution of cell-free cultures for 72 h. Control was NA medium. Data were mean ± SD of three replicates. Means in the column followed by the same letter did not differ significantly ($p = 0.05$) according to the LSD test.

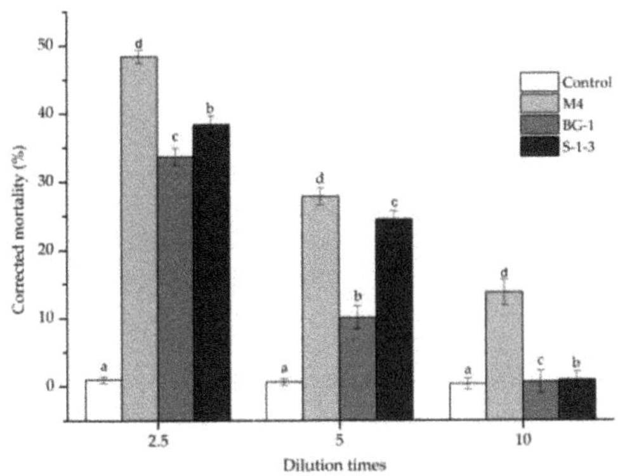

Figure 1. The nematicidal activity for the cell-free culture of strains M4, BG1, and S-1-3 diluted 10 times, 5 times, and 2.5 times, respectively. Data were given as mean. Means with the different letters were significantly different ($p < 0.05$) based on one-way ANOVA with multiple comparison analysis using the LSD test.

The colonies of the M4 strain were round with greenish-yellow color and entered margin. The cells were Gram-negative and motile with a short rod shape. Amylolysis, MR, and hydrogen sulfide tests were positive, while oil hydrolysis and Voges–Proskauer tests were negative. According to the comparison results of 16S rDNA sequences, the sequence from the M4 strain was homologous to that from *Lysinimonas* sp. F22 by 99.93%, which were both clustered in the same branch on the phylogenetic tree (Figure 2). Based on the morphological observation, analysis of physiological properties and 16S rDNA sequencing comparation, the M4 strain belonged to the genus *Lysinimonas*, and therefore was named *Lysinimonas* sp. M4.

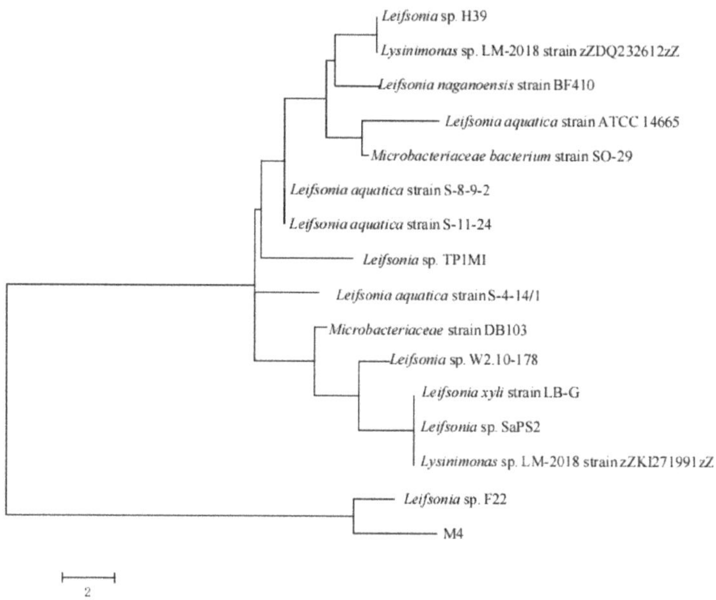

Figure 2. The phylogenetic tree constructed using a neighbor-joining method based on 16S rDNA gene sequence numbers at the nodes indicate the bootstrap values on neighbor-joining analysis.

3.2. Isolation and Identification of the Nematicidal Compounds from the M4 Strain

The nematicidal bioassay of ethyl acetate extract and the water phase of the *Lysinimonas* sp. M4 culture indicated that the ethyl acetate extract had significant nematicidal activity against PWN, while no obvious nematicidal activity was detected in the water phase (Table 2). Following a further isolation of ethyl acetate extract through silica gel column chromatography, five fractions (Fr. 1 to Fr. 5) were collected and tested for their nematicidal activity. The corrected mortality rate for Fr. 4 and Fr. 5 at a concentration of 200 μg/mL at 72 h was 91.87% and 70.68%, respectively, which was significantly higher than that for the other three fractions (Figure 3).

Table 2. The nematicidal activity of the aqueous phase and the ethyl acetate extract of the M4 strain culture for 72h.

Smples	Mean Corrected Mortality ± SD (%)		
	24 h	**48 h**	**72 h**
Water phase	1.00 ± 1.00 a	2.00 ± 1.00 a	2.00 ± 1.00 a
Ethyl acetate extract	37.33 ± 2.309 b	65.00 ± 2.00 c	87.33 ± 2.08 d

The mean corrected mortality of the nematodes was determined after exposure to the water phase and ethyl acetate extract at a different culture time. Data were mean ± SD of three replicates. Means in the column followed by the same letter did not differ significantly (p = 0.05) according to the LSD test.

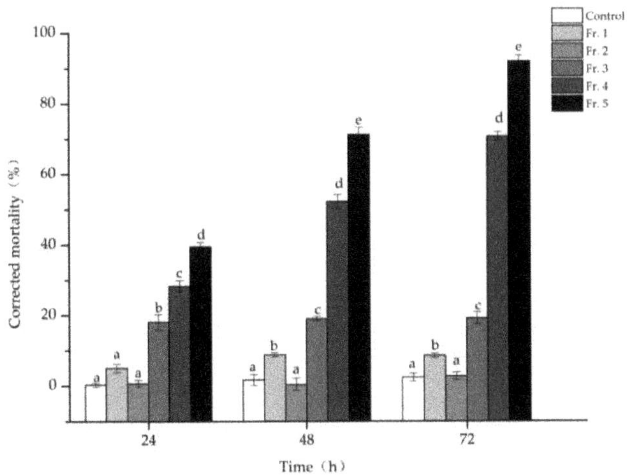

Figure 3. Corrected mortality of PWNs for different fractions. PWNs were treated with 200 μg/mL Fr. 1 - Fr. 5 effluent at 24 h, 48 h, and 72 h, respectively, and the corrected mortality was determined. Mean values were plotted with standard deviations (SD). The control group only contained NB medium. Means with the different letters were significantly different at $p < 0.05$ based on one-way ANOVA with multiple comparison analysis using the LSD test.

The nematicidal components in Fr.4 and Fr.5 were further purified and identified. The compound **1** derived from Fr. 4 was a pale-yellow crystal with a molecular weight of 134.037. Its molecular formula was determined to be $C_8H_6O_2$; ^{1}H NMR (600 MHz, $CDCl_3$): δ3.73 (2H, s, H-3), 7.25–7.31 (2H, m, H-4, 6), 7.08–7.14 (2H, m, H-5, 7); ^{13}C NMR ($CDCl_3$): δ33.07 (C-3), 110.88 (C-7), 123.13 (C-5), 124.18 (C-4), 124.72 (C-9), 128.98 (C-6), 154.82 (C-8), and 174.15 (C=O). Based on the spectrum analysis and comparison with the literature [33,34], compound **1** was identified as 2-coumaranone (Figure 4a).

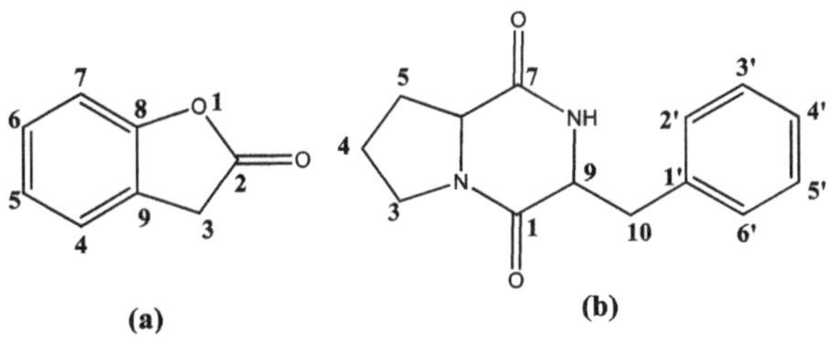

Figure 4. Chemical structures of 2-coumaranone (**a**) and cyclo-(Phe-Pro) (**b**).

Compound **2** purified from Fr.5 was a white powder with a relative molecular mass of 224.121, with the chemical formula of $C_{14}H_{16}N_2O_2$. ^{1}H NMR (600 MHz, $CDCl_3$): δ7.22–7.34 (5H, m, H-2, 3, 4, 5′, 6′), 4.26 (1H, m, H-6), 4.07 (1H, H-9), 3.65, 3.59 (2H, m, H-3), 3.54, 2.78 (2H, m, H-10), 2,33, 1.88 (each 1H, m, H-5), 2.01(2H, m, H-4). ^{13}C NMR ($CDCl_3$): δ169.50 (C-7), 165.15 (C-1), 136.01(C-1), 129.38 (C-3′, 5′), 129.20 (C-2′, 6′), 127.66 (C-4′), 59.22 (C-9), 56.28 (C-6), 45.55 (C-3), 36.87 (C-10), 28.44 (C-5), and 22.63 (C-4). According to EIMS and NMR data combined with the related literature, compound **2** was determined to be cyclo-(Pro-Phe) [35,36] (Figure 4b).

3.3. Nematicidal Activity against PWN of the Nematicidal Compounds

To determine the nematicidal activity, the semi-lethal concentrations against PWN at 24 h for 2-coumaranone and cyclo-(Pro-Phe) were detected. The LC_{50} values of 2-

coumaranone and cyclo-(Phe-Pro) were 196 and 425 mM, respectively, which indicated that 2-coumaranone had a stronger lethal effect against PWN (Table 3).

Table 3. LC_{50} values of 2-coumaranone and cyclo-(Phe-Pro) against PWN for 24 h.

	LC_{50} (mM)	95% Confidence Intervals	Probit Equation
Cyclo-(Pro-Phe)	0.425	0.374–0.483	y = 2.788x − 5.663
2-Coumaranone	0.196	0.167–0.233	y = 2.128x − 3.024

3.4. *Effect of the Nematicides on PWN Egg Hatching*

After being treated with 2-coumaranone and cyclo-(Phe-Pro) at concentrations of 0.1, 0.2, and 0.5 mM for 48 h, the hatching of eggs for PWN was significantly inhibited compared to the control ($p < 0.05$) (Figure 5). The hatching rates for eggs treated with 0.1 mM 2-coumaranone and cyclo-(Pro-Phe) were 50.77% and 59.92%, respectively.

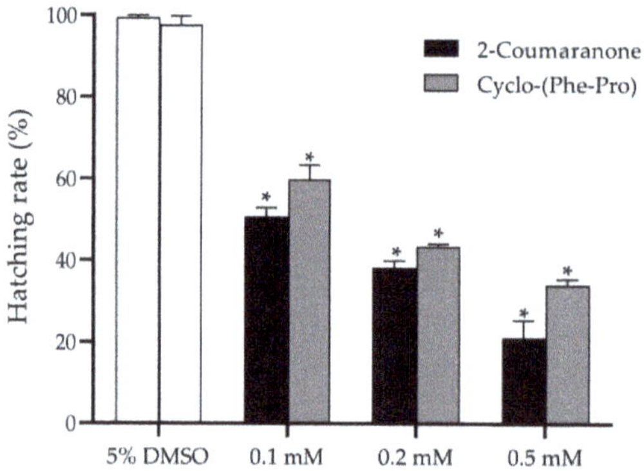

Figure 5. Effects of nematicidal compounds on cumulative hatching rates after 48 h. Mean values of four replications were plotted with SD. The asterisk denoted significant difference at $p < 0.05$ compared to controls according to Student's *t*-test.

3.5. *Influence of the Nematicides on PWN Feeding and Population*

To investigate the nematicidal effect on reproduction, PWNs were treated with 0.1 mM 2-coumaranone and cyclo-(Phe-Pro) for 24 h and then inoculated into *B. cinerea*. In contrast to the depleted mycelia in the control groups treated with 5% DMSO and sterile water at the end of 7 d, most of the mycelia was left in the treated groups (Figure 6). Specifically, the little mycelia of *B. cinerea* was fed in the 2-coumaranone-treated group. Therefore, the nematicidal compounds significantly reduced the feeding ability of the nematodes.

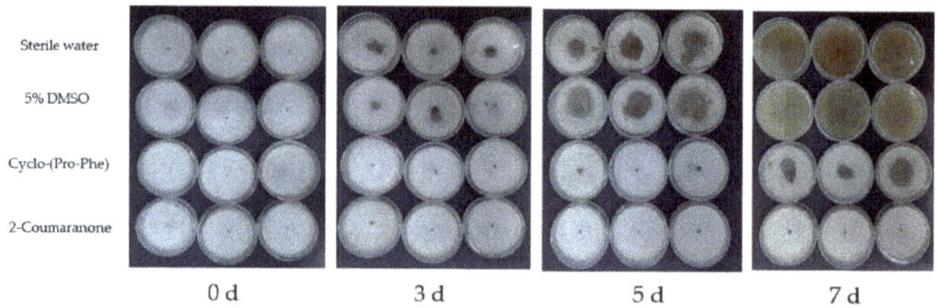

Figure 6. The feeding of *Botrytis cinerea* in Petri dishes after inoculation with different PWNs.

Moreover, the nematode populations were 270 and 3400 in the 2-coumaranone and cyclo-(Pro-Phe)-treated groups, respectively, which were significantly lower ($p < 0.01$) than that of 10,080 and 10,400 in the DMSO and sterile water-treated groups (Figure 7). The inhibition effect on reproduction was significantly higher for 2-coumaranone than that of cyclo-(Pro-Phe) ($p < 0.01$).

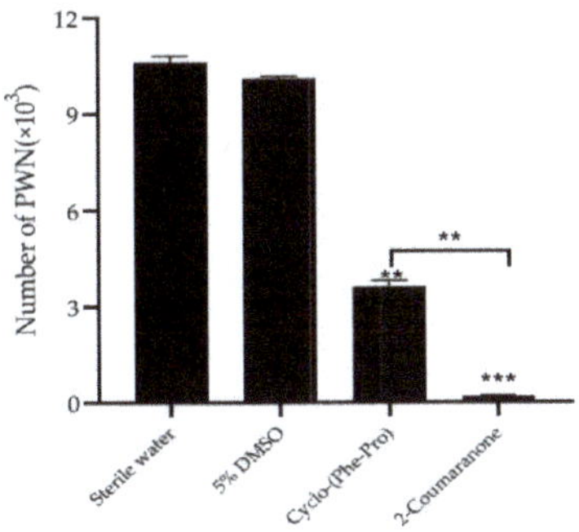

Figure 7. Numbers of PWN per dish treated with 2-coumaranone, cyclo-(Phe-Pro), 5% DMSO, and sterile water. Mean values of four replications were plotted with SD. ** and *** represent significant difference at $p < 0.01$ and $p < 0.001$, respectively, compared to control according to Student's *t*-test.

3.6. Detection of ROS and Antioxidant Enzymes in PWN

Fluorescence was revealed in the PWNs treated with 2-coumaranone and cyclo-(Pro-Phe) based on DCFH-DA fluorescence staining (Figure 8A–D), which indicated that the two nematicides stimulated the production of ROS in PWN. Furthermore, the activities of CAT and POD in nematodes treated with the nematicidal compounds were significantly lower, while the SOD activity was significantly higher, compared to that in the sterile water and 5% DMSO-treated groups (Figure 9). The results indicate that the accumulation of a large amount of H_2O_2 in the nematodes eventually led to their deaths.

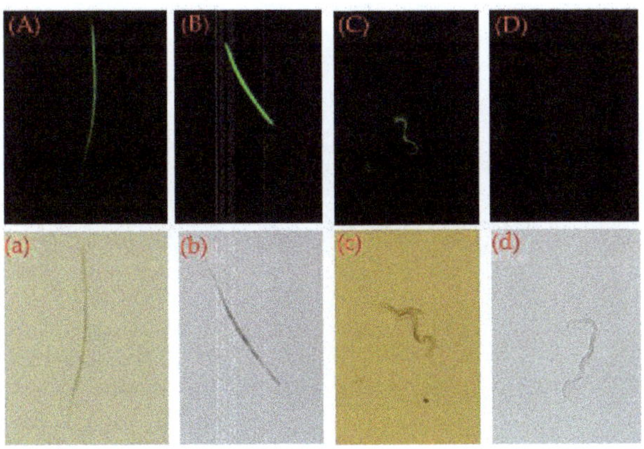

Figure 8. Reactive oxygen fluorescence staining of *B. xylophilus*. *B. xylophilus* was imaged in the darkfield (**A–D**) and the corresponding brightfield (**a–d**). The a–d groups were treated with 0.2 mM 2-coumaranone, 0.2 mM cyclo-(Phe-Pro), reactive oxygen hydrogen donor, and 5% DMSO, respectively.

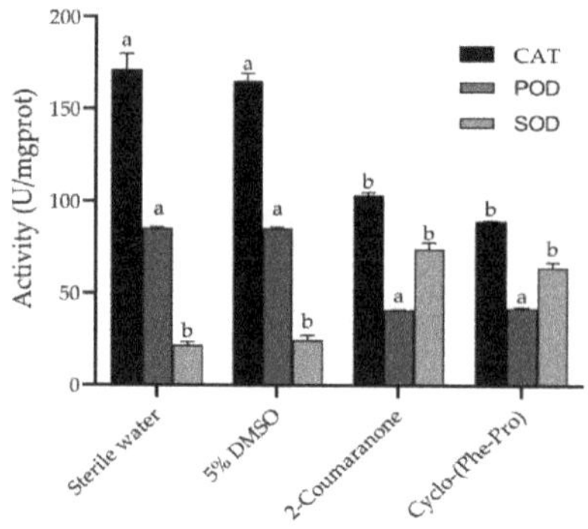

Figure 9. Activity of antioxidant enzymes of PWN. Means with the different superscript letters were significantly different at $p < 0.05$ according to LSD test.

3.7. Nematicidal Activity of the Nematicidal Compounds against PWN In Vivo

To understand the nematicidal effect of 2-coumaranone and cyclo-(Pro-Phe) in vivo, the PWN-infected Japanese black pine seedlings were treated with the two compounds. On the ninth day after inoculation with PWN, wilting symptoms of seedlings and dehydration of pine needles were observed in the group treated with PWN + 5% DMSO (Figure 10a(A)), while no obvious change was observed in the groups treated with PWN + 2-coumaranone and PWN + cyclo-(Pro-Phe) (Figure 10a(B,C)). On day 12, the pine seedlings treated with PWN + 5% DMSO almost dried up and died (Figure 10b(A)). However, the seedlings treated with PWN + nematicides only showed slight wilting (Figure 10b(B,C)), which indicated that the treatment with 2-coumaranone and cyclo-(Pro-Phe) effectively reduced the wilting symptoms of the Japanese pine seedlings caused by PWN. In contrast with the control group just treated with sterile water (Figure 10a(D),b(D)), no visible side effects were found in seedlings only treated with the nematicides (Figure 10a(E,F),b(E,F)).

Figure 10. Effect of 2-coumaranone and cyclo-(Phe-Pro) on inhibit PWN in vivo. A–F groups were treated with 5% DMSO + PWN, 2-coumaranone + PWN, cyclo-(Pro-Phe) + PWN, 5% DMSO, 50 mM 2-coumaranone, and 50 mM cyclo-(Pro-Phe), respectively. (**a**) Pine seedling symptoms on day 9 in different treatment groups; (**b**) pine seedling symptoms on day 12 in different treatment groups.

4. Discussion

Numerous species of microorganisms were tried in the biological control of PWD. Kang et al. [37] screened a nematicidal strain, *Streptomyces* sp. 680,560 among endophytic actinomycetes of Korean pines, and isolated a strongly active compound, teleocidin B4, from its metabolites, which had a significant inhibitory effect on the egg hatching and development of PWN. *Bacillus thuringiensis* was confirmed to produce endotoxins with strong nematicidal activity against PWN [38,39]. However, few of them were reported to be effective in pine forest tests, and no nematicidal bacteria were reported to be screened from the rhizosphere of pine trees. The microorganisms in the rhizosphere soil of pine trees differ greatly during different periods of infection with PWD, and some of these bacteria might have antagonistic effects against PWD [16]. Therefore, we tried to isolate bacteria with nematicidal activity from healthy Japanese black pine rhizosphere in the current study in order to obtain bioactive compounds as nematicidal candidates against PWN. A nematicidal bacterium, *Lysinimonas* sp M4, was isolated and the antagonistic effect against PWN of this genus was first reported. Several bacterial species were found to have nematicidal activity, among which *Bacillus* genus was the most reported. Some bacteria were confirmed to kill PWN through producing toxins or secreting extracellular proteases to digest nematode body surface tissues, and some others control PWN by interfering with nematode behavior or preying on nematodes [37–42]. We isolated two nematicides against PWN, 2-coumaranone, and cyclo-(Pro-Phe), from the fermentation broth of *Lysinimonas* sp. M4, and several strains of this genus were recorded to be resistant to metal ion stress [43–45], which implies the potential of this genus in PWN control.

The nematicidal compound cyclo-(Phe-Pro), belonging to the cyclic dipeptides (2,5-diketopiperazine), was identified in the present study. Park et al. [35] investigated the nematicidal metabolites of *Bacillus thuringiensis* JCK-1233, and found cyclo-(L-Pro-L-Phe) displayed an inhibitory effect on PWN in pine trees. However, the mechanism of the nematicidal effect of cyclic dipeptides was not further studied in their research. Lee et al. [46] revealed that cyclo-(Phe-Pro) could induce DNA damage in mammalian cells via ROS accumulation, which eventually causes apoptosis. In our research, cyclo-(Phe-Pro), as well as 2-coumaranone, were confirmed to cause the accumulation of ROS in nematodes, which could partially explain the nematicidal mechanism of the two nematicides. Cyclic dipeptides can be produced by bacteria, fungi, plants, and animals in nature [47,48], and have various biological functions, such as being antifungal, antibacterial, antiviral, antitumor, and cell signaling [49,50]. The cyclo-(Phe-Pro) secreted by the bacteria from the rhizosphere of pine trees might help pine trees to inhibit PWN reproduction and relieve pathogenicity symptoms of PWD [51]. Moreover, 2-coumaranone with strong nematicidal activity in the metabolites of *Lysinimonas* sp. M4 has nematicidal activity against PWN, which was not reported. The agent has potential to be developed as a nematicide or a lead compound for a more powerful nematicide by further modification in PWD management on a large scale.

ROS includes superoxide anion ($\cdot O^{2-}$), hydrogen peroxide (H_2O_2), and hydroxyl radicals ($\cdot OH$), which play key roles in cell metabolism, oxidative stress, signal transduction, and apoptosis [52]. H_2O_2 is mainly produced by SOD, while CAT and POD can remove H_2O_2 [53]. Analysis on the aforementioned antioxidant enzymes of the nematicide-treated nematodes in vivo revealed that the activities of CAT and POD were significantly reduced compared to the control, while the activity of SOD was significantly increased. According to the result, we inferred that excessive H_2O_2 production might lead to the death of pine nematodes. The levels of other antioxidant enzymes, such as glutathione peroxidases (GPX), are worthy of further study for further elucidating the mechanism of the nematicides isolated from *Lysinimonas* sp. M4 [54].

5. Conclusions

In this study, *Lysinimonas* sp. M4 with nematicidal activity was isolated from microorganisms from the rhizosphere soil of Japanese black pine, and two nematicidal compounds, 2-coumaranone and cyclo-(Phe-Pro), were purified from its culture. Both of the nematicidal compounds significantly inhibited egg hatching, feeding, and reproduction of PWN. The two nematicides caused the accumulation of ROS in PWNs, as well as a significant decrease in CAT and POS activity and an increase in SOD activity in PWNs, which might contribute to their nematicidal activity against PWN. Furthermore, both 2-coumaranone and cyclo-(Phe-Pro) could effectively reduce the wilt symptoms of Japanese black pine seedlings caused by PWN inoculation. This study provides a promising novel bacterial strain and two derived nematicidal compounds for controlling PWD.

Author Contributions: Conceptualization, Y.S., Q.G., Q.X., C.W., G.D. and R.L.; methodology, Y.S., C.W., G.D., R.L. and Q.G.; software, Y.S., R.L.; validation, Y.S., R.L., Q.X. and Q.G.; formal analysis, Y.S., G.D. and Q.G.; investigation, Y.S., W.D., H.Y.; resources, C.W., H.Y., G.D. and Q.G.; data curation, W.D., R.L.; writing—original draft preparation, Y.S., G.D., Q.G.; writing—review and editing, R.L., Q.G., W.D., Q.X.; visualization, C.W., G.D., Q.G., W.D.; supervision, Q.X., Q.G. and G.D.; project administration, Q.G. and Q.X.; funding acquisition, Q.G. and R.L. All authors have read and agreed to the published version of the manuscript.

Funding: This research was funded by Natural Science Foundation of Shandong Province, grant number ZR2020MC123.

Institutional Review Board Statement: Not applicable.

Informed Consent Statement: Not applicable.

Conflicts of Interest: The authors declare no conflict of interest.

References

1. Mamiya, Y. Pathology of the Pine Wilt Disease Caused by *Bursaphelenchus xylophilus*. *Annu. Rev. Phytopathol.* **1983**, *21*, 201–220. [CrossRef] [PubMed]
2. Proenca, D.N.; Grass, G.; Morais, P.V. Understanding pine wilt disease: Roles of the pine endophytic bacteria and of the bacteria carried by the disease-causing pinewood nematode. *MicrobiologyOpen* **2017**, *6*, e00415. [CrossRef] [PubMed]
3. Tóth, Á. *Bursaphelenchus xylophilus*, the pinewood nematode: Its significance and a historical review. *Acta Biol. Szeged.* **2011**, *55*, 213–217.
4. Kim, B.; Ji, H.; Ahn, J.; Kim, S.; Min, J. A short review of the pinewood nematode, *Bursaphelenchus xylophilus*. *Toxicol. Environ. Health Sci.* **2020**, *12*, 1–8. [CrossRef]
5. Lee, S.M.; Chung, Y.J.; Moon, Y.S.; Lee, S.G.; Lee, C.K. Insecticidal activity and fumigation conditions of several insecticides against Japanese pine sawyer (*Monochamus alternatus*) larvae. *J. Korean For. Soc.* **2003**, *92*, 191–198.
6. Liu, M.J.; Hwang, B.S.; Jin, C.Z.; Li, W.J.; Park, D.J.; Seo, S.T.; Kim, C.J. Screening, isolation and evaluation of a nematicidal compound from actinomycetes against the pine wood nematode, *Bursaphelenchus xylophilus*. *Pest Manag. Sci.* **2019**, *75*, 1585–1593. [CrossRef]
7. Choi, H.Y.; Van Minh, N.; Choi, J.M.; Hwang, J.Y.; Seo, S.T.; Lee, S.K.; Kim, W.G. Enzymatic synthesis of avermectin B(1a) glycosides for the effective prevention of the pine wood nematode *Bursaphelenchus xylophilus*. *Appl. Microbiol. Biotechnol.* **2018**, *102*, 2155–2165. [CrossRef] [PubMed]
8. Wang, H.; Wang, Y.; Yin, C.; Gao, J.; Sung, C. In vivo infection of *Bursaphelenchus xylophilus* by the fungus *Esteya vermicola*. *Pest Manag. Sci.* **2020**, *76*, 8254–8264. [CrossRef]
9. Li, Y.; Yu, H.; Araujo, J.; Zhang, X.; Ji, Y.; Hulcr, J. *Esteya floridanum* sp. nov.: An Ophiostomatalean Nematophagous Fungus and Its Potential to Control the Pine Wood Nematode. *Phytopathology* **2021**, *111*, 304–311. [CrossRef]
10. He, L.X.; Wu, X.Q.; Xue, Q.; Qiu, X.W. Effects of Endobacterium (*Stenotrophomonas maltophilia*) on Pathogenesis-Related Gene Expression of Pine Wood Nematode (*Bursaphelenchus xylophilus*) and Pine Wilt Disease. *Int. J. Mol. Sci.* **2016**, *17*, 778. [CrossRef]
11. Yu, J.; Du, G.C.; Li, R.G.; Li, L.; Li, Z.; Zhou, C.J.; Chen, C.C.; Guo, D.S. Nematicidal activities of bacterial volatiles and components from two marine bacteria, *Pseudoalteromonas marina* strain H-42 and *Vibrio atlanticus* strain S-16, against the pine wood nematode, *Bursaphelenchus xylophilus*. *Nematology* **2015**, *17*, 1011–1025. [CrossRef]
12. Ponpandian, L.N.; Rim, S.O.; Shanmugam, G.; Jeon, J.; Park, Y.H.; Lee, S.K.; Bae, H. Phylogenetic characterization of bacterial endophytes from four *Pinus* species and their nematicidal activity against the pine wood nematode. *Sci. Rep.* **2019**, *9*, 12457–12468. [CrossRef] [PubMed]
13. Ramadasappa, S.; Rai, A.K.; Jaat, R.S.; Singh, A.; Rai, R. Isolation and screening of phlD (+) plant growth promoting rhizobacteria antagonistic to *Ralstonia solanacearum*. *World J. Microbiol. Biotechnol.* **2012**, *28*, 1681–1690. [CrossRef] [PubMed]
14. Raaijmakers, J.M.; Weller, D.M. Exploiting genotypic diversity of 2,4-diacetylphloroglucinol-producing *Pseudomonas* spp.: Characterization of superior root-colonizing *P. fluorescens* strain Q8r1-96. *Appl. Environ. Microbiol.* **2001**, *67*, 2545–2554. [CrossRef] [PubMed]
15. Siddiqui, I.A.; Shaukat, S.S. Plant species, host age and host genotype effects on *Meloidogyne* incognita biocontrol by *Pseudomonas fluorescens* strain CHA0 and its genetically-modified derivatives. *J. Phytopathol.-Phytopathol. Z.* **2003**, *151*, 231–238. [CrossRef]
16. Liu, Z.; Budiharjo, A.; Wang, P.; Shi, H.; Fang, J.; Borriss, R.; Zhang, K.; Huang, X. The highly modified microcin peptide plantazolicin is associated with nematicidal activity of *Bacillus amyloliquefaciens* FZB42. *Appl. Microbiol. Biotechnol.* **2013**, *97*, 10081–10090. [CrossRef]

17. Deng, J.; Yu, D.; Zhou, W.; Zhou, L.; Zhu, W. Variations of phyllosphere and rhizosphere microbial communities of *Pinus koraiensis* infected by *Bursaphelenchus xylophilus*. *Microb. Ecol.* **2021**, *84*, 285–301. [CrossRef]
18. Dong, W.Y.; Du, G.C.; Zhang, T.T.; Yang, H.; Gou, Q.Q.; Li, R.G. Bacterial Diversity and Pathogenicity of Pine Wood Nematode (*Bursaphelenchus xylophilus*) from Two Diff erent Sources. *Chaing Mai J. Sci.* **2021**, *49*, 1–15. [CrossRef]
19. Elamary, R.; Salem, W. Optimizing and purifying extracellular amylase from soil bacteria to inhibit clinical biofilm-forming bacteria. *PeerJ* **2020**, *8*, e10288. [CrossRef] [PubMed]
20. Zhang, Y.; Yang, D.; Li, Z.; Hao, F.; Zhang, L.; Li, D. Screening and prevention effect of new and effective pesticides against pine wilt nematode *Bursaphelenchus xylophilus*. *Acta Phytophy. Sin.* **2017**, *44*, 856–862.
21. Wang, J.; Pan, X.; Han, Y.; Guo, D.; Guo, Q.; Li, R. Rosmarinic acid from eelgrass shows nematicidal and antibacterial activities against pine wood nematode and its carrying bacteria. *Mar. Drugs* **2012**, *10*, 2729–2740. [CrossRef] [PubMed]
22. Zhang, W.; Wu, X.; Wang, Y. *Bursaphelenchus xylophilus* and Its Fermentation and Culture Characteristics. *Biotechnol. Bull.* **2019**, *35*, 76–82.
23. Liu, D.W.; Liu, D.; Liu, L.; Wang, Y.; Zhang, Y.J. Screening and identification of antagonistic bacteria from vermicompost against *Fusarium oxysporum f.* sp. cucumerinum. *Acta Agric. Scand. Sect. B-Soil Plant Sci.* **2021**, *71*, 266–272. [CrossRef]
24. Sudhir, K.; Stecher, G.; Tamura, K. MEGA7: Molecular evolutionary genetics analysis version 7.0 for bigger datasets. *Mol. Biol. Evol.* **2016**, *33*, 1870–1874.
25. Wang, F.; Guo, Q.; Wang, L.; Ma, Y.; Zhang, T.; Li, R. Nematicidal Activities of Bacterial Volatiles from *Pseudoduganella violaceinigra* G5-3 and *Novosphingobium pokkalii* G8-2 against the Pine Wood Nematode *Bursaphelenchus xylophilus*. *Chiang Mai J. Sci.* **2019**, *46*, 236–246.
26. Jiang, L.; Ma, C.; Wang, S.; Zhou, M.; Wei, C. Effect of separation and preparation and of avermectin B2 by silica gel column chromatography. *Acta Agric. Shanghai* **2002**, *18*, 81–83.
27. Choi, I.H.; Kim, J.; Shin, S.C.; Park, I.K. Nematicidal activity of monoterpenoids against the pine wood nematode (*Bursaphelenchus xylophilus*). *Russ. J. Nematol.* **2007**, *15*, 35–40.
28. Xue, Q.; Wu, X.Q. Characteristics and function of a novel cystatin gene in the pine wood nematode *Bursaphelenchus xylophilus*. *Biol. Open* **2019**, *8*, bio042655. [CrossRef] [PubMed]
29. Wang, L.; Qin, Y.; Fan, Z.; Gao, K.; Zhan, J.; Xing, R.; Liu, S.; Li, P. Novel Lead Compound Discovery from Aspergillus fumigatus 1T-2 against *Meloidogyne incognita* Based on a Chemical Ecology Study. *J. Agric. Food Chem.* **2022**, *70*, 4644–4657. [CrossRef]
30. Mendez-Gomez, M.; Castro-Mercado, E.; Alexandre, G.; Garcia-Pineda, E. Superoxide anion production in the interaction of wheat roots and rhizobacteria *Azospirillum brasilense* Sp245. *Plant Soil* **2016**, *400*, 55–65. [CrossRef]
31. Bettaieb, T.; Mahmoud, M.; De Galarreta, J.I.R.; Du Jardin, P. Relation between the low temperature stress and catalase activity in gladiolus somaclones (Gladiolus grandiflorus Hort.). *Sci. Hortic.* **2007**, *113*, 49–51. [CrossRef]
32. Oku, H.; Shiraishi, T.; Ouchi, S.; Kurozumi, S.; Ohta, H. Pine wilt toxin, the metabolite of a bacterium associated with a nematode. *Naturwissenschaften* **1980**, *67*, 198–199. [CrossRef]
33. Lévai, A.; Silva, A.M.; Cavaleiro, J.A.; Patonay, T.; Silva., V.L. Dimethyldioxirane Oxidation of (E,E)-Cinnamylideneacetophenones. *Eur. J. Org. Chem.* **2001**, *17*, 3213–3219.
34. Venkateswarlu, S.; Panchagnula, G.K.; Guraiah, M.B.; Subbaraju, G.V. Isoaurones: Synthesis and stereochemical assignments of geometrical isomers. *Tetrahedron* **2006**, *62*, 9855–9860. [CrossRef]
35. Park, A.; Jeong, S.; Jeon, H.; Kim, J.; Kim, N.; Ha, M.T.; Mannaa, M.; Kim, J.; Lee, C.W.; Min, B.S.; et al. A Diketopiperazine,Cyclo-(L-Pro-L-Ile), Derived from *Bacillus thuringiensis* JCK-1233 Controls Pine Wilt Disease by Elicitation of Moderate Hypersensitive Reaction. *Front. Plant Sci.* **2020**, *11*, 1023–1037. [CrossRef]
36. Strom, K.; Sjogren, J.; Broberg, A.; Schnurer, J. Lactobacillus plantarum MiLAB 393 produces the antifungal cyclic dipeptides cyclo(L-Phe-L-Pro) and cyclo(L-Phe-trans-4-OH-L-Pro) and 3-phenyllactic acid. *Appl. Environ. Microbiol.* **2002**, *68*, 4322–4327. [CrossRef]
37. Bravo, A.; Gill, S.S.; Soberon, M. Mode of action of Bacillus thuringiensis Cry and Cyt toxins and their potential for insect control. *Toxicon* **2007**, *49*, 423–435. [CrossRef]
38. Wei, J.Z.; Hale, K.; Carta, L.; Platzer, E.; Wong, C.; Fang, S.C.; Aroian, R.V. Bacillus thuringiensis crystal proteins that target nematodes. *Proc. Natl. Acad. Sci. USA* **2003**, *100*, 2760–2765. [CrossRef]
39. Hu, H.J.; Gao, Y.; Li, X.; Chen, S.L.; Yan, S.Z.; Tian, X.J. Identification and Nematicidal Characterization of Proteases Secreted by Endophytic Bacteria *Bacillus cereus* BCM2. *Phytopathology* **2020**, *110*, 336–344. [CrossRef]
40. Luo, T.; Hou, S.; Yang, L.; Qi, G.; Zhao, X. Nematodes avoid and are killed by *Bacillus mycoides*-produced styrene. *J. Invertebr. Pathol.* **2018**, *159*, 129–136. [CrossRef]
41. Chan, S.Y.; Liu, S.Y.; Seng, Z.J.; Chua, S.L. Biofilm matrix disrupts nematode motility and predatory behavior. *ISME J.* **2021**, *15*, 260–269. [CrossRef]
42. Phani, V.; Somvanshi, V.S.; Rao, U. Silencing of a Meloidogyne incognita selenium-binding protein alters the cuticular adhesion of Pasteuria penetrans endospores. *Gene* **2018**, *677*, 289–298. [CrossRef]
43. Tan, W.; Li, Y.; Guo, F.; Wang, Y.; Zhang, X.W. Effect of *Leifsonia* sp. on retardation of uranium in natural soil and its potential mechanisms. *J. Environ. Radioact.* **2020**, *217*, 106202–106210. [CrossRef]
44. Trinh, T.; Wang, S.; Nguyen, V.B.; Tran, M.; Nguyen, A. A potent antifungal rhizobacteria *Bacillus velezensis* RB.DS29 isolated from black pepper (*Piper nigrum* L.). *Res. Chem. Intermed.* **2019**, *45*, 5309–5323. [CrossRef]

45. Egidi, E.; Wood, J.; Fox, E.; Liu, W.; Franks, A. Draft genome sequence of *Leifsonia* sp. strain NCR5, a rhizobacterium isolated from cadmium-contaminated soil. *Microbiol. Resour. Announc.* **2017**, *5*, e00520-17. [CrossRef]
46. Lee, K.; Jeong, J.; Kim, I.; Kim, K.; Ju, B. Cyclo(phenylalanine-proline) induces DNA damage in mammalian cells via reactive oxygen species. *J. Cell. Mol. Med.* **2015**, *19*, 2851–2864. [CrossRef]
47. Prasad, C. Bioactive cyclic dipeptides. *Peptides* **1995**, *16*, 151–164. [CrossRef]
48. Huang, R.; Zhou, X.; Xu, T.; Yang, X.; Liu, Y. Diketopiperazines from marine organisms. *Chem. Biodivers.* **2010**, *7*, 2809–2829. [CrossRef]
49. De Carvalho, M.P.; Abraham, W.R. Antimicrobial and biofilm inhibiting diketopiperazines. *Curr. Med. Chem.* **2012**, *19*, 3564–3577. [CrossRef]
50. Kumar, S.N.; Nambisan, B.; Mohandas, C. Purification and identification of two antifungal cyclic dipeptides from *Bacillus cereus* subsp. thuringiensis associated with a rhabditid entomopathogenic nematode especially against Fusarium oxysporum. *J. Enzyme Inhib. Med. Chem.* **2014**, *29*, 190–197. [CrossRef]
51. Cai, S.; Jia, J.; He, C.; Zeng, L.; Fang, Y.; Qiu, G.; Lan, X.; Su, J.; He, X. Multi-Omics of Pine Wood Nematode Pathogenicity Associated With Culturable Associated Microbiota Through an Artificial Assembly Approach. *Front Plant Sci.* **2021**, *12*, e798539. [CrossRef] [PubMed]
52. Fan, J.; Xie, Y.; Xue, J.; Li, B. Isolation and identification of toxins inhibiting Dentrolimus tabulaeformis from an antagonistic strain of Beauveria. *Wei Sheng Wu Xue Bao* **2008**, *48*, 596–601. [PubMed]
53. Zhang, W.; Zhao, L.; Zhou, J.; Yu, H.; Zhang, C.; Lv, Y.X.; Lin, Z.; Hu, S.N.; Zou, Z.; Sun, J.H. Enhancement of oxidative stress contributes to increased pathogenicity of the invasive pine wood nematode. *Philos. Trans. R. Soc. B-Biol. Sci.* **2019**, *374*, 20180323. [CrossRef]
54. Fu, Y.M.; Liu, H.B.; Wu, X.Q. Diversity and function of endo-bacteria in *Bursaphelenchus xylophilus* from *Pinus massoniana* Lamb. in different regions. *Forests* **2020**, *11*, 487. [CrossRef]

forests

MDPI

Article

Discovery and Biology of *Spathius verustus* Chao (Hymenoptera: Braconidae), a Potential Biological Agent on Two *Monochamus* Vectors of the Pinewood Nematode

Moo-Sung Kim [1], Jong-Kook Jung [2], Ki-Jeong Hong [3], Chang-Jun Kim [4], Bong-Woo Lee [1] and Il-Kwon Kim [1,*]

[1] Division of Forest Biodiversity, Korea National Arboretum, Pocheon 11186, Korea; cindy5138@korea.kr (M.-S.K.); eucosma@korea.kr (B.-W.L.)
[2] College of Forest and Environmental Science, Kangwon National University, Chuncheon 24341, Korea; jkjung@kangwon.ac.kr
[3] Department of Plant Medicine, Sunchon National University, Suncheon 57922, Korea; curcul@sunchon.ac.kr
[4] Research Planning and Coordination Team, Korea National Arboretum, Pocheon 11186, Korea; changjunkim@korea.kr
* Correspondence: ilkwons91@korea.kr; Tel.: +82-33-480-3021

Abstract: Pine wilt disease in Korea can be managed by reducing vector density through chemical application. To reduce the side effects from pesticides, effective natural enemies must be identified and used to reduce the vectors' natural density. Sentinel logs were used to investigate a parasitoid wasp parasitic to *Monochamus alternatus* and *Monochamus saltuarius*, the vectors of *Bursaphelenchus xylophilus*, which causes this disease. During 2016–2017, the parasitoid wasps distributed in the *Pinus densiflora* and *Pinus koraiensis* forests were surveyed using sentinel logs at six different areas. An ectoparasitoid wasp *Spathius verustus* Chao of *M. alternatus* and *M. saltuarius* was identified. We showed for the first time that *S. verustus* was parasitic to *M. alternatus* in South Korea and to *M. saltuarius* worldwide. The parasitism rates were affected by region and session but not by the sentinel log height in the tree and the distances between the trap and forest edge locations. Studies also showed that *S. verustus* appeared to prefer *M. alternatus* to *M. saltuarius* as a host. This study unveiled the ecological details of *S. verustus*. Further research on various environmental factors such as regional differences, host density differences, and the degree of damage from the pine wilt disease is required to understand the effects of environmental or ecological factors on parasitism rates.

Keywords: biological control agent; ectoparasitoid; Braconidae; *Monochamus* vector; pine wilt disease; parasitism rate

1. Introduction

Pinewood nematode (PWN), *Bursaphelenchus xylophilus* (Steiner & Buhrer), native to North America, causes pine wilt disease (PWD) in East Asia (Japan, China, Taiwan, and South Korea), Mexico, and Europe (Portugal and Spain), inflicting serious damage on pine forests around the world [1–8]. The first damage on pines in Korea was reported by Yi et al. [9]. The pinewood nematode is known to infest various species of *Pinus*, *Abies*, *Chamaecyparis*, *Cedrus*, *Larix*, *Picea*, and *Pseudotsuga*, all of which belonged to Pinales [10]. The hosts of the pinewood nematode in Korea include *P. densiflora* (Siebold & Zucc), *P. koraiensis* (Siebold & Zucc), and *P. thunbergii* (Parl), as well as *P. parviflora* Siebold & Zucc in some insular areas [11]. The pinewood nematode spreads among its hosts by using wood-boring beetles such as long-horned beetles from the genus *Monochamus* as vectors [12–14].

The representative insect vectors of PWN are, in particular, the species of *Monochamus* in Cerambycidae, and there are eight species known to the Korean insect fauna [15]. Of them, two native species, that have never had a relationship with PWN, have come to transmit PWN in the country since the invasion: *M. saltuarius* Gebler as the vector of PWN in the central region and *M. alternatus* Hope in the southern region of the country [16]. The pinewood nematode, transmitted by *M. alternatus*, has been causing immense damage to coniferous forests in Jeju Island since 2013 [17]. PWD causes a higher mortality rate than the pine caterpillar (*Dendrolimus spectabilis* Butler) or pine needle gall midge (*Thecodiplosis japonensis* Uchida & Inouye), causing great economic damage to the Korean forestry industry. The area affected by PWD has increased from 72 ha in 1988 to 6325 ha in 2016 [18].

Currently, the methods used in Korea to manage PWD include spraying insecticide using unmanned aerial vehicles, fumigation or incineration after harvesting damaged trees, and clear-felling around damaged trees [19,20]. However, these control methods negatively affect the environment and require a large budget and human capital for the processes of scientific forecasting with manned and unmanned aerial vehicles, as well as restoring forests [21,22]. Additionally, Jeon et al. [23] argued that clear-felling would further spread PWD. One of the ways to overcome these drawbacks to certain degrees is biological control using natural enemies to reduce the density of the insect vectors.

There have been many attempts to use natural enemies, such as fungus, parasitoids, or predators, to control populations of cerambycid vectors. For example, *M. alternatus* also transmits the pinewood nematode in China [24], where, to prevent the disease, *Sclerodermus harmandi* (Buysson) (Hymenoptera: Bethylidae) was used as a biological control agent (parasitism rate from 42.8 to 46.1%; average control effectiveness of 65.3%) [25]. There also has been a study on the indoor mass-rearing of *S. harmandi* using *Tenebrio molitor* Linnaeus as an alternative host [26]. As in South Korea, *M. alternatus* and *M. saltuarius* also transmit the pinewood nematode in Japan [27,28]. Several studies have addressed biological control methods such as using an entomopathogenic fungus, particularly *Beauveria bassiana* (Bals.—Criv.) Vuill., hymenopteran parasitoids, and coleopteran predators [29,30].

In Korea, research on the biological control of PWD has been conducted since 2004 [31] and highlighted the biological characteristics of *S. harmandi* as a natural enemy of *M. alternatus* and *M. saltuarius*. In indoor tests using small glass vials in a controlled laboratory environment, the parasitism on *M. saltuarius* larvae was as high as 98.6%, and it was selected for mass-rearing to reduce the density of *M. alternatus* and *M. saltuarius* [32]. However, there has been no report on the effectiveness of the bethylid wasp from outdoor release, if any, so far [33]. The reason for not having any report on actual usage of the bethylid wasps seems to be related to the host range. According to Lim et al. [34], the hosts of *S. harmandi* include more than 30 species of 9 insect families, which means the parasitoid is not a specialist parasitoid to the targeted cerambycids but a generalist. Further, Liu et al. [35] demonstrated the adverse effects on an important natural enemy (*Triaspis* sp. of Braconidae) of the pine weevil in China and suggested reconsideration of further release of *S. harmandi* for biological control programs. Since the first biological control attempt using *S. harmandi*, there has been little research on natural enemies or developing biological control methods on two *Monochamus* vectors in South Korea. Kim et al. (2010) [36] conducted an extensive study to find the natural enemies of various wood-boring insect pests, and of the surveyed parasitic wasps, two ichneumonid species were reared from prepupae and pupae of *M. saltuarius*. In Portugal, a relatively high diversity of parasitoids was also found associated with *M. galloprovincialis*, the pinewood nematode insect vector, but they are mainly idiobiont ectoparasitoids and generalists that attack a vast array of other insects living in dead and dying trees [37].

Great examples of the most active research on biological control development against wood-boring pests can be found in the U.S. and Europe, in which we can follow their natural enemy survey methods or strategies to develop a biological control program on the insect vectors, which are also wood-boring insects, of PWN. Three coleopteran invasive species inflict tremendous damage on hardwoods in those regions: *Anoplophora chinensis* (Förster) and *A. glabripennis* (Motschulsky) in Cerambycidae, and *Agrilus planipennis* Fairmaire in Buprestidae. Surveys for natural enemies, particularly parasitoids, have been conducted for over ten years to be applied to biological control programs (e.g., [38–41]). Especially, in the U.S., various studies have been carried out using sentinel logs to find natural enemies such as egg, larval, or pupal parasitoids that are suitable for biological control against the three wood-boring insect pests [42–45]. In particular, a biological control plan to stem the infestation of the emerald ash borer (*A. planipennis*) was initiated back in 2003, and groups of scientists from the U.S., China, and Russia searched for effective biological control agents in East Asian regions, from which the invasive pest was originated. Four parasitoids were proven to be effective and released to the fields for biological control since 2007. These parasitoids are *Oobius agrili* Zhang and Huang, an egg parasitoid in Encyrtidae; *Spathius agrili* Yang and S. *galinae* Belokobylskij & Strazanac, two larval parasitoids in Braconidae, and *Tetrastichus planipennisi* Yang, a larval parasitoid in Eulophidae. Several millions of them have been released, and the parasitoid species are found to be established in many states of the U.S. up to date (e.g., [46–49]).

It is imperative to find effective parasitoids that can be used with other control methods such as removing infested trees and injecting nematicides into trunks. Additionally, compared to other control methods, biological control can be cost-effective in the long term after parasitoids are successfully established in released areas. As the first step for biological control development, the present study aimed (1) to find potential parasitoids, particularly on eggs or larvae, of the two *Monochamus* insect vectors of the pinewood nematode using sentinel logs, and (2) to investigate ecological and biological characteristics of parasitoids in prior to the development of a biological control program and strategies.

2. Materials and Methods

2.1. Sentinel Log Preparation

To search for natural enemies of *M. alternatus* and *M. saltuarius*, we adopted the host egg-sentinel log method used by Duan et al. [50]. The sentinel logs used in our study are Korean pine (*P. koraiensis*) logs artificially infested with the two *Monochamus* species [51]. The reasons for selecting the Korean pine for the sentinel logs were that the larvae of *M. saltuarius* were known to have high survival in the Korean pine, and that *M. alternatus* preferred it over the Korean red pine (*P. densiflora*) for oviposition [52,53]. Preparation of the sentinel logs was carried out by (1) cutting freshly chopped Korean pine wood into split logs with bark intact (width 15 × length 25 × thickness 5 cm). The sentinel logs of this size were found to be advantageous for the field survey in terms of transportation and installation. (2) Two Korean pine blocks and ten pairs of adult vectors were placed in a plastic box (50 × 40 × 20 cm) and induced to lay eggs for three to four days. This was performed separately for each *Monochamus* species, avoiding them infesting the same split logs. (3) A hole was drilled near the top of a sentinel log for hanging. It is crucial that the sentinel logs should be installed in the field right after preparation to catch egg or larval parasitoids because the egg stage of *M. alternatus* and *M. saltuarius* only lasts about five to ten days according to Ochi [54].

2.2. Survey Site

Survey areas were chosen according to three criteria: (1) affected by PWD; (2) no aerial pesticide application; and (3) recorded presence of the insect vectors. The information of PWN-infested areas was provided by the National Institute of Forest Science, South Korea. *M. alternatus* is mainly distributed in the *Pinus densiflora* forests from the central to the southern region, including Jeju Island; *M. saltuarius* is mainly distributed in the *P. densiflora*, the Korean red pine, and *P. koraiensis*, the Korean pine, in forests from the central to the northern region [16]. Based on this information, the search for parasitoids of *M. alternatus* was first conducted in three Korean red pine stands of three southern locations in 2016: Jinju (N35.130196 E128.100493) and Geoje (N34.563059 E128.402835) in Gyeongsangnam-do, and Suncheon (N34.582965 E127.283301) in Jeollanam-do. In 2017, the parasitoid survey of *M. saltuarius* was conducted based on a more detailed survey plan in three Korean pine stands in three northern locations: Hongcheon (N37.454955 E127.533691) in Gangwon-do, and Yangpyeong (N37.372783 E127.390890) and Gapyeong (N37.511048 E127.285610) in Gyeonggi-do (Figure 1).

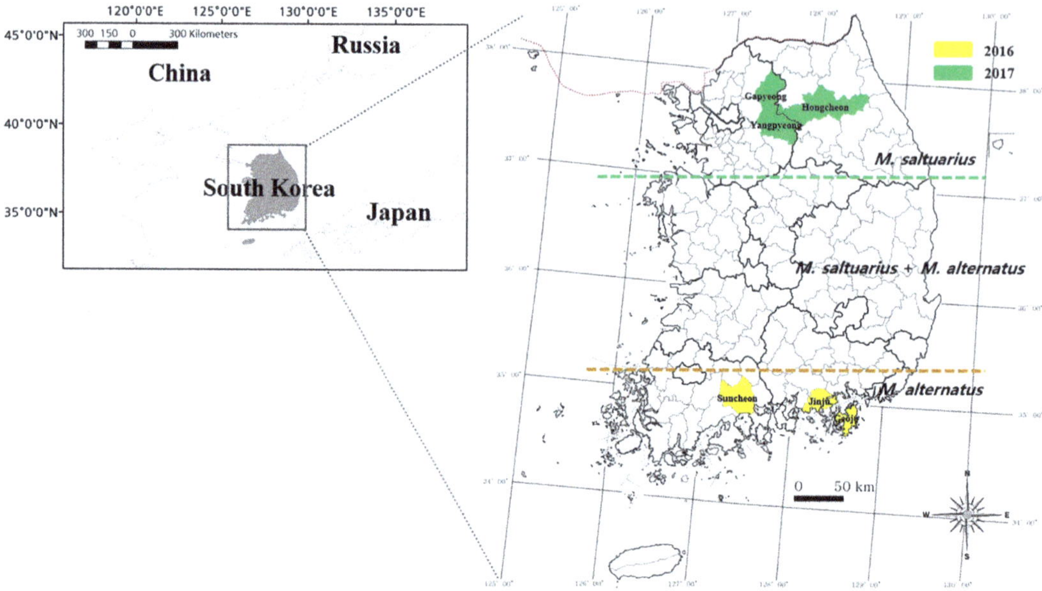

Figure 1. Parasitoid survey sites in South Korea in 2016 (yellow; Suncheon, Jinju, and Geoje) and 2017 (green; Gapyeong, Yangpyeong, and Hongcheon).

2.3. Field Survey

The first field survey was carried out as a pilot test in August 2016. The surveys were conducted twice, each in 14 days, at three sites in the southern region of South Korea (Figure 1). Each survey was conducted with three repeats of the *M. alternatus* sentinel log installations at each site. Along the edge of a Korean red pine stand, two sentinel logs (i.e., one sentinel log installation) were attached to the trunk of a pine tree at each of two different heights (0 and 1.8 m above the ground), adopting the method proposed by Duan et al. (Figure 2A) [48]. In 2017, the survey plan was revised to include more repeats during a longer period than the previous year. The surveys were made four times from mid-May to late July, which is the flight period for *M. saltuarius* according to Han et al. [55], at three sites in the northern region of South Korea (Figure 1). This was designed to investigate the differences in parasitoids composition in the southern and northern parts of Korea. In both 2016 and 2017, the sentinel logs installed were replaced every 14 days after the initiation until the end of the surveys, and after the 14-day exposure, the sentinel logs were brought back to the Korea National Arboretum lab for dismantling (Figure 2B). The barks of the collected sentinel logs were carefully peeled off using a small knife to find host eggs or larvae that were parasitized while recording all the number of host eggs and larvae that were found under the barks to calculate the parasitism rate.

Figure 2. Sampling parasitoids of a vector with a sentinel log of *Monochamus* sp. and indoor oviposition observation setup for *Spathius verustus*: (**A**) hanging two sentinel logs at two different heights (0 m and 1.8 m) from the base of a pine tree; (**B**) peeling off the bark of the collected sentinel logs after 14-day exposure to find parasitoid larvae. (**C**) A whole sentinel log with *S. verustus* in a glass cage. (**D**) Four females (black arrows) gathered in one place, drilling for oviposition.

2.4. Rearing and Identification of the Parasitoid

Collected larvae of the parasitic wasps were reared in small plastic holders ($2 \times 2 \times 2$ cm) under room temperature conditions (25 ± 1 °C) until adults emerged. The emerged parasitoids were killed with alcohol, dried using hexamethyldisilazane (HMDS), and card-mounted. Photographs were taken with a Leica DFC 495 camera through a Leica M205A Stereozoom stereomicroscope (Leica, Microsystems, Solms, Germany), and LAS software (version 4.1.0., Leica Microsystems, Switzerland) was used for image stacking. The images were adjusted using Adobe Photoshop CS6 (Adobe Systems Incorporated, San Jose, CA, USA). The parasitic wasp was identified according to the identification key by Tang et al. [56].

2.5. *Indoor Oviposition Observation for the Parasitoid*

For indoor mass-rearing of the parasitoid wasp in the future, it is crucial to study the host-seeking and oviposition behavior of the parasitoid, which is to find semiochemical or physical cues for oviposition. A simple oviposition test was performed with female wasps in the laboratory. First, early instar larvae of the vectors (= host larvae) were prepared under three different conditions to see to which condition the parasitoid reacts: (1) host larvae completely exposed; (2) larvae placed in a sandwich made of two *P. koraiensis* bark pieces (4 × 4 cm) with oviposition scars made by the vectors; and (3) larvae placed in a sandwich made of two *P. koraiensis* bark pieces (4 × 4 cm) without oviposition scars of the vectors. Each of the larvae was exposed to a pair of the parasitoids, which were fed with diluted honey (5%) for seven days after emergence, in a Petri dish (ø10 cm) to observe the oviposition behavior. Oviposition scars have traces of secretions during the oviposition of the vectors, and these substances (such as long-chain hydrocarbons, ketones of fatty acids, esterified cholesterols, and proteins) act as kairomones to attract parasitoids [57]. We also set up another test in which a dozen female wasps were exposed to a whole sentinel log (width 15 × length 25 × thickness 5 cm) where young host larvae were growing and supposedly feeding actively under the bark (Figure 2C). According to Wang et al. [58], *Spathius agrili* Yang, which is a braconid parasitoid of the emerald ash borer, responded to vibration signals from host-feeding and movement.

2.6. *Data Analysis*

Correlation analysis, *t*-test, and ANOVA were performed using the rstatix package [59] in R [60] to compare the relationships between variables as follows: (1) size of parasitized host larvae × number of *S. verustus* (the head capsule size of parasitized host larvae was measured, and the number of parasitoid progeny on a host was counted); (2) size of parasitized host larvae × size of female *S. verustus* (the head capsule size of parasitized host larvae was measured, and the hind tibia length of female parasitoid progeny on a host was measured); (3) host density × sex ratio (the number of hosts in a sentinel log and sex ratio of parasitoid progeny on a host were counted); and (4) host density × the number of parasitized host larvae (the number of hosts in a sentinel log and the number of parasitized host larvae were counted).

We also analyzed the effect of environmental variables on the parasitism rate of *S. verustus* with the generalized linear mixed model (GLMM) using the lme4 package [61] in R. The GLMM analysis method was used because the parasitism distribution was suggested to be binomial [62]. In the data of the present study, the fixed effect is the parasitism rate, and the random effects are site, height of sentinel log, forest depth, and session. Wilcoxon test was performed using the stats package [60] to analyze the statistical significance of the relationship between the parasitism rate of *S. verustus* and *Monochamus* species.

2.7. *Depository*

All materials examined are deposited in the Entomological Collection of the Korea National Arboretum (KNAE), Pocheon, Republic of Korea.

3. Results

3.1. *Taxonomic Account of Spathius verustus*

Family Braconidae

Genus *Spathius* Nees, 1819

Spathius verustus Chao, 1977

Diagnosis (based on [63,64]). Head behind eyes weakly convex anteriorly and evenly narrowed posteriorly. Transverse diameter of eye 1.1–1.3 times as long as temple. Ocelli arranged in triangle with base 1.2–1.3 times as long as its sides. Palpi light reddish brown or yellow. Mesopleuron always smooth in lower half. Forewing with darkened areas along the veins; vein 2-SR about 1.5 times as long as vein 3-SR. Hind tibia with long pale near the base. Ovipositor slightly shorter than metasoma.

Male (Figure 3A). Body length 2.64 ± 0.29 mm, fore wing length 1.6 ± 0.21 mm (n = 20). Antennae with 20–23 flagellar segments. Body saddle brown, legs pale yellow, tarsal claw blackish, metasoma evenly blackish posteriorly.

Figure 3. *Spathius verustus*: (**A**) male habitus, lateral view; (**B**) female habitus, lateral view; (**C**) female genitalia; (**D**) mature larvae of *S. verustus* on a larva of *M. alternatus*; (**E**) cocoons near the dead host in the tunnel dug by the host; (**F**) Pupa, female.

Female (Figure 3B, C). Body length 4.21 ± 0.68 mm, fore wing length 2.2 ± 0.27 mm, ovipositor length 1.99 ± 0.25 mm (n = 20). Antennae with 22–29 flagellar segments. Body saddle brown, tarsal claw blackish, metasoma evenly blackish posteriorly. Tergite 3 of metasoma to ovipositor black. Mesoscutum, axilla, and scutellum narrow and black. Fore and middle legs pale yellow. Hind femur pale yellow to blackish brown.

Distribution. China, Japan, Korea, Mongolia, Russia (Far East), Central and Western Europe.

3.2. Biology of Spathius verustus

Previously, *S. verustus* was known to attack only two coleopteran hosts, including *M. alternatus* (Cerambycidae) and *Pissodes obscurus* Roelofs (Curculionidae) [65,66], but the present study identified *M. saltuarius* as a new host. *S. verustus* was found to be an idiobiont ectoparasitoid (Figure 3D). The wasp mostly parasitized the second instar of *M. alternatus* and the first instar of *M. saltuarius*. Host instar was determined from the average head capsule width (HCW), as per [67,68]: 1.3 mm for *M. alternatus* second instar larvae and 1.2 mm for *M. saltuarius* first instar larvae. Additionally, *M. alternatus* third (HCW 1.6 mm), fourth (HCW 2.3 mm), and fifth (HCW 3.0 mm) instar larvae were found parasitized. The number of progenies on a larva of each host was slightly different. Up to four parasitoids were found on a larva of *M. alternatus* and up to three parasitoids on a larva of *M. saltuarius*. The average number of the parasitoid larvae was 3.15 ± 1.64 on *M. alternatus* and 1.48 ± 0.62 on *M. saltuarius* (Table 1).

Table 1. Biological information of *Spathius verustus* by host.

Host	Larval Stages of Parasitized Hosts (n)	Ave. No of Parasitoids on a Host Larva	Ave. Pupation Period (Day)	Sex Ratio (♂: Total)
M. alternatus	2nd (7) 3rd (1) 4th (3) 5th (1)	3.15 ± 1.64	6.6 ± 1.39	0.16
M. saltuarius	1st (33) 2nd (3)	1.48 ± 0.62	7.76 ± 3.06	0.37

Additionally, the ANOVA test results showed only the relationship between host density and number of parasitized host larvae statistically significant, when the host was *M. alternatus*, although the sample numbers were too small for a consistent analytic result (Figure 4, Table 2). On the other hand, all the other relationships such as size of host larvae and number of *S. verustus* were not statistically significant.

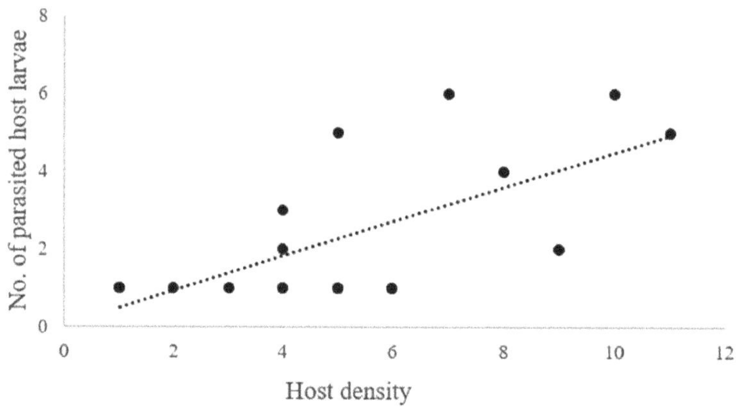

Figure 4. Relationship between host density (=number of host larvae per sentinel log) and number of parasitized *M. alternatus* larvae ($y = 0.4445x + 0.043$, $r = 0.66$, $df = 34$, $t = -1.64$, $p < 0.002$).

Table 2. Correlation analysis of host characteristics and *Spathius verustus* ecology in 2016 and 2017.

Host	Variable	n	r	df	t	p
M. alternatus	Size of parasitized host larvae × no. of *S. verustus*	12	0.35	10	1.16	0.27
	Size of parasitized host larvae × size of female *S. verustus*	25	0.26	23	1.3	0.21
	Host density × sex ratio	10	−0.55	8	−1.87	0.1
	Host density × no. of parasitized host larvae	18	0.66	16	3.56	0.002 **

Table 2. *Cont.*

Host	Variable	*n*	*r*	*df*	*t*	*p*
M. saltuarius	Size of parasitized host larvae × no. of *S. verustus*	36	−0.27	34	−1.64	0.1
	Size of parasitized host larvae × size of female *S. verustus*	57	0.09	55	0.6	0.55
	Host density × sex ratio	29	−0.27	27	−1.44	0.16
	Host density × no. of parasitized host larvae	45	0.17	43	1.11	0.27

** $p < 0.01$.

At the time of dismantling the sentinel logs, the early instar larvae of the parasitoid were white, and as they developed into mature larvae, fat bodies were distributed inside the body. The body size of fully matured larvae, with numerous fat bodies distributed rather sparsely throughout the body, was approximately 2–3 mm (Figure 3D). They then detached themselves from the body of the host before pupation. They spun silks and formed cocoons in the tunnel made by the host larvae. Upon this time, the host larvae lost all their body fluid and completely shrunk, with only the head remaining (Figure 3E). The parasitoid pupa was an exarate pupa, which was first white and then gradually turned darker over time (Figure 3F). The pupal stage lasted about 6.6 ± 1.39 d on *M. alternatus*, and 7.76 ± 3.06 d on *M. saltuarius* at room temperature 25 ± 3 °C (Table 1).

3.3. Indoor Oviposition Observation of S. verustus

In simple oviposition tests, *S. verustus* did not respond at all to either the exposed host larva or two types of wood sandwiches, with or without a host oviposition scar, even after several hours of exposure. On the other hand, the parasitoids exposed to a whole sentinel log approached certain areas within a few minutes after the exposure. A few of them were observed to have gathered in a certain area starting antennal drumming for a few minutes, then lowered their abdomen to start drilling (Figure 2D). After about ten minutes of drilling, the parasitoids started ovipositor probing as if trying to locate an exact or suitable point for oviposition and then put the ovipositor down and stayed still for a while. After this, the female withdrew its ovipositor completely and left from the site.

3.4. Parasitism Rates of S. verustus from the Field Surveys in 2016 and 2017

In 2016, among the survey sites, the average parasitism rate of Jinju was higher than Geoje and Suncheon and was statistically significant (Figure 5A, Table 3). The average parasitism rate by tree height was higher at 1.8 m than at 0 m, but there was no statistical significance (Figure 5B, Table 3).

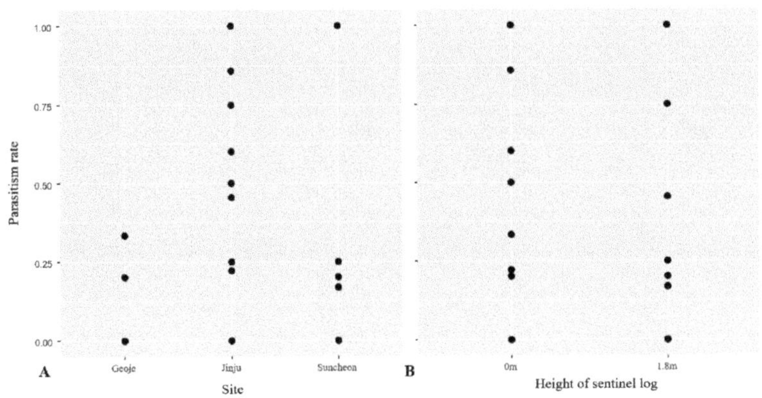

Figure 5. Scatter plot of *Spathius verustus* in 2016: (**A**) relationship between site and parasitism rate; (**B**) relationship between height of sentinel log and parasitism rate.

Table 3. Parasitism rate analysis of *Spathius verustus* by ecological variables using GLMM in 2016 and 2017.

Year	Variable	Parameter	Mean (%) ± SD	*n*	*df*	*p*
2016	Site	Geoje Suncheon Jinju	1.66% ± 6.61 5.59% ± 18.2 17.62% ± 31.26	96 †	2	4.279×10^{-12} ***
	Height of sentinel log	0 m 1.8 m	7.75% ± 21.18 8.83% ± 23.33		1	0.94
2017	Site	Gapyeong Yangpyeong Hongcheon	0.68% ± 0.03 1.25% ± 0.04 3.01% ± 0.09	480 †	2	0.001501 **
	Height of sentinel log	0 m 1.8 m	1.2% ± 0.04 1.4% ± 0.05		1	0.38
	Forest depth	0 m 10 m 20 m 30 m 40 m	1.43 ± 0.05 0.7% ± 0.03 0.89% ± 0.03 1.6% ± 0.04 1.91% ± 0.04		4	0.52
	Session	1st (5/17–31) 2nd (6/1–15) 3rd (6/15–30) 4th (7/1–15)	3.35% ± 0.07 0.28% ± 0.01 0.64% ± 0.02 2.12% ± 0.06		3	2.849×10^{-7} ***

** $p < 0.01$, *** $p < 0.001$. † total number of sentinel logs used for the survey.

In 2017, among the survey sites, the average parasitism rate of Hongcheon was higher than Yangpyeong and Gapyeong, and it was statistically significant (Figure 6A, Table 3). The average parasitism rate by height of sentinel log was higher at 1.8 m than at 0 m, but there was no statistical significance (Figure 6B, Table 3). The average parasitism rate (0 m, 10 m, 20 m, 30 m, 40 m) appeared to increase with forest depth, but it was not statistically significant (Figure 6C, Table 3). The average parasitism by session was high in the order of second, third, fourth, and first, and it was statistically significant (Figure 6D, Table 3). In addition, as a result of comparing the parasitism rate of *M. alternatus* and *M. saltuarius* as the hosts of *S. verustus*, the parasitism rate of *M. alternatus* was high, which showed a statistical significance ($p < 0.01$).

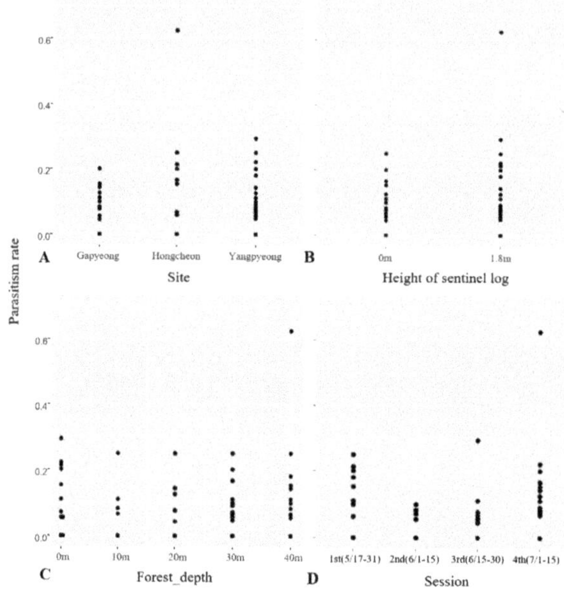

Figure 6. Scatter plot of *Spathius verustus* in 2017: (**A**) relationship between site and parasitism rate;

(**B**) relationship between the height of sentinel log and parasitism rate; (**C**) relationship between forest depth and parasitism rate; (**D**) relationship between session and parasitism rate.

4. Discussion

4.1. Indoor Oviposition Observation of *S. verustus*

Spathius verustus females were observed to display a series of ovipositional behaviors such as antennal drumming, drilling, and oviposition only when exposed to sentinel logs. This may indicate that *S. verustus* also responds to the same or similar ovipositional cues as *S. agrilis* did in a previous study [58]. According to them, *S. agrili*, used as the biological control of *A. planipennis*, the emerald ash borer, was reported to be attracted by the volatile substance from the leaves of ash trees when harmed by *A. planipennis* and the vibration generated when the beetle larvae gnawed the tree with its mandibles. Then, this may explain the reasons why *S. verustus* females did not respond to the hosts in the other conditions: (1) the parasitoids females were not attracted to naked hosts since no chemical cues are expected to be released and act as kairomone [69], and (2) the host larvae could scratch the wall to make a noise, but its body was not tightly fixed in place due to the space in the bark sandwich and was not able to generate vibrations strong enough to attract *S. verustus*. As an exact oviposition cue of *S. verustus* was not identified at the present time, it will be ideal to supply whole sentinel logs upon mass-rearing until an alternative host or artificial diet, which is more convenient than a whole sentinel log, is identified or developed.

4.2. The Ecological Relationship between the Host and *S. verustus*

There was no statistical significance in the relationship between the size of host larvae and the number of *S. verustus* in 2016 and 2017. Häckermann et al. [70] reported that the clutch size of parasitoids was not influenced by the host size. This suggests that the number of eggs was not proportional to the larval size of the host (Table 2). Nicol and Mackauer [71] reported that the host size did not affect parasitoid size. However, the size of female *S. verustus* was affected by the size of host larvae in 2016 and 2017 (Table 2). This will require further investigation with larger samples. Studies of the relationship between the sex ratio of *Bracon hebetor* and the density of the host have shown that as the host density increases, the male ratio tends to decrease and the female ratio tends to increase [72]. In 2016 and 2017, the relationship between host density and the sex ratio of *S. verustus* was similar to the result of Jamil et al. [72] but was not statistically significant (Table 2). This can be addressed by increasing the sample size in future studies. We observed a relationship between host density and the number of parasitized host larvae in 2016 and 2017. These results suggest that the number of parasitized host larvae increases with an increase in host density [73]. However, only the 2016 results are statistically significant, and further study on *M. saltuarius* is needed due to the small sample size (Table 2).

4.3. In Spatial Effect of Parasitism Rate of *S. verustus* According to Environmental Variables

In 2016 and 2017, parasitism rates for *S. verustus* were investigated in three regions, respectively, in the northern and southern regions of South Korea. Although the host was different, there was a statistically significant difference in the parasitism rate by region (Table 3). This may be related to the difference in host density among the sites. According to Furlong and Zalucki [74], the density of parasitoids changes with the density of the host. However, there has not been any study on host density in infected or healthy pine forests in the country. Therefore, further research is needed to investigate the density of the host and other factors so as to determine which factor is the cause.

During the investigation period, installation heights of sentinel logs were both 0 and 1.8 m, and the installation by forest depth was divided into 5 stages (0 m, 10 m, 20 m, 30 m, 40 m) from the edge to the inside of the forest in only 2017. However, the parasitism rate of *S. verustus* was not affected by installation height and forest depth. Chung et al. [75] investigated the within-tree distribution of larval entrance of *M. laternatus* and reported that larvae of *M. laternatus* were distributed most at 6 to 7 m and at least 0 to 3 m height of Korean red pines. This may suggest that the parasitism of *S. verustus* by sentinel log height was not statistically significant because the sentinel logs were installed at a height lower than where the hosts might have mostly been distributed in a tree. According to Ma et al. [76], when the density of *M. alternatus* was investigated using a pheromone trap outdoors, it was highest at 0 to 50 m from the forest edge toward the center. In the present study, all of the sentinel logs were installed within the forest depth range where the vector might be mostly distributed. Therefore, the parasitism rate did not show differences along the forest depth.

In 2017, the parasitism rate of *S. verustus* was affected by session as the emergence time of the host (Table 3). The parasitism rate of the 1st (5/17–31) and 4th (7/1–15) is higher than the 2nd (6/1–15) and 3rd (6/15–30) of the session. There are about 40 days between the 1st and 4th parasitism peaks. It is thought that the occurrence of a natural enemy is closely related to the life cycle of the host. According to Han et al. [77], the first emergence time of *M. saltuarius* is in early May and lasts until mid-June in South Korea. The life cycle *S. agrili*, which is a braconid parasitoid of the emerald ash borer, takes about 40 days from one generation to the next [78]. If *S. verustus* has a life cycle similar to that of *S. agrili*, the interval between two parasitism peaks in the 2017 survey result could be due to the developmental period of *S. verustus*. In addition, *S. verustus* appears to prefer *M. alternatus* to *M. saltuarius*, which shows which host to use for future indoor breeding.

4.4. Possibility of *S. verustus* as a Biological Control Agent on the Insect Vectors of PWN

Developing a biological control project requires investigating natural enemies, especially specialists, their availability in nature, biological traits, etc. The present study can be considered as the very first step, in finding natural enemies and exploring basic biological traits, toward developing biological control against the spread of PWD.

Two insect vectors in South Korea had never been problematic before the PWN invasion, but now they are a key factor in the spread of PWD. It would be another viable biological control option to cut the link between the *Monochamus* vectors and PWN by utilizing natural enemies of the vectors so as to slow down or restrain the spread of PWD. In regard to restraining PWN transmission, there is ongoing research on chemical interactions among PWN, insect vectors, and host trees to break the link in the transmission from the infected wood to the insect vectors and later from the insect vectors to the healthy host trees during feeding [69,79].

Spathius verustus appears to be distributed nationwide, from northern to southern regions, and up to date, the wasp seems to be a generalist but with a narrow range of host species, now including two Korean native *Monochamus* species. This trait alone may make the parasitoid a potential candidate as a biological control agent against both of the PWN insect vectors; controlling two pest species with one natural enemy species anywhere in the country would be very cost-effective.

Secondly, the parasitoid was most parasitic on the second instar larva of *M. alternatus* and the first instar larva of *M. saltuarius*. This may be explained by the ovipositor length, about 2 mm, which means that the parasitoid cannot parasitize host larvae living deeper than 2 mm from the bark. Previously, Pershing and Linit [80] reported that the fourth instar larvae of both *Monochamus* species move into the sapwood area, and from then on, the probability of being attacked by their natural enemies would be significantly reduced, which makes it difficult to control them. Therefore, using the parasitoid to attack the early-stage larvae would be an effective measure to control the insect vectors.

Thirdly, the parasitoid can be considered as gregarious, laying an average of 3.15 ± 1.64 eggs on the larvae of *M. alternatus* and more than 1.48 ± 0.62 eggs on *M. saltuarius*, which is advantageous in terms of mass-rearing of any parasitoid. Compared to this, *S. agrili*, a biological control agent on the emerald ash borer, has been reported to lay an average of eight eggs on a host [81]. Although the number of eggs laid by a parasitic wasp is different depending on host size, it can be a great advantage in increasing the number of *S. verustus* as long as host larvae are well managed, at the instar stages preferred by the parasitoid, during mass-rearing [82]. Further indoor tests are required to investigate the accurate life cycle of *S. verustus* to figure out biological characteristics such as developmental period, life span, daily fecundity, and sex ratio.

Lastly, the flight period of the parasitoid appears to coincide with the flight period of the two *Monochamus* species, as the parasitoid constantly parasitized the hosts throughout the survey period. Therefore, the parasitoid can be augmented into a certain area during the peak flight period of the vectors, when their oviposition period can be assumed to be from mid-May to late June according to Han et al. [77].

4.5. Compensate the Defect and Future Study

In this study, the large difference in the parasitism rate in the two vectors and the parasitism rate by region needs to be addressed in the future. According to Choi et al. [18], the intraspecific competition among the parasitic wasps resulted in an increase in the density of the parasitic wasps as the density of the vectors decreased over time. Referring to this and considering the history of the area affected by PWD, some regional characteristics may have affected the parasitism rate. For example, the Jinju site has been an area devastated by PWD for the longest time next to Busan where PWD was first identified [83]. In addition to that, we were told that the Jinju site of the present study has never been treated with aerial spray except for nematicide injection for experimental purposes. Insecticide spraying reduces the vector's density, as well as the parasitoid's density [84]. Therefore, this area seems to have been stably accommodating numbers of *S. verustus* parasitic to *M. alternatus* [85]. In future studies, it is necessary to investigate other infested areas with parasitoids.

Lastly, to use *S. verustus*, which has been identified to be parasitic to the vectors of PWN, as a biological control agent, it is necessary to address several limitations. For example, the host range of *S. verustus* requires further research. If this species is only parasitic to the wood-boring beetles that only inhabit live pines, it would be a highly interesting option in terms of forest protection. Unlike the chemical or physical control methods, such a control method for various wood-boring insect pests would enable pest control without a great impact on the environment [86]. Recently, with increased negative awareness of chemical control agents and continuous reports on environmental risks such as the harm to honeybees [87], it is critical to explore, select, and utilize the means of biological control for large-scale forests. Therefore, continuous research on other natural enemies, including parasitoids such as *S. verustus*, as well as predators is required.

Author Contributions: Conceptualization, I.-K.K. and B.-W.L.; methodology, M.-S.K. and J.-K.J.; validation, I.-K.K. and B.-W.L.; formal analysis, M.-S.K. and J.-K.J.; investigation, M.-S.K. and K.-J.H.; resources, C.-J.K. and B.-W.L.; data curation, M.-S.K. and J.-K.J.; writing—original draft preparation, M.-S.K.; writing—review and editing, K.-J.H., I.-K.K., and J.-K.J.; visualization, M.-S.K. and J.-K.J.; supervision, I.-K.K.; project administration, C.-J.K. and I.-K.K.; funding acquisition, I.-K.K. and B.-W.L. All authors have read and agreed to the published version of the manuscript.

Funding: This research was fully funded by the Korea National Arboretum (of the Korea Forest Service), grant number KNA1-1-24, 18-2, and the APC was also funded by the Korea National Arboretum.

Data Availability Statement: The data presented in this study are available on request from the corresponding author. The data are not publicly available due to "the regulations of research management" of the Korea National Arboretum.

Acknowledgments: The authors thank Min Chul Kim and Myeong-Ho Kim (the Korea National Arboretum, South Korea) for their hard work as research assistants. Special thanks go to Wonil Choi and Youngwoo Nam (the National Institute of Forest Science, South Korea) for providing valuable advice and comment about the analyses. This study was supported by the Korea National Arboretum (Project No. KNA1-1-24, 18-2).

Conflicts of Interest: The authors declare no conflict of interest.

References

1. Choi, W.I.; Park, Y.S. Dispersal patterns of exotic forest pests in South Korea. *Insect Sci.* **2012**, *19*, 535–548. [CrossRef]
2. Dwinell, L.D. First report of pinewood nematode (*Bursaphelenchus xylophilus*) in Mexico. *Plant Dis.* **1993**, *77*, 846. [CrossRef]
3. Enda, N. The damage of pine wilt disease in Taiwan. *For. Pests* **1988**, *37*, 161–166.
4. Malek, R.B.; Appleby, J.E. Epidemiology of pine wilt in Illinois. Disease distribution. *Plant Dis.* **1984**, *68*, 180–186. [CrossRef]
5. Mamiya, Y. History of pine wilt disease in Japan. *J. Nematol.* **1988**, *20*, 219–226. [PubMed]
6. Mota, M.M.; Braasch, H.; Bravo, M.A.; Penas, A.C.; Burgermeister, W.; Metge, K.; Sousa, E. First report of *Bursaphelenchus xylophilus* in Portugal and in Europe. *Nematology* **1999**, *1*, 727–734. [CrossRef]
7. Zamora, P.; Rodríguez, V.; Renedo, F.; Sanz, A.V.; Domínguez, J.C.; Pérez-Escolar, G.; Miranda, J.; Álvarez, B.; González-Casas, A.; Mayor, E.; et al. First report of *Bursaphelenchus xylophilus* causing pine wilt disease on *Pinus radiata* in Spain. *Plant Dis.* **2015**, *99*, 1449. [CrossRef]
8. Zhao, B.G. Pine wilt disease in China. In *Pine Wilt Disease*; Springer: Tokyo, Japan, 2008; pp. 18–25.
9. Yi, C.K.; Byun, B.H.; Park, J.D.; Yang, S.I.; Chang, K.H. First finding of the pinewood nematode, *Bursaphelenchus xylophilus* (Steiner et Buhrer) nickle and its insect vector in Korea. *Res. Rep. For. Res. Inst.* **1989**, *38*, 141–149.
10. Tóth, Á. *Bursaphelenchus xylophilus*, the pinewood nematode: Its significance and a historical review. *Acta Biol. Szegediensis* **2011**, *55*, 213–217.
11. Shin, S.C. Pine wilt disease in Korea. In *Pine Wilt Disease*; Springer: Tokyo, Japan, 2008; pp. 32–36.
12. Mamiya, Y.; Enda, N. Transmission of *Bursaphelenchus lignicolus* (nematoda: Aphelenchoididae) by *Monochamus alternatus* (coleoptera: Cerambycidae). *Nematologica* **1972**, *18*, 159–162. [CrossRef]
13. Linit, M.J. Nemtaode-vector relationships in the pine wilt disease system. *J. Nematol.* **1988**, *20*, 227–235. [PubMed]
14. Sousa, E.; Bravo, M.A.; Pires, J.; Naves, P.; Penas, A.C.; Bonifácio, L.; Mota, M.M. *Bursaphelenchus xylophilus* (nematoda; aphelenchoididae) associated with *Monochamus galloprovincialis* (coleoptera; Cerambycidae) in Portugal. *Nematology* **2001**, *3*, 89–91.
15. National Institute of Biological Resources. *National Species List of Korea. III. Insects (Hexapoda)*; Designzip: Seoul, Korea, 2019; Volume 988.
16. Kwon, T.S.; Lim, J.H.; Sim, S.J.; Kwon, Y.D.; Son, S.; Lee, K.Y.; Kim, Y.T.; Shin, C.H.; Ryu, S.B.; Lee, C.K.; et al. Distribution patterns of *Monochamus alternatus* and *M. saltuarius* (Coleoptera: Cerambycidae) in Korea. *J. Korean For. Soc.* **2006**, *95*, 543–550.
17. Kwon, S.H.; Go, M.S.; Park, J.; Han, T.W.; Kim, K.B.; Shin, C.H.; Kim, D.S. The bimodal adult activity of *Monochamus alternatus* (coleoptera: Cerambycidae) caught in pheromone traps in Jeju can be explained by the competitive attractiveness of dying pine trees. *Entomol. Res.* **2019**, *49*, 172–178. [CrossRef]
18. Choi, W.I.; Jeon, M.J.; Park, Y.S. Structural dynamics in the host-parasitoid system of the pine needle gall midge (*Thecodiplosis japonensis*) during invasion. *PeerJ* **2017**, *5*, e3610. [CrossRef] [PubMed]
19. Kwon, T.S.; Shin, J.H.; Lim, J.H.; Kim, Y.K.; Lee, E.J. Management of Pine Wilt Disease in Korea Through Preventative Silvicultural Control. *Forest Ecol. Manag.* **2011**, *261*, 562–569. [CrossRef]
20. Suh, D.Y.; Jung, J.K.; Lee, S.K.; Seo, S.T. Effect of aerial spraying of thiacloprid on pine sawyer beetles (*Monochamus alternatus*) and honey bees (*Apis mellifera*) in pine forests. *Entomol. Res.* **2021**, *51*, 83–89. [CrossRef]
21. Kim, J.B.; Kim, D.Y.; Park, N.C. Development of an aerial precision forecasting techniques for the pine wilt disease damaged area based on GIS and GPS. *J. Korean Assoc. Geogr. Inf.* **2010**, *13*, 28–34. [CrossRef]
22. Park, Y.S.; Park, Y.K.; Yang, H.M. Effects of clear-cutting on forest arthropod communities at two different vertical levels (crown and ground surface). *Korean J. Ecol. Environ.* **2016**, *49*, 271–278. [CrossRef]
23. Jeon, K.S.; Kim, C.S.; Park, N.C.; Hur, T.C.; Hong, S.C. Effects on control of pine wilt disease (*Bursaphelenchus xylophilus*) by thinning methods in red pine (*Pinus densiflora*) forest. *J. Korean For. Soc.* **2011**, *100*, 165–171.
24. Wang, Y.; Chen, F.; Wang, L.; Li, M. Investigation of beetle species that carry the pinewood nematode, *Bursaphelenchus xylophilus* (Steiner and Buhrer) nickle, in China. *J. For. Res.* **2020**, *32*, 1745–1751. [CrossRef]
25. Li, L.; Wei, W.; Liu, Z.; Sun, J. Host Adaptation of a gregarious parasitoid *Sclerodermus harmandi* in artificial rearing. *BioControl.* **2010**, *55*, 465–472. [CrossRef]
26. Tian, S.P.; Xu, Z.Q. Effects of different temperatures on the development of *Scleroderma guani* reared with *Tenebrio molitor*. *Entomol. Knowl.* **2003**, *40*, 356–359.
27. Morimoto, K.; Iwasaki, A. Role of *Monochamus alternatus* (Coleoptera: Cerambycidae) as a vector of *Bursaphelenchus lignicolus* (Nematoda: Aphelenchoididae). *J. Jpn. For. Soc.* **1972**, *54*, 177–183. [CrossRef]
28. Takizawa, Y.; Shoji, T. Distribution of *Monochamus saltuarius* gebler, and its possible transmission of pinewood nematodes in Iwate prefecture. *For. Pests* **1982**, *31*, 4–6.
29. Kobayashi, F.; Yamane, A.; Ikeda, T. The Japanese pine sawyer beetle as the vector of pine wilt disease. *Annu. Rev. Entomol.* **1984**, *29*, 115–135. [CrossRef]
30. Togashi, K. A Field experiment on dispersal of newly emerged adults of *Monochamus alternatus* (coleoptera: Cerambycidae). *Res. Popul. Ecol.* **1990**, *32*, 1–13. [CrossRef]
31. Hong, J.I.; Koh, S.H.; Chung, Y.J.; Shin, S.C.; Kim, G.H.; Choi, K.S. Biological characteristics of *Sclerodermus harmandi* (Hymenoptera: Bethylidae) parasitized on cerambycid. *Korean J. Appl. Entomol.* **2008**, *47*, 133–139. [CrossRef]
32. Sin, S.C. Biological control of tree insect pest. *Landscaping Tree* **2012**, *129*, 28–34.

33. Choi, W.I.; Nam, Y.; Lee, C.Y.; Choi, B.K.; Shin, Y.J.; Lim, J.H.; Koh, S.H.; Park, Y.S. Changes in major insect pests of pine forests in Korea over the last 50 years. *Forests* **2019**, *10*, 692. [CrossRef]
34. Lim, J.O.; Lyu, D.P.; Choi, G.S.; Jeong, Y.J.; Shin, S.C.; Lee, S.H. A taxonomic note on *Sclerodermus harmandi*, ectoparasite of stem and wood boring insect larvae (Hymenoptera: Chrysidoidea: Bethylidae) in South Korea. *J. Asia Pac. Entomol.* **2006**, *9*, 115–119. [CrossRef]
35. Liu, Y.; Zhang, B.; Xie, Y.; Yang, S. Potential non-target effects of *Sclerodermus harmandi* (Hymenoptera: Bethylidae) on *Triaspis* sp. (Hymenoptera: Braconidae). *Biocontrol Sci. Technol.* **2017**, *27*, 909–917. [CrossRef]
36. Kim, J.K.; Won, D.S.; Park, Y.C.; Koh, S.H. Natural Enemies of Wood Borers and Seasonal Occurrence of Major Natural Enemies of *Monochamus saltuarius* on Pine Trees. *J. Korean Soc. For. Sci.* **2010**, *99*, 439–445.
37. Petersen-Silva, R.; Pujade-Villar, J.; Naves, P.; Sousa, E.; Belokobylskij, S. Parasitoids of *Monochamus galloprovincialis* (Coleoptera, Cerambycidae), vector of the pinewood nematode, with identification key for the palaearctic region. *ZooKeys* **2012**, *251*, 29. [CrossRef]
38. Brabbs, T.; Collins, D.; Hérard, F.; Maspero, M.; Eyre, D. Prospects for the use of biological control agents against *Anoplophora* in Europe. *Pest Manag. Sci.* **2015**, *71*, 7–14. [CrossRef]
39. Duan, J.J.; Aparicio, E.; Tatman, D.; Smith, M.T.; Luster, D.G. Potential new associations of north american parasitoids with the invasive asian longhorned beetle (Coleoptera: Cerambycidae) for biological control. *J. Econ. Entomol.* **2016**, *109*, 699–704. [CrossRef]
40. Hérard, F.; Bon, M.C.; Maspero, M.; Cocquempot, C.; Lopez, J. Survey and evaluation of potential natural enemies of *Anoplophora glabripennis* and *A. chinensis*. In Proceedings of the 16th US Department of Agriculture Interagency Research Forum on Gypsy Moth and Other Invasive Species, Newton Square, PA, USA, 18–21 January 2005; pp. 18–21.
41. Wang, X.; Aparicio, E.M.; Murphy, T.C.; Duan, J.J.; Elkinton, J.S.; Gould, J.R. Assessing the host range of the north american parasitoid *Ontsira mellipes*: Potential for biological control of asian longhorned beetle. *Biol. Control* **2019**, *137*, 104028. [CrossRef]
42. Delvare, G.; Bon, M.C.; Hérard, F.; Cocquempot, C.; Maspero, M.; Colombo, M. Description of *Aprostocetus anoplophorae* n. sp. (Hymenoptera: Eulophidae), a new egg parasitoid of the invasive pest *Anoplophora chinensis* (Förster) (Coleoptera: Cerambycidae). *Ann. Société Entomol. Fr.* **2004**, *40*, 227–233. [CrossRef]
43. Kim, M.S.; Lee, H.L.; Ku, D.S.; Hérard, F.; Gould, J.R.; Williams, D.W.; Kim, I.K.; Hong, K.J. Discovery of *Spathius ibarakius* Belokobylskij et Maeto (Hymenoptera: Braconidae) as a larval ectoparasitoid of citrus longhorned beetle in Korea. *Korean J. Appl. Entomol.* **2016**, *55*, 285–291. [CrossRef]
44. Yang, Z.Q.; Strazanac, J.S.; Marsh, P.M.; Van Achterberg, C.; Choi, W.Y. First Recorded Parasitoid from China of *Agrilus planipennis*: A New Species of *Spathius* (Hymenoptera: Braconidae: Doryctinae). *Ann. Entomol. Soc. Am.* **2005**, *98*, 636–642. [CrossRef]
45. Yao, W.J.; Yang, Z.Q. Studies on biological control of *Anoplophora glabripennis* (Coleoptera: Cerambycidae) with a parasitoid, *Sclerodermus guani* (Hymenoptera: Bethylidae). *J. Environ. Entomol.* **2008**, *30*, 127–134.
46. Abell, K.J.; Duan, J.J.; Shrewsbury, P.M. Determining Optimal parasitoid release timing for the biological control of emerald ash borer (Coleoptera: Buprestidae). *Fla. Entomol.* **2020**, *102*, 691–694. [CrossRef]
47. Gould, J.S.; Bauer, L.S.; Lelito, J.; Duan, J. *Emerald Ash Borer Biological Control Release and Recovery Guidelines*; US Department of Agriculture. Animal Plant Health Inspection Service; Forest Service Northern Research Station and Agricultural Research Service: Riverdale, MD, USA, 2012; pp. 1–76.
48. Gould, J.R.; Bauer, L.S.; Duan, J.J.; Williams, D.; Liu, H. History of emerald ash borer biological control. In *Biology and Control of Emerald Ash Borer. FHTET-2014-09*; Van Driesche, R.G., Reardon, R.C., Eds.; United States Department of Agriculture, Forest Service, Forest Health Technology Enterprise Team: Morgantown, WV, USA, 2015; Volume 5, pp. 83–95.
49. Wang, X.; Wang, X.Y.; Kenis, M.; Cao, L.M.; Duan, J.J.; Gould, J.R.; Hoelmer, K.A. Exploring the potential for novel associations of generalist parasitoids for biological control of invasive woodboring beetles. *BioControl* **2021**, *66*, 97–112. [CrossRef]
50. Duan, J.J.; Bauer, L.S.; Hansen, J.A.; Abell, K.J.; Van Driesche, R. An improved method for monitoring parasitism and establishment of *Oobius Agrili* (Hymenoptera: Encyrtidae), an egg parasitoid introduced for biological control of the emerald ash borer (Coleoptera: Buprestidae) in North America. *Biol. Control* **2012**, *60*, 255–261. [CrossRef]
51. Abell, K.J.; Bauer, L.S.; Miller, D.L.; Duan, J.J.; Van Driesche, R.G.V. Monitoring the establishment and flight phenology of parasitoids of emerald ash borer (Coleoptera: Buprestidae) in Michigan by using sentinel eggs and larvae. *Fla. Entomol.* **2016**, *99*, 667–672. [CrossRef]
52. Futai, K.; Shirakikawa, S.; Nakai, I. The suitability of korean pine (*Pinus koraiensis* Sieb. et Zucc.) and japanese red pine (*P. densiflora* Sieb. et Zucc.) as a host of the japanese pine sawyer, *Monochamus alternatus* hope (Coleoptera: Cerambycidae). *Appl. Entomol. Zool.* **1994**, *29*, 167–177. [CrossRef]
53. Hwang, I.C.; Kim, J.H.; Park, J.B.; Shin, S.C.; Chung, Y.J.; Cho, S.; Park, Y.C. Growth, development, and reproduction of *Monochamus saltuarius* (Coleoptera: Cerambycidae) on conifers fed to larvae. *Korean J. Appl. Entomol.* **2008**, *47*, 385–394. [CrossRef]
54. Ochi, K. Ecological studies on cerambycid injurious to pine trees (II) biology of two monochamus (Coleoptera, Cerambycidae). *J. Jpn. For. Soc.* **1969**, *51*, 188–192. [CrossRef]
55. Han, J.H.; Yoon, C.; Shin, S.C.; Kim, G.H. Seasonal occurrence and morphological measurements of pine sawyer, *Monochamus saltuarius* adults (Coleoptera: Cerambycidae). *J. Asia Pac. Entomol.* **2007**, *10*, 63–67. [CrossRef]
56. Tang, P.; Belokobylskij, S.; Chen, X.X. *Spathius* Nees, 1818 (Hymenoptera: Braconidae, Doryctinae) from China with a key to species. *Zootaxa* **2015**, *3960*, 1–132. [CrossRef]

57. Rutledge, C.E. A survey of identified kairomones and synomones used by insect parasitoids to locate and accept their hosts. *Chemoecology* **1996**, *7*, 121–131. [CrossRef]
58. Wang, X.Y.; Yang, Z.Q.; Gould, J.R.; Wu, H.; Ma, J.H. Host-seeking behavior and parasitism by *Spathius agrili* yang (Hymenoptera: Braconidae), A parasitoid of the emerald ash borer. *Biol. Control* **2010**, *52*, 24–29. [CrossRef]
59. Kassambara, A. *rstatix: Pipe-Friendly Framework for Basic Statistical Tests*. R Package Version 0.7.0. Available online: http://CRAN.R-project.org/package=rstatix (accessed on 14 May 2022).
60. R Core Team. *R: A Language and Environment for Statistical Computing*; R Foundation for Statistical Computing: Veinna, Asutria, 2021; Available online: https://www.R-project.org/ (accessed on 14 May 2022).
61. Douglas, B.; Martin, M.; Ben, B.; Steve, W. Fitting linear mixed-effects models using lme4. *J. Stat. Softw.* **2015**, *67*, 1–48. [CrossRef]
62. Murdoch, W.W.; Reeve, J.D.; Huffaker, C.B.; Kennett, C.E. Biological control of olive scale and its relevance to ecological theory. *Am. Nat.* **1984**, *123*, 371–392. [CrossRef]
63. Ku, D.S.; Belokobylskij, S.A.; Cha, J.Y. Family Braconidae. Economic insects of Korea 16. *Insecta Koreana* **2001**, *23*, 1–281.
64. Lee, H.R.; An, T.H.; Ku, D.S.; Byun, B.K. A newly recorded species of the genus *Spathius* nees (Hymenoptera: Braconidae: Doryctinae) with a checklist in Korea. *Korean J. Appl. Entomol.* **2017**, *56*, 295–299. [CrossRef]
65. Belokobylskij, S.A.; Maeto, K. *Doryctinae (Hymenoptera: Braconidae) of Japan*; Natura Optima Dux Foundation: Warszawa, Poland, 2009; pp. 1–806.
66. Kishi, Y. *Pine Wood Nematode and the Japanese Pine Sawyer*; Thomas Company Limited: Tokyo, Japan, 1995; pp. 1–302.
67. Go, M.S.; Kwon, S.H.; Kim, S.B.; Kim, D.S. The developmental characteristics for the head capsule width of *Monochamus alternatus* (Coleoptera: Cerambycidae) larvae and determination of the number of instars. *J. Insect Sci.* **2019**, *19*, 26. [CrossRef]
68. Li, M.; Dai, Y.; Wang, Y.; Wang, L.; Sun, S.; Chen, F. New insights in to the life history of *Monochamus saltuarius* (Cerambycidae: Coleoptera) can enhance surveillance strategies for pine wilt disease. *J. For. Res.* **2021**, *32*, 2699–2707. [CrossRef]
69. Gonçalves, E.; Figueiredo, A.C.; Barroso, J.G.; Millar, J.G.; Henriques, J.; Sousa, E.; Bonifácio, L. Characterization of cuticular compounds of the cerambycid beetles *Monochamus galloprovincialis*, *Arhopalus syriacus*, and *Pogonocherus perroudi*, potential vectors of pinewood nematode. *Entomol. Exp. Appl.* **2021**, *169*, 183–194. [CrossRef]
70. Häckermann, J.; Rott, A.S.; Dorn, S. How two different host species influence the performance of a gregarious parasitoid: Host size is not equal to host quality. *J. Anim. Ecol.* **2007**, *76*, 376–383. [CrossRef]
71. Nicol, C.M.Y.; Mackauer, M. The scaling of body size and mass in a host-parasitoid association: Influence of host species and stage. *Entomol. Exp. Appl.* **1999**, *90*, 83–92. [CrossRef]
72. Jamil, A.; Ul Abdin, Z.; Arshad, M.; Falabella, P.; Abbas, S.K.; Tahir, M.; Jamil, A.; Manzoor, A.; Shaina, H. Multiple effects of host density on egg density and the sex ratio of progeny of *Bracon hebetor* (Say.) (Hymenoptera: Braconidae). *Pak. J. Zool.* **2015**, *47*, 455–460.
73. Yu, S.-H.; Ryoo, M.I.; Na, J.H.; Choi, W.I. Effect of host density on egg dispersion and the sex ratio of progeny of *Bracon hebetor* (Hymenoptera: Braconidae). *J. Stored Prod. Res.* **2003**, *39*, 385–393. [CrossRef]
74. Furlong, M.J.; Zalucki, M.P. Climate change and biological control: The consequences of increasing temperatures on host–parasitoid interactions. *Curr. Opin. Insect Sci.* **2017**, *20*, 39–44. [CrossRef]
75. Chung, Y.J.; Choi, K.S.; Lee, S.M.; Kim, D.S.; Lee, S.G.; Park, J.G. Measurement and within-tree distribution of larval entrance and adult emergence holes of japanese pine sawyer, *Monochamus alternatus* (Coleoptera: Cerambycidae). *Korean J. Appl. Entomol.* **2003**, *42*, 315–321.
76. Ma, T.; Shi, X.; Shen, J.; Wang, C.; Zhang, S.; Lu, X.; Sun, Z.; Chen, X.; Wang, C.; Xie, W.; et al. Field evaluation of commercial attractants and trap placement for monitoring pine sawyer beetle, *Monochamus alternatus* (Coleoptera: Cerambycidae) in Guangdong, China. *J. Econ. Entomol.* **2018**, *111*, 239–246. [CrossRef]
77. Han, J.H.; You, J.H.; Koo, C.D.; Yoon, C.M.; Choi, K.S.; Shin, S.C.; Kim, G.H. Emergence timing of the pine sawyer beetle, *Monochamus saltuarius* (Coleoptera: Cerambycidae) by tree species. *Korean J. Appl. Entomol.* **2009**, *48*, 189–195. [CrossRef]
78. Yang, Z.Q.; Wang, X.Y.; Gould, J.R.; Reardon, R.C.; Zhang, Y.N.; Liu, G.J.; Liu, E.S. Biology and behavior of *Spathius Agrili*, a parasitoid of the emerald ash borer, *Agrilus planipennis*, in China. *J. Insect Sci.* **2010**, *10*, 30. [CrossRef]
79. Gonçalves, E.; Figueiredo, A.C.; Barroso, J.G.; Henriques, J.; Sousa, E.; Bonifácio, L. Effect of *Monochamus galloprovincialis* feeding on *Pinus pinaster* and *Pinus pinea*, oleoresin and insect volatiles. *Phytochemistry* **2020**, *169*, 112–159. [CrossRef]
80. Pershing, J.C.; Linit, M.J. Biology of *Monochamus carolinensis* (Coleoptera: Cerambycidae) on scotch pine in Missouri. *J. Kans. Entomol. Soc.* **1986**, *59*, 706–711.
81. Gould, J.R.; Ayer, T.; Fraser, I. Effects of rearing conditions on reproduction of *Spathius Agrili* (Hymenoptera: Braconidae), a parasitoid of the emerald ash borer (Coleoptera: Buprestidae). *J. Econ. Entomol.* **2011**, *104*, 379–387. [CrossRef] [PubMed]
82. Le Masurier, A.D. A comparative study of the relationship between host size and brood size in *Apanteles* spp. (Hymenoptera: Braconidae). *Ecol. Entomol.* **1987**, *12*, 383–393. [CrossRef]
83. Seo, B.S.; Jeon, H.S.; Kim, J.B.; Cho, G.S. Development of pine wilt disease (*Bursaphelenchus Xylophilus*) prevention system. In *Proceedings of the Korean Society of Surveying Geodesy Photogrammetry Cartography Conference, Jeju, Korea*; Korean Society of Surveying, Geodesy, Photogrammetry and Cartography: Seoul, Korea, 2007; pp. 283–287.
84. Rebek, E.J.; Sadof, C.S. Effects of pesticide applications on the euonymus scale (Homoptera: Diaspididae) and its parasitoid, *Encarsia citrina* (Hymenoptera: Aphelinidae). *J. Econ. Entomol.* **2003**, *96*, 446–452. [CrossRef] [PubMed]
85. Hassell, M.P.; Waage, J.K. Host-parasitoid population interactions. *Annu. Rev. Entomol.* **1984**, *29*, 89–114. [CrossRef]

86. Frank, S.D. Biological control of arthropod pests using banker plant systems: Past progress and future directions. *Biol. Control* **2010**, *52*, 8–16. [CrossRef]
87. Sánchez-Bayo, F.; Goulson, D.; Pennacchio, F.; Nazzi, F.; Goka, K.; Desneux, N. Are bee diseases linked to pesticides? A brief review. *Environ. Int.* **2016**, *89–90*, 7–11. [CrossRef]

forests

MDPI

Article

Identification of Olfactory Genes in *Monochamus saltuarius* and Effects of *Bursaphelenchus xylophilus* Infestation on Their Expression

Sufang Zhang [1,*,†], Xizhuo Wang [1,†], Yanlong Zhang [1,†], Yanan Zheng [2], Zhizhi Fan [1] and Rong Zhang [1]

[1] Key Laboratory of Forest Protection of National Forestry and Grassland Administration, Research Institute of Forest Ecology, Environment and Nature Conservation, Chinese Academy of Forestry, Beijing 100091, China; wavingsnow@126.com (X.W.); zhangyl@caf.ac.cn (Y.Z.); 18234494745@163.com (Z.F.); 13625650836@163.com (R.Z.)

[2] College of Forestry, Shenyang Agricultural University, Shenyang 110866, China; rockyya@163.com

* Correspondence: zhangsf@caf.ac.cn

† These authors contributed equally to this work.

Abstract: The pine wood nematode (PWN) *Bursaphelenchus xylophilus* has caused disastrous losses of pine forests in many countries, and the success of PWN depends strongly on interactions with its insect vectors. *Monochamus saltuarius* is a newly recorded vector in Northeast China. Feeding (i.e., immature) and egg-laying (i.e., mature) *Monochamus* spp. target different host plants, and olfactory cues play important roles regarding host choice. Whether infestation with PWN affects olfactory mechanisms in *M. saltuarius* related to feeding and oviposition is of interest as this may affect the spread of nematodes to new healthy hosts. However, little is known about molecular mechanisms of the olfactory system of *M. saltuarius*. We identified chemosensory-related genes in adult *M. saltuarius* and examined the influence of *B. xylophilus* on the respective expression patterns. Fifty-three odorant-binding proteins (OBPs), 15 chemosensory proteins, 15 olfactory receptors (ORs), 10 gustatory receptors, 22 ionotropic receptors (IRs), and two sensory neuron membrane proteins were identified, and sex bias among non-infested beetles was mainly found with respect to expression of OBPs. Interestingly, OBPs and ORs were markedly down-regulated in male *M. saltuarius* infested with *B. xylophilus*, which may reduce olfactory sensitivity of male *M. saltuarius* and affect the spreading of *B. xylophilus* to new hosts. Our results will help understand the interactions between *B. xylophilus* and *M. saltuarius*, which may lead to the identification of new control targets in the olfactory system of *M. saltuarius*.

Keywords: odorant-binding proteins; olfactory receptors; pine wood nematode; gene expression; *Monochamus saltuarius*

1. Introduction

Pine is the common name of *Pinus* species (family Pinaceae), and it is the largest genus of conifers, which is widely distributed throughout the world [1]. Pines are not only pioneer species in afforestation, but also have very high economic value owing to their by-products such as pine resin, rosin, and turpentine, which are important raw materials for producing medicinal compounds, organic solvents, and fuels. Currently, pine forests are threatened by pine wood nematodes (PWNs; *Bursaphelenchus xylophilus*) (Steiner and Buhrer, 1934) Nickle (1970) [2]. More than 60 pine species can be infected with *B. xylophilus*, and vast areas of pine forests in several countries have been devastated as a result of pine wilt disease. In China alone, an estimated 50 million trees were killed in an area of 80,000 ha between 1982 and 2000 [3].

In 1982, PWN was first found in Nanjing, China, and in only a few decades, it spread rapidly from Nanjing to vast areas of China (Li and Zhang, 2018). According to the announcement of the National Forestry and Grassland Administration in 2021, there were 721 PWN-epidemic areas at county level, which were widely distributed in 16 provinces, thus seriously threatening pine forests in northern China.

The spread of *B. xylophilus* relies on insects of the genus *Monochamus* [4–7]. *B. xylophilus* is attracted by newly emerged *Monochamus* adults, and when infested beetles feed on shoots of healthy trees, the nematode can disperse to these trees [8,9]. With the spread of PWN from South Central China to Northeast China, its main vector insects changed from *Monochamus alternatus* (Coleoptera: Cerambycidae) to *Monochamus saltuarius* [10]. The interactions among trees, vector insects, and *B. xylophilus* are complex, considering previous studies on *M. alternatus* [8,11]; moreover, elucidating such interactions may be of importance for controlling the spread of PWN. *M. saltuarius* has been identified as a vector of *B. xylophilus* in Japan and Korea [12]; however, only limited information is available regarding its biology, pheromones, and potential control measures of *M. saltuarius* [12–15]. Few studies so far focused on the interactions between nematodes and *M. saltuarius* [16].

Sexual maturity of *Monochamus* spp. lasts for almost one month, and these insects undergo an immature (nutrition) period and mature reproduction period, which in females includes oviposition [17]. Nematodes in the spiracles of *Monochamus* spp. depart to invade new healthy trees through feeding wounds inflicted to the trees [9] or through oviposition incisions [18]. *Monochamus* spp. preferably feed on healthy trees and lay eggs in weak or dying trees [17]. Volatile organic compounds of plants [19–21] and insect olfactory systems [22] play important roles regarding insect host choice for feeding and oviposition; for example, terpene volatiles (such as α-pinene) are the main attractants for *M. alternatus* to feed [23,24], and ethanol exuded from weak or dying trees can attract female longicorn beetles to lay eggs [25–27]. However, it is unknown whether the olfactory recognition sensitivity of insect vectors may be affected by infestation with nematodes, which may then also influence the spreading of nematodes. For example, the spread of nematodes to new healthy hosts mainly depends on the feeding behavior of *Monochamus* spp.; if the olfactory sensitivity of vectors to healthy trees is affected by infestation with nematodes, the spreading efficiency of PWN may also be affected by such mechanisms.

The olfactory system is critical for insect feeding and reproduction (including mate selection and oviposition site searching) [28]. Several classes of genes are associated with insect olfactory performance, including two classes of binding proteins (odorant-binding proteins, [OBPs] and chemosensory proteins [CSPs]) [29], three receptor gene classes (olfactory receptors [ORs], ionotropic receptors [IRs], and gustatory receptors [GRs]), and sensory neuron membrane proteins (SNMPs) [30]. The binding proteins can preliminarily screen and transport organic hydrophobic volatile molecules from the environment (such as host volatiles and pheromones) to the olfactory neurons through hydrophilic sensillum lymph [31,32]. These receptors can detect and discriminate olfactory signals and translate them into bioelectric signals, and different receptor types frequently respond to different signals. For example, ORs mainly respond to olfactory signals [33], and GRs respond to taste signals such as sugars, bitter compounds, and other contact stimuli [34].

Identification and expression analysis of olfactory-related genes are the basis for further exploration of olfactory mechanisms in insects. However, information on olfactory-related genes in Coleoptera is scarce. Before 2013, members of the major chemosensory gene families had been identified only in *T. castaneum* [35] and in two bark beetles, *Ips typographus* and *Dendroctonus ponderosae* [36]. Chemosensory genes of *M. alternatus* were identified in 2014, and this is also the first longicorn beetle species whose olfactory recognition was examined [3]. Molecular olfactory reception mechanisms have received more attention [37,38], however, very little is known about the molecular mechanisms of olfactory recognition by *M. saltuarius*, which is a new and important vector of *B. xylophilus* in northeast China. This limits our general understanding of the olfactory cues underlying the spread of this pest.

Here, we aim to identify the chemosensory genes of *M. saltuarius*, and detect the influence of *B. xylophilus* infestation on chemosensory gene expression in *M. saltuarius*. This work may provide a basis for further research on the chemosensory mechanisms of

M. saltuarius and help us to further understand the interactions between *B. xylophilus* and *M. saltuarius* further.

2. Materials and Methods

2.1. Insects

M. saltuarius at the mating stage were collected on the Dahuofang Experimental Forest Farm, Fushun City, Shenyang Province, China, in May 2019. Samples (n = 40) were collected in the forest. We searched *M. saltuarius* carefully in the area where some of the host trees were weak, and once mating couples were found (a male and a female *M. saltuarius* were mating, usually located on the trunk), they would be put it in a breathable plastic tube and taken back to the laboratory. Antennae were collected as follows: both antennae of each *M. saltuarius* were cut off at the base, placed in a marked centrifuge tube, and were immediately frozen in liquid nitrogen. Then, the respective donor individual was examined for nematodes using the Baermann funnel method. Four categories were used, i.e., males and females without nematodes (here termed 'healthy') and with nematodes ('infested'), respectively. Five biological replicates were used per category, and each of the five biological replicates consisted of antennae coming from one beetle.

2.2. RNA-Seq Library Preparation and Sequencing

Total RNA was isolated from *M. saltuarius* antennae using TRIzol reagent (Invitrogen, Carlsbad, CA, USA) [39,40], and genomic DNA was removed using DNase I (TaKara, Dalian, Liaoning, China). Integrity and purity of the extracted RNA were determined using a 2100 Bioanalyzer (Agilent Technologies, Inc., Santa Clara, CA, USA), and RNA was quantified using a ND-2000 device (NanoDrop Thermo Scientific, Wilmington, DE, USA). Only high-quality RNA samples (OD260/280 = approximately 1.8–2.2, OD260/230 $\geq$ 2.0, RIN $\geq$ 8.0, 28S:18S $\geq$ 1.0) were used to prepare sequencing libraries. One microgram total RNA was used to synthesize cDNA [41,42]. Sequencing libraries were prepared using the Illumina TruSeq™ RNA Sample Preparation Kit (Illumina, San Diego, CA, USA). Libraries were selected for cDNA target fragments of 200–300 bp, followed by PCR amplification using Phusion DNA polymerase (New England Biolabs, Boston, MA, USA) with 15 amplification cycles. After quantification, the sequencing of 2 $\times$ 150 bp paired-end reads was performed on an Illumina HiseqXTen sequencer (Illumina).

2.3. De-Novo Sequence Assembly

Clean data were obtained from raw sequencing data by trimming and quality control using SeqPrep (https://github.com/jstjohn/SeqPrep, accessed on 11 November 2019) and Sickle (https://github.com/najoshi/sickle, accessed on 6 January 2022) with default parameters. Clean data were then used to perform *de novo* assembly with Trinity software (http://trinityrnaseq.sourceforge.net/, accessed on 20 November 2019) [43], and redundancies were removed using TGICL [44].

2.4. Annotation

All assembled transcripts were searched against six public databases (NR, Swiss-Prot, Pfam, COG, GO, and KEGG) using BLAST to identify proteins with the highest sequence similarity with the produced transcripts to produce functional annotations (cut-off e-value < 1^{-5}). BLAST2GO (http://www.blast2go.com/b2ghome, accessed on 25 November 2019) [45,46]. Software was used to produce gene ontology (GO) annotations of unique assembled transcripts to describe biological processes, molecular functions, and cellular components [47,48]. Metabolic pathway analysis was performed using the Kyoto Encyclopedia of Genes and Genomes (KEGG; http://www.genome.jp/kegg/, accessed on 26 November 2019).

Chemosensory genes were identified in two ways: first, contig tBLASTx searches were performed using an *M. saltuarius* assembled unigene database with query sequences as chemosensory genes from other insects, such as *Drosophila melanogaster*, *Bombyx mori*, and other long-horned beetles. Second, annotation information of the unigenes was used. Open reading frames of the candidate genes from the two ways were identified and used to as confirmation through BLAST (http://blast.ncbi.nlm.nih.gov/Blast.cgi, accessed on 6 January 2022).

2.5. Gene Expression Quantification

The expression level of each transcript was calculated according to the fragments per kilobase of exon per million mapped reads (FRKM) method [49]. The R software package EdgeR vision 2.12 (http://www.bioconductor.org/packages/2.12/bioc/html/edgeR.html, accessed on 10 October 2019) was used for differential expression analysis [50]. A compatible-hits-norm model was used to normalize unigene expression levels and DEGs differentially expressed genes [51].

2.6. Phylogenetic Analyses

IQ-TREE web service was used to construct phylogenetic trees of the predicted protein sequences and orthologous genes from other insects [52]. The sequences were first aligned using BioEdit software with the ClustalW multiple alignment tool. Bootstrap phylogenetic trees were generated using 1000 replicates, and the best models were selected using the model fit result of IQ-TREE. iTOL was used to annotate the phylogenetic trees [53].

3. Results

3.1. *M. saltuarius* Antennae Transcriptomes

More than 7 Gbp clean sequence data were produced from most samples (except for one sample from an infested male, which produced 6.98 Gb data), and the Q30 value was > 94%. *De-novo* transcriptome assembly resulted in 38,198 contigs with an N50 of 1798 bp, and the largest unigene was 32,017 bp long. The length distribution of the transcriptome assembly is shown in Figure S1. The mapping rates of clean data were >78.16% for each sample. Gene annotation based on the NR, Swiss-Prot, Pfam, COG, GO, and KEGG databases yielded annotations for 20,931transcripts.

3.2. *M. saltuarius* Chemosensory Genes

We identified 53 OBPs, 15 CSPs, 15 ORs, 10 GRs, 22 IRs, and 2 SNMPs (Table S1). The identified 117 chemosensory genes were submitted to NCBI under the accession numbers MT008336–MT008452.

3.3. OBPs of *M. saltuarius* with and without *B. xylophilus*

We analyzed sex-specific expression patterns of OBPs in non-*B. xylophilus*-infested *M. saltuarius*, and 15 OBPs were expressed differentially between sexes. Most differentially expressed OBPs (14/15) showed higher expression in females (Figure 1A). Among OBPs with female-biased expression in healthy beetles, OBP28, OBP52, OBP38, and OBP39 had high expression levels, and the other female-biased OBPs (OBP32, OBP18, OBP19, OBP31, OBP43, OBP3, OBP6, OBP7, OBP40, and OBP51) showed low expression. OBP14 expression was male-biased among healthy beetles, and, the expression of OBP14 was the highest in males and very low in females (Figure 1A). Thus, OBP14 may be related to sex pheromone recognition in male *M. saltuarius*. Further studies are needed to understand the reason of its differential expression between males and females. Effects of *B. xylophilus* on the expression patterns of OBPs in male and female beetles was examined, and none of the OBPs showed altered expression levels in females after one month of infestation with *B. xylophilus* (Figure 1B). However, expressions of five OBPs (including OBP14, OBP30, OBP52, OBP44, and OBP35) were significantly lower in infested male beetles than in non-infested ones (Figure 1C). Thus, OBP expression in male *M. saltuarius* was affected by infestation with *B. xylophilus*.

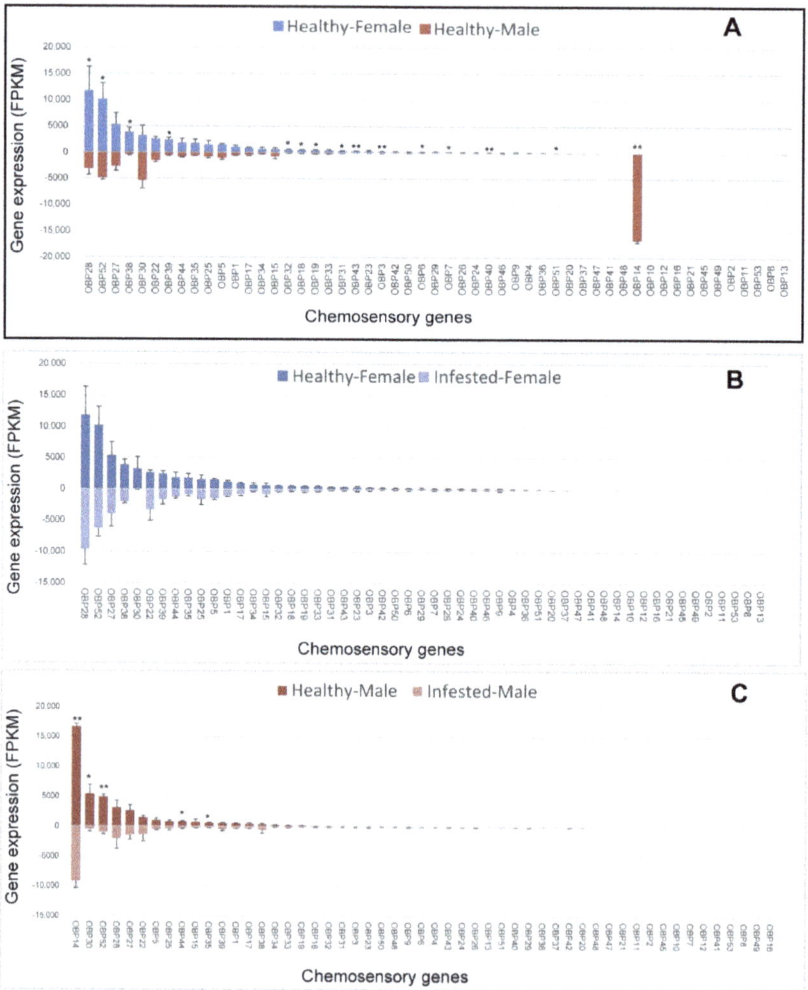

Figure 1. Expression patterns of odorant-binding protein in healthy and *Bursaphelenchus xylophilus*-infested *Monochamus saltuarius*. (**A**) Differences between sexes of healthy *M. saltuarius*; (**B**) Differences between healthy and infested female *M. saltuarius*; (**C**) Differences between healthy and infested male *M. saltuarius*. Transcript levels are expressed as reads per kilobase of exon per million mapped reads (RPKM), and the expression data were shown as mean ±SE. The stars above each bar of figure indicate significant differences ($p < 0.05$ * and $p < 0.05$ **).

A phylogenetic tree was constructed to analyze the characteristics of OBPs from *M. saltuarius*, based on orthologs to OBP sequences from several other Coleoptera insects *Semanotus bifasciatus* [54], *Tribolium castaneum* [55], *Anomala corpulenta* [56], *Agrilus planipennis*, and *Dendroctonus ponderosae* [57] (Figure 2). OBPs from *M. saltuarius* seems to be homologous to OBPs of Coleopteran insects whose genomic information are available, and indicates the relative complete identification of OBPs from *M. saltuarius*. MsalOBP14, which expressed at an especially high level in male adults, seems not to be a typical PBPs. We did not get a special PBP clade in Figure 2 (MsalOBP14 was clustered together with other *M. saltuarius* OBPs) and MsalOBP14 was not crusted in PBP clade in another phylogenetic tree containing more model insect species (Figure S2). However, Pfam search hits

PBP/GOBP family (PF01395.24, E-value is 4.8×10^{-18}) (http://pfam.xfam.org, accessed on 2 February 2019). The function of MsalOBP14 deserves further study.

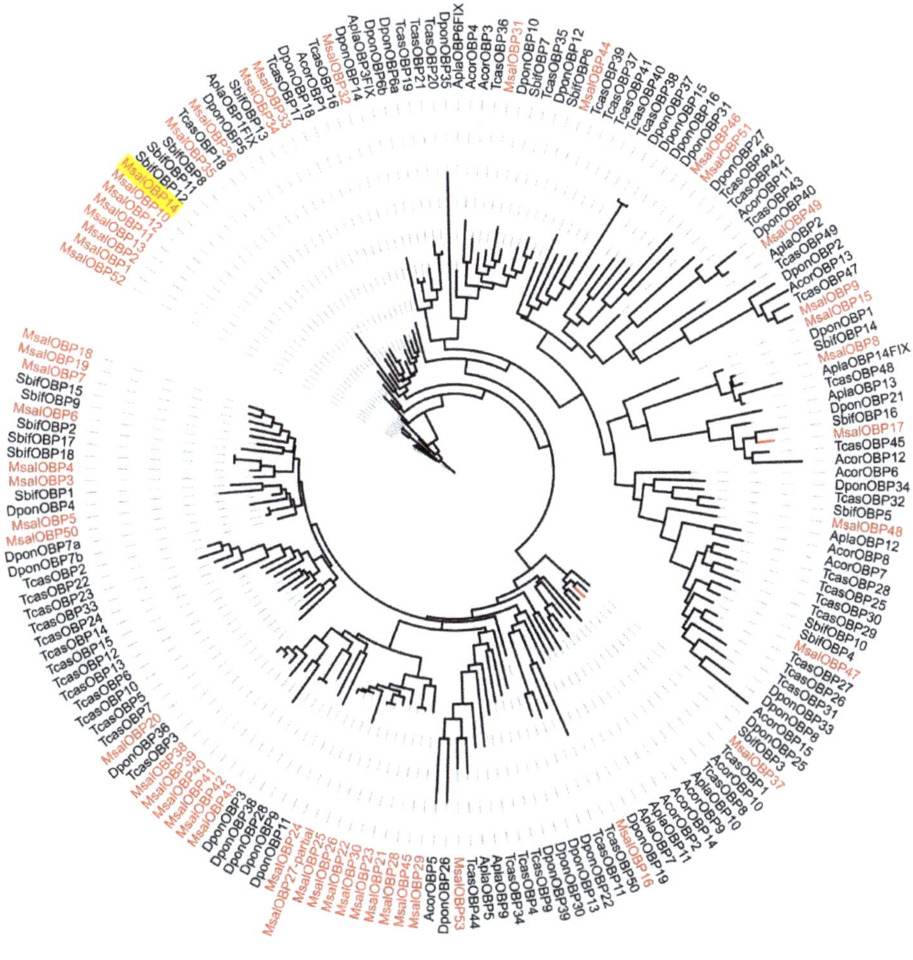

Figure 2. Maximum likelihood dendrogram based on protein sequences of candidate odorant-binding proteins (OBP) in *M. saltuarius* and orthologs from other Coleopteran insects.JTT + F + R6 model based on the best fit result of IQ-tree model finding, and bootstrap consensus tree inferred from 1000 replicates were used. *Monochamus saltuarius* (Msal) (red), *Semanotus bifasciatus* (Sbif), *Tribolium castaneum* (Tcas), *Anomala corpulenta* (Acor), *Agrilus planipennis* (Apla), and *Dendroctonus ponderosae* (Dpon). MsalOBP14 expressed at an especially high level in male antennae of *M. saltuarius* was marked with yellow background.

3.4. CSPs of *M. saltuarius* with and without *B. xylophilus*

Sex-specific expression of CSPs was assessed in healthy *M. saltuarius* (Figure 3A), and three CPSs were expressed differently between sexes; two of these (CSP1 and CSP13) showed higher expression in females, and one (CSP5) showed extremely high expression in males. Infestation with *B. xylophilus* slightly affected CSP expression, only CSP13 in female (Figure 3B), and CSP9 and CSP13 male (Figure 3C) were down-regulated in the infested beetles (please see supplementary Table S2).

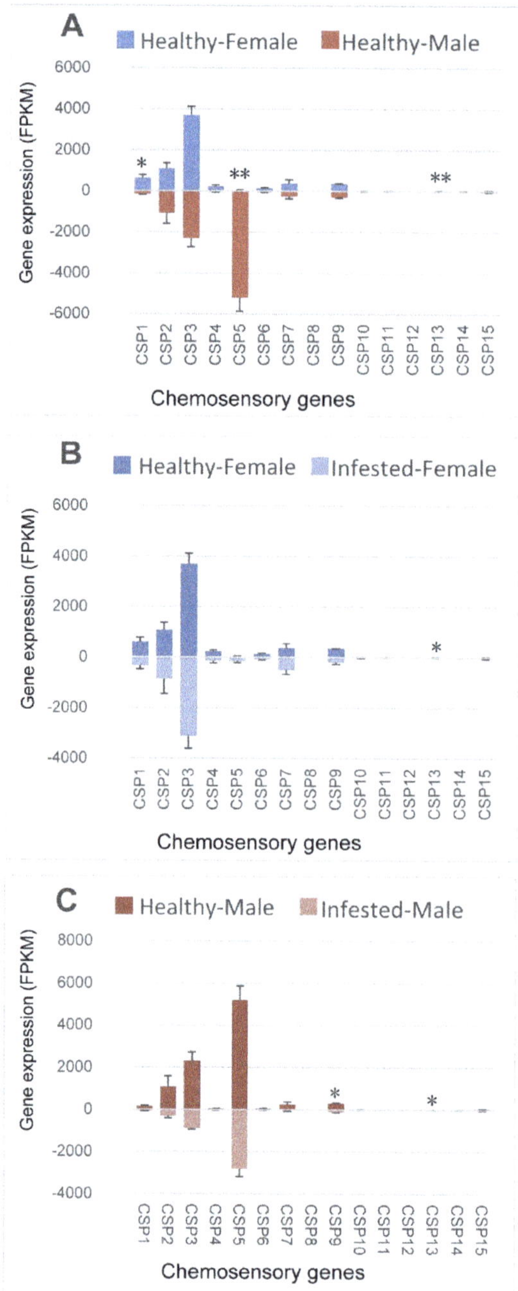

Figure 3. Expression patterns of chemosensory genes in healthy and *B. xylophilus*-infested *M. saltuarius*. (**A**) Differences between sexes of healthy *M. saltuarius*; (**B**) Differences between healthy and infested female *M. saltuarius*; (**C**) Differences between healthy and infested male *M. saltuarius*. Transcript levels are expressed as reads per kilobase of exon per million mapped reads (RPKM), and the expression data were shown as mean±SE. The stars above each bar of figure indicate significant differences ($p < 0.05$ * and $p < 0.05$ **).

A phylogenetic tree of CSPs was constructed based on orthologs to CPS sequences from *M. saltuarius*, *S. bifasciatus*, *T. castaneum*, *A. corpulenta*, *A. planipennis*, and *D. ponderosae* (Figure 4). The identified CSPs from *M. saltuarius* were homologous to CSPs of Coleopteran insects whose genomic information were available, indicating relative complete identification of CSPs from *M. saltuarius*.

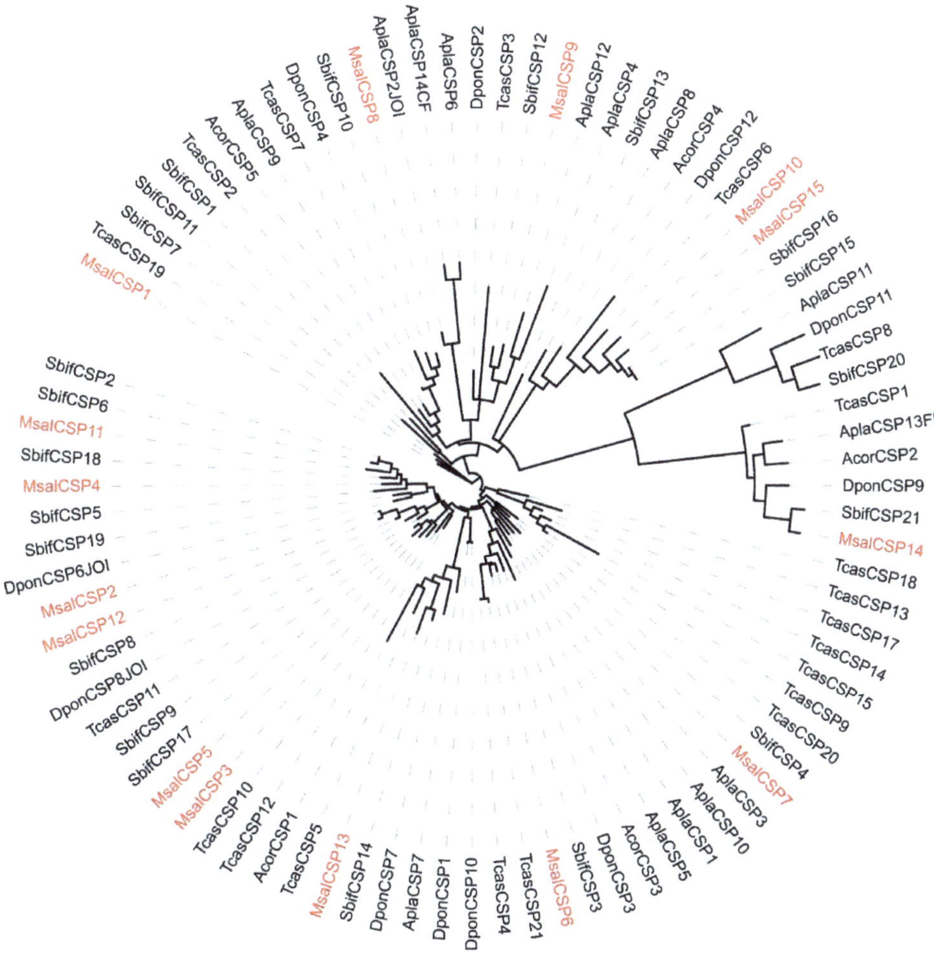

Figure 4. Maximum likelihood dendrogram based on protein sequences of candidate chemosensory genes (CSP) in *M. saltuarius* and orthologs from other Coleopteran insects. LG + I + G4 model based on the best fit result of IQ-tree model finding, and bootstrap consensus tree inferred from 1000 replicates were used. *Monochamus saltuarius* (Msal) (red), *Semanotus bifasciatus* (Sbif), *Tribolium castaneum* (Tcas), *Anomala corpulenta* (Acor), *Agrilus planipennis* (Apla), and *Dendroctonus ponderosae* (Dpon).

3.5. ORs of M. saltuarius with and without B. xylophilus

No sex bias was observed regarding OR expression in healthy *M. saltuarius* (Figure 5A). Infestation with *B. xylophilus* did not affect OR expression in female *M. saltuarius* (Figure 5B). Marked changes in OR expression were detected in male *M. saltuarius* carrying *B. xylophilus* compared with non-infested individuals (Figure 5C). OR1, OR5, OR10, OR12, OR14, and OR16 showed higher expression in healthy *M. saltuarius* than those that were infested. Thus, *B. xylophilus* can reduce olfactory sensitivity of male *M. saltuarius*.

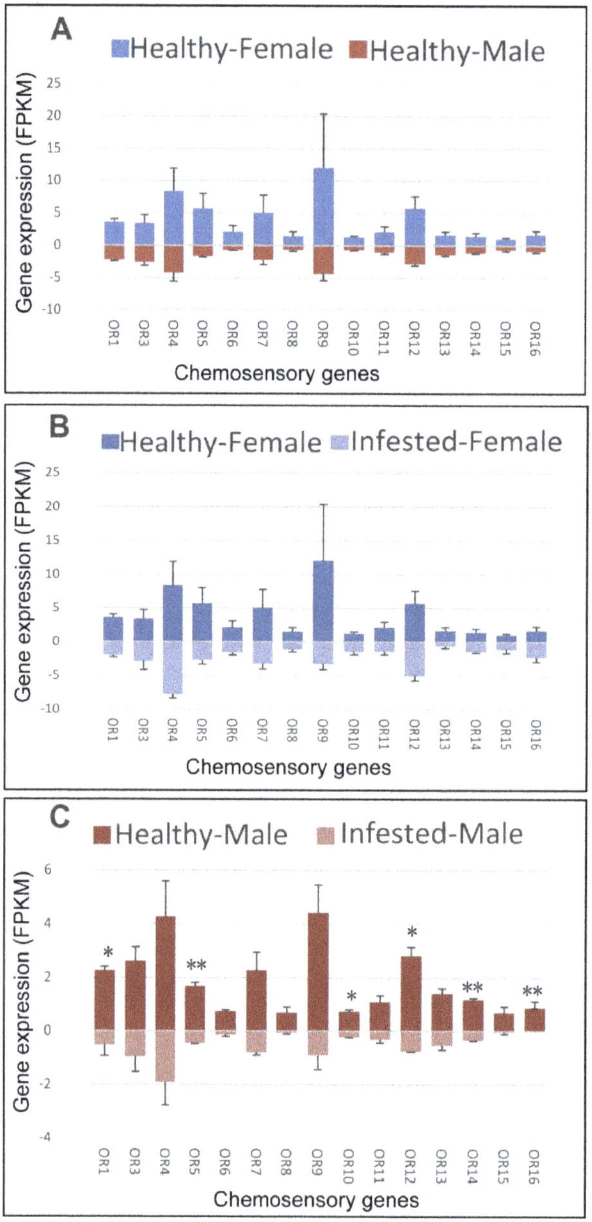

Figure 5. Expression patterns of olfactory receptors in healthy and *B. xylophilus*-infested *M. saltuarius*. (**A**) Differences between sexes of healthy *M. saltuarius*; (**B**) Differences between healthy and infested female *M. saltuarius*; (**C**) Differences between healthy and infested male *M. saltuarius*. Transcript levels are expressed as reads per kilobase of exon per million mapped reads (RPKM), and the expression data were shown as mean±SE. The stars above each bar of figure indicate significant differences ($p < 0.05$ * and $p < 0.05$ **).

A phylogenetic analysis of ORs was performed based on the OR sequence orthologs from *M. saltuarius*, *S. bifasciatus*, *A. corpulenta*, *A. planipennis*, and *D. ponderosae* (Figure 6).

Generally, almost all *M. saltuarius* ORs found orthology to other Coleoptera ORs, including the insect OR obligatory co-receptor, Orco. However, the total number of *M. saltuarius* ORs may have not been recovered due to its integrity in this transcriptomic study. This is obvious after comparison with genomic data from *A. planipennis* and *D. ponderosae*.

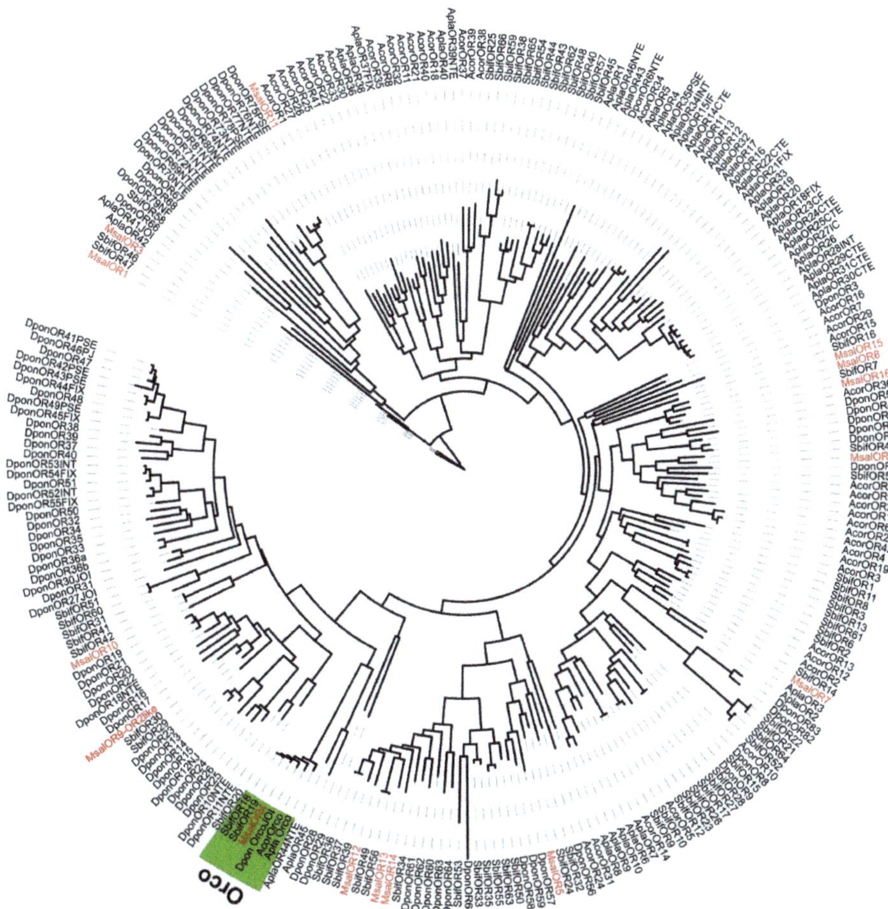

Figure 6. Maximum likelihood dendrogram based on protein sequences of candidate olfactory receptors (OR) in *M. saltuarius* and orthologs from other Coleopteran insects. JTT + F + R6 model based on the best fit result of IQ-tree model finding, and bootstrap consensus tree inferred from 1000 replicates were used. *Monochamus saltuarius* (Msal) (red), *S. bifasciatus* (Sbif), *A. corpulenta* (Acor), *A. planipennis* (Apla), and *D. ponderosae* (Dpon). Orco clade marked with green background.

3.6. IRs, GRs, and SNMPs in Healthy and Infested M. saltuarius

Regarding IRs, IR10 and IR12 were expressed higher in healthy females compared to healthy males, IR4 expressed higher in healthy than infested females, while IR25 and IR21 were expressed higher in healthy than infested males (Figure 7(A1–A3)). Regarding GRs, no sex bias was observed in healthy adults, and only GR8 showed higher expression in female *M. saltuarius* infested with *B. xylophilus* (Figure 7(B1–B3)). GR8 was the only up-regulated chemosensory gene in *M. saltuarius* infested with *B. xylophilus*. Regarding SNMP, no sex bias was observed in healthy *M. saltuarius* and SNMP2 showed lower expression in infested males than in healthy ones (Figure 7(C1–C3)).

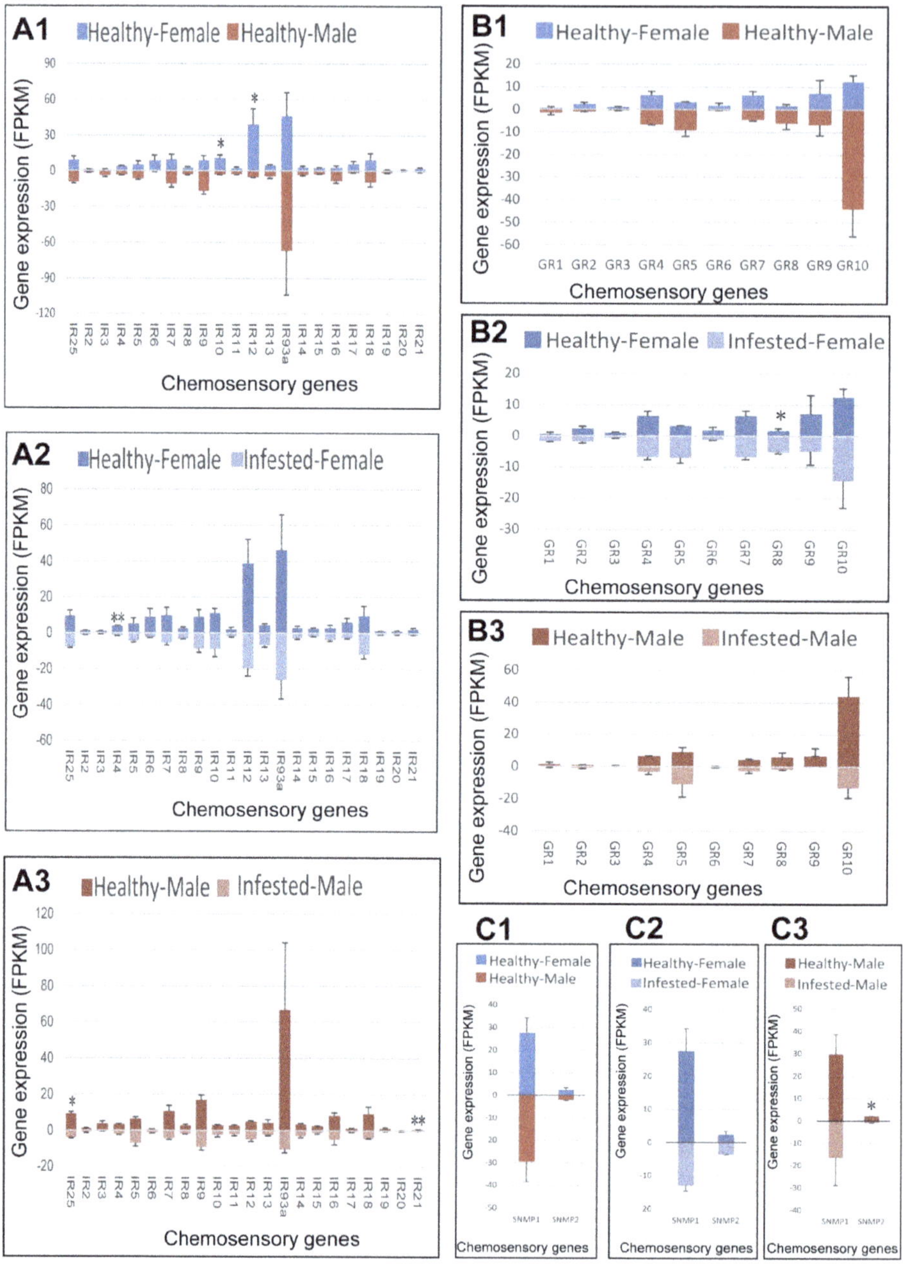

Figure 7. Expression patterns of ionotropic receptors (IRs), gustatory receptors (GRs), and sensory neuron membrane proteins (SNMPs) in healthy and *B. xylophilus*-infested *M. saltuarius*. (**A1**), IRs expression differences between sexes of healthy *M. saltuarius*; (**A2**), IRs expression differences between healthy and infested female *M. saltuarius*; (**A3**), IRs expression differences between healthy and infested male *M. saltuarius*. (**B1**), GRs expression differences between sexes of healthy *M. saltuarius*;

(**B2**), GRs expression differences between healthy and infested female *M. saltuarius*; (**B3**), GRs expression differences between healthy and infested male *M. saltuarius*. (**C1**), SNMPs expression differences between sexes of healthy *M. saltuarius*; (**C2**), SNMPs expression differences between healthy and infested female *M. saltuarius*; (**C3**), SNMPs expression differences between healthy and infested male *M. saltuarius*. Transcript levels are expressed as reads per kilobase of exon per million mapped reads (RPKM), and the expression data were shown as mean±SE. The stars above each bar of figure indicate significant differences ($p < 0.05$ * and $p < 0.05$ **).

4. Discussion

Olfactory cues play important roles in the host choice of immature and mature *M. saltuarius*. The spread of nematodes to new healthy hosts mainly relies on the feeding activity of immature *M. saltuarius*. If infestation with *B. xylophilus* for approximately one month can affect olfactory recognition in longicorn beetles it could also affect the nematode's spreading interval duration. In the present study, we identified 53 OBPs, 15CSPs, 15 ORs, 10 GRs, 22 IRs, and 2 SNMPs in *M. saltuarius*, a new vector of *B. xylophilus* in northern China. Sex-specific expression patterns in healthy beetles occurred mainly regarding OBPs. Interestingly, we found that OBPs and ORs were down-regulated in male *M. saltuarius* infested with *B. xylophilus*, which may reduce olfactory sensitivity and delay their maturation; consequently, this may prolong the time period, during which *B. xylophilus* can be transmitted through infested beetles.

The numbers of OBPs and CSPs in *M. saltuarius* are consistent with other Coleopteran insects whose genomic information are available [55,57], indicating relative complete identification of these two chemosensory families in *M. saltuarius*. However, the number of OR genes in *M. saltuarius* was lower than the number of other Coleopteran insects, this may be due to some missing genes down-regulated in the adults, especially in sexually mature adults. Another reason maybe that these insects have mated and some of the chemosensory genes were down-regulated. However, *S. bifasciatus* ORs were also identified from antennae transcriptome data, and its numbers were much higher than *M. saltuarius* [54]. To elucidate the underlying causation, we reviewed olfactory gene identifications of other longicorn beetles. To date, olfactory-related gene identification and respective studies have only been performed on a few longicorn beetles. Fifty-two OBPs and 132 ORs were identified in the genome of *Anoplophora glabripennis* [58], and 57 ORs were identified in *Megacyllene caryae* [59]. However, 31 OBPs and only nine ORs were identified in the antenna transcriptome of *M. alternatus* [3]. Both *Monochamus* species possess considerably fewer ORs than other longicorn beetles, which may be a particular characteristic of this genus. However, further studies on the olfactory system of *Monochamus* spp. based on genome data are needed to confirm this prediction.

The present study is the first to confirm that infestation with *B. xylophilus* can influence olfactory gene expression in the vector *M. saltuarius*. Interactions between *B. xylophilus* and their vector insects are complex throughout all stages of development. Before accumulation in its vector, *B. xylophilus* can secrete ascarosides to promote beetle pupation. After developing into the adult stage, the beetle secretes ascarosides to attract the *B. xylophilus* 4th instar larvae [8]. *B. xylophilus* 4th instar larvae development relies on C16 and C18 fatty acid ethyl esters, which are abundantly produced by beetle pupae at emergence and newly emerged adults [60]. Thus, *B. xylophilus* can promote and use signals from vector beetles to facilitate its accumulation on its vector before the newly emerged beetles leave their old host plants. When the beetles disperse for foraging, the time span for *B. xylophilus* to spread on healthy trees is critical; a longer foraging time increases the probability that *B. xylophilus* will spread to healthy trees. Our results indicate that *B. xylophilus* can reduce the olfactory sensitivity of male *M. saltuarius*, which may delay sexual maturity and further prolong the beetle's feeding activity, thus increasing the likelihood of spreading PWNs to new healthy hosts. Conversely, the olfactory system of female *M. saltuarius* was not affected by long-time *B. xylophilus* infestation, which may be a compromise for *B. xylophilus* as it requires female *M. saltuarius* to lay eggs, thereby initiating a new spreading cycle. However, further research is required to test whether GR8 upregulation in female *M. saltuarius* infested with *B. xylophilus* (Figure 7(B1–B3)) can facilitate feeding activity in female beetles and therefore enhance egg maturity and efficient oviposition. Further studies should be undertaken to test whether GR up regulation in infested mated females may help discriminate adequate hosts for oviposition and thereby facilitate the spread of nematode to damaged trees.

On the other hand, the blocking of chemosensation in infested *M. saltuarius* may be for the self-preservation of the insect. Sexually mature males need to find dead logs and trees where females are waiting for mating. Chemosensory gene down-regulation in males would reduce their sensitivity and influence mating success, and thus transfer less nematodes to its host trees. Anyway, whether these hypotheses are correct requires further physiological and gene function researches, and clarification of the complex relationships between nematodes-vectors-trees will imagine new pest control strategies.

Genes affected by *B. xylophilus* infestation may be used as new molecular targets for *M. saltuarius* control in the future. However, the functions of the regulated ORs and OBPs genes in male *M. saltuarius* are unclear, and further research is needed to elucidate the respective interactions between nematodes and their beetle vectors. This work extends our understanding of the complex relationship between nematodes-vectors-trees; this is critical for us to grasp new control targets and develop new control methods to protect large areas of pine forest.

Conclusion: firstly, we identified 53 OBPs, 15 CSPs, 15 ORs, 10 GRs, 22 IRs, and two SNMPs in *M. saltuarius*. Secondly, sex expression bias of the chemosensory genes was analyzed. Finally, we found that *B. xylophilus* infestation mainly down-regulated the OBPs and ORs expression levels in male *M. saltuarius* at the mating stage.

Supplementary Materials: The following supporting information can be downloaded at: https://www.mdpi.com/article/10.3390/f13020258/s1. Figure S1. Sequence length distribution of the *Monochamus saltuarius* antennal transcriptomes assembly. Figure S2. Maximum likelihood dendrogram based on protein sequences of candidate odorant binding protein (OBP) in *M. saltuarius* and orthologs from other insects. IQ-tree best fit model (VT+R6) was used and bootstrap consensus tree inferred from 1000 replicates. *Monochamus saltuarius* (Msal) (red), *S. bifasciatus* (Sbif), *T. castaneum* (Tcas), *A. corpulenta* (Acor), *A. planipennis* (Apla), *D. ponderosae* (Dpon), *Drosophila melanogaster* (Dmel), and *Bombyx mori* (Bmor). Orco clade was marked with green background. Table S1. Candidate chemosensory genes of *Monochamus saltuarius*. Table S2. FPKM and p values of the differently expressed chemosensory genes of *Monochamus saltuarius*.

Author Contributions: S.Z., X.W. and Y.Z. (Yanlong Zhang) designed the experiments; S.Z., X.W., Y.Z. (Yanlong Zhang) and Y.Z. (Yanan Zheng) collected samples, S.Z., X.W., Z.F. and R.Z. examined the samples; S.Z. analyzed the data and drafted the manuscript. Y.Z. (Yanlong Zhang), Y.Z. (Yanlong Zhang) and X.W. revised the manuscript. All authors have read and agreed to the published version of the manuscript.

Funding: This research was funded by Fundamental Research Funds for the Central Non-profit Research Institution of CAF: CAFYBB2018SZ006, CAFYBB2020QC001.

Data Availability Statement: The identified chemosensory genes were submitted to NCBI under the accession numbers MT008336–MT008452.

Acknowledgments: This research was funded by the Fundamental Research Funds for the Central Non-profit Research Institution of CAF (CAFYBB2018SZ006, CAFYBB2020QC001).

Conflicts of Interest: The authors declare no conflict of interest.

References

1. Shen, H.; Wang, F. The most plants in *Pinus*. *Plants* **1991**, *6*, 40.
2. Nickle, W.R. A taxonomic review of the genera of the *Aphelenchoidea* (Fuchs, 1937) Thorne, 1949 (Nematoda: Tylenchida). *J. Nematol.* **1970**, *2*, 375–392.
3. Wang, J.; Li, D.Z.; Min, S.F.; Mi, F.; Zhou, S.S.; Wang, M.Q. Analysis of chemosensory gene families in the beetle *Monochamus alternatus* and its parasitoid *Dastarcus helophoroides*. *Comp. Biochem. Physiol. Part D Genom. Proteom.* **2014**, *11*, 1–8. [CrossRef] [PubMed]

4. Morimoto, K.; Iwasaki, A. Role of *Monochamus alternatus* (Coleoptera: Cerambycidae) as a vector of *Bursaphelenchus lignicolus* (Nematoda: Aphelenchoididae). *Nihon Ringakkai Shi J. Jpn. For. Soc.* **1972**, *54*, 177–183.
5. Zhang, J.; Zhang, R.; Chen, J. Species and their dispersal ability of *Monochamus* as vectors to transmit *Bursaphelenchus xylophilus*. *J. Zhejiang For. Coll.* **2007**, *24*, 350–356.
6. Linit, M.J.; Kondo, E.; Smith, M.T. Insects associated with the pinewood nematode, *Bursaphelenchus xylophilus* (Nematoda: Aphelenchoididae), in missouri. *Environ. Entomol.* **1983**, *12*, 467–470. [CrossRef]
7. Kobayashi, F.; Yamane, A.; Ikeda, T. The japanese pine sawyer beetle as the vector of pine wilt disease. *Annu. Rev. Entomol.* **1984**, *29*, 115–135. [CrossRef]
8. Zhao, L.; Zhang, X.; Wei, Y.; Zhou, J.; Zhang, W.; Qin, P.; Chinta, S.; Kong, X.; Liu, Y.; Yu, H.; et al. Ascarosides coordinate the dispersal of a plant-parasitic nematode with the metamorphosis of its vector beetle. *Nat. Commun.* **2016**, *7*, 12341. [CrossRef]
9. Enda, N.; Mamiya, Y. Transmission of *Bursaphelenchus lignicolus* (Nematoda: Aphelenchoididae) by *Monochamus alternatus* (Coleoptera: Cerambycidae). *Nematologica* **1972**, *18*, 159–162.
10. Meng, Y. Monitoring, identification and control of pine wood nematode in Dandong City, Liaoning Province. *Agric. Technol.* **2018**, *38*, 62–63.
11. Wu, Y.; Wickham, J.D.; Zhao, L.; Sun, J. Co_2 drives the pine wood nematode off its insect vector. *Curr. Biol.* **2019**, *29*, R619–R620. [CrossRef]
12. Cho, W.S.; Koo, H.N.; Yun, S.H.; Lee, J.S.; Kim, G.H. Electron beam-induced sterility and inhibition of ovarian development in the sakhalin pine longicorn, *Monochamus saltuarius* (Coleoptera: Cerambycidae). *J. Econ. Entomol.* **2018**, *2*, 725–731. [CrossRef]
13. Li, Y.; Zhang, X. Analysis on the trend of invasion and expansion of *Bursaphelenchus xylophilus*. *For. Pest Dis.* **2018**, *37*, 1–4.
14. Jung, C.S.; Koh, S.-H.; Nam, Y.; Ahn, J.J.; Lee, C.Y.; Choi, W.I. A model for predicting spring emergence of *Monochamus saltuarius* (Coleoptera: Cerambycidae) from Korean white pine, *pinus koraiensis*. *J. Econ. Entomol.* **2015**, *108*, 1830–1836. [CrossRef] [PubMed]
15. Lee, H.-R.; Lee, S.-C.; Lee, D.H.; Choi, W.-S.; Jung, C.-S.; Jeon, J.-H.; Kim, J.-E.; Park, I.-K. Identification of the aggregation-sex pheromone produced by male *Monochamus saltuarius*, a major insect vector of the pine wood nematode. *J. Chem. Ecol.* **2017**, *43*, 670–678. [CrossRef] [PubMed]
16. Li, M.; Li, H.; Sheng, R.-C.; Sun, H.; Sun, S.-H.; Chen, F.-M. The first record of *Monochamus saltuarius* (Coleoptera; Cerambycidae) as vector of *Bursaphelenchus xylophilus* and its new potential hosts in china. *Insects* **2020**, *11*, 636. [CrossRef] [PubMed]
17. Webb, J.L. The southern pine sawyer. *USDA Bur. Entomol. Bull.* **1909**, *58*, 41–56.
18. Wingfield, M.J. Transmission of pine wood nematode to cut timber and girdled trees. *Plant Dis.* **1983**, *67*, 35–37. [CrossRef]
19. Skoczek, A.; Piesik, D.; Wenda-Piesik, A.; Buszewski, B.; Bocianowski, J.; Wawrzyniak, M. Volatile organic compounds released by maize following herbivory or insect extract application and communication between plants. *J. Appl. Entomol.* **2017**, *141*, 630–643. [CrossRef]
20. Piesik, D.; Rochat, D.; Delaney, K.J.; Marion-Poll, F. Orientation of European Corn Borer first instar larvae to synthetic green leaf volatiles. *J. Appl. Entomol.* **2013**, *137*, 234–240. [CrossRef]
21. Piesik, D.; Rochat, D.; van der Pers, J.; Marion-Poll, F. Pulsed odors from maize or spinach elicit orientation in European Corn Borer neonate larvae. *J. Chem. Ecol.* **2009**, *35*, 1032–1042. [CrossRef] [PubMed]
22. De Bruyne, M.; Baker, T.C. Odor detection in insects: Volatile codes. *J. Chem. Ecol.* **2008**, *34*, 882–897. [CrossRef]
23. Toshiya, I.; Nobuo, E.; Akiomi, Y.; Katsuo, O.; Takaaki, T. Attractants for the Japanese pine sawyer, *Monochamus alternatus* hope (Coleoptera: Cerambycidae). *Nihon Ringakkai Shi J. Jpn. For. Soc.* **1980**, *15*, 358–361.
24. Fan, J.; Zhang, D.; Zhang, Z.; Meng, J.; Wang, Y. Feeding behavior of *Monochamus alternatus* and its relationship with the host volatiles. *J. Zhejiang AF Univ.* **2014**, *31*, 78–82.
25. Hao, D.; Fenglin, M.A.; Wang, Y.; Zhang, Y.; Dai, H. Electroantennogram and behavioral responses of *Monochamus alternatus* to the volatiles from pinus thunbergii with different physiological status. *Chin. J. Appl. Ecol.* **2006**, *17*, 1070.
26. Fettköther, R.; Reddy, G.V.P.; Noldt, U.; Dettner, K. Effect of host and larval frass volatiles on behavioural response of the old house borer, *Hylotrupes bajulus* (L.) (Coleoptera: Cerambycidae), in a wind tunnel bioassay. *Chemoecology* **2000**, *10*, 1–10. [CrossRef]
27. Phillips, T.W.; Wilkening, A.J.; Atkinson, T.H.; Nation, J.L.; Wilkinson, R.C.; Foltz, J.L. Synergism of turpentine and ethanol as attractants for certain pine-infesting beetles (Coleoptera). *Environ. Entomol.* **1988**, *17*, 456–462. [CrossRef]
28. Wynand, V.D.G.V.N.; Carlson, J.R. Insects as chemosensors of humans and crops. *Nature* **2006**, *444*, 302–307.
29. Pelosi, P.; Zhou, J.-J.; Ban, L.P.; Calvello, M. Soluble proteins in insect chemical communication. *Cell. Mol. Life Sci.* **2006**, *63*, 1658–1676. [CrossRef]
30. Leal, W.S. Odorant reception in insects: Roles of receptors, binding proteins, and degrading enzymes. *Annu. Rev. Entomol.* **2013**, *58*, 373–391. [CrossRef]
31. Pelosi, P.; Calvello, M.; Ban, L. Diversity of odorant-binding proteins and chemosensory proteins in insects. *Chem. Senses* **2005**, *30*, i291–i292. [CrossRef]
32. Rützler, M.; Zwiebel, L. Molecular biology of insect olfaction:Recent progress and conceptual models. *J. Comp. Physiol. A* **2005**, *191*, 777–790. [CrossRef]
33. Su, C.Y.; Menuz, K.; Carlson, J.R. Olfactory perception: Receptors, cells, and circuits. *Cell* **2009**, *139*, 45–59. [CrossRef]
34. Vosshall, L.B.; Stocker, R.F. Molecular architecture of smell and taste in *Drosophila*. *Annu. Rev. Neurosci.* **2007**, *30*, 505–533. [CrossRef]
35. Abdel-Latief, M. A family of chemoreceptors in *Tribolium castaneum* (Tenebrionidae: Coleoptera). *PLoS ONE* **2007**, *2*, e1319. [CrossRef]
36. Andersson, M.N.; Grosse-Wilde, E.; Keeling, C.I.; Bengtsson, J.M.; Yuen, M.M.S.; Li, M.; Hillbur, Y.; Bohlmann, J.; Hansson, B.S.; Schlyter, F. Antennal transcriptome analysis of the chemosensory gene families in the tree killing bark beetles, *Ips typographus* and *Dendroctonus ponderosae* (Coleoptera: Curculionidae: Scolytinae). *BMC Genom.* **2013**, *14*, 198. [CrossRef]

37. Ali, S.; Ahmed, M.Z.; Li, N.; Ali, S.A.I.; Wang, M.Q. Functional characteristics of chemosensory proteins in the sawyer beetle *Monochamus alternatus* Hope. *Bull. Entomol. Res.* **2018**, *109*, 34–42. [CrossRef]
38. Hu, P.; Wang, J.; Cui, M.; Tao, J.; Luo, Y. Antennal transcriptome analysis of the asian longhorned beetle *Anoplophora glabripennis*. *Sci. Rep.* **2016**, *6*, 26652. [CrossRef] [PubMed]
39. Zhang, S.; Zhang, Z.; Wang, H.; Kong, X. Antennal transcriptome analysis and comparison of olfactory genes in two sympatric defoliators, *Dendrolimus houi* and *Dendrolimus kikuchii* (lepidoptera: Lasiocampidae). *Insect Biochem. Mol. Biol.* **2014**, *52*, 69–81. [CrossRef] [PubMed]
40. Zhang, S.-F.; Liu, H.-H.; Kong, X.-B.; Wang, H.-B.; Liu, F.; Zhang, Z. Identification and expression profiling of chemosensory genes in *Dendrolimus punctatus* walker. *Front. Physiol.* **2017**, *8*, 471. [CrossRef] [PubMed]
41. Bogdanova, E.A.; Shagin, D.A.; Lukyanov, S.A. Normalization of full-length enriched cdna. *Mol. BioSystems* **2008**, *4*, 205–212. [CrossRef] [PubMed]
42. Zhulidov, P.A.; Bogdanova, E.A.; Shcheglov, A.S.; Vagner, L.L.; Khaspekov, G.L.; Kozhemyako, V.B.; Matz, M.V.; Meleshkevitch, E.; Moroz, L.L.; Lukyanov, S.A.; et al. Simple cDNA normalization using Kamchatka Crab duplex-specific nuclease. *Nucleic Acids Res.* **2004**, *32*, e37. [CrossRef] [PubMed]
43. Grabherr, M.G.; Haas, B.J.; Yassour, M.; Levin, J.Z.; Thompson, D.A.; Amit, I.; Adiconis, X.; Fan, L.; Raychowdhury, R.; Zeng, Q.; et al. Trinity: Reconstructing a full-length transcriptome without a genome from rna-seq data. *Nat. Biotechnol.* **2011**, *29*, 644–652. [CrossRef] [PubMed]
44. Pertea, G.; Huang, X.; Liang, F.; Antonescu, V.; Sultana, R.; Karamycheva, S.; Lee, Y.; White, J.; Cheung, F.; Parvizi, B.; et al. TIGR gene indices clustering tools (TGICL): A software system for fast clustering of large EST datasets. *Bioinformatics* **2003**, *19*, 651–652. [CrossRef] [PubMed]
45. Conesa, A.; Götz, S.; García-Gómez, J.M.; Terol, J.; Talón, M.; Robles, M. Blast2GO: A universal tool for annotation, visualization and analysis in functional genomics research. *Bioinformatics* **2005**, *21*, 3674–3676. [CrossRef]
46. Götz, S.; García-Gómez, J.M.; Terol, J.; Williams, T.D.; Nagaraj, S.H.; Nueda, M.J.; Robles, M.; Talón, M.; Dopazo, J.; Conesa, A. High-throughput functional annotation and data mining with the blast2GO suite. *Nucleic Acids Res.* **2008**, *36*, 3420–3435. [CrossRef] [PubMed]
47. Krieger, J.; Grosse-Wilde, E.; Gohl, T.; Dewer, Y.M.E.; Raming, K.; Breer, H. Genes encoding candidate pheromone receptors in a moth (*Heliothis virescens*). *Proc. Natl. Acad. Sci. USA* **2004**, *101*, 11845–11850. [CrossRef] [PubMed]
48. Ashburner, M.; Ball, C.A.; Blake, J.A.; Botstein, D.; Butler, H.; Cherry, J.M.; Davis, A.P.; Dolinski, K.; Dwight, S.S.; Eppig, J.T.; et al. Gene ontology: Tool for the unification of biology. *Nat. Genet.* **2000**, *25*, 25–29. [CrossRef]
49. Trapnell, C.; Williams, B.A.; Pertea, G.; Mortazavi, A.; Kwan, G.; van Baren, M.J.; Salzberg, S.L.; Wold, B.J.; Pachter, L. Transcript assembly and quantification by rna-seq reveals unannotated transcripts and isoform switching during cell differentiation. *Nat. Biotechnol.* **2010**, *28*, 511–515. [CrossRef]
50. Anders, S.; Huber, W. *Differential Expression of RNA-Seq Data at the Gene Level—The DESeq Package*; European Molecular Biology Laboratory: Heidelberg, Germany, 2012.
51. Robinson, M.D.; McCarthy, D.J.; Smyth, G.K. Edger: A bioconductor package for differential expression analysis of digital gene expression data. *Bioinformatics* **2010**, *26*, 139–140. [CrossRef]
52. Trifinopoulos, J.; Nguyen, L.-T.; von Haeseler, A.; Minh, B.Q. W-IQ-TREE: A fast online phylogenetic tool for maximum likelihood analysis. *Nucleic Acids Res.* **2016**, *44*, W232–W235. [CrossRef] [PubMed]
53. Letunic, I.; Bork, P. Interactive Tree Of Life (iTOL) v5: An online tool for phylogenetic tree display and annotation. *Nucleic Acids Res.* **2021**, *49*, W293–W296. [CrossRef] [PubMed]
54. Zhang, S.; Shen, S.; Zhang, S.; Wang, H.; Kong, X.; Liu, F.; Zhang, Z. Chemosensory characteristics of two semanotus bifasciatus populations. *Forests* **2019**, *10*, 655. [CrossRef]
55. Vieira, F.G.; Rozas, J. Comparative genomics of the odorant-binding and chemosensory protein gene families across the arthropoda: Origin and evolutionary history of the chemosensory system. *Genome Biol. Evol.* **2011**, *3*, 476–490. [CrossRef]
56. Xiao, L.; Qian, J.; Jie, W.; Fei, L.; Jiang, X.; Hu, J.; Qu, M. Chemosensory gene families in adult antennae of *Anomala corpulenta* motschulsky (Coleoptera: Scarabaeidae: Rutelinae). *PLoS ONE* **2015**, *10*, e0144214.
57. Andersson, M.N.; Keeling, C.I.; Mitchell, R.F. Genomic content of chemosensory genes correlates with host range in wood-boring beetles (*Dendroctonus ponderosae*, *Agrilus planipennis*, and *Anoplophora glabripennis*). *BMC Genom.* **2019**, *20*, 690. [CrossRef]
58. McKenna, D.D.; Scully, E.D.; Pauchet, Y.; Hoover, K.; Kirsch, R.; Geib, S.M.; Mitchell, R.F.; Waterhouse, R.M.; Ahn, S.-J.; Arsala, D.; et al. Genome of the Asian longhorned beetle (*Anoplophora glabripennis*), a globally significant invasive species, reveals key functional and evolutionary innovations at the beetle–plant interface. *Genome Biol.* **2016**, *17*, 227. [CrossRef]
59. Mitchell, R.F.; Hughes, D.T.; Luetje, C.W.; Millar, J.G.; Soriano-Agatón, F.; Hanks, L.M.; Robertson, H.M. Sequencing and characterizing odorant receptors of the cerambycid beetle *Megacyllene caryae*. *Insect Biochem. Mol.* **2012**, *42*, 499–505. [CrossRef]
60. Zhao, L.; Zhang, S.; Wei, W.; Hao, H.; Zhang, B.; Butcher, R.; Sun, J. Chemical signals synchronize the life cycles of a plant-parasitic nematode and its vector beetle. *Curr. Biol.* **2013**, *23*, 2038. [CrossRef]

forests
MDPI

Article

Long-Term Projections of the Natural Expansion of the Pine Wood Nematode in the Iberian Peninsula

Begoña de la Fuente * and Santiago Saura

MONTES (ETSI de Montes, Forestal y del Medio Natural), Universidad Politécnica de Madrid, Ciudad Universitaria s/n, 28040 Madrid, Spain; santiago.saura@upm.es
* Correspondence: fuemarbego@hotmail.com

Abstract: The invasive pine wood nematode (PWN), *Bursaphelenchus xylophilus*, causal agent of pine wilt disease, was first reported in Europe, near Lisbon, in 1999, and has since then spread to most of Portugal. We here modelled the spatiotemporal patterns of future PNW natural spread in the Iberian Peninsula, as dispersed by the vector beetle *Monochamus galloprovincialis*, using a process-based and previously validated network model. We improved the accuracy, informative content, forecasted period and spatial drivers considered in previous modelling efforts for the PWN in Southern Europe. We considered the distribution and different susceptibility to the PWN of individual pine tree species and the effect of climate change projections on environmental suitability for PWN spread, as we modelled the PWN expansion dynamics over the long term (>100 years). We found that, in the absence of effective containment measures, the PWN will spread naturally to the entire Iberian Peninsula, including the Pyrenees, where it would find a gateway for spread into France. The PWN spread will be relatively gradual, with an average rate of 0.83% of the total current Iberian pine forest area infected yearly. Climate was not found to be an important limiting factor for long-term PWN spread, because (i) there is ample availability of alternative pathways for PWN dispersal through areas that are already suitable for the PWN in the current climatic conditions; and (ii) future temperatures will make most of the Iberian Peninsula suitable for the PWN before the end of this century. Unlike climate, the susceptibility of different pine tree species to the PWN was a strong determinant of PWN expansion through Spain. This finding highlights the importance of accounting for individual tree species data and of additional research on species-specific susceptibility for more accurate modelling of PWN spread and guidance of related containment efforts.

Keywords: pine wood nematode; pine wilt disease; forest pests; Iberian Peninsula; invasive alien species

1. Introduction

In 1999, the invasive alien species pine wood nematode (PWN; *Bursaphelenchus xylophilus* Steiner & Buhrer, 1934 (Nickle, 1970)), causal agent of pine wilt disease (PWD), was first reported in Europe, on *Pinus pinaster* Ait. trees in the district of Setubal, in southwestern Portugal, near Lisbon [1]. Since then, it has quickly spread through Portugal, producing large-scale damage in Portuguese forests [2]. The area now affected by the PWN covers most of the Portuguese mainland territory (http://www.icnf.pt/portal/florestas/prag-doe/ag-bn/nmp/infgeo, accessed on 10 September 2019) and has reached Spain, with several and so-far isolated foci close to the border with Portugal [3–9].

Under suitable climatic conditions that trigger PWD, the PWN is able to rapidly kill susceptible pine trees [10]. Though this pest causes limited damage in its native range in North America, it has caused massive mortality in conifer forests elsewhere, most recently in Portugal but also in Eastern Asia since its accidental introduction in Japan and neighboring Asian countries at the beginning of the 20th century [11]. The mortality patterns caused by PWD are already gradually changing the structure and composition of pine stands in Portugal, rendering them less dense and with fewer mature and old trees [12].

As is typical for epidemic processes, early detection plays a vital role for the eradication and containment of the PWD. Knowing which areas are most likely to be infected at a particular point in time is essential to control the disease. The widely distributed cerambycid beetle *Monochamus galloprovincialis* Oliver is the only known PWN vector in the Iberian Peninsula [13]. Consequently, the natural spread of the PWN in Europe depends on the dispersal abilities of this vector beetle that can infect pine trees when feeding or breeding on them [14,15]. Understanding the behavior of *M. galloprovincialis* is therefore a key to predicting the dynamics of the PWN invasion front.

Coniferous forests cover almost 1 Mha in Portugal and about 8 Mha in Spain, or ca. 30% of the total forest area in these countries, reflecting their major ecological and economic importance in the Iberian Peninsula. Pine forests are ubiquitous on the Iberian Peninsula, with pine species ranging from the coastal areas (Aleppo pine *Pinus halepensis* M. and Stone pine *Pinus pinea* L.) to the upper areas of the mountain ranges (Scots pine *Pinus sylvestris* L. and Mountain pine *Pinus uncinata* Ram.). As *Pinus* species make up 95% of coniferous forests in Portugal and Spain, further spread of the PWN would have a strong impact on the forest-based economy [16–19], on biodiversity and on other important ecosystem services provided by coniferous forests inside and outside protected areas [20–22].

There is evidence that the resistance of pine species to the PWN may be related not only to the possibility of migration of the nematodes through the pine tissues after their entrance into the tree, but also to other specific intrinsic factors that repel, immobilize or disrupt the life cycle of nematodes, including their reproduction [23,24]. Although all European pine tree species appear susceptible to some degree to the nematode infection [25], inoculation tests in controlled conditions have shown that they differ in their susceptibility to PWD [26,27]. According to the results of the inoculation tests carried out by Menéndez-Gutierrez et al. [28], *Pinus sylvestris* and *Pinus pinaster* are the European native pine species most susceptible to the PWN, while *Pinus pinea* and *Pinus halepensis* appear to be less susceptible to the PWN. So far in Portugal, *Pinus pinea* trees seem not to be affected in areas where *Pinus pinaster* [1,12] and *Pinus nigra* Arn. [29] trees are killed by the PWN. This agrees with the results reported by Nunes da Silva et al. [26]. Among the non-native pine species that are planted for timber production in the northern part of the Iberian Peninsula, *Pinus radiata* showed considerable susceptibility to the PWN [28]. Thus far, however, the different susceptibility of the European pine species to the PWN has not been considered in PWN spread predictive modelling [30,31].

PWN transmission to a host tree does not necessarily lead to PWD and tree decline. PWD expression depends on environmental factors [32], such as the role of water stress caused by seasonal drought and high temperature in disease development [33]. When an area is not climatically suitable for the development of the PWD, the lack of declining trees prevents further spread of the pest to other potentially susceptible trees or areas. According to Gruffudd et al. [34], the risk of conifer trees infected with PWN becoming symptomatic is low if the mean summer temperature (June–August) is below 19.31 °C, and high if it is above 20 °C. Locations between 19.31 °C and 20 °C are considered at some risk of PWD [34]. Therefore, the northern part of the Iberian Peninsula, and some high-altitude areas in certain Iberian mountain chains, are currently not climatically suitable for PWD expression [30,35]. However, climate warming may progressively change this [36,37]. Therefore, considering both current and future climate in the Iberian Peninsula is important for long-term projections of the potential PWN expansion.

In this study, we modelled the spatiotemporal patterns of future natural spread of the PWN in the Iberian Peninsula, as dispersed by the longhorn beetle *Monochamus galloprovincialis*. We improved the accuracy, informative content, forecasted period and number of factors considered in previous modelling efforts for the PWN in Southern

Europe [30,31,38]. Unlike in previous studies, we jointly considered the distribution and different susceptibility of pine tree species, the effect of climate change projections on environmental suitability for PWN spread and the dynamics of PWN natural expansion toward the end of the century. We applied the spatio-temporal network model described in De la Fuente et al. [30], which was validated with actual PWN spread data in Portugal and was shown to have a high predictive accuracy [30]. We applied this model to predict the areas of the Iberian Peninsula through which the PWN is most likely to spread by natural means and in absence of dedicated containment measures, and the most likely year of infection in each location. We also evaluated the potential influence of climate warming on PWN spread and the effect of the susceptibility to the PWN of different pine tree species in the expected patterns and magnitude of spread. Our results can support management efforts by forecasting when areas are at greater risk of vector-mediated invasion. They also indicate how climate and tree species distribution may mediate the long-term patterns of PWN spread and inform the related pest containment and forest planning measures that may be applied in the Iberian Peninsula.

2. Materials and Methods

2.1. Study Area

In our analysis, we considered the entire Iberian Peninsula as study area, which has an extent of about 597,000 km^2. We described the study area using a grid of 1 km × 1 km cells (referred to as nodes in the modelling). This spatial resolution was considered sufficiently high given the extent of the study area and the >1 km dispersal range of the vector beetle *Monochamus galloprovincialis* [30].

2.2. Pine Tree Species Susceptibility to Pine Wood Nematode

We considered two different scenarios (A and H) regarding which pine species are susceptible to infection and which species can be a source of newly infected vector beetles that can further transmit the PWN to other pine trees:

- Scenario A: All pine tree species on the Iberian Peninsula are equally susceptible to the PWN. Under this scenario, the PWN spread modelling does not distinguish between pine species.
- Scenario H: Only pine species with high (*P. pinaster* and *P. radiata*) or very high (*P. sylvestris*) susceptibility to the PWN, following Menéndez-Gutiérrez et al. [28], can get infected and infect other pine trees. The other pine species were considered as neither targets nor sources of infection in the PWN spread modelling.

These two scenarios covered the range of plausible responses of different pine tree species to the PWN, according to the available studies [26–28], as well as the range of modelling scenarios within which the actual spread of the PWN in the Iberian Peninsula may be contained. They also allowed to evaluate the impact that not considering the different susceptibility to the PWN of pine tree species may have had on the results of previous modelling work on this invasive tree pest in Southern Europe.

2.3. Forest Tree Species Maps

For each cell i, we calculated the area covered by pine forest in the cell (a_i), either considering all pine tree species (Scenario A above) or only those pine tree species with high susceptibility to the PWN (Scenario H above). Pine forest area in the cell was calculated considering the area of pure pine forest ($a_{c,i}$) and of mixed pine–broadleaved ($a_{m,i}$) forest in the available maps (see below) as $a_i = a_{c,i} + F \times a_{m,i}$, where F indicates the fraction of pine tree cover in mixed stands, which was taken from the maps when available and otherwise set to 0.5.

For this purpose, we considered the spatial distribution of the different pine forest species as reported in the most detailed forest maps available in our study area: the Forest Map of Spain developed by the Spanish Ministry of Environment in coordination with the Third Spanish National Forest Inventory (scale 1:50,000, minimum mapping unit of

2.25 ha) and the 2007 land cover map of continental Portugal (COS2007) developed by the Portuguese Geographic Institute (1:25,000 scale, minimum mapping unit of 1 ha).

The Spanish Forest Map differentiates the distribution of each individual pine tree species. The Portuguese land cover map only differentiates *Pinus pinaster* (Maritime pine forests) and *Pinus pinea* (Stone pine forests), which are by far the pine species more widely distributed in Portugal. A total of 94.6% of the coniferous forest territory of continental Portugal is covered by these two species according to the results of the 6th National Forest Inventory of Portugal (http://www2.icnf.pt/portal/florestas/ifn/ifn6, accessed on 10 September 2019). Therefore, for Portugal, we used only the spatial distribution of Maritime pine forests when running Scenario H. For Scenario A, we used the distribution of the following three map categories: Stone pine forests, Maritime pine forests and forests of other coniferous species.

This resulted, for Scenario A, in a network of 210,338 nodes (1 km^2 cells) describing the distribution of all pine forests on the Iberian Peninsula (Scenario A). For Scenario H, we obtained a network of 121,200 nodes describing the spatial distribution of the three most susceptible pine species (*P. pinaster*, *P. sylvestris* and *P. radiata*) on the Iberian Peninsula.

2.4. Climatic Suitability to Pine Wilt Expression

We considered the climatic suitability of different areas in Spain and Portugal for the development of PWD in PWN-infected areas in current and future climates. The current climate data for Spain and Portugal were obtained from Ninyerola et al. [39]. For future climates, we examined the climate change projections provided by the ENSEMBLES project (http://ensembles-eu.metoffice.com/, accessed on 5 March 2020), considering the IPCC A1B scenario. These projections indicate that by 2100 all the areas with pine forests in the Iberian Peninsula will be climatically suitable for the PWN, except some isolated locations around a few mountain peaks that cover less than 1% of the pine forest distribution in the Iberian Peninsula. Therefore, we modelled PWN spread under the following two climate scenarios:

- Current climate (Scenario C): Only 1 km^2 cells in which the current mean summer temperature is above 19.31 °C are considered climatically suitable for wilt expression [31]. Excluded from the climatically suitable areas was 23% of the total pine forest extent in the Iberian Peninsula when considering all pine tree species (Scenario A) or 38% of the area covered in the Iberian Peninsula by the pine tree species that are highly susceptible to the PWN (Scenario H). In Scenario C, we simulated PWN spread assuming no dispersal of infected beetles from the areas where the disease does not express due to climate constraints.
- Future climate (Scenario F): All the Iberian Peninsula, except the isolated locations mentioned above, is climatically suitable for the expression of PWD. All 1 km^2 cells can act as spreaders of the PWN once they are invaded by the disease.

These two scenarios covered, therefore, the most climatically restrictive scenario (C) and the least restrictive scenario (F) that can be considered for the PWN expansion modelling.

2.5. PWN Spread Model

We simulated the PWN expansion using the process-based network model presented in De la Fuente et al. [30], which predicts the natural spread of the PWN through the dispersal of the vector beetle *Monochamus galloprovincialis*. This model was validated with data on the past PWN expansion patterns in Portugal and showed a high predictive accuracy [30].

In this model, nodes represent locations with pine forest (here 1 km^2 cells), and links (edges) reflect the probability of disease spread between two nodes. The model predictions are calculated based on two main analytical steps. First, the model calculates the weight of the links in the network as the probability that a sufficient number of vector beetles is able to disperse between two nodes during the yearly flight period so as to transmit the disease to the destination node (if the source node was infected). Second, using the distribution of infected and non-infected nodes in a given year and the link weights, it calculates, each year, the probability that individual nodes actually become infected, considering the contribution of all possible infection source nodes and dispersal pathways. Each of the nodes could be, in any year, in an infected or non-infected state: for the purpose of this study, we considered a cell infected when the infection probability is higher than 0.9 and non-infected when the infection probability is lower than 0.1. Intermediate probability values (between 0.1 and 0.9) were considered separately but represent a considerably small number of cells given that the infection probabilities predicted by the model are almost binary, i.e., very close to 0 or 1 [30].

The model has the following five parameters:

- d, the mean dispersal distance of the vector beetle, *Monochamus galloprovincialis*.
- N, the number of infected beetles emerging from each infected tree and thus available for dispersal.
- k, the number of infected beetles that need to reach an uninfected pine tree to actually transmit the disease to that tree.
- β, a coefficient determining the tail fatness (kurtosis) of the vector beetle dispersal kernel.
- c, the number of infected trees per hectare of pine forest in newly infected areas. This parameter was used to convert the pine forest area in a cell i (a_i) to the number of pine trees expected to be infected in that cell when PWN transmission occurs.

Based on available empirical and modelling studies on the PWN (see De la Fuente et al. [30]), we set $d = 1500$ m, $N = 9$, $k = 2$, $\beta = 1$ and $c = 0.5$, which were the best supported and validated parameter values and hence those used in all the predictions presented in this study. Further model details and equations are provided in De la Fuente et al. [30].

2.6. PWN Spread Modelling Scenarios

We predicted future patterns of natural PWN spread through the Iberian Peninsula at yearly intervals up to year 2150 in the absence of effective containment measures, starting the model runs from the PWN distribution observed in Portugal in June 2018 (http://www.icnf.pt/portal/florestas/prag-doe/ag-bn/nmp/infgeo, accessed on 10 September 2019). Model runs were conducted in the four scenarios combining those on pine tree species susceptibility to the PWN and those related to climate described above:

- Scenario AC: all pine tree species (Scenario A) and current climate (Scenario C) considered.
- Scenario AF: all pine tree species (Scenario A) and future climate (Scenario F) considered.
- Scenario HC: only highly susceptible tree pine species (Scenario H) and current climate (Scenario C) considered.
- Scenario HF: only highly susceptible tree species (Scenario H) and future climate (Scenario F) considered.

Of these scenarios, AF has the greatest potential for higher rates of PWN spread, because any pine forest stand of any pine tree species and in any location in the Iberian Peninsula can be infected by the nematode and thus be a source of PWD. Conversely, HC is the most restrictive scenario for the modelled spread.

The models were implemented and run in the R programming language [40], as were all related analyses.

3. Results

3.1. PWN Spread in Scenario AF: All Pine Forests and Future Climate

In Scenario AF, which is the least restrictive scenario for the PWN expansion (all pine forests considered and no climatic restrictions), the PWN spread from Portugal into Spain is predicted to occur uninterrupted through the expansive tracts of host vegetation (Figure 1). The spread in scenario AF is faster than in the other three scenarios (Figures 1 and 2). Modelling results indicate that by 2050 the PWN will be present throughout the region of Galicia, entering the north of the region of Asturias through the coastal area and the south of Asturias through the Cantabrian Range (Figure 1). By 2050, the disease will have reached the pine forests of the Central Plateau of the region of Castilla y León through Sierra de la Culebra, in the province of Zamora. Through the pathway of the Central Range, the infection will have reached by 2050 the central part of Spain up to the border of the region of Madrid (Figure 1). In the south of Spain, the mountains of Sierra Morena are predicted to start to be affected by the PWN by 2050, and through the coast the PWN will have reached the pine forests on the Guadalquivir Valley by that date (Figure 1).

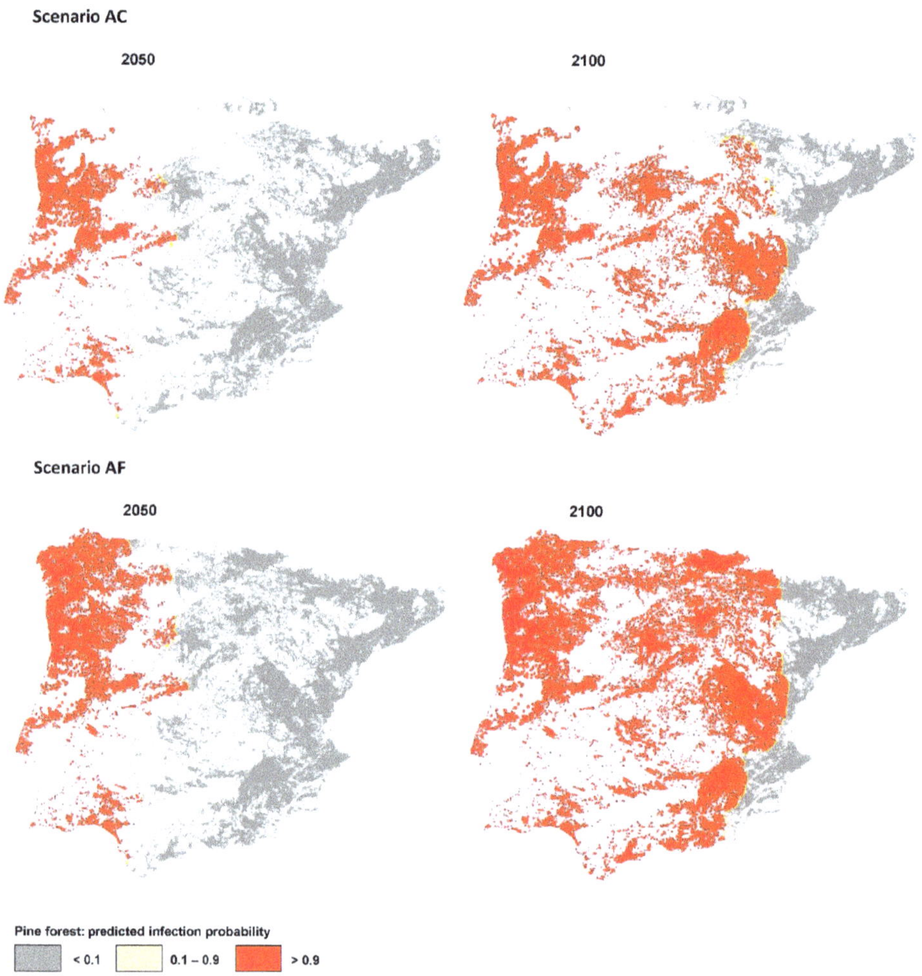

Figure 1. PWN spread in the Iberian Peninsula for the years 2050 and 2100, considering all pine tree species as susceptible of being infected. Both the current climate (Scenario AC) and future climate (Scenario AF) are considered.

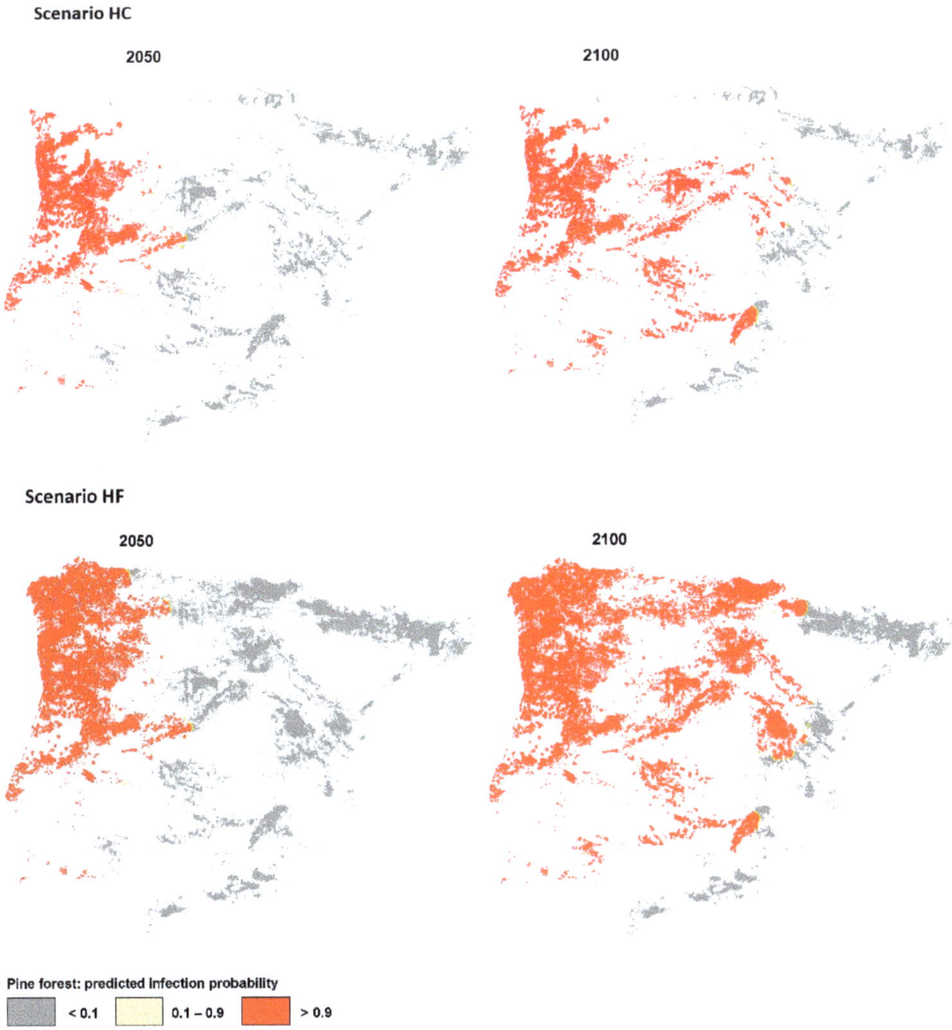

Figure 2. PWN spread in the Iberian Peninsula for the years 2050 and 2100, considering only the forests composed by pine tree species with high susceptibility to infection. Both the current climate (Scenario HC) and future climate (Scenario HF) are considered.

The rate of PWN spread is predicted to considerably increase after 2050 in Scenario AF (Figure 3). The average rate of infection increases from 43,763 ha per year in the period 2020–2050 to 68,038 ha per year in the period 2050–2100 (Figure 3), which is due to the higher continuity of host pine trees in the eastern half of the Iberian Peninsula (Figure 1).

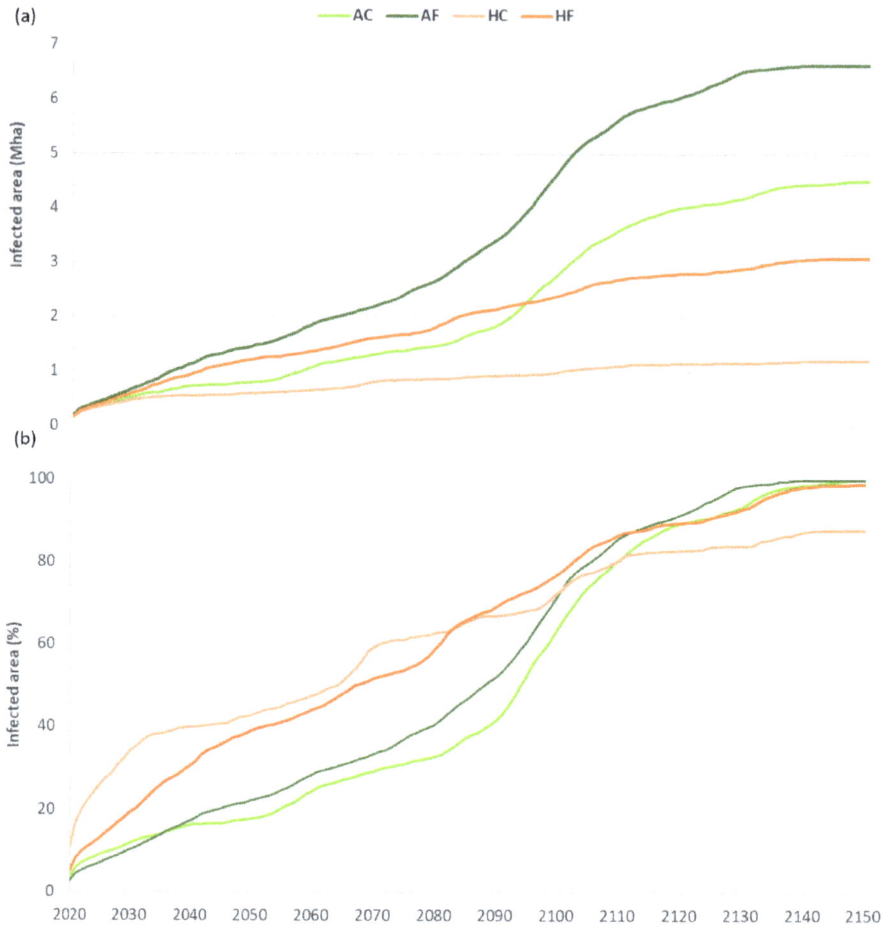

Figure 3. Infected area projections in the Iberian Peninsula for the four considered scenarios (AC, AF, HC and HF) up to the year 2150. Results are expressed in (**a**) as the forest area infected (Mha) and in (**b**) as the percent of the total forest area in the Iberian Peninsula that is infected. Projections consider all pine tree forest for Scenarios AC and AF (under current and future climate, respectively) and only the pine forest composed of highly susceptible species for Scenarios HC and HF (under current and future climate, respectively). Infected area corresponds to the cells with a probability of infection higher than 0.9 for a given year in the spread modelling. The asymptotes reached by the projections extend indefinitely beyond 2150 (all values remain constant up to 300 years of model runs).

In 2100, most of the Iberian Peninsula will be invaded by the PWN. Only the regions of Cataluña, Valencia and Murcia are predicted to be mostly free of infection by that year (Figure 1). All pine forests in the Iberian Peninsula are predicted to be infected by 2130 in Scenario AF (Figure 3).

3.2. PWN Spread in Scenario AC: All Pine Forests and Current Climate

When considering the current climate (Scenario AC), large tracts of forest in the northern part of the Iberian Peninsula, and particularly in the western half of this northern part, are not suitable to be infected or to transmit the disease (Figure 1). This results in less area that can be infected in any year compared to Scenario AF (Figure 3) and in a smaller portion of infected forest in the northeastern part of the Iberian Peninsula by year 2100 compared to Scenario AF, due to the lower continuity of susceptible forest in the north (Figure 1). This delay in the PWN expansion due to the lower forest continuity in the north does not occur, however, in the invasion front through the central and southern parts of Spain (Figure 1), with patterns of spread that show similarities to those described above for Scenario AF (Figure 1). Finally, the PWN is able to spread, in the AC scenario, to all the pine forest in the Iberian Peninsula, even if a few years later than in Scenario AF (Figure 3).

The climatic constraints to PWN expression affect mostly the northwestern part of Spain, which is the one closer to the PWN sources in Portugal, and there is also a higher continuity of host pine trees in the eastern half of the Iberian Peninsula (Figure 1). For these reasons, the speed of PWN spread is considerably lower in the period 2020–2050 than in the period 2050–2100, with average rates of infection of 21,365 ha/year and of 43,146 ha/year, respectively (Figure 3).

3.3. PWN Spread in Scenario HF: Highly Susceptible Pine Forests and Future Climate

When only the pine tree species with high or very high susceptibility to the PWN (*P. pinaster*, *P. radiata* and *P. sylvestris*) are considered, the host vegetation distribution is considerably less continuous (Figure 2). In addition, the distribution of host vegetation (highly susceptible pine trees) is concentrated towards the western half of the Iberian Peninsula in this scenario (compare Figures 1 and 2).

This different distribution gives much less host forest area that can be infected on the Iberian Peninsula in Scenario HF compared to Scenarios AF and AC (Figure 3). The rate of spread (quantified as the increase in the percent of total forest area that is infected), however, is faster in the first two decades (2020–2040) in Scenario HF than in Scenarios AF or AC (Figure 3) because a larger proportion of the considered forest is distributed closer to the infection sources in Portugal (Figure 2). However, in later decades this rate of spread does not exceed that of scenarios AF and AC, as the invasion front moves towards the eastern half of the Iberian Peninsula, where the distribution of highly susceptible pine forest is sparser (Figures 2 and 3).

In Scenario HF, there is only one small patch of highly susceptible pine forest that is predicted to remain free from infection even in the long run: Serranía de Ronda in the Sistema Bético (south of Spain), which represents only about 0.2% of the total area of highly susceptible pine forest on the Iberian Peninsula. All the other highly susceptible pine forests on the Iberian Peninsula are predicted to get infected by the PWN eventually, although it takes longer than in Scenario AF (Figure 3).

3.4. PWN Spread in Scenario HC: Highly Susceptible Pine Forests and Current Climate

Scenario HC, which considers only highly susceptible pine tree species together with the unsuitable areas for pine wilt expression in the current climatic conditions, is the most restrictive one for PWN expansion. Scenario HC has the lowest amount of pine forest in the Iberian Peninsula of the four scenarios, has the lowest proportion of the Iberian Peninsula infected by 2100 of all four scenarios and is the only scenario in which a significant proportion of the considered forest area never gets infected, even in the long run. Our model predicts that 12.2% of the forest area will never be reached by the PWN in the HC scenario (Figure 3). These non-infected areas in the long run include the entire Pyrenees, some northern coastal areas, the coastal mountain ranges in Catalonia (Cordillera Costero Catalana) and, as in Scenario HF, Serranía de Ronda (south of Spain).

The speed of spread in Scenario HF considerably decreases with time (Figure 3), with a rate of infection of 14,946 ha/year up to 2050 and of 8280 ha/year from 2050 to 2100.

4. Discussion

4.1. The Large Spread Potential of the PWN in the Iberian Peninsula

Our results indicate that, in the absence of effective containment measures, the PWN will naturally invade, through the dispersal flights of the vector beetle *Monochamus galloprovincialis*, all the pine forests of the Iberian Peninsula. This will not be, however, a fast invasion process. Our model predicts that the natural PWN spread to the entire Iberian Peninsula will take a century or longer, progressing at an average rate (average for all the four considered scenarios) of 0.83% of the total pine forest area per year over the next century, which increases to 0.90% per year if considering only the scenarios under future climate (AF and HF).

Our findings stress the high capacity of the PWN to spread through the rest of Portugal and through Spain via the dispersal flights of the vector beetle. Even large spatial discontinuities in the distribution of suitable areas for the PWN, caused by the lack of susceptible host vegetation or by climatic restrictions (Figures 1 and 2), do not generally block the spread of the PWN to other regions of the Iberian Peninsula, and at best delay to some extent the PWN expansion (with the only relative exception of Scenario HC discussed in the next section). These discontinuities are either circumvented through other available pathways or simply jumped over thanks to the considerable capacity of the PWN natural spread associated with the dispersal flights of the vector beetle populations. These vector populations are large in the Iberian pine forests, and only a few individuals that succeed in long-distance dispersal events are needed to further spread the disease to relatively distant areas [30].

4.2. The Impact of Current and Future Climate on PWN Spread

Of the four considered scenarios, only Scenario HC (high susceptibility pine tree species and current climate) gives some significant percentage of total forest area free from PWN invasion in the long run (about 12% of the total pine forest area considered in this scenario). Importantly, the areas excluded from PWN reach under this scenario include the entire Pyrenees, which would hence halt the natural spread of the PWN from Spain to France and the rest of Europe, naturally containing the PWN spread to the Iberian Peninsula only.

The inability of the PWN to spread to the Pyrenees under Scenario HC is because of two reasons. First, the coastal mountain ranges in Catalonia, running parallel to the Mediterranean Sea, are dominated by *P. halepensis*, a non-highly susceptible species, which makes these ranges lack sufficient density and continuity of host vegetation to provide a corridor for PWN spread towards the Pyrenees. Second, the potential west-to-east PWN dispersal to the Pyrenees through the north of Spain is impeded by the natural discontinuity due to the climatically unsuitable areas there in current conditions. The first reason is a structural one that will remain in the long run, impeding the PWN spread through that coastal pathway. The second reason, however, is unlikely to last: climate warming makes increasing portions of the currently climatically unsuitable areas adequate for pine wilt expression, making Scenario HC not representative of the longer-term PWN distribution and spread potential. Climate change will therefore pave the way for PWN spread through this northern corridor even when considering only the highly susceptible pine tree species as hosts (Scenario HF).

Current climatic constraints mean that an important portion of the Iberian Peninsula is not suitable for pine wilt expression and hence for the PWN spread: this happens in about 23% of the total pine forest area and 38% of the highly susceptible pine forest area, mainly in the north of Spain. These climatically unsuitable areas for the PWN will, however, progressively vanish with time, and all the pine forest area will be able to host and further transmit the wilt disease to other pine forest patches not later than by the end of the century, as predicted by the considered climate change models. Although we here base our climate forecasts on the IPCC A1B scenario, this may be in fact a relatively conservative or intermediate scenario that may underestimate the actual progression of climate warming [41], which would hence in any case result in the abovementioned significant lack of climatically suitable areas for PWN transmission by or before the end of the century. This will make, in summary, the natural spread of the PWN to the Pyrenees being delayed but not prevented in practice, in this or other realistic climate change scenarios. The expected climatic conditions do not offer, therefore, a significant limitation or barrier for the long-term spread and for the total pine forest area that can be invaded by the PWN in the Iberian Peninsula. Our results are consistent with the results obtained by Haran et al. [35] on the southern side of the Pyrenean chain, where they found that western and eastern hillsides may represent corridors favoring natural spread of the nematode from the Iberian Peninsula to France, and that temperature rise due to climate change may significantly reduce the extent of the barrier formed by the highest elevations.

4.3. The Importance of Pine Tree Species Susceptibility to the PWN

Unlike climate, the different susceptibility of pine tree species to the PWN is a structural and major long-term determinant of the invasion potential of the PWN; i.e., of the total pine forest area that can be invaded by the PWN in the Iberian Peninsula. The pine forest area that will be potentially affected by the PWN is less than half when considering only the highly susceptible pine tree species as PWN targets than when considering all pine tree species (compare Scenarios HF and AF in Figure 3). While several studies have examined and estimated the susceptibility of different pine tree species to PWD [26–28], this is probably the area of PWN research in which more additional efforts and studies are needed. More solid and detailed estimates of the responses of individual species to the PWN would considerably improve our ability to accurately predict the short-term and long-term spatiotemporal patterns of PWN spread, given the crucial role that tree species susceptibility plays in the modelling outcomes, as examined and determined for the first time in this study. A related and important avenue of further research is to understand and eventually quantify the potential adaptability of the genetic diversity of the domestic PWN populations and of the host tree species, which may alter the PWN capacity to infect and damage different pine forests in the Iberian Peninsula.

4.4. Measures to Contain PWN Spread

If no effective containment measures are taken, the PWN will invade all the pine forests in Portugal and Spain, resulting in large-scale ecological and economic impacts throughout the Iberian Peninsula. These impacts include, among others, wood and wood production losses [17–19]; changes in soil properties and pine forest communities; altered forest succession and regeneration; loss of the oldest and biggest pines [20,42], which frequently have a higher conservation and biodiversity value; and potential disruption of protected area networks [22].

Avoiding this damage to Iberian pine forests requires the combined application of several measures for PWN containment. These measures mainly consist of reducing the vector beetle population through mass trapping during the flight period (spring–summer) and the intensive surveillance of pine forests to allow the early detection and removal of infected trees before the PWN-infected beetles emerge and spread the disease to other pine trees or patches [30,43–47]. However, the effectiveness of this early detection and removal measure in containing the PWN spread may be compromised to some extent by the role of asymptomatic tree infections in the expansion patterns [48]. Other longer-term measures that could promote the capacity of the pine forests to resist or cope with the PWN infection include forest management treatments, such as thinning, that increase tree vigor or tree species diversity [49–51], or establishing pine half-sib families with higher genetic tolerance to the PWN [28,52].

It needs to be emphasized that our model and PWN spread projections only consider the natural spread of the PWN by its vector beetle. Human-mediated dispersal is not considered in our analysis and may, in case of uncontrolled transport of infected wood, drastically increase the speed and modify the spatiotemporal patterns of the PWN spread throughout the Iberian Peninsula and the rest of Europe. Therefore, our PWN expansion projections will only be valid, and the potential management measures outlined above will only be effective, under a rigorous application of the regulations and restrictions on timber treatment and transport. The control of the origin and sanitation of pine wood for transportation is essential to avoid larger-scale and uncontrolled human-mediated dispersal through the Iberian Peninsula [12,53]. Modelling anthropogenic PWN dispersal is complex, involving a considerable stochastic component, and is out of the scope of this study; however, it could be further explored in the future, by incorporating in our modelling framework some of the methods and ideas by Robinet et al. [54], Pukkala et al. [38] and Hudgins et al. [55].

5. Conclusions

Our modelling approach and the results for the set of considered scenarios allow us to reach the following four main conclusions:

- In the absence of effective containment measures, the PWN will spread naturally, through the dispersal of its vector beetle, to the entire Iberian Peninsula, including the Pyrenees, which would provide a gateway for further PWN expansion into France and the rest of Europe.
- The natural spread of the PWN will be relatively gradual, with an average rate of 0.83% of the total Iberian forest area infected yearly, so that it will take a century or longer for all the susceptible pine forests in Spain to be invaded by the PWN.
- Climate is not an important limiting factor for long-term spread of the PWN, because (i) there is ample availability of alternative pathways for PWN dispersal through the areas that are already suitable for the PWN in the current climatic conditions; and (ii) climate change projections, even conservative ones, indicate that future temperatures will make the whole Iberian Peninsula suitable for PWN transmission by or before the end of this century, except for some isolated mountain peaks.
- Unlike climatic conditions, the susceptibility of different pine tree species to PWN is by far the main determinant of PWN spread rates and of the extent of forest area affected by the wilt disease. Our findings highlight the need and importance of integrating data on individual pine tree species into predictive models on the spread and magnitude of damage caused by the PWN. Additional research that can refine our knowledge on tree species susceptibility is hence of particular importance for more accurate modelling of the PWN spread and for better guiding related containment efforts.

Author Contributions: Conceptualization and methodology, B.d.l.F. and S.S.; modelling and formal analysis, B.d.l.F.; writing—original draft preparation, B.d.l.F.; writing—review and editing, B.d.l.F. and S.S. All authors have read and agreed to the published version of the manuscript.

Funding: This research received no external funding.

Data Availability Statement: The results of the PWN spread modelling runs presented in this study are available on request from the corresponding author.

Acknowledgments: We thank Portugal's DG Território and the Instituto da Conservação da Natureza e das Florestas for providing their land cover map (COS2007) and PWN range data, respectively. We thank Pieter Beck for his helpful comments on this article and Aitor Gastón for his advice on climate change models.

Conflicts of Interest: The authors declare no conflict of interest.

References

1. Mota, M.M.; Braasch, H.; Bravo, M.A.; Penas, A.C.; Burgermeister, W.; Metge, K.; Sousa, E. First report of *Bursaphelenchus xylophilus* in Portugal and in Europe. *Nematology* **1999**, *1*, 727–734. [CrossRef]
2. Rodrigues, J.M. National eradication programme for the pinewood nematode. In *Pine Wilt Disease: A Worldwide Threat to Forest Ecosystems*; Springer: Dordrecht, The Netherlands, 2008; pp. 5–14.
3. EPPO. *Reporting Service (2010/051), Reporting Service (2010/058): Detected on a Single Pinus Tree Near Caceres (Extremadura) in 2008; Reporting Service (2010/202): Detected in November 2010 on Seven Trees (P. pinaster) in a Forest Area in the Municipality of 'As Neves' (Province of Pontevedra), in Galicia*; European and Mediterranean Plant Protection Organization (EPPO): Paris, France, 2010. Available online: https://gd.eppo.int/taxon/BURSXY/reporting (accessed on 3 May 2021).

4. EPPO. *Reporting Service (2012/047): Detected in One Tree (P. pinaster) in a Forest of the 'Monte Barroco Toiriña' (Province of Cáceres in Extremadura;* European and Mediterranean Plant Protection Organization (EPPO): Paris, France, 2012. Available online: https://gd.eppo.int/taxon/BURSXY/reporting (accessed on 3 May 2021).
5. EPPO. *Reporting Service (2014/020): Detected in December 2013 in a Sample Collected from Symptomatic Pine Trees in a Forest Stand Located in the Municipality of 'Sancti-Spíritus', Province of Salamanca (Castilla y León);* European and Mediterranean Plant Protection Organization (EPPO): Paris, France, 2014. Available online: https://gd.eppo.int/taxon/BURSXY/reporting (accessed on 3 May 2021).
6. EPPO. *Reporting Service (2018/140): Detected in one Pine Tree in a Forest Area in the Municipality of Lagunilla (Salamanca Province, Castilla y Léon);* European and Mediterranean Plant Protection Organization (EPPO): Paris, France, 2018. Available online: https://gd.eppo.int/taxon/BURSXY/reporting (accessed on 3 May 2021).
7. Abelleira, A.; Picoaga, A.; Mansilla, J.P.; Aguin, O. Detection of *Bursaphelenchus xylophilus*, causal agent of pine wilt disease on *Pinus pinaster* in northwestern Spain. *Plant Dis.* **2011**, *95*, 776. [CrossRef]
8. Zamora, P.; Rodríguez, V.; Renedo, F.; Sanz, A.V.; Domínguez, J.C.; Pérez-Escolar, G.; Miranda, J.; Álvarez, B.; González-Casas, A.; Mayor, E.; et al. First report of *Bursaphelenchus xylophilus* causing pine wilt disease on *Pinus radiata* in Spain. *Plant Dis.* **2015**, *99*, 1449. [CrossRef]
9. Xunta de Galicia. *Resolución de 28 de Diciembre de 2018 de la Dirección General de Ganadería Agricultura e Industrias Agroalimentarias por la que se Declara en el Territorio de la Comunidad Autónoma de Galicia la Presencia de cinco Nuevos Positivos del Organismo de Cuarentena Bursaphelenchus xylophilus (Steiner et Buhrer) Nickle et al. (Nematodo del pino) y se Ordena Comenzar las Medidas Para su Erradicación;* Diario Oficial de Galicia: Galicia, Spain, 2019; p. 1159.
10. Zhao, B.G.; Futai, K.; Sutherland, J.R.; Takeuchi, Y. Vector–Host Tree Relationships and the Abiotic Environment. In *Pine Wilt Disease;* Springer: Tokyo, Japan, 2008; pp. 144–161.
11. Kobayashi, F.; Yamane, A.; Ikeda, T. The Japanese pine sawyer beetle as the vector of pine wilt disease. *Annu. Rev. Entomol.* **1984**, *29*, 115–135. [CrossRef]
12. Sousa, E.; Rodrígues, J.M.; Bonifácio, L.F.; Naves, P.M.; Rodrígues, A. Management and control of the pine Wood nematode, *Bursaphelenchus xylophilus*, in Portugal. In *Nematodes: Morphology, Functions and Management Strategies;* Nova Science Publishers Inc.: New York, NY, USA, 2011; Chapter 6.
13. Sousa, E.; Bravo, M.A.; Pires, J.; Naves, P.; Penas, A.C. *Bursaphelenchus xylophilus* (Nematoda; Aphelenchoididae) associated with *Monochamus galloprovincialis* (Coleoptera; Cerambycidae) in Portugal. *Nematology* **2001**, *3*, 89–91.
14. Linit, M.J. Nematode-vector relationships in the pine wilt system. *J. Nematol.* **1988**, *20*, 227–235.
15. Naves, P.M.; Camacho, S.; De Sousa, E.M.; Quartau, J.A. Transmission of the pine wood nematode *Bursaphelenchus xylophilus* through feeding activity of *Monochamus galloprovincialis* (Col., Cerambycidae). *J. Appl. Entomol.* **2007**, *131*, 21–25. [CrossRef]
16. Pimentel, D.; Lach, L.; Zuniga, R.; Morrison, D. Environmental and economic costs of nonindigenous species in the United States. *BioScience* **2000**, *50*, 53–65. [CrossRef]
17. Pimentel, D.; Zuniga, R.; Morrison, D. Update on the environmental and economic costs associated with alien-invasive species in the United States. *Ecol. Econ.* **2005**, *52*, 273–288. [CrossRef]
18. Webster, J.; Mota, M. *Pine Wilt Disease: Global Issues, Trade and Economic Impact;* Springer: Berlin/Heidelberg, Germany, 2008; pp. 315–316.
19. Soliman, T.; Mourits, M.C.; Van Der Werf, W.; Hengeveld, G.M.; Robinet, C.; Lansink, A.O. Framework for modelling economic impacts of invasive species, applied to pine wood nematode in Europe. *PLoS ONE* **2012**, *7*, e45505.
20. Hu, G.; Xu, X.; Wang, Y.; Lu, G.; Feeley, K.J.; Yu, M. Regeneration of different plant functional types in a Masson pine forest following pine wilt disease. *PLoS ONE* **2012**, *7*, e36432. [CrossRef]
21. Boyd, I.L.; Freer-Smith, P.H.; Gilligan, C.A.; Godfray, H.C.J. The consequence of tree pests and diseases for ecosystem services. *Science* **2013**, *342*, 1235773. [CrossRef] [PubMed]
22. De la Fuente, B.; Beck, P.S. Invasive species may disrupt protected area networks: Insights from the pine wood nematode spread in Portugal. *Forests* **2018**, *9*, 282. [CrossRef]
23. Mori, Y.; Miyahara, F.; Tsutsumi, Y.; Kondo, R. Relationship between resistance to pine wilt disease and the migration or proliferation of pine wood nematodes. *Eur. J. Plant Pathol.* **2018**, *122*, 529–538. [CrossRef]
24. Zhang, F.; Kajiwara, J.; Mori, Y.; Ohira, M.; Tsutsumi, Y.; Kondo, R. Metabolites from resistant and susceptible *Pinus thunbergii* after inoculation with pine wood nematode. *Am. J. Plant Sci.* **2013**, *4*, 512–518. [CrossRef]
25. Evans, H.F.; McNamara, D.G.; Braasch, H.; Chadoeuf, J.; Magnusson, C. Pest risk analysis (PRA) for the territories of the European Union (as PRA area) on *Bursaphelenchus xylophilus* and its vectors in the genus *Monochamus*. *EPPO Bull.* **1996**, *26*, 199–249. [CrossRef]
26. Da Silva, M.N.; Solla, A.; Sampedro, L.; Zas, R.; Vasconcelos, M.W. Susceptibility to the pinewood nematode of four pine species involved in potential range expansion across Europe. *Tree Physiol.* **2015**, *35*, 987–999. [CrossRef] [PubMed]
27. Pimentel, C.S.; Gonçalves, E.V.; Firmino, P.N.; Calvão, T.; Fonseca, L.; Abrantes, I.; Máguas, C. Differences in constitutive and inducible defences in pine species determining susceptibility to pinewood nematode. *Plant Pathol.* **2017**, *66*, 131–139. [CrossRef]
28. Menéndez-Gutiérrez, M.; Alonso, M.; Jiménez, E.; Toval, G.; Mansilla, P.; Abelleira, A.; Abelleira-Sanmartín, A.; Díaz, R. Interspecific variation of constitutive chemical compounds in *Pinus* spp. xylem and susceptibility to pinewood nematode (*Bursaphelenchus xylophilus*). *Eur. J. Plant Pathol.* **2018**, *150*, 939–953. [CrossRef]

29. Inácio, M.L.; Nóbrega, F.; Vieira, P.; Bonifácio, L.; Naves, P.; Sousa, E.; Mota, M. First detection of *Bursaphelenchus xylophilus* associated with *Pinus nigra* in Portugal and in Europe. *For. Pathol.* **2017**, *45*, 235–238. [CrossRef]
30. De la Fuente, B.; Saura, S.; Beck, P.S.A. Predicting the spread of an invasive tree pest: The pine wood nematode in Southern Europe. *J. Appl. Ecol.* **2018**, *55*, 2374–2385. [CrossRef]
31. Robinet, C.; Van Opstal, N.; Baker, R.; Roques, A. Applying a spread model to identify the entry points from which the pine wood nematode, the vector of pine wilt disease, would spread most rapidly across Europe. *Biol. Invasions* **2011**, *13*, 2981–2995. [CrossRef]
32. Rutherford, T.A.; Webster, J.M. Distribution of pine wilt disease with respect to temperature in North America, Japan, and Europe. *Can. J. For. Res.* **1987**, *17*, 1050–1059. [CrossRef]
33. Evans, H.F. Plant Health Risk and Monitoring Evaluation (PHRAME) Final Report. In PHRAME Final Report and Conclusions. 2007. Available online: https://www.forestry.gov.uk/fr/infd-7xrfx9 (accessed on 25 April 2020).
34. Gruffudd, H.R.; Jenkins, T.A.R.; Evans, H.F. Using an evapo-transpiration model to predict the risk and expression of symptoms of pine wilt disease across Europe. *Biol. Invasions* **2016**, *18*, 2823–2840. [CrossRef]
35. Haran, J.; Roques, A.; Bernard, A.; Robinet, C.; Roux, G. Altitudinal Barrier to the Spread of an Invasive Species: Could the Pyrenean Chain Slow the Natural Spread of the Pinewood Nematode? *PLoS ONE* **2015**, *10*, e0134126. [CrossRef] [PubMed]
36. Pérez, G.; Díez, J.J.; Ibeas, F.; Pajares, J.A. Assessing pine wilt disease risk under a climate change scenario in Northwestern Spain. In *Managing Forest Ecosystems: The Challenge of Climate Change*; Springer: Dordrecht, The Netherlands, 2008; pp. 269–282.
37. Roques, A.; Zhao, L.; Sun, J.; Robinet, C. Pine wood nematode, pine wilt disease, vector beetle and pine tree: How a multiplayer system could reply to climate change. *Clim. Chang. Insect Pests* **2015**, *7*, 220–234.
38. Pukkala, T.; Möykkynen, T.; Robinet, C. Comparison of the potential spread of pinewood nematode (*Bursaphelenchus xylophilus*) in Finland and Iberia simulated with a cellular automaton model. *For. Pathol.* **2014**, *44*, 341–352. [CrossRef]
39. Ninyerola, M.; Pons, X.; Roure, J.M. *Atlas Climático Digital de la Península Ibérica: Metodología y Aplicaciones en Bioclimatología y Geobotánica*; Universidad Autónoma de Barcelona: Barcelona, Spain, 2005.
40. R Core Team. *R: A Language and Environment for Statistical Computing*; R Foundation for Statistical Computing: Vienna, Austria, 2017.
41. Sheffield, J.; Wood, E.F. Projected changes in drought occurrence under future global warming from multi-model, multi-scenario, IPCC AR4 simulations. *Clim. Dyn.* **2008**, *31*, 79–105. [CrossRef]
42. Gao, R.; Shi, J.; Huang, R.; Wang, Z.; Luo, Y. Effects of pine wilt disease invasion on soil properties and Masson pine forest communities in the Three Gorges reservoir region, China. *Ecol. Evol.* **2015**, *5*, 1702–1716. [CrossRef] [PubMed]
43. Gallego, D.; Sánchez-García, F.J.; Mas, H.; Campo, M.T.; Lencina, J.L. Estudio de la capacidad de vuelo a larga distancia de *Monochamus galloprovincialis* en un mosaico agro-forestal. *Boletín Sanid. Veg. Plagas* **2012**, *38*, 109–123.
44. Álvarez, G.; Etxebeste, I.; Gallego, D.; David, G.; Bonifacio, L.; Jactel, H.; Pajares, J.A. Optimization of traps for live trapping of pine wood nematode vector *Monochamus galloprovincialis*. *J. Appl. Entomol.* **2015**, *139*, 618–626. [CrossRef]
45. Etxebeste, I.; Sanchez-Husillos, E.; Álvarez, G.; Gisbert, H.M.i.; Pajares, J. Dispersal of *Monochamus galloprovincialis* as recorded by mark–release–recapture using pheromone traps. *J. Appl. Entomol.* **2015**, *140*, 485–499. [CrossRef]
46. Firmino, P.N.; Calvão, T.; Ayres, M.P.; Pimentel, C.S. *Monochamus galloprovincialis* and *Bursaphelenchus xylophilus* life history in an area severely affected by pine wilt disease: Implications for forest management. *For. Ecol. Manag.* **2017**, *389*, 105–115. [CrossRef]
47. De la Fuente, B.; Beck, P.S. Management measures to control pine wood nematode spread in Europe. *J. Appl. Ecol.* **2019**, *56*, 2577–2580. [CrossRef]
48. Zhao, B.G.; Futai, K.; Sutherland, J.R.; Takeuchi, Y. Pine Wilt in Japan: From First Incidence to the Present. In *Pine Wilt Disease*; Springer: Tokyo, Japan, 2008; pp. 5–12.
49. Jactel, H.; Menassieu, P.; Vetillard, F.; Gaulier, A.; Samalens, J.C.; Brockerhoff, E.G. Tree species diversity reduces the invasibility of maritime pine stands by the bast scale, *Matsucoccus feytaudi*. *Can. J. For. Res.* **2006**, *36*, 314–323. [CrossRef]
50. Muzika, R.M.; Liebhold, A.M. A critique of silvicultural approaches to managing defoliating insects in North America. *Agric. For. Entomol.* **2000**, *2*, 97–105. [CrossRef]
51. Liebhold, A.M. Forest pest management in a changing world. *Int. J. Pest Manag.* **2012**, *58*, 289–295. [CrossRef]
52. Gaspar, D.; Trindade, C.; Usié, A.; Meireles, B.; Barbosa, P.; Fortes, A.M.; Ramos, A.M. Expression profiling in *Pinus pinaster* in response to infection with the pine wood nematode *Bursaphelenchus xylophilus*. *Forests* **2017**, *8*, 279. [CrossRef]
53. Vicente, C.; Espada, M.; Vieira, P.; Mota, M. Pine wilt disease: A threat to European forestry. *Eur. J. Plant Pathol.* **2012**, *133*, 89–99. [CrossRef]
54. Robinet, C.; Roques, A.; Van Opstal, N.; Baker, R.; Pan, H.; Fang, G.; Sun, J. Anthropogenic pathways in the spread of the pinewood nematode and predictions of future expansion. In *International Congress on Biological Invasions*; HAL-INRAE: Fuzhou, China, 2009; p. 1.
55. Hudgins, E.J.; Liebhold, A.M.; Leung, B. Predicting the spread of all invasive forest pests in the United States. *Ecol. Lett.* **2017**, *20*, 426–435. [CrossRef]

forests

MDPI

Article

Invasion of Emerald Ash Borer *Agrilus planipennis* and Ash Dieback Pathogen *Hymenoscyphus fraxineus* in Ukraine—A Concerted Action

Kateryna Davydenko [1], Yuriy Skrylnyk [1], Oleksandr Borysenko [1], Audrius Menkis [2,*], Natalia Vysotska [1], Valentyna Meshkova [1], Åke Olson [2], Malin Elfstrand [2] and Rimvys Vasaitis [2]

1 Ukrainian Research Institute of Forestry and Forest Melioration, Pushkinska 86, UA-61024 Kharkiv, Ukraine; kateryna.davydenko@slu.se (K.D.); yuriy.skrylnik@gmail.com (Y.S.); xalekter@gmail.com (O.B.); vysotska_n@ukr.net (N.V.); valentynameshkova@gmail.com (V.M.)

2 Department of Forest Mycology and Plant Pathology, Swedish University of Agricultural Sciences, P.O. Box 7026, SE-75007 Uppsala, Sweden; ake.olson@slu.se (Å.O.); malin.elfstrand@slu.se (M.E.); rimvys.vasaitis@slu.se (R.V.)

* Correspondence: audrius.menkis@slu.se

Abstract: Emerald Ash Borer (EAB), *Agrilus planipennis*, is a beetle that originates from East Asia. Upon invasion to North America in the early 2000s, it killed untold millions of ash trees. In European Russia, EAB was first detected in Moscow in 2003 and proved to have the potential to also kill native European ash (*Fraxinus excelsior*). The beetle has since spread in all geographic directions, establishing itself in eastern Ukraine by 2019 and possessing potential for further westward spread towards the EU. Apart from the approaching EAB, *F. excelsior* is currently threatened by the dieback disease (ADB) caused by the invasive ascomycete fungus *Hymenoscyphus fraxineus*. The infestation by EAB combined with ADB infection is expected to be more lethal than either of them alone, yet the potential consequences are unknown. To date, eastern Ukraine represents the geographic area in which both invasions overlap, thus providing the opportunity for related investigations. The aims of the study were to investigate: (i) the EAB expansion range in Ukraine, (ii) the relative susceptibility of *F. excelsior* and American ash (*Fraxinus pennsylvanica*) to EAB and ADB, and (iii) the combined effect/impact on ash condition imposed by both the pest and disease in the area subjected to the invasion. The results have demonstrated that (i) the invasion of EAB is currently expanding both in terms of newly infested trees and invaded geographic area; (ii) *F. excelsior* is more resistant to EAB than *F. pennsylvanica*, while *F. excelsior* is more susceptible to ADB than *F. pennsylvanica*; and (iii) the infection by ADB is likely to predispose *F. excelsior* to the infestation by EAB. It was concluded that inventory and mapping of surviving *F. excelsior*, affected by both ADB and EAB, is necessary to acquire genetic resources for the work on strategic, long-term restoration of *F. excelsior* in devastated areas, thereby tackling a possible invasion of EAB to the EU.

Keywords: emerald ash borer; ash dieback; *Fraxinus* spp.; forest pests; invasive populations; eastern Ukraine

1. Introduction

The Emerald Ash Borer (EAB) *Agrilus planipennis* Fairmaire, 1888 (Coleoptera: Buprestidae), is a buprestid beetle that originates from East Asia. In its native range, EAB is a minor pest colonizing dying ash (*Fraxinus* spp.) trees and causing insignificant damage to viable ones [1]. It was first detected in North America (southeast Michigan) in 2002 (introduced with wood trade) and has since killed millions of trees in forest, riparian, and urban areas. It is estimated that EAB could virtually eliminate *Fraxinus* spp., one of North America's most widely distributed tree genera, with devastating economic and ecological impacts [2].

In European Russia, EAB was first detected in Moscow in 2003 (transported with wood) and proved to have the potential to kill native European ash (*Fraxinus excelsior*

Linnaeus, 1753). The beetle has since spread in all geographic directions, but most notably towards the west and southwest, by 2019 crossing the border of Belarus and entering and establishing in eastern Ukraine [3]. Notably, ash is continuously present in areas from the eastern borders of Ukraine and Belarus towards the west: apart from natural woodlands, in both countries, *F. excelsior* and (highly EAB-susceptible) North American green ash (*Fraxinus pennsylvanica* Marshall, 1785) have historically been extensively planted along roads, railways, field shelter belts, and in urban greenings [4]. These plantings and woodlands potentially provide an excellent route pathway for the spread of EAB towards EU countries, e.g., into Romania, Hungary, Slovakia, and Poland.

Importantly in this respect, EAB beetles are strong flyers. For example, in the Great Lakes region of North America, EAB adults were reported to be capable of long dispersal flights, and gravid females are estimated to fly more than 10 km in 24 h [5]. Moreover, EAB beetles are efficient "hitchhikers": they can easily travel by cars, being hidden behind flanges of the car body; the insect can even stay on a tree twig pressed by a wiper to a windshield of a car driving at speeds of up to 120 km/h [6]. Railway cars also often serve as means for traveling EAB [7]. Consequently, all this makes the invasion of EAB to the EU highly probable [8]. The possibility therefore cannot be excluded that such situation will also put *F. excelsior* under the threat of extinction, as American ash species currently are. Therefore, to monitor and predict the pace at which EAB is approaching the EU is an important task, e.g., for plant quarantine authorities and other stakeholders, for the prognosis of future developments, planning and imposing preventive measures, and to prepare for the mitigation of eventual impacts.

Apart from the approaching EAB, another threat to *F. excelsior* is the dieback disease (ADB) caused by the invasive ascomycete fungus *Hymenoscyphus fraxineus* (T. Kowalski) Baral et al. that has impacted ash forests all over Europe [9]. Yet, there is the evidence that a certain proportion of trees exhibits resistance to ADB, providing the basis for future propagation [10,11]. However, the proportion showing resistance to ADB is low, namely, 1–5% of individuals [12,13]. As EAB infestation combined with ADB infection is expected to be more lethal than either of them alone [13], the following questions arise: (1) what fraction of the 1–5% remaining viable *F. excelsior* individuals in ADB-devastated areas in Europe will survive subsequent infestation by EAB? (2) Will the number of survivors of the ADB epidemic combined with EAB invasion be enough for initiating sustainable restoration of *F. excelsior*?

Currently, eastern parts of Ukraine represent a geographic area over which the invasive ranges of EAB and ADB overlap, thus providing the opportunity to investigate the impact imposed over populations of *Fraxinus* spp. by both the invasive pest and disease simultaneously. Moreover, such investigations would allow one to compare the relative susceptibility of two ash species to EAB and ADB, namely, native *F. excelsior* vs. introduced *F. pennsylvanica*. In particular, the data regarding the consequences of infestation of EAB to *F. excelsior* are scarce and fragmented. Thus, during the survey of EAB in western Russia and eastern Ukraine, an overwhelming majority of the infestations were found on the highly susceptible *F. pennsylvanica*, and all observed cases of infestation of the native species *F. excelsior* occurred in artificial plantings [14]. On the other hand, despite such certain indications that *F. excelsior* might be more resistant to EAB, e.g., along roadsides and city plantings [15,16], a recent study conducted in the Moscow Province provided certain evidence that EAB is susceptible to massive outbreaks causing significant damage also in forest stands [17]. Consequently, more detailed studies are needed to elucidate the relative susceptibility of *F. excelsior* compared to *F. pennsylvanica*. The aims of the present study were to investigate the following: (1) the EAB expansion range in Ukraine; (2) the relative susceptibility of *F. excelsior* and *F. pennsylvanica* to EAB and ADB; and (3) the combined effect/impact on ash conditions imposed by both the pest and disease in the area subjected to the invasion.

2. Materials and Methods

To estimate the extent of spread of EAB and to assess the health condition of ash, visual surveys have been carried out in forest stands of *F. excelsior* and shelterbelts composed of *F. pennsylvanica* in eastern Ukraine, in the years 2019 to 2021. Symptoms of EAB infestation (canopy decline, woodpecker attacks, D-shaped exit holes, and epicormic sprouting) were preliminary recorded and mapped. Simultaneously, the frequency of occurrence of ADB-symptomatic trees has been assessed based on the presence of typical bark necroses on branches and shoots, shoot dieback, and wilting leaves. The presence of ADB (*Hymenoscyphus fraxineus*, formerly *H. pseudoalbidus*) was confirmed earlier in the research regions using fungal culturing followed by ITS rDNA sequencing [18,19] and using ITS rDNA sequencing directly from the infected tree tissues [9,20]. Based on information acquired during the surveys, locations of 10 monitoring plots of the present study were selected (Figure 1). They were located in areas where the earliest infestation by EAB in Ukraine was observed, as well as in its advancing front and in the areas where the EAB presence could have been expected. All monitoring plots were subjected to ash dieback disease for at least a decade.

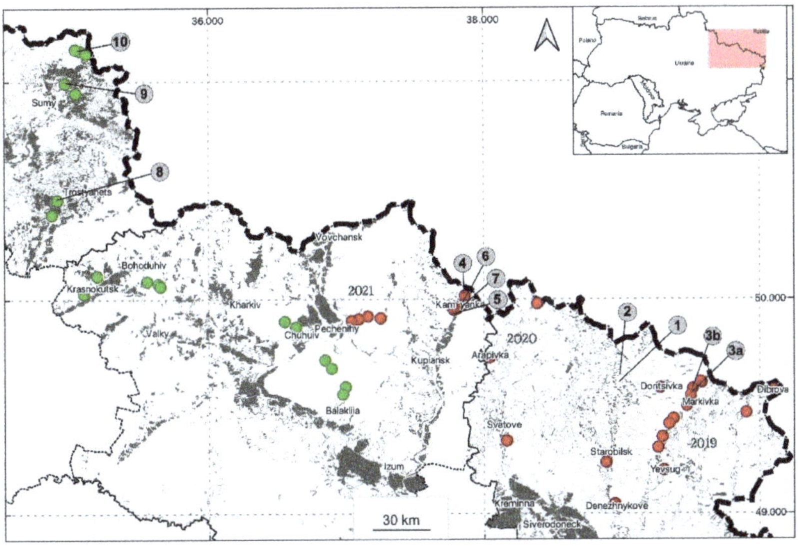

Figure 1. Map of eastern Ukraine showing locations of survey sites and monitoring plots of the present study. Red circles indicate survey sites where the emerald ash borer has been detected, and green indicates where the beetle has not been detected. Numbers of the monitoring plots correspond to those in the Tables 1 and 2.

There were two distinct types of the monitoring plots, depending on ash species investigated. All monitoring plots for *F. pennsylvanica* were located in field shelterbelts, where the trees were positioned in rows. Monitoring plots for *F. excelsior* were located in the interior of forest stands (Plot 3b in the urban forest) and were comprised of compact groups of trees within the area of approx. 15 × 50 m. In each plot, ash trees were investigated in a simple systematic manner: after the inspection of the first tree (located at the edge of a plot), the next nearest tree was inspected, and so on. Each examined tree was assigned a specific number and mapped. During the detailed investigations in the monitoring plots, apart from examining crowns (for woodpecker activity and decline) the lower part of tree stems was visually inspected for incidence of bark loosening and cracks and for the direct signs of EAB infestation: exit holes, dead beetles, larval galleries, and larvae (Figure 2). Simultaneously, the presence of galleries of ash bark beetles (genus *Hylesinus* Fabricius, 1801) and symptoms of ADB were recorded based on the disease symptoms listed above.

The upper part of the stems and thick branches were inspected using binoculars. Numbers of trees examined on each monitoring plot are shown in the Table 1.

Chi-squared tests for comparison of proportions were calculated, applying Yates correction.

Table 1. Infestations of the ash dieback disease fungus *Hymenoscyphus fraxineus* (ADB) and emerald ash borer *Agrilus planipennis* (EAB) on *Fraxinus pennsylvanica* and *F. excelsior* trees in eastern Ukraine. The plots are as shown in Figure 1.

Plot	*Fraxinus* spp.	Trees Monitored, no. (%)										
		All	Year 2020					Year 2021				
			ADB	EAB	Of those, ADB and EAB	Dead [a]	Visually Healthy	ADB	EAB	Of Those, ADB and EAB	Dead [a]	Visually Healthy
Luhansk region (LH)												
1LH	*F. pen.* [b]	38	1 (3)	19 (50)	0	12 (32)	18 (47)	2 (5)	37 (97)	1 (3)	29 (76)	0
2LH	*F. pen.*	25	0	16 (64)	0	7 (28)	9 (36)	1 (4)	22 (88)	0	14 (56)	2 (8)
3aLH	*F. pen.*	25	0	21 (84)	0	9 (36)	4 (16)	0	25 (100)	0	17 (68)	0
All LH *F. pen.*		88	1 (1)	56 (64)	0	28 (32)	31 (35)	3 (3.5)	84 (95)	1 (1)	60 (68)	2 (2)
3bLH, all LH *F. ex.* [c]		16	4 (25)	3 (19)	2 (13)	2 (13)	11 (69)	6 (38)	7 (44)	3 (19)	7 (44)	6 (38)
χ^2 test *F. pen.* vs. *F. ex.* [d]			***	**		n.s.	*	****	****	**	n.s.	****
Kharkiv region (KH, northwest from LH)												
4KH	*F. ex.*	60	-	-	-	-	-	15 (25)	17 (28)	9 (15)	9 (15)	37 (62)
5KH	*F. ex.*	55	-	-	-	-	-	18 (33)	12 (22)	6 (11)	9 (16)	31 (56)
6KH	*F. pen.*	52	-	-	-	-	-	7 (13)	31 (60)	4 (8)	23 (44)	18 (35)
7KH	*F. pen.*	45	-	-	-	-	-	3 (7)	24 (53)	2 (4)	19 (42)	20 (44)
All KH *F. pen.*		97	-	-	-	-	-	10 (10)	55 (57)	6 (6)	42 (43)	38 (39)
All KH *F. ex.*		115	-	-	-	-	-	33 (29)	29 (25)	15 (13)	18 (16)	68 (59)
χ^2 test *F.pen.* vs. *F.ex.*								**	****	n.s.	****	**
Sumy region (SU, northwest from KH)												
8SU	*F. ex.*	50	-	-	-	-	-	32 (64)	-	-	31 (62)	18 (36)
9SU	*F. ex.*	50	-	-	-	-	-	27 (54)	-	-	19 (38)	23 (46)
10SU, all SU *F. pen.*		25	-	-	-	-	-	8 (32)	-	-	5 (20)	17 (68)
All SU *F. ex.*		100	-	-	-	-	-	59 (59)	-	-	50 (50)	41 (41)
χ^2 test *F.pen.* vs. *F.ex.*								*			**	*
All plots (LH + KH + SU)												
All *F. pen.*		210	-	-	-	-	-	21 (10)	139 (66)	7 (3)	107 (51)	57 (27)
All *F. ex.*		231	-	-	-	-	-	98 (42)	36 (16)	18 (8)	75 (32)	115 (50)
χ^2 test *F.pen.* vs. *F.ex.*								****	****	*	****	****
Plots infested by the emerald ash borer (LH + KH)												
LH + KH *F. pen.*		185						13 (7)	139 (75)	7 (4)	102 (55)	40 (22)
LH + KH *F. ex.*		131						39 (30)	36 (27)	18 (14)	25 (19)	74 (56)
χ^2 test *F.pen.* vs. *F.ex.*								****	****	**	****	****

[a] All showing ADB and/or EAB infestations. [b] *Fraxinus pennsylvanica* (field shelterbelts). [c] *Fraxinus excelsior* (forest interior). [d] Significance of χ^2 tests: * $p < 0.05$; ** $p < 0.01$; *** $p < 0.001$; **** $p < 0.0001$; n.s. not significant.

Table 2. Colonization by the ash bark beetles (*Hylesinus* spp.) on trees of *Fraxinus excelsior* and *F. pennsylvanica* infected by the ash dieback disease (ADB) fungus *Hymenoscyphus fraxineus*. The plots are as shown in Figure 1.

Plot	*Fraxinus* spp.	Trees with ADB Symptoms, No. (%)		
		All (100%)	Colonized by *Hylesinus* spp. [a]	Dead (% of Colonized by *Hylesinus* spp.)
4KH	*F. excelsior*	15	4 (27)	4 (100)
5KH	*F. excelsior*	18	7 (39)	5 (71)
6KH	*F. pennsylvanica*	7	3 (43)	1 (33)
7KH	*F. pennsylvanica*	3	1 (33)	1 (100)
8SU	*F. excelsior*	32	31 (97)	31 (100)
9SU	*F. excelsior*	27	22 (81)	19 (86)
10SU	*F. pennsylvanica*	8	5 (63)	5 (100)
All	*F. excelsior*	92	64 (70)	59 (92)
	F. pennsylvanica	18	9 (50)	7 (78)
χ^2 test *F. excelsior* vs. *F. pennsylvanica* [b]			n.s.	*

[a] Galleries of *H. fraxini* and *H. crenatus*. [b] Significance of $\chi2$ tests: * $p < 0.05$; n.s. not significant.

Figure 2. The emerald ash borer, *Agrilus planipennis* on ash trees: (**A**) Exit holes, shown by white arrows; (**B**) Old galleries on a dry-sided stem; (**C**) Larva.

3. Results

According to the results of surveys, in the Luhansk region with the longest EAB invasion history, in 2019, approx. 10–30% of ash trees were infested by the beetle, and in the years 2020–2021, this proportion had increased up to 60 to 90%. Typically, many of those trees exhibited old EAB exit holes, bark cracks, and dry-sided stems (Figure 2A,B), indicating that the initial invasion took place at the eastern Russian–Ukrainian border several years previously, e.g., in the years 2016–2017 at the latest. Further northwest in the Kharkiv region, no EAB attacks had been observed in 2019, but already in 2020 up to 10–30% of trees were infested, and in 2021, this proportion increased up to 60 to 90%. Neither of the surveys accomplished during the period of 2019–2021 revealed the presence of EAB in the neighboring Sumy region (Figure 1).

Investigations in the monitoring plots of the Luhansk region (1, 2, 3a; LH) demonstrated that 56% of *F. pennsylvanica* trees were infested by EAB in year 2020, and in 2021, the proportion had increased to 68%. For *F. excelsior* (plot 3bLH), the corresponding numbers were 19% and 44%. A similar trend was observed in 2021 in the plots of Kharkiv region (4, 5, 6, 7; KH), where EAB infested 57% *F. pennsylvanica* and 25% *F. excelsior* trees. Pooled data from all plots where EAB was observed (LH + KH) showed that out of 185 *F. pennsylvanica* trees, 139 (75%) were EAB-infested, while the corresponding proportion for *F. excelsior* was 36 out of 131 (31%). In this respect, all comparisons between the two tree species were statistically significant (Table 1).

The situation with frequencies of infections by ADB was the reverse. Thus, in 2020 in the LH plots, ADB symptoms were observed on 1% of *F. pennsylvanica* and on 25% of *F. excelsior* trees, and in 2021, the respective proportions increased up to 3.5% and 38%. Consistently, in 2021 in the KH plots, ADB symptoms were observed on 10% and 29% of *F. pennsylvanica* and *F. excelsior* trees, respectively, and in the Sumy plots (8, 9, 10; SU), on 32% and 59% of *F. pennsylvanica* and of *F. excelsior* trees, respectively. Pooled data from all study plots (LH + KH + SU) had shown that out of 210 *F. pennsylvanica* trees, 21 (10%) were ADB-infected, while the corresponding proportion for *F. excelsior* was 98 out of 231 (42%). In this respect, all comparisons between the two tree species were statistically significant. In the plots invaded by both EAB and ADB (LH + KH), 7 (4%) *F. pennsylvanica* and 18 (14%) *F. excelsior* trees were affected by both the pest and disease, and the difference in proportions was significant (Table 1).

The proportion of dead *F. pennsylvanica* trees in LH plots between the years 2020 and 2021 increased significantly from 32% to 68% (χ^2 test; $p < 0.00001$), although the increase in dead *F. excelsior* (13% to 44%) was statistically insignificant ($p = 0.12$). The differences in proportions of dead trees between the two species were insignificant ($p = 0.20$ and $p = 0.06$; Table 1). This was in contrast with the KH and SU plots, where significant differences in interspecific mortality rates have been noted, but for different reasons: in KH, it was higher in *F. pennsylvanica* due to EAB, while in SU, in *F. excelsior* due to ADB. The mortality in EAB-infested plots (LH + KH), as well as in the whole study area, was to a significant extent higher in *F. pennsylvanica* due to prevailing infestations by EAB (Table 1).

In reverse, the proportions of visually healthy (no signs neither of EAB nor ADB) *F. pennsylvanica* trees in LH plots between the years 2020 and 2021 decreased significantly, from 35% to 2% (χ^2 test; $p < 0.00001$), although the decrease in visually healthy *F. excelsior* (69% to 38%) was statistically insignificant ($p = 0.16$). In each accomplished comparison, the proportion of visually healthy *F. pennsylvanica* was significantly lower than that of *F. excelsior*, except in the SU plots, where the proportion was higher (68% vs. 41%) due to the absence of EAB infestation (Table 1).

Galleries of ash bark beetles (*Hylesinus* spp.) were observed only on trees with ADB symptoms and only in the KH and SU plots. In all, 64 out of 92 (70%) ADB-symptomatic *F. excelsior* trees and 9 of 18 (50%) *F. pennsylvanica* trees were colonized by the beetles, yet the difference in proportions was not significant ($p = 0.11$). Among the trees colonized by *Hylesinus* spp., 59 (92%) of *F. excelsior* and 7 (78%) of *F. pennsylvanica* were dead. Of all trees with ADB symptoms, this comprised 64% and 39% of *F. excelsior and F. pennsylvanica*, respectively, and the difference in the proportions was (marginally) statistically significant ($p = 0.046$; Table 2).

4. Discussion

Rapid geographic expansion of EAB and a steady increase in its killed trees (for *F. pennsylvanica*, very significant) in 2020–2021 point out that the ecological conditions are indeed suitable for the development of EAB populations in eastern Ukraine. This is different from the situation observed in populations of the beetle approx. 1200 km north in Saint Petersburg, where, following the invasion, further spread of EAB was slow and locally restricted, likely due to climatic conditions characterized by cool and wet summers and freezing winter temperatures [8]. Preceding studies have shown that the life cycle of EAB in eastern Ukraine fells into two cohorts: (1) a "spring" cohort, comprised of larvae hatching in May to June, which develop to prepupae, overwinter in pupal chambers, and adult beetles emerge in May to June the next year (one-year generation); (2) a "summer" cohort, comprised of larvae that hatch later in the summer and overwinter twice—first winter as larvae and the second as prepupae; the ratio of the cohorts is about 1:1 [21]. For comparison, in European Russia, the life cycle of the EAB lasts more than 1 year for most individuals in Moscow [22] and at least 2 years in Saint Petersburg [8].

The current study demonstrated significantly higher susceptibility to EAB of *F. pennsylvanica* as compared to *F. excelsior* (Table 1). This is in agreement with the results of the previous study in eastern Ukraine (the same study area), wherefrom it has been reported that EAB inhabits mainly *F. pennsylvanica*, and in *F. excelsior* it occurs more occasionally, preferring sprouts, trees in the stands with low relative stocking density, and trees at the edges and along the perimeter of small forests or forest belts. Here, despite that the density of EAB larvae in colonized branches did not differ significantly between the two ash species (approx. 0.6 and 0.7 larvae/dm^2, respectively), the numbers of viable EAB larvae in branches collected from attacked *F. pennsylvanica* (91.4%) were significantly higher than in *F. excelsior* (76.1%) [23]. Differently, a study from Saint Petersburg (1200 km north) showed that the EAB exhibited a slightly more successful development in *F. excelsior* than in *F. pennsylvanica*: larval densities, numbers of larval galleries, exit holes, viable larvae, and emerged adult beetles were slightly higher in *F. excelsior* than in *F. pennsylvanica*; larval densities were 0.1–0.5 larvae/dm^2 for *F. pennsylvanica* and 0.3–0.8 for *F. excelsior* [8]. Yet, the cited work was conducted in the city environment and consisted of small and fragmented empirical data.

On the other hand, the current study demonstrated significantly higher susceptibility to ADB of *F. excelsior* as compared to *F. pennsylvanica* (Table 1). This is the first report of its kind and is supported not only by the data acquired in SU monitoring plots, where invasion of EAB was still absent, but also from the whole study area, as in all comparisons the proportions of *F. excelsior* trees with ADB symptoms was significantly higher than that of *F. pennsylvanica*. In monitoring plots located in EAB-invaded areas (LH + KH), the proportion of trees affected by both ADB and EAB simultaneously was also higher for *F. excelsior*, indicating that its trees diseased with ADB are to a certain extent also predisposed to EAB infestation. In case of *F. pennsylvanica*, this demonstrates such vulnerability of the species to EAB that the beetle infests and kills trees indiscriminately of their health condition (e.g., despite presence/absence of ADB). Indeed, the previous study in the eastern Ukraine has reported regular attacks of EAB on vigorous *F. pennsylvanica* trees and the aggressiveness of the pest was emphasized [23].

The results of our work demonstrated a common occurrence of *Hylesinus* bark beetles on ADB-diseased ash (slightly preferring native *F. excelsior*), and the prevailing majority of the colonized trees were already dead. Similarly, the report from ADB-infected ash stands in southeastern Germany had shown that breeding galleries of *Hylesinus* were only found in ADB-diseased trees that have recently died, concluding that the beetle is not able to colonize vigorous trees and is acting as a secondary opportunistic pest [24].

Finally, it must be noted that only a minority of *F. excelsior* trees in affected sites by EAB and ADB (LH + KH) have died, while over a half remained visually healthy. Therefore, those over 100 of *F. excelsior* trees that survived or were not susceptible to attacks by EAB (Table 1) constitute a source of material for further monitoring and eventual propagation, for which the current work represents a starting point. As it was emphasized previously [8], inventory and mapping of surviving ash trees, focusing on European native *F. excelsior*, affected by both ADB and EAB is necessary to acquire genetic resources for the work on strategic, long-term restoration of devastated areas, thereby tackling a possible invasion of EAB to the EU.

5. Conclusions

1. Invasion of EAB to Ukraine occurred 2–3 years previously to its first records in 2019, and is currently expanding both in terms of newly infested trees and invaded geographic area.
2. *Fraxinus excelsior* (in interior of forest stands) is more resistant to EAB than *F. pennsylvanica* (in field shelterbelts).
3. *Fraxinus excelsior* is more susceptible to ADB than *F. pennsylvanica*.
4. Infection by ADB is likely to predispose *F. excelsior* to infestation by EAB.
5. Ash trees infected by ADB are predisposed for the colonization by ash bark beetles *Hylesinus* spp.
6. Inventory and mapping of surviving *F. excelsior*, affected by both ADB and EAB, is necessary to acquire genetic resources for the work on strategic, long-term restoration of devastated areas, thereby tackling a possible invasion of EAB to the EU.

Author Contributions: Conceptualization, R.V., V.M. and O.B.; methodology, K.D., Y.S., O.B. and N.V.; validation, R.V. and N.V.; formal analysis and investigation, K.D., Y.S., O.B., V.M. and A.M.; resources, Y.S., R.V., N.V. and Å.O.; data curation, K.D. and Y.S.; writing—original draft preparation, K.D., R.V. and O.B.; writing—review and editing, K.D., R.V., A.M., Å.O., M.E. and V.M.; visualization, Y.S., O.B. and A.M.; supervision, K.D., A.M. and R.V.; project administration, N.V., Å.O. and A.M.; funding acquisition, K.D., Å.O., M.E. and R.V. All authors have read and agreed to the published version of the manuscript.

Funding: K.D. was funded by the FAO project TCP/RER/3801 "Strengthening the resilience of forests to beetle outbreaks and associated dieback", the Swedish Foundation for Strategic Research (SSF) project UKR22-0060, Knut and Alice Wallenbergs foundation project KAW 2022.0211 and United States Agency for International Development (USAID) project Ukraine Economic Resilience PUR-SEV-21-0001. K.D., A.M. and A.O. were funded by the Marie-Claire Cronstedts foundation. A.M. was funded by the Swedish Research Council Formas (grant no. 2019-00597). This project was also supported by the SLU Forest Damage Center, Swedish University of Agricultural Sciences.

Data Availability Statement: Data are available upon email request to the corresponding author.

Conflicts of Interest: The authors declare no conflict of interest.

References

1. Baranchikov, Y.; Mozolevskaya, E.; Yurchenko, G.; Kenis, M. Occurrence of the emerald ash borer, *Agrilus planipennis* in Russia and its potential impact on European forestry. *EPPO Bull.* **2008**, *38*, 233–238. [CrossRef]
2. Herms, D.A.; McCullough, D.G. Emerald ash borer invasion of North America: History, biology, ecology, impacts, and management. *Annu. Rev. Entomol.* **2014**, *59*, 13–30. [CrossRef] [PubMed]
3. Drogvalenko, A.N.; Orlova-Bienkowskaja, M.J.; Bienkowski, A.O. Record of the emerald ash borer (*Agrilus planipennis*) in Ukraine is confirmed. *Insects* **2019**, *10*, 338. [CrossRef] [PubMed]
4. Meshkova, V.L. Emerald ash borer-newly arriving to our territories. *Lisoviy Visn.* **2019**, *6*, 8–11. (In Ukrainian)
5. Taylor, R.A.J.; Bauer, L.S.; Poland, T.M.; Windell, K.N. Flight performance of *Agrilus planipennis* (Coleoptera: Buprestidae) on a flight mill and in free flight. *J. Insect Behav.* **2010**, *23*, 128–148. [CrossRef]
6. Gninenko, Y.I.; Klyukin, M.S.; Khegay, I.V. Distribution speed of emerald ash borer in Russia. In *Emerald Ash Borer—Distribution and Protection Measures in the USA and Russia*; Gninenko, Y.I., Ed.; VNIILM: Pushkino, Russia, 2016; pp. 57–62. (In Russian)
7. Short, M.T.; Chase, K.D.; Feeley, T.E.; Kees, A.M.; Wittman, J.T.; Aukema, B.H. Rail transport as a vector of emerald ash borer. *Agric. For. Entomol.* **2020**, *22*, 92–97. [CrossRef]
8. Selikhovkin, A.V.; Musolin, D.L.; Popovichev, B.G.; Merkuryev, S.A.; Volkovitsh, M.G.; Vasaitis, R. Invasive Populations of the Emerald Ash Borer *Agrilus planipennis* Fairmaire, 1888 (Coleoptera: Buprestidae) in Saint Petersburg, Russia: A Hitchhiker? *Insects* **2022**, *13*, 191. [CrossRef]
9. Vasaitis, R.; Enderle, R. *Dieback of European Ash (Fraxinus spp.)–Consequences and Guidelines for Sustainable Management*; Swedish University of Agricultural Sciences: Uppsala, Sweden, 2017; p. 320. Available online: http://www.slu.se/globalassets/ew/org/inst/mykopat/forskning/stenlid/dieback-of-european-ash.pdf (accessed on 31 August 2021).
10. Menkis, A.; Bakys, R.; Åslund, M.S.; Davydenko, K.; Elfstrand, M.; Stenlid, J.; Vasaitis, R. Identifying *Fraxinus excelsior* tolerant to ash dieback: Visual field monitoring versus a molecular marker. *For. Pathol.* **2019**, *50*, e12572. [CrossRef]
11. Enderle, R.; Stenlid, J.; Vasaitis, R. An overview of ash (*Fraxinus* spp.) and ash dieback disease in Europe. *CAB Rev.* **2019**, *14*, 025. [CrossRef]
12. McKinney, L.V.; Nielsen, L.R.; Collinge, D.B.; Thomsen, I.M.; Hansen, J.K.; Kjær, E.D. The ash dieback crisis: Genetic variation in resistance can prove a long-term solution. *Plant Pathol.* **2014**, *63*, 485–499. [CrossRef]
13. Semizer-Cuming, D.; Krutovsky, K.V.; Baranchikov, Y.N.; Kjær, E.D.; Williams, S.G. Saving the world's ash forests calls for international cooperation now. *Nat. Ecol. Evol.* **2019**, *3*, 141–144. [CrossRef] [PubMed]
14. Orlova-Bienkowskaja, M.J.; Drogvalenko, A.N.; Zabaluev, I.A.; Sazhnev, A.S.; Peregudova, E.Y.; Mazurov, S.G.; Komarov, E.V.; Struchaev, V.V.; Martynov, V.V.; Nikulina, T.V.; et al. Current range of *Agrilus planipennis* Fairmaire, an alien pest of ash trees, in European Russia and Ukraine. *Ann. For. Sci.* **2020**, *77*, 29. [CrossRef]
15. Baranchikov, Y.N.; Seraya, L.G.; Grinash, M.N. All European ash species are susceptible to emerald ash borer *Agrilus planipennis* Fairmaire (Coleoptera: Buprestidae)–A Far Eastern invader. *Sib. For. J.* **2014**, *6*, 80–85, (In Russian with English summary).
16. Musolin, D.L.; Selikhovkin, A.V.; Shabunin, D.A.; Zviagintsev, V.B.; Baranchikov, Y.N. Between ash Dieback and Emerald Ash Borer: Two Asian Invaders in Russia and the Future of Ash in Europe. *Baltic For.* **2017**, *23*, 316–333. Available online: https://www.balticforestry.mi.lt/bf/PDF_Articles/2017-23%5B1%5D/Baltic%20Forestry%202017.1_316-333.pdf (accessed on 31 August 2021).
17. Volkovitsh, M.G.; Bieńkowski, A.O.; Orlova-Bienkowskaja, M.J. Emerald ash borer approaches the borders of the European union and Kazakhstan and is confirmed to infest European ash. *Forests* **2021**, *12*, 691. [CrossRef]
18. Davydenko, K.V.; Borysova, V.; Shcherbak, O.; Kryshtop, Y.; Meshkova, V. Situation and perspectives of European ash (*Fraxinus* spp.) in Ukraine: Focus on eastern border. *Balt. For.* **2019**, *25*, 193–202. [CrossRef]
19. Meshkova, V.; Samoday, V.; Davydenko, K. Ash dieback and contributing factors of forest weakening in provenance tests in the Sumy region. *Cent. Eur. For. J.* **2021**, *67*, 113–121. [CrossRef]
20. Davydenko, K.; Vasaitis, R.; Stenlid, J.; Menkis, A. Fungi in foliage and shoots of *Fraxinus excelsior* in eastern Ukraine: A first report on *Hymenoscyphus Pseudoalbidus*. *For. Path.* **2013**, *43*, 462–467. [CrossRef]
21. Kucheryavenko, T.V.; Skrylnik, Y.E.; Davydenko, K.V.; Zinchenko, O.V.; Meshkova, V.L. The first data on the biological characteristics of *Agrilus planipennis* Fairmare, 1988 (Coleoptera: Buprestidae) in Ukraine. *Ukrainian Entomol. J.* **2020**, *18*, 58–66. [CrossRef]
22. Orlova-Bienkowskaja, M.J.; Bienkowski, A.O. The life cycle of the emerald ash borer *Agrilus planipennis* in European Russia and comparisons with its life cycles in Asia and North America. *Agric. For. Entomol.* **2016**, *18*, 182–188. [CrossRef]
23. Meshkova, V.L.; Kucheryavenko, T.V.; Skrylnik, Y.E.; Zinchenko, O.V.; Borysenko, A.I. Beginning of the spread of *Agrilus planipennis* Fairmaire (Coleoptera: Buprestidae) on the territory of Ukraine. *Proc. St. Petersburg For. Tech. Acad.* **2021**, *236*, 163–184. [CrossRef]
24. Lenz, H.D.; Bartha, B.; Strasser, L.; Lemme, H. Development of ash dieback in south-eastern Germany and the increasing occurrence of secondary pathogens. *Forests* **2016**, *7*, 41. [CrossRef]

forests

MDPI

Article

Evidence for the Widespread Occurrence of Bacteria Implicated in Acute Oak Decline from Incidental Genetic Sampling

Louise A. P. Gathercole [1,2], Gabriele Nocchi [1,2], Nathan Brown [3], Timothy L. R. Coker [2], William J. Plumb [1,2], Jonathan J. Stocks [2], Richard A. Nichols [1], Sandra Denman [4] and Richard J. A. Buggs [1,2,*]

[1] School of Biological and Behavioural Sciences, Faculty of Science and Engineering, Mile End Campus, Queen Mary University of London, Mile End Road, London E1 4NS, UK; l.a.p.gathercole@qmul.ac.uk (L.A.P.G.); g.nocchi@qmul.ac.uk (G.N.); plumbwj01@gmail.com (W.J.P.); r.a.nichols@qmul.ac.uk (R.A.N.)

[2] Science Directorate, Royal Botanic Gardens, Kew, Richmond TW9 3AE, UK; tim@coker-online.co.uk (T.L.R.C.); jjstocks@gmail.com (J.J.S.)

[3] Woodland Heritage, P.O. Box 1331, Cheltenham GL50 9AP, UK; nathan@woodlandheritage.org

[4] Forest Research, Centre for Ecology, Biosecurity and Society, Farnham GU10 4LH, UK; sandra.denman@forestresearch.gov.uk

* Correspondence: r.buggs@qmul.ac.uk; Tel.: +44-(0)-2083325755

Abstract: Acute Oak Decline (AOD) is complex syndrome affecting Britain's keystone native oak species, (*Quercus robur* L. and *Q. petraea* L. (Matt.) Liebl.), in some cases causing mortality within five years of symptom development. The most distinguishable symptom is weeping stem lesions, from which four species of bacteria have been isolated: *Brenneria goodwinii*, *Gibbsiella quercinecans*, *Lonsdalea britannica* and *Rahnella victoriana*. We do not yet know where else these bacteria exist, and little is known about the relationship of the wider oak leaf microbiome (phyllosphere) to acute oak decline. Here we investigate whether incidental evidence from a large oak genome re-sequencing dataset could be used to detect these bacteria in oak foliage, and whether bacterial incidence co-varied with AOD status or location. Oak leaves and buds were sampled from 421 trees at five sites in England. Whole genomic DNA from these samples was shot-gun sequenced with short reads. Non-oak reads were extracted from these data and queried to microbial databases. Reads uniquely matching AOD-associated bacterial genomes were found to be present on trees from all five sites and included trees with active lesions, trees with historic lesions and trees without AOD symptoms. The abundance of the AOD-associated bacteria did not differ between tree health categories but did differ among sites. We conclude that the AOD-associated bacteria may be members of the normal oak microbiome, whose presence on a tree is not sufficient to cause AOD symptoms.

Keywords: *Quercus*; acute oak decline; phyllosphere; *Brenneria goodwinii*; *Gibbsiella quercinecans*; *Lonsdalea britannica*; *Rahnella victoriana*

1. Introduction

The native oak trees, *Quercus robur* L. and *Q. petraea* (Matt.) Liebl. are the most common native broadleaf species in Britain [1]. They have historical, cultural, and economic importance and are ecological keystone species for other organisms [1–3]. A recent study found 2300 macro-species of arthropods, fungi and lichens, mosses and liverworts, birds and mammals associated with British native oaks, of which 326 were obligate (found only on *Q. robur/petraea)* [4]. Oak in the UK is currently experiencing damage or elevated mortality from oak processionary moth (*Thaumetopoea processionea* L.), powdery mildew (*Erysiphe alphitoides* (Griffon & Maubl.) [5,6], and complex multi-factorial decline disorders [7]. Decline disorders are complex interactions of multiple factors such as long- and short-term adverse environmental conditions and the actions of secondary pests and diseases [8]. With no clear universal key factors, Manion defined tree decline disorders as "an interaction of interchangeable, specifically ordered abiotic and biotic factors to produce a gradual general deterioration, often ending in death of trees" [9]. In the UK, two separate decline disorders have been described: Chronic Oak Decline (COD) and Acute Oak Decline (AOD) [10]. Chronic Oak Decline is a slow decline syndrome thought to be related to root health with root rot fungi as a key biotic factor [10].

Acute Oak Decline has been described as a distinctive decline condition that can rapidly kill trees [7,11]. The key observable symptoms of AOD are cracks in the bark plate and stem bleeds, together with signs of the presence of the beetle *Agrilus biguttatus* Fabricius (larval galleries, and adult exit holes in the bark). It is a complex episodic decline syndrome associated with several predisposing environmental factors causing stress to the trees. Affected sites are more likely to be found in warmer areas of Britain and areas of low rainfall [11,12]. Soil pH, and the rhizosphere microbiome, have been found to differ between symptomatic and asymptomatic trees at some sites [13,14]. A study that included Hatchlands Park and Sheen Wood (both of which are studied here, see below) found that the bacterial composition of the rhizosphere differed between sites and tree health conditions [13]. Observations at AOD affected sites show a fluctuating pattern of visible symptoms, which may relate to weather patterns. In many affected trees, callus tissue overgrows lesions and the trees enter remission, where external stem bleeds are no longer visible. However, some trees die within 3 to 5 years of the first infection [11,12].

The biotic cause of the lesions characteristic of Acute Oak Decline is a subject of ongoing investigation. A complex polyspecies microbiome has been found in necrotic lesion tissue. Three bacterial species, *Gibbsiella quercinecans, Brenneria goodwinii*, and *Rahnella victoriana*, have been consistently found, with *Lonsdalea britannica* sometimes present [15–23]. Comparisons of the genomes of these species with orthologous genes in canonical phytopathogens and non-pathogenic symbionts suggest that *B. goodwinii* and *L. britannica* have the genomic potential to cause tissue necrosis, with *B. goodwinii* the most likely to be the key causal necrogenic phytopathogen in AOD [16]. The role of the apparently less pathogenic *G. quercinecans* may be to release necrotizing enzymes that contribute to tissue necrosis or to help the more pathogenic *B. goodwinii* to activate [16].

Below the bark, lesions are associated with the galleries of larvae of a wood-boring beetle, the two-spotted buprestid *Agrilus biguttatus*, which may play a role in tree death through girdling and disrupting vascular flow [24–26]. Co-inoculations of logs with *A. biguttatus* eggs and Acute Oak Decline associated bacteria found that the necrogenic bacterium *B. goodwinii* appears to be triggered by the presence of the larvae, with upregulation of damaging genes increased 10-fold [27]. Whilst colonisation by *Agrilus* species can occur without AOD symptoms, the lesion symptoms of AOD are not found outside of the range of *A. biguttatus*, of which Britain is the northernmost limit [25].

A significant difference in microbial composition has been found in the cultivable bacteria isolated from diseased stem tissue compared with healthy stem tissue. Diseased tissue yielded a greater abundance of bacteria in general, and higher *G. quercinecans* and *B. goodwinii* abundance in particular [28]. Microbial isolation and cultivation from stem tissue from both healthy and symptomatic trees at Acute Oak Decline affected and unaffected sites, with identification of bacteria using DNA amplicon sequencing, showed that *G. quercinecans* and *B. goodwinii* were significantly associated with each other, but were not found in AOD free sites and were low in abundance in the stems of healthy trees at AOD sites. *Rahnella* species were identified at AOD sites, but *R. victoriana* was found only in diseased trees and not at all sites [15], although it has previously been found in both healthy and diseased trees [28]. *Lonsdalea britannica* was not widespread [15]. *Gibbsiella quercinecans* and *B. goodwinii* have recently been isolated from stem bleeds in *Q. robur* in Spain [29], whilst *Brenneria* and *Rahnella* species have been isolated from AOD-like stem bleeds in *Q. castaneifolia, Q. brantii* and *Carpinus betulus* in Iran [30] suggesting that these genera may be involved in similar decline syndromes in other tree species in different parts of the world.

At present little is known about the wider occurrence of the four Acute Oak Decline-associated bacteria outside of AOD lesions. Studies to date have focused on cultures from oak stems [15,28,31]. However, little data is available about other parts of the oak phyllosphere: The above-ground parts of the plant that can be colonised by micro-organisms. Leaf surfaces form the largest component of the phyllosphere, providing a highly variable environment where bacteria are estimated to be the most abundant colonisers [32–34]. In deciduous plants like *Q. robur* and *Q. petraea*, leaves are a temporary habitat, available for only part of the year. Laboratory experiments assessing the ability of *B. goodwinii* and *G. quercinecans* to survive in water and soil showed that *G. quercinecans*, although not culturable from soil, was detectable using qPCR up to 84 days after inoculation in both rainwater and soil. *Brenneria goodwinii* was not able to survive in either environment suggesting that it is an endophytic or epiphytic plant pathogen, whereas *G. quercinecans* is more generalist and able to survive in environmental reservoirs [35]. Recent unpublished research into potential transmission mechanisms of AOD found *B. goodwinii* and *G. quercinecans* on samples of leaves of both healthy and symptomatic trees and at AOD affected and unaffected sites (Emma Bonham, personal communication 20 June 2019). It is therefore possible that these species are usually present on the leaves of the oak and other forest trees. In contrast, Meaden et al. [31] found no evidence of *B. goodwinii* in healthy trees in a woodland with no AOD symptoms.

Metagenomics can be applied to analyse microbiomes from high-throughput sequencing data. Originally proposed as a way of capturing the unculturable genomes and chemistry of microbes in soil samples [36], metagenomics has increasingly been used as a culture-independent method to characterise the microbial communities of a range of environments [37–39]. A marker gene approach, such as that used in the analysis of the oak rhizosphere at sites with Acute Oak Decline, is one method [13]. An alternative approach analyses shotgun sequence data from whole sample DNA, either assembling the raw reads into contigs for functional gene annotation and taxonomic profiling or, by taking an assembly free approach, using raw reads for reference-based classification. Metagenomic classification software exists to rapidly provide taxonomic identification of raw reads using a database of microbial reference genomes. The recent publication of whole genome assemblies of the four bacteria implicated in AOD [16] means that they can be detected using Kraken2, an ultrafast classification tool designed to assign taxonomic labels to sequences of DNA using exact alignments [40].

Here, we analyse a large short-read dataset based on high-throughput sequencing of DNA extracted from oak leaves and buds from five sites, including Acute Oak Decline symptomatic, asymptomatic and 'in remission' native oaks. This dataset was generated for the study of oak genomic variation [41], but here we examine the incidental evidence it may provide about the phyllosphere of the sequenced trees. We aim to discover whether the four AOD-associated bacteria, previously identified in stem bleeds, could be identified in these foliar data using metagenomic classification of reads that do not map to the oak genome. We wish to understand whether read numbers suggest that these bacteria differ in presence or abundance between affected and unaffected trees, or between geographical locations. We also aim to identify the composition of the wider microbiome of the oak phyllosphere and to ascertain whether there are differences between geographical locations and between health status groups.

2. Materials and Methods

This study makes use of data initially collected for an investigation of oak tree genomes in relation to Acute Oak Decline and a wider population study of oak trees at four sites where research and monitoring of AOD was already taking place [41]. These sites are Attingham Park in Shropshire, Langdale Wood in Herefordshire, Hatchlands Park in Surrey and Sheen Wood in Richmond Park, Greater London. In addition, samples and limited data were available from a set of trees at Chestnuts Wood in the Forest of Dean, Gloucestershire. Samples from Chestnuts Wood were either healthy or showing signs of Chronic Oak Decline.

Although AOD is present in parts of the Forest of Dean, none of the sampled trees from Chestnuts Wood had AOD symptoms and these symptoms are not present within 50m of the sampled area. Table 1 summarises the features of each site, and the sample sizes of trees in each health category, based on annual Forest Research and Woodland Heritage monitoring data. Some of the trees at Attingham Park were planted in the 1980s and therefore are less likely to be affected by AOD which mainly affects mature trees [10].

Table 1. Summary of sites and numbers of trees in each health category at each site. n = 421. In addition, 15 technical replicates were sequenced. These were used in the NMDS and PERMANOVA analysis but not in any subsequent analysis.

			Health Groups Sampled at Each Site (Asymptomatic [1], Remission [2], Acute Oak Decline [3] or Chronic Oak Decline [4] Symptoms)-Number of Trees in Each Health Category				
	Features	**AOD First Recorded [42]**	**A [1]**	**R [2]**	**AOD [3]**	**COD [4]**	**Total**
Attingham Park	Small area of closely planted oak dominated amenity woodland within parkland. Area of new planting with trees approximately 30 years old	2007	58	12	15	0	85
Hatchlands Park	Landscaped parkland with shelterbelts	After 2001	37	26	19	0	82
Langdale Wood	Managed open forest, not closely planted, mowed grass below	Before 2006	77	30	20	0	127
Sheen Wood	Oak dominated woodland in a Royal Park, trees not closely planted. Location as a boundary belt on the edge of the park in an urban area	1991	17	60	25	0	102
Chestnuts Wood	Area of plantation 50–60 years old in the Forest of Dean. Close to natural woodland. COD present. No AOD in the woodland sampled, but AOD is present in the wider area	N/A	14	0	0	11	25
Total			203	128	79	11	421

Samples of small branches containing leaves and buds were collected by the Forest Research Technical Services Unit and placed into plastic bags, which were delivered the same day, or the day after, to RBG Kew. Sampling took place from 1 November 2017 to 14 November 2017. At RBG Kew, samples were placed in a cold room upon receipt. A small amount of leaf and/or bud material from each sample was placed into a small zip-lock bag with silica gel. Another portion of each sample was pressed in a plant press and dried using blotting paper. Another portion was placed in a −80 °C freezer.

Health status, categorised as Acute Oak Decline or Chronic Oak Decline symptomatic, in remission or asymptomatic, was recorded at the time of sampling. Trees were also allocated an age category of semi-mature, mature, or over-mature based on a combination of a guide diameter, physiological stages based on a model by Raimbault for early tree development [43], and whether the tree is perceived to be in the first, second, or third stage of life.

The sampling for this study was not originally designed for a metagenomic study and therefore did not follow all the best practice protocols for microbiome and metagenomic studies as outlined by Knight, Vrbanac et al. [44]. Specifically, the samples were taken from two tissue types and were dried using two different methods. Samples were not collected and processed in the sterile conditions normally associated with studies of microbiomes, but there was little opportunity for cross contamination from other samples in the field as each sample was cut and bagged immediately. Detailed information was not collected about the exact position in the canopy or the part of the leaf from which each sample was taken, both of which are associated with variability in microbial community abundance and composition [34,45]. However, since samples were collected by hand or long-arm pruner the maximum height is likely to be 3m from the ground.

Whole genomic DNA was extracted in most cases from leaf or bud tissues that had been dried using silica gel, and in a few cases from bud material dried in a plant press. Extractions were done using the Qiagen DNeasy Plant Mini Kit 250 (Qiagen, Germantown, Maryland) following the standard protocol. Whole genomic DNA was sent to Novogene (Hong Kong) for library preparation and whole-genome shotgun sequencing. A total of 436 DNA samples with full phenotypic information were successfully sequenced, including 15 technical replicates from separate DNA extractions. Of these 330 were from leaf tissue dried on silica gel, 21 from bud tissue dried on silica gel, and 78 from bud tissues dried in plant presses. Additionally, four samples were extracted from both bud and leaf tissues dried in silica gel and 3 samples were a combination of press dried bud and silica gel dried leaf. Of the successfully sequenced individual trees, 351 were *Q. robur*, 10 were *Q. petraea* and 16 were hybrids, according to genome-based analyses. A further 45 trees not included in Nocchi et al. 2021 [41] were identified by morphology, 43 as *Q. robur*, and one as *Q. petraea*, making a total of 394 *Q. robur*, 11 *Q. petraea*, and 16 hybrids.

Sequencing libraries were prepared with NEBNext DNA Library Prep Kit (New England Biolabs Inc., Ipswich, Massachusetts, USA). Whole genome short read sequencing data was generated with the Illumina Novaseq 6000 platform (Illumina Inc., San Diego, CA, USA) with read lengths of 2 × 150 base pairs (bp) to produce 16.5 Gbp of data per sample. All subsequent analysis utilised Queen Mary University of London's Apocrita HPC facility, supported by QMUL Research-IT [46]. The whole genome sequencing (WGS) data was cleaned and aligned to the reference *Q. robur* genome V2.3 [47] using the bwa mem algorithm from bwa version 0.7.15 with the -M option [48]. Reads that did not align to the reference genome were extracted from the bwa output with samtools v1.9 using the fastq function with flag (−f) 12 [49] and written to separate files. To remove potential human tissue contamination, these non-oak reads were aligned to the human genome build37 using bowtie2 v2.3.4 with the default parameters [50]. The non-human reads were separated and returned as paired read files using samtools v1.9 and bedtools v2.28 with default parameters. Raw reads for each sample are available in the European Nucleotide Archive, accession PRJEB30573.

The non-oak, non-human reads from each sample were taxonomically classified with the bioinformatics software Kraken2, as follows. A custom library was created from all available fungal (1562 genomes), bacterial (56,460 genomes), viral (13,466 genomes), protozoal (11,151 genomes), and archaeal (614 genomes) genomes along with the UniVec_Core library of common contaminants downloaded from the NCBI refseq database [51] on 9/5/21. Three of the four Acute Oak Decline-associated genomes were present in the library and a *Rahnella victoriana* genome was downloaded separately and added to the library. The library contained a total of 5 *Brenneria* species (*B. goodwinii, B. nigrifluens, B. rubrifaciens, B. sp. IR50* and *B. sp. EniD312*), 2 *Lonsdalea* species (*L. britannica* and *L. populi*), and 5 *Rahnella* species (*R. britannica, R. aquatilis, R. aceris, R. sp. Y9602*, and *R. sp. ERMR1:05*). The only *Gibbsiella* species in the library was *G. quercinecans*. From this library, kraken2 created substrings (or *k-mers*), each 31 bases in length.

This custom Kraken database was used to identify the biotic sources of the reads that had not aligned to the *Q. robur* or *H. sapiens* genomes. Kraken2 queried all the possible 31-mers in our paired reads against the 31-mers in the custom database and matched them to the lowest common ancestor (i.e., the highest taxonomic level within which all lower levels contained that 31-mer). Only exact matches for full 31-mers were classified. Each matched read was then allocated an appropriate taxonomic classification using the set of taxa matched to the 31-mers, and a report produced, showing the number and percentage of reads classified at each clade and taxon level for each sample. Kraken-biom 1.0.1 (https://pypi.org/project/kraken-biom/ accessed on 13 December 2019) was used to create BIOM files [52] for downstream analysis. One BIOM file included classification of reads to all possible taxonomic levels for each sample. A second BIOM file classified reads only to phylum level. Statistical analysis was carried out using R version 3.5.3, in RStudio [53,54]. Package citations for packages not included in the main text is available in Supplementary Materials.

To compare the whole microbiomes among sampled trees, read counts for each classified taxon at all taxonomic levels were normalised by dividing by the total number of all classified non-oak reads per sample. We ran a Non-metric Multidimensional Scaling analysis (NMDS) on these data using the vegan package in R [55] to identify any potential patterns in variation across the whole microbiome in issue type, drying method, tree species, sampling site, or tree health status as well as examining the technical replicates. We further analysed the data using Permutational Multivariate Analysis of Variance (PERMANOVA) implemented in the vegan adonis and adonis2 functions using Bray-Curtis dissimilarity and 999 permutations. The analysis in adonis analyses the variables sequentially, passing on the unexplained variability to the next variable. The order of the variables was decided by first using the adonis2 function with the by = "margin" option, which reports the variance not explained by the other variables, as described in Hill et al. [56]. The vegan betadisper function was also used for Permutational Analysis of Multivariate Dispersions (PERMADISP) to test whether data dispersion was uniform for each variable.

The software Bracken (Bayesian Reestimation of Abundance after Classification with Kraken) [57] can be used with Kraken classifications to estimate abundance at species level by re-classifying higher level taxonomic reads at species level. A Bracken analysis was run on the Kraken allocations. No additional reads were allocated to the AOD-associated bacterial classifications, so this approach was not pursued further.

Each DNA sample was scored as having presence or absence of the Acute Oak Decline-associated bacterial according to the Kraken results using two thresholds: (1) presence of one or more reads assigned to an AOD-associated bacterium; (2) presence of ten or more reads assigned to an AOD-associated bacterium. These allocations of the DNA samples are referred to hereafter as the one-read and the ten-read scorings respectively. The value of ten, the default cut-off for Bracken re-estimation, is somewhat arbitrary. It eliminates two thirds of the samples, hence the analysis was carried out on both scorings. The percentage of trees with AOD-associated bacteria scored as present at each site and in each health group (AOD Symptomatic, COD Symptomatic, Remission, Asymptomatic) was calculated for each scoring. Chi-squared tests were carried out to assess whether the differences in the percentage of trees with each AOD bacteria present was significant between sites or health categories.

In addition to these analyses of presence/absence we analysed the abundance of each Acute Oak Decline-associated taxon (where present) as follows. A relative abundance score (a_{ij}) was obtained from the counts of each AOD-associated taxon (x_{ij}) expressed as a proportion of the total number of bacterial reads in sample ($t_{i,}$). A log(log) transformation was applied after inspection of the residuals in a preliminary analysis: $a_{ij} = \log(\log(x_{ij}/t_{i,}) + 20)$. For the one-read data set, 305 of the original 421 trees were included in this analysis once Chestnuts Wood and trees with none of the AOD-associated bacteria were removed. For the ten-read data set, 97 trees were included in the analysis. To investigate whether the relative abundance score varied as a function of multiple factors, we ran mixed effects models using the lme function from the nlme package in R [58]. In all models, the random effect was the sample identifier allowing for sample specific effects on the counts of up to 4 taxa for each sample. The response variable was the proportioned and log(log) transformed count of reads classified by Kraken as *G. quercinecans*, *B. goodwinii*, *L. britannica*, or *R. victoriana*. Fixed effects were bacterial species, site (population) and health status. Trees were categorised as asymptomatic, in remission, or AOD symptomatic—according to the surveyed health status in 2017, which was the year in which the leaves were collected. Previous work has shown an association between tree size as an approximation of age, being a predictor of bacterial community composition [31]. Diameter at breast height (DBH) could be used as a continuous variable to approximate age, but it was not available for all samples. Instead, age was included in the models as a categorical variable with three levels (semi-mature, mature, or over-mature). An initial additive model was run as follows:

Random effects: individual

Abundance_score ~ (fixed effects): drying_method + tissue + bacterial_species + age + health + site

The initial model was run three times. First with the health status as three separate categories, secondly with health status as two categories-remission trees were included with symptomatic trees, thirdly with remission trees included with asymptomatic trees. There was little difference in outcome, so health was included as a three-category variable for the final model. (See Supplementary Materials for results).

Health status, having been found to be not significant, was removed for a final model. Tissue type and drying method were significant, but closely correlated, therefore tissue type was included in the final model. The final model for the one-read scoring included tissue, age, bacterial species, site, and the interaction between bacterial species and site, as follows:

Random effects: individual

Abundance_score ~ (fixed effects): tissue + bacterial_species + age + health*site

The interaction between species and site was included to investigate whether the combination of a particular bacterial species and a site had an effect on the proportion of reads classified as Acute Oak Decline-associated bacteria. Both of these independent variables had been found to be significant in earlier models.

The missing values for some of the bacteria in the smaller ten-read scoring meant that it was not possible to include the interaction between bacterial species and site, therefore an additive model was run, as follows:

Random effects: individual

Abundance_score ~ (fixed effects): tissue + bacterial_species + age + health + site

In order to further confirm the classifications made by Kraken of some reads to the four Acute Oak Decline-associated bacteria on the basis of 31-mers, we took the full-length reads and queried them against a custom BLAST database of the four AOD-associated genomes using magic-BLAST [59]. We calculated the percentage of these reads, classified by Kraken as being from AOD-related bacteria, that had a 95% or greater alignment with the genome of the species that Kraken had classified them to.

3. Results

Within each of the short-read sequence datasets for each oak tree, a mean of 96.44% (standard deviation = 4.4) of reads mapped to the *Q. robur* genome. A further 0.22% (standard deviation = 1.13) of reads mapped to the human genome. This left a mean of 3.33% (standard deviation = 3.97) of non-oak, non-human reads. Of these reads, the mean percentage of reads classified by Kraken was 6.05% (range = 0.49% to 51.14%, sd = 5.25): thus between 2724 and 6,626,088 reads were classified per sample. Of these, across all samples, 44.55% were classified as Proteobacteria, 33.9% as Actinobacteria, 12.13% as Bacteroidetes, 6.2% as Ascomycota, and 1.28% as Firmicutes. The relative abundance of reads classified as fungi differed among sites (Figure 1). Notably, ascomycetes were unusually common in the samples from Hatchlands Park and amongst the bacteria, proteobacteria was unusually common at Chestnuts Wood. Basidiomycota only accounted for 1% of the classified reads across all reads. These were in greater abundance at Hatchlands Park (2.4%) where the predominant environment is wood pasture. There was a relatively low percentage (0.1%) of the reads classified as Basidiomycota at Chestnuts Wood, which is a more closely planted woodland with COD, often associated with the Basidiomycota genus *Armillaria* causing problems with the tree roots. These seems not to have translated into the presence of Basidiomycota on the leaves.

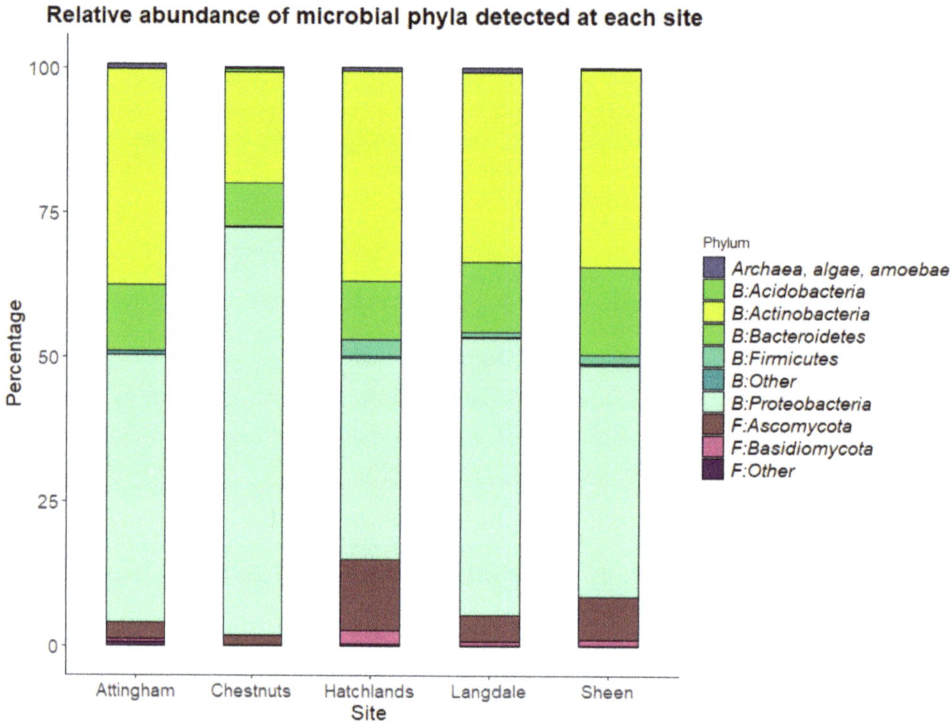

Figure 1. The relative abundance of microbial phyla detected by Kraken classification software at each site. Fungi (F) are at the bottom of the bars and bacterial phyla (B) at the top. Archaea and other eukaryotic microbes comprise less than 0.05% of the total classified. The most abundant phylum is the proteobacteria to which the AOD-associated bacteria belong.

The bacterial phyla with the highest percentages of the total reads classified were Proteobacteria (44.55%), Actinobacteria (33.9%), Bacteroidetes (12.13%), and Firmicutes (1.28%). The fungal phylum Ascomycota accounted for 6.2% of classified reads and the fungal Basidiomycota just 1.02%. The AOD-associated all belong to the Proteobacteria phylum.

A Non-metric Multidimensional Scaling Analysis based on all samples (including technical replicates) and all classified microbiome read counts across all taxonomic levels is shown in Figure 2. Each panel of this figure shows NMDS1 and NMDS2 for the same analysis, but with different labels. In Figure 2A, the different species of oak are labelled; showing little clustering due to oak species, whether identified morphologically or genetically. In Figure 2B, the sample drying method and tissue source are labelled. Samples dried in the plant press and those dried in silica gel fall separately, the pattern being very similar to that of the tissue source and these appear to be confounded. Most bud samples were dried in the press and most leaf samples in silica gel. In Figure 2C, the health status of the trees is labelled. From the plot, there appears to be little clustering of trees with different health statuses. In Figure 2D, the sample sites are labelled, where there appears to be a minimal amount of clustering by site. Chronic Oak Decline is confounded with site as it was only found in Chestnuts Wood. Figure 2E shows the same NMDS with just the pairs of technical replicates visible. Most technical replicates appeared close to each other, although some differences are to be expected.

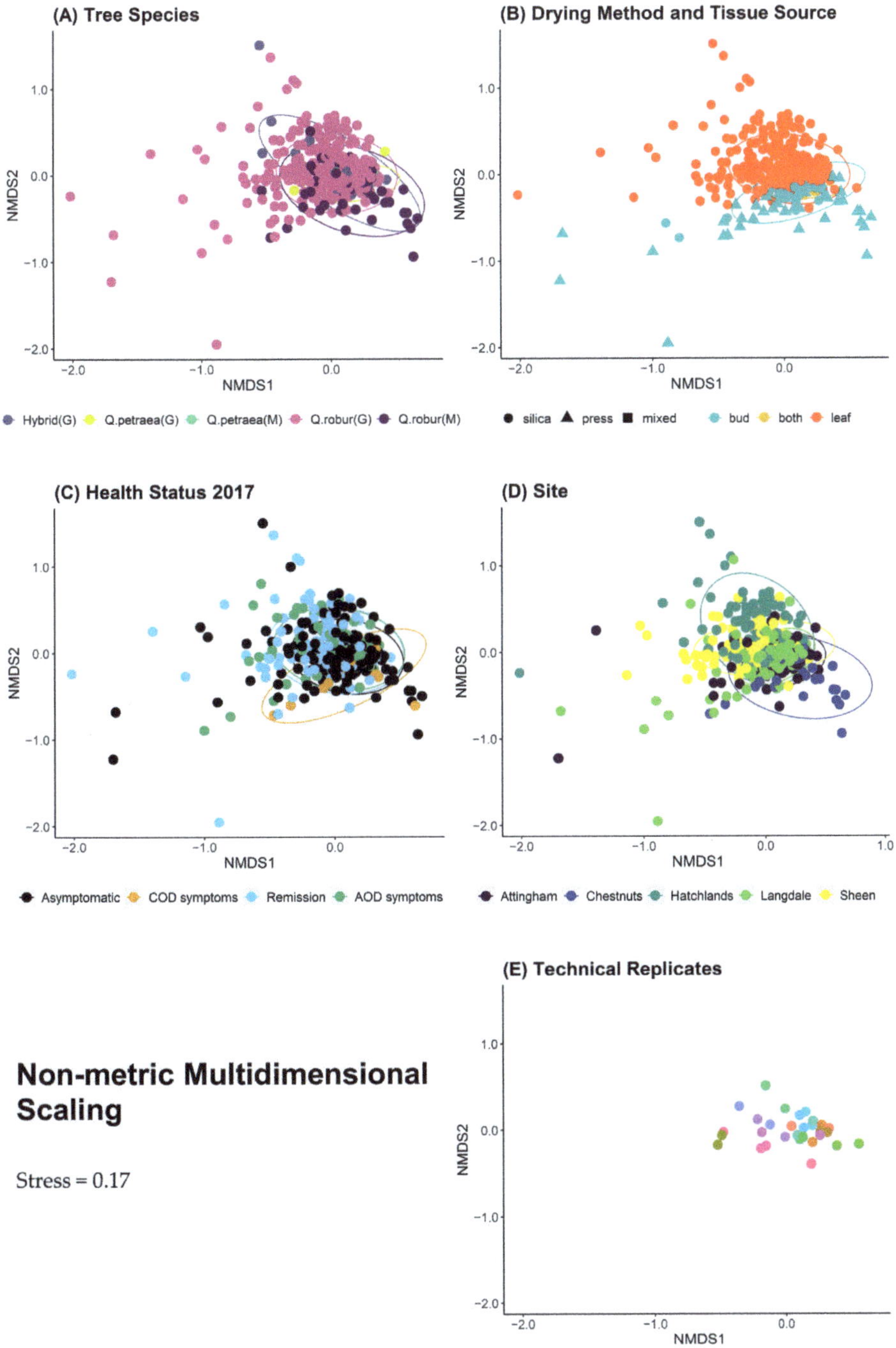

Figure 2. Non-metric Multidimensional Scaling plot based on all samples (including technical replicates) and all classified microbiome read counts across all taxonomic levels. Each panel shows NMDS1 and NMDS2 for the same NMDS, but with

different labels. (**A**) oak tree species; showing little clustering. (Species are categorised as Hybrid, Q. robur or Q. petraea (according to genome-based analysis, or morphologically categorised as Q. robur or Q. petraea). (**B**) the sample drying method and tissue source. Samples dried in the plant press and those dried on silica gel fall on different sides of the plot and are very similar to tissue source. Most bud samples were dried in the press and leaf samples in silica gel. The differences in microbial classification may be due to differences between the bud and open leaf, or be from differences in drying method. (**C**) Health Status: There is minimal visibly discernible clustering of trees by health status, especially amongst the trees from the AOD sites. (**D**) Sample site: There is a small visible clustering by site. COD health status is confounded with site as COD was only found in Chestnuts Wood. (**E**) Showing only the technical replicates, which mostly appear close to their partner.

Table 2 shows the PERMANOVA results indicating significant differences in the microbiome between the microbiomes according to tissue source, site, and health status. However, the difference between sites may not be a reliable result, as the PERMADISP analysis found that dispersion was uniform for all the tested variables, except site possibly because of the difference in sample sizes between the sites. The health status differences in microbial composition are not clear in the NMDS visualisation with the ellipse for the COD symptomatic trees being set slightly apart from the other categories.

Table 2. Results of the statistical tests on Bray-Curtis dissimilarity matrices of all of the Kraken2 classified microbial taxa visualised in the NMDS.

		Permanova				Permdisp
		Adonis 2		Adonis		
Variable	**DF**	**Marginal R^2**	***p*-Value**	**R^2**	***p*-Value**	***p***
Tissue	2	0.018	**0.001**	0.084	**0.001**	0.172
Site	4	0.1	**<0.001**	0.118	**0.001**	**<0.001**
Health Status	3	0.01	**0.011**	0.01	**0.009**	0.31
Tree Species	4	0.007	0.52	0.007	0.506	0.052
Drying Method	2	0.002	0.84	0.002	0.84	0.37

Note: Statistically significant results ($p < 0.05$) are highlighted in **bold**.

Low numbers of reads classified by Kraken as *Brenneria goodwinii*, *Gibbsiella quercinecans*, *Lonsdalea britannica*, and *Rahnella victoriana* were found from both leaf and bud samples at all sites and across all health categories. Total reads for all four taxa per tree (including technical replicates) ranged from 0 to 1,607,824 with a mean of 3955 and a median of 3 classified reads. There were two very high outliers, and apart from these the highest count was 7961. Only 35 samples in total had 100 reads or more.

In the one-read scoring, *B. goodwinii* was detected in 42 trees (10%), *G. quercinecans* in 132 (31%), *L. britannica* in 195 (46%), and *R. victoriana* in 237 (56%). In the more stringent ten-read scoring *B. goodwinii* was detected in just 5 trees (1%), *G. quercinecans* in 12 (3%), *L. britannica* in 48 (11%), and *R. victoriana* in 66 (16%).

Figure 3 shows the proportion of trees at each site where each species of bacteria was detected in each scoring. It is notable that reads from all four Acute Oak Decline-associated bacteria were found at Chestnuts Wood where Acute Oak Decline is not present, but Chronic Oak Decline is. The lowest incidence of the four bacteria was at Hatchlands, where AOD is present. Figure 4 shows the proportion of samples in the four sites (grouped together) where AOD is present. The bars show the proportion of trees in the different health statuses where AOD-bacterial reads were found in their samples. This shows no noticeable difference between symptomatic, asymptomatic, and remission trees.

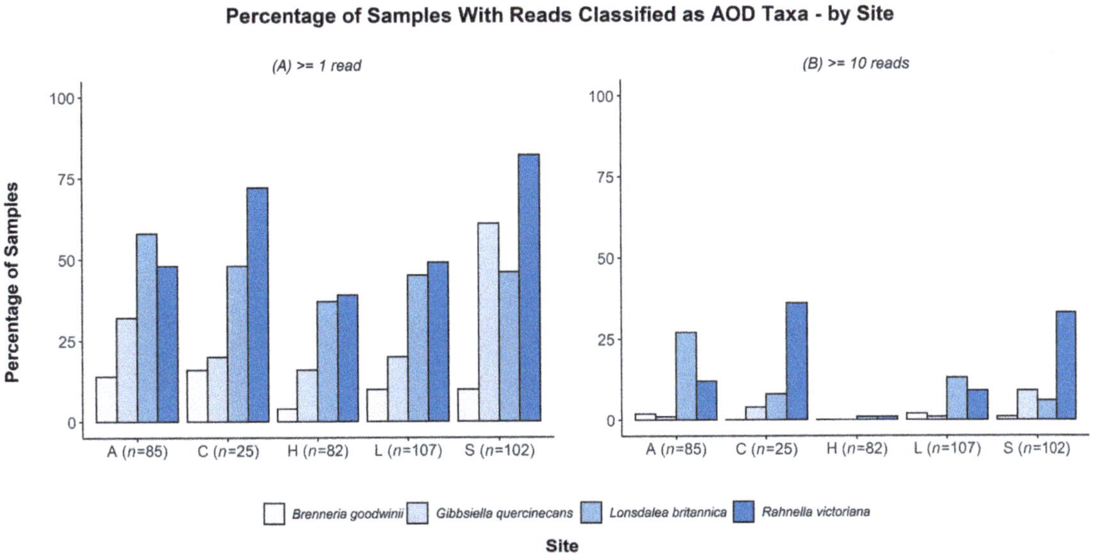

Figure 3. Percentage of samples with AOD-associated bacteria present at each of the sites. A = Attingham Park, C = Chestnuts Wood, H = Hatchlands Park, L = Langdale Wood, S = Sheen Wood. (**A**) shows the percentage of trees where 1 or more reads were classified by Kraken2. (**B**) shows the percentage where at least 10 reads were classified. Hatchlands Park has very few trees in which more than 10 reads were classified as AOD-associated bacterial genomes. Reads that matched AOD genomes were found at all sites, including Chestnuts Wood which did not have any AOD symptomatic trees.

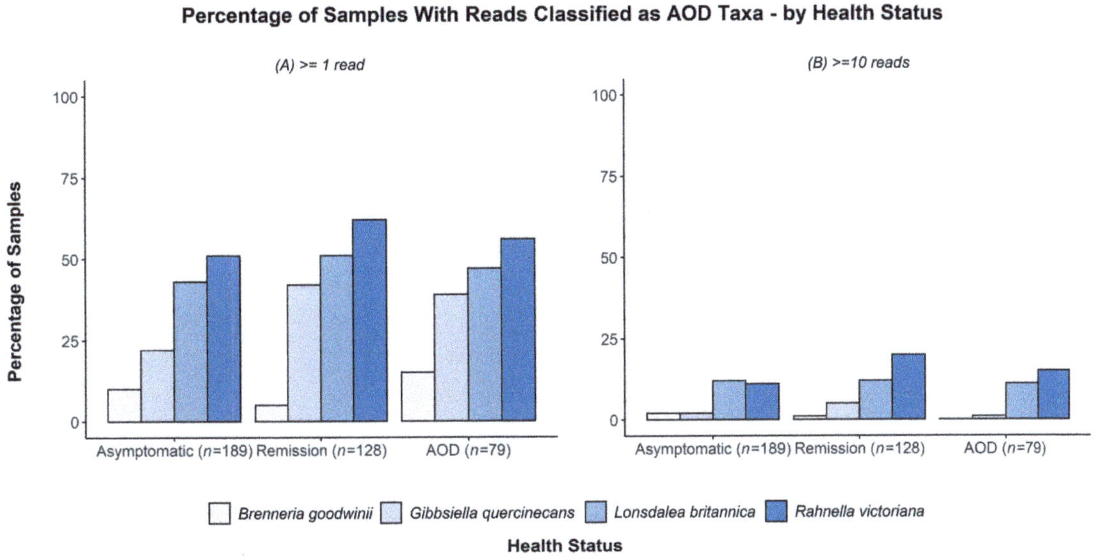

Figure 4. Percentage of samples with Acute Oak Decline-associated bacteria present in each health group across all sites with AOD-symptomatic trees. (**A**) shows the percentage of trees where 1 or more reads were classified by Kraken2. (**B**) shows the percentage where at least 10 reads were classified. Reads that matched AOD genomes were found on Asymptomatic, AOD symptomatic trees and trees in remission. Brenneria goodwinii, the putative necrotic agent in AOD stem lesions was detected on the leaves of very few trees in any health group.

We carried out chi-squared tests on the differences in the proportion of trees in which each bacterial species was detected across the four sites with Acute Oak Decline. These found significant differences across sites for all of the bacteria except *B. goodwinii*, which was detected on the lowest number of trees. Hatchlands Park had few trees with more than 10 reads classified as each AOD bacteria. Removing Hatchlands reduced the differences between sites, but this still remained significant (Table 3).

Table 3. Pearson's chi-squared test of differences between the proportion of trees showing presence of each bacteria across the four sites with Acute Oak Decline symptomatic trees. Presence is considered to be one or more classified reads or 10 or more classified reads. There are significant differences across all bacterial species except B. goodwinii, which was detected on a low number of trees. Hatchlands Park had few trees with more than 10 reads classified as each AOD bacteria. Removing Hatchlands reduced the differences between sites, but results still remained significant. Chestnuts Wood was not included in this analysis.

	Presence = One-Read Scoring				**Presence = 10-Read Scoring**			
	Pearson's chi-squared test with Hatchlands (df = 3)		Pearson's chi-squared test with Hatchlands removed (df = 2)		Pearson's chi-squared test with Hatchlands (df = 3)		Pearson's chi-squared test with Hatchlands removed (df =2)	
	X^2	*p*-Value	X^2	*p*-Value	X^2	*p*-Value	X^2	*p*-Value
G. quercinecans	57.448	**<0.001**	42.359	**<0.001**	18.818	<0.001	12.673	**0.002**
B. goodwinii	5.400	0.145	1.049	0.591	2.024	0.568	0.558	0.757
L. britannica	7.619	0.054	3.745	0.154	31.762	**<0.001**	17.456	**<0.001**
R. victoriana	42.861	**<0.001**	32.608	**<0.001**	44.24	**<0.001**	24.952	**<0.001**

Note: Statistically significant results ($p < 0.05$) are highlighted in **bold**.

The abundance of the Acute Oak Decline-associated bacteria, measured by the number of reads assigned to the taxon by Kraken, was analysed for any patterns in abundance related to health status, geographical site, tissue, and drying method. In an initial additive mixed effect models on the samples from the four AOD sites we found no significant differences between symptomatic, remission and healthy trees in the proportion of genomic sequences classified as AOD-associated bacteria. This was the case when samples were categorised in three groups (symptomatic, remission, and healthy), or when trees in remission were included with healthy trees, and when trees in remission were included with AOD symptomatic trees (see Supplementary Materials Table S1). However, we did find a significant effect by including site in the models of the one-read scoring ($p = 0.024$), but not the lower powered comparison for the ten-read scoring ($p = 0.967$) (see Supplementary Materials Table S1).

In a second mixed-effects model, health status was not included, and we included interactions between bacterial species and site. We found a highly significant interaction between bacterial species abundance and site in the one-read scoring (Table 4). Site and age category were both significant even if semi-mature (too young to be affected by Acute Oak Decline) trees were removed from the dataset. Site and age also remained significant if the site Attingham, which has a large number of young trees, was removed from the dataset. In the ten-read scoring, tree age had a significant effect on the read counts but not site. In all analyses tissue type was a significant factor.

Figure 5 shows the abundance score for each bacterial species by health status and by population in both the one- and ten-read scorings. Figure 5A,C show the abundance scores for each species in each health category. There are scores for all of the bacteria in all health categories in the one-read scoring, and also in the ten-read scoring, with the exception of *B. goodwinii* which had no samples with 10 or more reads in the Acute Oak Decline symptomatic category. Figure 5B,D show the scorings by site, where differences in abundance can be seen. The ten-read scoring has no scores for *B. goodwinii* or *G. quercinecans* at Hatchlands Park. *L. britannica* and *R. victoriana* have greater abundance scores in the more stringent ten-read scoring than *B. goodwinii* or *G.quercinecans*. In the BLAST searches of the reads identified by Kraken2 as being from AOD-related bacteria, 57% of *B. goodwinii* classified reads, 60% of *G. quercinecans*, 57% of *L. britannica*, and 61% of

R. victoriana matched their respective genomes with 95% or more of the full read aligning to the genome.

Table 4. (A) ANOVA results of mixed effect model of abundance scores for Acute Oak Decline-associated bacteria using one- (A) and ten-read scoring (B). Random effect was the sample number, fixed effects were tissue (leaf or bud), age category (semi-mature, mature, over-mature), bacterial species, and site. Model (A) included an interaction between bacterial species and site, which was lacking in (B) due to different numbers of samples in each category. In addition to the significant effect of site on the abundance of reads classified in the (A) (1 or more reads classified), an interaction was found between the abundance of the different bacterial species and the site. Site was not significant in (B) after accounting for the other effects.

	Fixed Effect	Numerator DF	Denominator DF	*F*-Value	*p*-Value
(A) One-read scoring(n = 305)	Tissue	2	297	28.46	**<0.001**
	Age	2	297	16.95	**<0.001**
	Bacterial Species	3	250	25.58	**<0.001**
	Site	3	297	2.98	**0.0316**
	Bacterial Species: Site	9	250	7.34	**<0.001**
(B) Ten-read scoring(n = 97)	Tissue	2	89	10.274	**<0.001**
	Age	2	89	0.107	0.899
	Bacterial Species	3	19	3.870	**0.0257**
	Site	3	89	0.835	0.478

Note: Statistically significant effects ($p < 0.05$) are highlighted in **bold**.

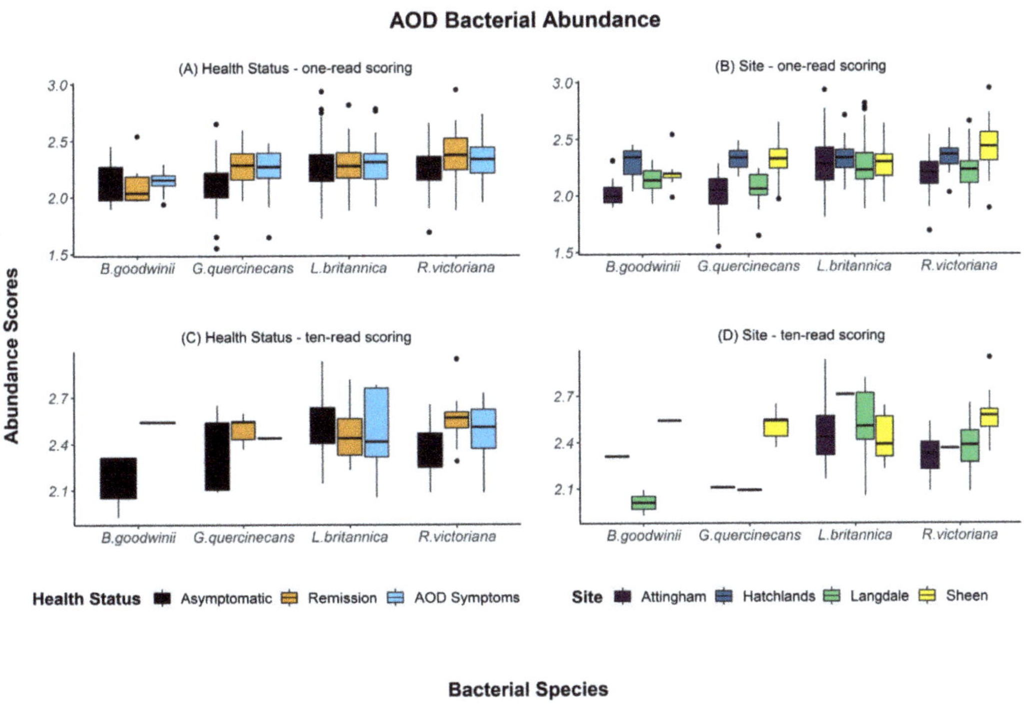

Figure 5. The abundance score for each bacterial species by health status and by population in both the one- and ten-read scorings. (**A**,**C**) show the abundance scores for each species in each health category. There are scores for all of the bacteria in all health categories in the one-read scoring, and also in the ten-read scoring, with the exception of *B. goodwinii* which had no samples with 10 or more reads in the Acute Oak Decline symptomatic category. (**B**,**D**) show the scorings by site, where differences in abundance can be seen. The ten-read scoring has no scores for *B. goodwinii* or *G. quercinecans* at Hatchlands Park. *L. britannica* and *R. victoriana* have greater abundance scores in the more stringent ten-read scoring than *B. goodwinii* or *G.quercinecans*.

4. Discussion

In this study we were able to use incidental read data and the metagenomic classifier Kraken2, to characterise the foliar microbiome, find evidence for the presence of Acute Oak Decline-associated bacteria across five geographical sites, and all AOD health categories.

In total, Kraken2 classified reads were assigned to 56 phyla. The bacterial phyla *Proteobacteria* and *Actinobacteria* accounted for over 78% of the read classifications and *Bacteroidetes* a further 12%. Previous 16s RNA taxonomic studies of *Q. robur* metagenomes from the trunk and the rhizosphere (root microbiome) have found that *Acidobacteria* were highly abundant [31,60]. The present study found that *Acidobacteria* made up only 0.07% of the total reads classified, but this is unsurprising as *Acidobacteria* are abundant in soil and less likely to be dispersed in the phyllosphere. A similar pattern was found in a study of bacterial communities in *Ginkgo biloba* which found that *Acidobacteria* were abundant in trunk tissue, but not in leaf tissue, with *Bacteroidetes* also one of the four most abundant taxon across the different tissues [61]. Therefore, our findings seem to be in line with previous bacterial microbiome studies. Fungi made up only a small percentage of the classified reads compared with bacteria. The two major fungal phyla, *Ascomycota* and *Basidiomycota* are both represented in our classifications, with *Ascomycota* being the most abundant phylum. This is a similar finding to an ITS1 amplicon sequencing study of several leaf samples from a single *Q. robur,* which found that 50% of the operational taxonomic units (OTU)s were *Ascomycota* and 11% *Basidiomycota* [62]. Visualisations of Non-metric multidimensional analysis of the full microbial classifications at all taxonomic levels did not show any obvious clustering by health status or site. However, the PERMANOVA statistical analysis did identify a significant difference in the microbial composition between the different health statuses, which may be of interest for further study. A significant difference was also found between sites, but this may not be a secure finding as the PERMDISP analysis identified a non-uniform distribution amongst sites (Table 2).

The bacteria associated with Acute Oak Decline had previously been identified from isolations taken from lesions in the stem tissue of affected trees [17,19–21,23]. Higher abundances of these bacteria have been found in lesion tissue than unaffected stem tissue from symptomatic trees and stem tissue from healthy trees [28]. We found evidence of their genome sequences in the leaf and leaf-bud tissues, in low and variable abundances, suggesting that these bacteria are in fact ubiquitous. These results imply that the presence of the bacteria on an individual is not a sufficient cause for Acute Oak Decline symptoms to appear. It is of particular interest that in our more stringent ten-read scoring, there was no evidence for the necrogenic species *B.goodwinii* in symptomatic trees. It may therefore be the case that the proliferation of these bacteria at AOD lesion sites is triggered by abiotic factors and/or boring by the beetle *A. biguttatus*, as has been suggested by Doonan et al. [27]. Factors such as low rainfall or poor soil conditions may cause stress in the oak tree, allowing *A.biguttatus,* which usually lives on dying trees, greater opportunities to colonise oaks. However, the nature of the interactions between the bacteria and the beetle larvae are yet to be fully understood. Current work by the BAC-STOP project is investigating whether secondary metabolites in the beetle larvae trigger virulence gene expression in *B. goodwinii* (https://bacterialplantdiseases.uk/bac-stop/, accessed on 24 November 2021).

The detection of these bacterial sequence data in DNA extracted from leaves does not distinguish between bacteria living as endophytes, epiphytes or present due to rain or insect dispersal. Pettifor et al. [35] found that rainwater, and to a lesser extent, soil, could act as a reservoir for *G. quercinecans* but not for *B. goodwinii.* This evidence indirectly suggests that *B. goodwinii* is an endosymbiont, but *G. quercinecans* may have a different niche [35]. We found *R. victoriana* and *L. britannica* to be more widespread than either *G. quercinecans* or *B. goodwinii,* which could suggest a wider dispersal in the environment for these two species.

Our analysis of the abundance of Acute Oak Decline-associated bacteria showed differences between the geographical sites. This is in line with a study of the oak rhizosphere, which found differences in the bacterial composition of the root microbiome between sites [13]. It is also in line with phyllosphere studies on other tree species, reviewed by Rastogi et al. [63] in which geographical location has been shown to be a driver of bacterial community composition. To our knowledge, this has not previously been shown in the oak phyllosphere.

One site, Hatchlands Park, had a lower proportion of trees where the bacteria were present than the other sites. The ecology of the selected trees at Hatchlands Park is different from the other sites in that they are spaced further apart in wood pasture and hedgerow settings rather than in closer planted woodland. The samples were also taken from a wider geographical area in the site. This may have led to greater differences in soil condition or exposure to rainwater, sunlight, and insect activity. Differences in the microbiome and the abundance of Acute Oak Decline-associated bacteria are likely to be partly due to environmental and climactic factors, but there may also be an effect of genetics if the trees were naturally grown or planted from the same seed source at each plantation site. Genotype has been shown to have an effect on the microbiome in poplar, particularly of fungal foliar communities [64,65], and may have an impact on the geographical differences in the full microbiome. AOD symptomatic trees at some of these sites have previously been found in physical clusters [42], which could suggest a mechanism passing the bacteria between the trees. However, the presence of AOD-associated bacteria across many trees in the woodland or parkland could suggest that these clusters may be related more to the action of adult *A. biguttatus*, or to soil or other abiotic conditions in specific parts of the site.

The number of reads classified at species level for the Acute Oak Decline-associated bacteria of interest were low. The developers of Kraken tested the original Kraken1 program against a simulated Hi-seq metagenome and found that precision at genus level, defined as the proportion of correctly classified reads from all the classification attempts, was 99.2% [66]. Although this precision is high, and Kraken and Kraken2 have been found by other authors to be precise and accurate and in line with other metagenomic pipelines [67,68], the low numbers of total AOD-associated bacterial reads classified may put our results within the potential margin of error for false positives. The trends we detected in the species level comparisons should therefore be verified, for example using qPCR studies on the oak phyllosphere. Our classifications at the phylum level analysis are more robust due to the greater likelihood of a correct classification at this level [68] and the higher numbers of reads we classified in each phylum.

What is not clear from this study is whether the AOD-associated bacteria are part of the leaf microbiome only at sites where Acute Oak Decline is present. The bacteria were found in Chestnuts Wood, which is not currently showing any AOD symptoms, although AOD is present in other parts of the Forest of Dean. A microbiome study by Meaden et al. [31] of the *Q. robur* trunk microbiome in 64 trees at Wytham Woods, which has no cases of AOD, found no 16S RNA signals for *Brenneria goodwinii*, but did not look for the other AOD-associated bacteria. It is therefore possible that presence of the AOD-associated bacteria indicates a symptomatic site, rather than symptomatic individuals, and that these bacteria are not a normal part of the oak microbiome outside of these sites. A useful further study would be to collect leaf samples from trees in woodlands with no AOD present and use the recently developed real-time PCR assay to look for AOD-associated bacteria at these sites [69]. In addition, a metagenomic study of oak trees and other species from a variety of sites could shed further light on the range and habitats of these bacteria. The presence of bacteria on the leaves in non-AOD woodlands would confirm that these bacteria are indeed to be part of the normal oak microbiome. An absence could indicate that AOD is a site-wide infection in which trees with predisposing factors become symptomatic. A systematic survey across the *Q. robur* and *Q. petraea* species range in Britain or Europe, including outside of the range of *A. biguttatus*, could shed greater light on whether these species are symbionts of the oak tree.

5. Conclusions

In this study, we were able to identify the most abundant fungal and bacterial phyla in the oak phyllosphere, and detect signatures of Acute Oak Decline-associated bacterial genomes over a much larger number of samples than are typically used for metagenomic studies. This was achieved using incidental read data from a whole-genome-sequencing population genomics project on oak trees in England. We found sequence-read evidence for the presence of AOD-associated bacteria on leaf and bud material in four Acute Oak Decline affected sites and additionally at a site where AOD symptoms are not present in the immediate vicinity. We found evidence that the AOD-associated bacteria were present on trees in all health categories. There were no significant differences in either the proportion of trees with AOD-associated bacteria present, or in the abundance of any of the four AOD-associated bacteria between the symptomatic, remission, and healthy trees. There were significant differences between geographical sites in both the proportion of trees with the bacteria present and the abundance of bacteria, although the abundance trend could not be demonstrated in the lower powered analysis of the more stringent ten-read scoring. Interestingly, *Brenneria goodwinii*, the bacterial species believed to have a necrogenic affect in AOD lesions was not detected on the leaves of symptomatic trees in the more stringent ten-read scoring, but was found on healthy trees and those in remission suggesting that its presence on the tree is not sufficient for infection to occur.

Supplementary Materials: The following are available online at https://www.mdpi.com/1999-4907/12/12/1683/s1, Table S1: ANOVA results of additive mixed effect models of abundance scores for AOD-associated bacteria with samples in three health categories. Table S2: List of R packages. [70–88].

Author Contributions: Conceptualization, L.A.P.G., N.B., R.A.N., R.J.A.B.; methodology, L.A.P.G., G.N., N.B., R.A.N.; formal analysis, L.A.P.G., R.A.N.; investigation, L.A.P.G., N.B., S.D., T.L.R.C., J.J.S., W.J.P., Forestry Commission Technical Services Unit; resources, Forest Research, RBG Kew, QMUL ITS Research Team.; data curation, L.A.P.G., G.N., N.B.; writing—original draft preparation, L.A.P.G.; writing—review and editing, R.J.A.B., R.A.N., S.D., N.B., G.N.; visualization, L.A.P.G.; supervision, R.J.A.B., R.A.N.; project administration, R.J.A.B.; funding acquisition, R.J.A.B., S.D. All authors have read and agreed to the published version of the manuscript.

Funding: This research was funded by the Department for Environment, Food and Rural Affairs (UK Government) in association with Action Oak; the Department for Environment, Food and Rural Affairs (UK Government) Future Proofing Plant Health scheme; Forest Research (UK Government); and Woodland Heritage (UK Registered Charity No. 1041611).

Data Availability Statement: DNA sequences: Trimmed sequencing reads have been deposited in the European Nucleotide Archive under the Project Accession. No. PRJEB30573, which can be found at: https://www.ebi.ac.uk/ena/browser/view/PRJEB30573 (accessed on 23 November 2019).

Acknowledgments: Samples were collected by the Forestry Commission Technical Services Unit. Thanks to Raphaella Jackson for helpful advice on microbial analysis and to James Crowe for help with scripting for bioinformatics tools. We also acknowledge the assistance of the ITS Research team at Queen Mary University of London.

Conflicts of Interest: The authors declare no conflict of interest. The funders had no role in the design of the study; analyses, or interpretation of data; in the writing of the manuscript, or in the decision to publish the results.

References

1. Quine, C.P.; Atkinson, N.; Denman, S.; Desprez-Loustau, M.-L.; Jackson, R.; Kirby, K. *Action Oak Knowledge Review: An Assessment of the Current Evidence on Oak Health in the UK, Identification of Evidence Gaps and Prioritisation of Research Needs*; Action Oak: Haslemere, UK, 2019.

2. Leroy, T.; Plomion, C.; Kremer, A. Oak symbolism in the light of genomics. *New Phytol.* **2019**, *226*, 1012–1017. [CrossRef]
3. Rackham, O. *Trees and Woodland in the British Landscape: The Complete History of Britain's Trees, Woods & Hedgerows*; Phoenix Press: London, UK, 2001.
4. Mitchell, R.; Bellamy, P.; Ellis, C.; Hewison, R.; Hodgetts, N.; Iason, G.; Littlewood, N.; Newey, S.; Stockan, J.; Taylor, A. Collapsing foundations: The ecology of the British oak, implications of its decline and mitigation options. *Biol. Conserv.* **2019**, *233*, 316–327. [CrossRef]
5. Tomlinson, I.; Potter, C.; Bayliss, H. Managing tree pests and diseases in urban settings: The case of Oak Processionary Moth in London, 2006–2012. *Urban For. Urban Green.* **2015**, *14*, 286–292. [CrossRef]
6. Lonsdale, D. Review of oak mildew, with particular reference to mature and veteran trees in Britain. *Arboric. J.* **2015**, *37*, 61–84. [CrossRef]
7. Denman, S.; Brown, N.; Kirk, S.; Jeger, M.; Webber, J. A description of the symptoms of Acute Oak Decline in Britain and a comparative review on causes of similar disorders on oak in Europe. *Forestry* **2014**, *87*, 535–551. [CrossRef]
8. Manion, P.D.; Lachance, D. *Forest Decline Concepts*; The American Phytopathological Society: St. Paul, MN, USA, 1992.
9. D Manion, P. *Tree Disease Concepts/Paul D. Manion*; Prentice-Hall: Englewood Cliffs, NJ, USA, 1981.
10. Denman, S.; Webber, J. Oak declines—New definitions and new episodes in Britain. *Q. J. For.* **2009**, *103*, 285–290.
11. Brown, N.; Bosch, F.V.D.; Parnell, S.; Denman, S. Integrating regulatory surveys and citizen science to map outbreaks of forest diseases: Acute oak decline in England and Wales. *Proc. Biol. Sci.* **2017**, *284*, 20170547. [CrossRef] [PubMed]
12. Brown, N.; Vanguelova, E.; Parnell, S.; Broadmeadow, S.; Denman, S. Predisposition of forests to biotic disturbance: Predicting the distribution of Acute Oak Decline using environmental factors. *For. Ecol. Manag.* **2018**, *407*, 145–154. [CrossRef]
13. Pinho, D.; Barroso, C.; Froufe, H.; Brown, N.; Vanguelova, E.; Egas, C.; Denman, S. Linking Tree Health, Rhizosphere Physicochemical Properties, and Microbiome in Acute Oak Decline. *Forests* **2020**, *11*, 1153. [CrossRef]
14. Scarlett, K.; Denman, S.; Clark, D.R.; Forster, J.; Vanguelova, E.; Brown, N.; Whitby, C. Relationships between nitrogen cycling microbial community abundance and composition reveal the indirect effect of soil pH on oak decline. *ISME J.* **2020**, *15*, 623–635. [CrossRef]
15. Denman, S.; Doonan, J.; Ransom-Jones, E.; Broberg, M.; Plummer, S.; Kirk, S.; Scarlett, K.; Griffiths, A.R.; Kaczmarek, M.; Forster, J.; et al. Microbiome and infectivity studies reveal complex polyspecies tree disease in Acute Oak Decline. *ISME J.* **2018**, *12*, 386–399. [CrossRef] [PubMed]
16. Doonan, J.; Denman, S.; Pachebat, J.; McDonald, J.E. Genomic analysis of bacteria in the Acute Oak Decline pathobiome. *Microb. Genom.* **2019**, *5*, e000240. [CrossRef] [PubMed]
17. Brady, C.; Arnold, D.; McDonald, J.; Denman, S. Taxonomy and identification of bacteria associated with acute oak decline. *World J. Microbiol. Biotechnol.* **2017**, *33*, 1–11. [CrossRef] [PubMed]
18. Brady, C.; Cleenwerck, I.; Denman, S.; Venter, S.; Rodriguez-Palenzuela, P.; Coutinho, T.; De Vos, P. Proposal to reclassify Brenneria quercina (Hildebrand and Schroth 1967) Hauben et al. 1999 into a new genus, Lonsdalea gen. nov., as Lonsdalea quercina comb. nov., descriptions of Lonsdalea quercina subsp. quercina comb. nov., Lonsdalea quercina subsp. iberica subsp. nov. and Lonsdalea quercina subsp. britannica subsp. nov., and emendation of the description of Dickeya dadantii. *Int. J. Syst. Evol. Microbiol.* **2012**, *62*, 1592–1602. [PubMed]
19. Brady, C.; Denman, S.; Kirk, S.; Venter, S.; Rodríguez-Palenzuela, P.; Coutinho, T. Description of Gibbsiella quercinecans gen. nov., sp. nov., associated with Acute Oak Decline. *Syst. Appl. Microbiol.* **2010**, *33*, 444–450. [CrossRef]
20. Brady, C.; Hunter, G.; Kirk, S.; Arnold, D.; Denman, S. Rahnella victoriana sp. nov., Rahnella bruchi sp. nov., Rahnella woolbedingensis sp. nov., classification of Rahnella genomospecies 2 and 3 as Rahnella variigena sp. nov. and Rahnella inusitata sp. nov., respectively and emended description of the genus Rahnella. *Syst. Appl. Microbiol.* **2014**, *37*, 545–552.
21. Li, Y.; Xue, H.; Guo, L.-M.; Koltay, A.; Palacio-Bielsa, A.; Chang, J.; Xie, S.; Yang, X. Elevation of three subspecies of Lonsdalea quercina to species level: Lonsdalea britannica sp. nov., Lonsdalea iberica sp. nov. and Lonsdalea populi sp. nov. *Int. J. Syst. Evol. Microbiol.* **2017**, *67*, 4680–4684. [CrossRef] [PubMed]
22. Soutar, C.D.; Stavrinides, J. Phylogenetic analysis supporting the taxonomic revision of eight genera within the bacterial order Enterobacterales. *Int. J. Syst. Evol. Microbiol.* **2020**, *70*, 6524–6530. [CrossRef] [PubMed]
23. Denman, S.; Brady, C.; Kirk, S.; Cleenwerck, I.; Venter, S.; Coutinho, T.; De Vos, P. Brenneria goodwinii sp. nov., associated with acute oak decline in the UK. *Int. J. Syst. Evol. Microbiol.* **2012**, *62*, 2451–2456. [CrossRef]
24. Vansteenkiste, D.; Tirry, L.; Van Acker, J.; Stevens, M. Predispositions and symptoms of Agrilus borer attack in declining oak trees. *Ann. For. Sci.* **2004**, *61*, 815–823. [CrossRef]
25. Brown, N.; Inward, D.J.; Jeger, M.; Denman, S. A review of Agrilus biguttatus in UK forests and its relationship with acute oak decline. *Forestry* **2014**, *88*, 53–63. [CrossRef]
26. Reed, K.; Denman, S.; Leather, S.R.; Forster, J.; Inward, D.J.G. The lifecycle of Agrilus biguttatus: The role of temperature in its development and distribution, and implications for Acute Oak Decline. *Agric. For. Èntomol.* **2017**, *20*, 334–346. [CrossRef]
27. Doonan, J.M.; Broberg, M.; Denman, S.; McDonald, J.E. Host–microbiota–insect interactions drive emergent virulence in a complex tree disease. *Proc. R. Soc. B: Boil. Sci.* **2020**, *287*, 20200956. [CrossRef]
28. Denman, S.; Plummer, S.; Kirk, S.; Peace, A.; McDonald, J.E. Isolation studies reveal a shift in the cultivable microbiome of oak affected with Acute Oak Decline. *Syst. Appl. Microbiol.* **2016**, *39*, 484–490. [CrossRef] [PubMed]

29. González, A.J.; Ciordia, M. Brenneria goodwinii and Gibbsiella quercinecans isolated from weeping cankers on Quercus robur L. in Spain. *Eur. J. Plant Pathol.* **2019**, *156*, 965–969. [CrossRef]
30. Moradi-Amirabad, Y.; Rahimian, H.; Babaeizad, V.; Denman, S. Brenneria spp. and Rahnella victoriana associated with acute oak decline symptoms on oak and hornbeam in Iran. *For. Pathol.* **2019**, *49*, e12535. [CrossRef]
31. Meaden, S.; Metcalf, C.J.E.; Koskella, B. The effects of host age and spatial location on bacterial community composition in the English Oak tree (Quercus robur). *Environ. Microbiol. Rep.* **2016**, *8*, 649–658. [CrossRef]
32. Stone, B.W.; Weingarten, E.A.; Jackson, C.R. The Role of the Phyllosphere Microbiome in Plant Health and Function. *Annu. Plant Rev. Online* **2018**, *1*, 533–556.
33. Leveau, J. Microbial Communities in the Phyllosphere. In *Biology of the Plant Cuticle*; Riederer, M.M.C., Ed.; Blackwell: Oxford, UK, 2006; pp. 334–367.
34. Vorholt, J.A. Microbial life in the phyllosphere. *Nat. Rev. Microbiol.* **2012**, *10*, 828. [CrossRef] [PubMed]
35. Pettifor, B.J.; Doonan, J.; Denman, S.; McDonald, J.E. Survival of Brenneria goodwinii and Gibbsiella quercinecans, associated with acute oak decline, in rainwater and forest soil. *Syst. Appl. Microbiol.* **2020**, *43*, 126052. [CrossRef] [PubMed]
36. Handelsman, J.; Rondon, M.R.; Brady, S.F.; Clardy, J.; Goodman, R.M. Molecular biological access to the chemistry of unknown soil microbes: A new frontier for natural products. *Chem. Biol.* **1998**, *5*, R245–R249. [CrossRef]
37. Müller, T.; Ruppel, S. Progress in cultivation-independent phyllosphere microbiology. *FEMS Microbiol. Ecol.* **2014**, *87*, 2–17. [CrossRef]
38. Zolfo, M.; Asnicar, F.; Manghi, P.; Pasolli, E.; Tett, A.; Segata, N. Profiling microbial strains in urban environments using metagenomic sequencing data. *Biol. Direct* **2018**, *13*, 9. [CrossRef]
39. Roumpeka, D.D.; Wallace, R.J.; Escalettes, F.; Fotheringham, I.; Watson, M. A Review of Bioinformatics Tools for Bio-Prospecting from Metagenomic Sequence Data. *Front. Genet.* **2017**, *8*, 23. [CrossRef]
40. Wood, D.E.; Lu, J.; Langmead, B. Improved metagenomic analysis with Kraken 2. *Genome Biol.* **2019**, *20*, 257. [CrossRef]
41. Nocchi, G.; Brown, N.; Coker, T.; Plumb, W.; Stocks, J.; Denman, S.; Buggs, R.J.A. Genomic structure and diversity of oak populations in British parklands. *Plants People Planet* **2021**, 1–15. [CrossRef]
42. Brown, N.; Jeger, M.; Kirk, S.; Xu, X.; Denman, S. Spatial and temporal patterns in symptom expression within eight woodlands affected by Acute Oak Decline. *For. Ecol. Manag.* **2016**, *360*, 97–109. [CrossRef]
43. Raimbault, P. *Physiological Diagnosis*; Societe Francaise d'Arboriculture: Versailles, France, 1995.
44. Knight, R.; Vrbanac, A.; Taylor, B.C.; Aksenov, A.; Callewaert, C.; Debelius, J.; Gonzalez, A.; Kosciolek, T.; McCall, L.-I.; McDonald, D.; et al. Best practices for analysing microbiomes. *Nat. Rev. Genet.* **2018**, *16*, 410–422. [CrossRef] [PubMed]
45. Herrmann, M.; Geesink, P.; Richter, R.; Küsel, K. Canopy Position Has a Stronger Effect than Tree Species Identity on Phyllosphere Bacterial Diversity in a Floodplain Hardwood Forest. *Microb. Ecol.* **2020**, *81*, 157–168. [CrossRef] [PubMed]
46. King, T.; Butcher, S.; Zalewski, L. *Apocrita—High Performance Computing Cluster for Queen Mary University of London*; Zenodo: London, UK, 2017.
47. Plomion, C.; Aury, J.-M.; Amselem, J.; Leroy, T.; Murat, F.; Duplessis, S.; Faye, S.; Francillonne, N.; Labadie, K.; Le Provost, G.; et al. Oak genome reveals facets of long lifespan. *Nat. Plants* **2018**, *4*, 440–452. [CrossRef] [PubMed]
48. Li, H.; Durbin, R. Fast and accurate short read alignment with Burrows-Wheeler transform. *Bioinformatics* **2009**, *25*, 1754–1760. [CrossRef]
49. Li, H.; Handsaker, B.; Wysoker, A.; Fennell, T.; Ruan, J.; Homer, N.; Marth, G.; Abecasis, G.; Durbin, R.; 1000 Genome Project Data Processing Subgroup. The Sequence Alignment/Map format and SAMtools. *Bioinformatics* **2009**, *25*, 2078–2079. [CrossRef]
50. Langmead, B.; Salzberg, S.L. Fast gapped-read alignment with Bowtie 2. *Nat. Methods* **2012**, *9*, 357–359. [CrossRef]
51. NCBI. *The NCBI Handbook*; National Center for Biotechnology Information (US): Bethesda, MD, USA, 2013.
52. McDonald, D.; Clemente, J.C.; Kuczynski, J.; Rideout, J.R.; Stombaugh, J.; Wendel, D.; Wilke, A.; Huse, S.; Hufnagle, J.; Meyer, F.; et al. The Biological Observation Matrix (BIOM) format or: How I learned to stop worrying and love the ome-ome. *GigaScience* **2012**, *1*, 7. [CrossRef]
53. R Core Team. *R: A Language and Environment for Statistical Computing*; R Foundation for Statistical Computing: Vienna, Austria, 2013.
54. RStudio Team. *RStudio: Integrated Development for R*; RStudio, Inc.: Boston, MA, USA, 2019.
55. Oksanen, J.; Guillaume Blanchet, F.; Friendly, M.; Kindt, R.; Legendre, P.; McGlinn, D.; Minchin, P.R.; O'Hara, R.B.; Simpson, G.L.; Solymos, P.; et al. Vegan: Community Ecology Package; R package version 2.5-7. 2020. Available online: https://CRAN.R-project.org/package=vegan (accessed on 30 November 2021).
56. Hill, R.; Llewellyn, T.; Downes, E.; Oddy, J.; MacIntosh, C.; Kallow, S.; Panis, B.; Dickie, J.B.; Gaya, E. Seed Banks as Incidental Fungi Banks: Fungal Endophyte Diversity in Stored Seeds of Banana Wild Relatives. *Front. Microbiol.* **2021**, *12*, 508. [CrossRef]
57. Lu, J.; Breitwieser, F.P.; Thielen, P.; Salzberg, S.L. Bracken: Estimating species abundance in metagenomics data. *PeerJ Comput. Sci.* **2017**, *3*, e104. [CrossRef]
58. Pinheiro, J.; Bates, D.; DebRoy, S.; Sarkar, D.; R Core Team. _ nlme: Linear and Nonlinear Mixed Effects Models; R package version 3.1-148. 2020. Available online: https://CRAN.R-project.org/package=nlme (accessed on 30 November 2021).
59. Boratyn, G.M.; Thierry-Mieg, J.; Thierry-Mieg, D.; Busby, B.; Madden, T.L. Magic-BLAST, an accurate RNA-seq aligner for long and short reads. *BMC Bioinform.* **2019**, *20*, 405. [CrossRef] [PubMed]

60. Uroz, S.; Buée, M.; Murat, C.; Frey-Klett, P.; Martin, F. Pyrosequencing reveals a contrasted bacterial diversity between oak rhizosphere and surrounding soil. *Environ. Microbiol. Rep.* **2010**, *2*, 281–288. [CrossRef] [PubMed]
61. Leff, J.W.; Del Tredici, P.; Friedman, W.E.; Fierer, N. Spatial structuring of bacterial communities within individual Ginkgo biloba trees. *Environ. Microbiol.* **2015**, *17*, 2352–2361. [CrossRef] [PubMed]
62. Jakuschkin, B. Genetic architecture of the interactions between English oak (*Quercus robur* L.) and the microbial community of its phyllosphere. Ph.D. Thesis, Université de Bordeaux, Bordeaux, France, 2015.
63. Rastogi, G.; Coaker, G.L.; Leveau, J.H. New insights into the structure and function of phyllosphere microbiota through high-throughput molecular approaches. *FEMS Microbiol. Lett.* **2013**, *348*, 1–10. [CrossRef] [PubMed]
64. Bálint, M.; Tiffin, P.; Hallström, B.; O'Hara, R.B.; Olson, M.S.; Fankhauser, J.D.; Piepenbring, M.; Schmitt, I. Host Genotype Shapes the Foliar Fungal Microbiome of Balsam Poplar (*Populus balsamifera*). *PLoS ONE* **2013**, *8*, e53987. [CrossRef] [PubMed]
65. Cregger, M.A.; Veach, A.; Yang, Z.K.; Crouch, M.J.; Vilgalys, R.; Tuskan, G.A.; Schadt, C.W. The Populus holobiont: Dissecting the effects of plant niches and genotype on the microbiome. *Microbiome* **2018**, *6*, 31. [CrossRef] [PubMed]
66. Wood, D.E.; Salzberg, S.L. Kraken: Ultrafast metagenomic sequence classification using exact alignments. *Genome Biol.* **2014**, *15*, R46. [CrossRef] [PubMed]
67. Kibegwa, F.M.; Bett, R.C.; Gachuiri, C.K.; Stomeo, F.; Mujibi, F.D. A Comparison of Two DNA Metagenomic Bioinformatic Pipelines While Evaluating the Microbial Diversity in Feces of Tanzanian Small Holder Dairy Cattle. *BioMed Res. Int.* **2020**, *2020*, 2348560. [CrossRef] [PubMed]
68. Menzel, P.; Ng, K.; Krogh, A. Fast and sensitive taxonomic classification for metagenomics with Kaiju. *Nat. Commun.* **2016**, *7*, 11257. [CrossRef] [PubMed]
69. Crampton, B.G.; Plummer, S.J.; Kaczmarek, M.; McDonald, J.E.; Denman, S. A multiplex real-time PCR assay enables simultaneous rapid detection and quantification of bacteria associated with acute oak decline. *Plant Pathol.* **2020**, *69*, 1301–1310. [CrossRef]
70. Arnold, J.B. ggthemes: Extra Themes, Scales and Geoms for 'ggplot2'; R package version 4.2.0. 2019. Available online: https://CRAN.R-project.org/package=ggthemes (accessed on 30 November 2021).
71. Auguie, B. GridExtra: Miscellaneous Functions for "Grid" Graphics; R package version 2.3. 2017. Available online: https://CRAN.R-project.org/package=gridExtra (accessed on 30 November 2021).
72. Wilke, C.O. Cowplot: Streamlined Plot Theme and Plot Annotations for 'ggplot2'; R package version 1.1.1. 2020. Available online: https://CRAN.R-project.org/package=cowplot (accessed on 30 November 2021).
73. Henry, L.; Wickham, H. Purrr: Functional Programming Tools; R package version 0.3.4. 2020. Available online: https://CRAN.R-project.org/package=purrr (accessed on 30 November 2021).
74. Müller, K.; Wickham, H. Tibble: Simple Data Frames; R package version 3.1.0; 2021. Available online: https://CRAN.R-project.org/package=tibble (accessed on 30 November 2021).
75. Wickham, H. The split-apply-combine strategy for data analysis. *J. Stat. Softw.* **2011**, *40*, 1–29. [CrossRef]
76. Wickham, H. Forcats: Tools for Working with Categorical Variables (Factors); R package verion 0.5.1; 2021. Available online: https://CRAN.R-project.org/package=forcats (accessed on 30 November 2021).
77. Wickham, H. Stringr: Simple, Consistent Wrappers for Common String Operations; R package version 1.4.0; 2019. Available online: https://CRAN.R-project.org/package=stringr. (accessed on 30 November 2021).
78. Wickham, H. Tidyr: Tidy Messy Data; R package version 1.1.3. 2021. Available online: https://CRAN.R-project.org/package=tidyr. (accessed on 30 November 2021).
79. Wickham, H.; Averick, M.; Bryan, J.; Chang, W.; McGowan, L.D.A.; François, R.; Grolemund, G.; Hayes, A.; Henry, L.; Hester, J. Welcome to the Tidyverse. *J. Open Source Softw.* **2019**, *4*, 1686. [CrossRef]
80. Wickham, H.; Chang, W.; Henry, L.; Pedersen, T.L.; Takahashi, K.; Wilke, C.; Woo, K.; Yutani, H.; Dunnington, D. ggplot2: Create Elegant Data Visualisations Using the Grammar of Graphics. 2020. Available online: https://ggplot2.tidyverse.org/ (accessed on 30 November 2021).
81. Wickham, H.; François, R.; Henry, L.; Müller, K. dplyr: A Grammar of Data Manipulation; R package version 1.0.5; 2021. Available online: https://CRAN.R-project.org/package=dplyr (accessed on 30 November 2021).
82. Wickham, H.; Hester, J. Readr: Read Rectangular Text Data; R package version 1.4.0; 2020. Available online: https://CRAN.R-project.org/package=readr (accessed on 30 November 2021).
83. Wickham, H.; Seidel, D. Scales: Scale Functions for Visualization; R package version 1.1.0; 2019. Available online: https://CRAN.R-project.org/package=scales. (accessed on 30 November 2021).
84. Hope, R.M. Rmisc: Ryan Miscellaneous; R package version 1.5; 2013. Available online: https://CRAN.R-project.org/package=RMisc (accessed on 30 November 2021).
85. Sarkar, D. *Lattice: Multivariate Data Visualization with R*; Springer: New York, NY, USA, 2008.
86. Kassambara, A. ggpubr: 'ggplot2' Based Publication Ready Plots; R package version 0.4.0. 2020. Available online: https://CRAN.R-project.org/package=ggpubr. (accessed on 30 November 2021).
87. Simpson, G.L. Permut: Functions for Generating Restricted Permutations of Data: R package version 0.9-5. Available online: https://CRAN.R-project.org/package=permute. (accessed on 30 November 2021).
88. Wolfe, B.E. Signs: Insert proper Minus Signs; R package version 0.1.2. Available online: https://CRAN.R-project.org/package=signs. (accessed on 30 November 2021).

forests

MDPI

Article

Chemical Control of *Corythucha arcuata* (Say, 1832), an Invasive Alien Species, in Oak Forests

Flavius Bălăcenoiu [1,2,*], Constantin Nețoiu [2], Romică Tomescu [1], Dieter Carol Simon [1], Andrei Buzatu [1,2], Dragoș Toma [1,2] and Ion Cătălin Petrițan [1]

[1] Faculty of Silviculture and Forest Engineering, Transilvania University of Brașov, Sirul Beethoven 1, 500123 Brașov, Romania; ro_tomescu2003@yahoo.fr (R.T.); disim@unitbv.ro (D.C.S.); andrei.buzatu@yahoo.com (A.B.); dragost93@gmail.com (D.T.); petritan@unitbv.ro (I.C.P.)

[2] National Institute for Research and Development in Forestry "Marin Drăcea", Eroilor 128, 077190 Voluntari, Romania; c_netoiu@yahoo.com

* Correspondence: flavius.balacenoiu@icas.ro

Abstract: In addition to the classic biotic and abiotic factors that have disrupted the health of forests throughout history, lately, the balance of forest ecosystems has been disturbed by different phenomena such as climate change, pollution, and, especially, biological invasions of invasive alien species. One of the alien species that has invaded Europe relatively quickly is an insect species of North American origin, the oak lace bug (*Corythucha arcuata* Say, 1832 Heteroptera: Tingidae). In the context of the rapid spread of infestations and the severity of attacks on oak trees in infested forests, this paper aims to assess measures to manage this species in the future. Namely, the effect of aerial chemical treatments on oak lace bug has been investigated with two influencing factors: the mode of insecticide action (contact and systemic) and the treatment volume (low volume and ultra-low volume). The experiment was conducted in two forests over a total area of 350 hectares. The results show that the reduction of the nymph population varied from 91% to 96%. However, the residual population was sufficient to allow differentiated re-infestations over time, more quickly after contact insecticide sprays (22 days after treatment) and slowly after systemic insecticide sprays (more than a month after treatment). This re-infestation time difference had implications on attack intensity as well, with stronger leaf discoloration observed in areas treated with a contact insecticide compared with those treated with a systemic insecticide.

Keywords: forest health; forest pests; chemical control; invasive alien species; oak lace bug

1. Introduction

The capacity of forests to accomplish multiple functions can be reduced by the harmful action of biotic and abiotic factors. Forests urgently need protection against the harmful action of these factors, especially in the light of the risks posed by global climate change, which is manifesting increasingly intensely in the last decade. People rely on healthy forest ecosystems for energy, building materials, and food, as well as services such as carbon storage, biodiversity management, and climate regulation [1]. During the lifespan of a forest, its health can be disrupted by biotic or abiotic factors, especially insect outbreaks, phytopathogens, wind, rainfall, or fire. Lately, in addition to these destabilizing factors, the health of forest ecosystems is endangered by different phenomena such as climate change and pollution, and, especially, biological invasions of invasive alien species (IAS) [1].

IAS are defined as organisms that have been introduced by humans, deliberately or accidentally, out of their natural range, that have multiplied and begun to exert negative effects on the new ecosystem [2–4]. The phenomenon by which they establish population, spread, and negatively affect native species is called biological invasion [2,3,5–7].

Against the background of increasingly intensive globalization, biological invasions have caused multiple negative effects on the economy and health, but especially on the environment [7–16], becoming one of the factors implicated in the loss of biodiversity and changes in ecosystem services [4].

One of the alien species that invaded Europe relatively quickly is an insect species of North American origin, the oak lace bug (*Corythucha arcuata* Say, 1832 Heteroptera: Tingidae), which was first reported in Europe in 2000 in Northern Italy [17]. Two years later, in 2002, it was also reported in Switzerland [18] and Turkey [19]. Meanwhile, oak lace bug (OLB) expanded its range, reaching Eastern Europe. In 2012, it was first observed in Bulgaria [20], and in 2013, in Hungary [21], Croatia [22], and Serbia [23–25]. Since then, with the presence of the appropriate climatic conditions in Europe as well as a lack of natural enemies, OLB has experienced explosive growth. Thus, at present, the species has been reported in Russia [26], Romania [27,28], Albania [29], Slovenia [30], Bosnia and Herzegovina [31,32], France [33], Ukraine [29], Greece [29], Slovakia [34], and Austria [35].

After the first detection of the insect in many of the countries mentioned above, such as Turkey, Hungary, Bulgaria and Romania, the further invasion and expansion of the distribution of OLB was reported on the territory of the given country [36–39].

Both nymphs and adults of this species feed on the undersides of the leaves of host trees, usually oaks, by piercing the epidermis and drawing out the cellular sap material. The typical characteristics of OLB feeding are the spots resulting from small, separate stings (1–3 mm), which then grow and merge into gray-yellow spots on the upper surface of the leaf, resulting in chlorotic discoloration of the leaf [36,40,41].

In its area of origin, the OLB is not considered a significant pest, especially because there are a noteworthy number of native predators, such as *Hyaliodes vitripennis* (Say, 1832), *Deraeocoris nebulosus* (Uhler, 1832), *Erythmelus klopomor* (Triapitsyn), *Orius insidiosus* (Say, 1832), and syrphid larvae, which are used in integrated pest management [42–44]. In Europe, the way OLB has invaded the continent, as well as the extremely high population of OLB in consecutive years, suggests a high potential risk both ecologically and economically [41].

Given the wide range of varied host plants available to OLB, which feed on almost all Eurasian oaks [29], at least 30 million hectares of oak forest can be a suitable habitat in Europe [41].

Considering that thus far in Europe, no natural enemies have been identified that would significantly limit the population of OLB and thus limit the invasion of the species [41], it is necessary to study possible removal methods to prescribe experimentally tested control measures. In North America, where OLB is endemic, some insecticides are recommended as being effective [45–47]; however, no research has yet been performed using these. In Europe, small-scale experiments were carried out in a Serbian study [48] in which an 8 years old oak culture infested with OLB was treated with four chemical insecticides based on the active substances bifenthrin, buprofezin, thiamethoxam, and abamectin. The results showed that bifenthrin and thiamethoxam insecticides, which are neonicotinoids and pyrethroids, respectively, are highly efficient for OLB control. Based on these results, we set out to answer questions such as (i) whether the method chemical control used at the nursery level could be replicated at the forest level and (ii) of the two recommended insecticides, which is more effective, the neonicotinoid or the pyrethroid?

This research aims to assess the extent to which chemical control can limit the population of *Corythucha arcuata* and therefore limit the invasion of the species in Europe.

2. Materials and Methods

2.1. Study Site and Data Collection

The experiment was carried out in two forests (Forest A and B) in Romania, Giurgiu County (Figure 1). The forests are managed by the Bolintin Forest District and consist of Turkey oak (*Quercus cerris* L.), Hungarian oak (*Quercus frainetto* Ten.), and pedunculate oak

(*Quercus robur* L.) stands or a mixture of these species, with small numbers of secondary species such as Norway maple, field maple, and European ash. The characteristics of the two forests were extracted from the forest management plans of each and are summarized in Table 1.

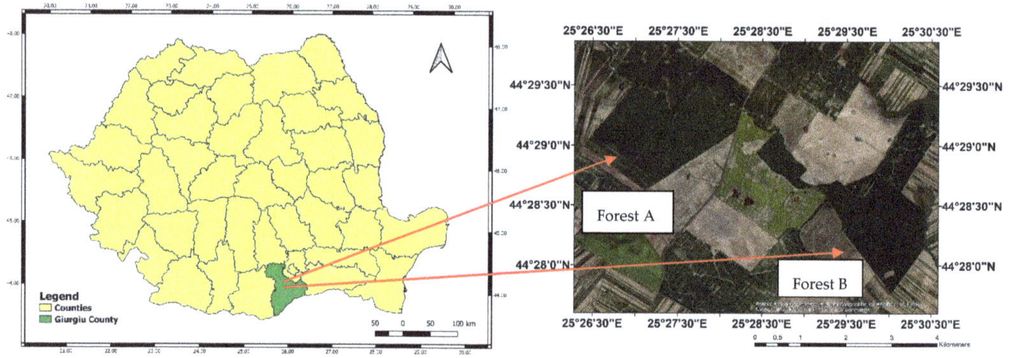

Figure 1. Location of the study area (Esri, Maxar, GeoEye, Earthstar Geographics, CNES/Airbus DS, USDA, USGS, Aero Grid, IGN, and the GIS User Community).

Table 1. Description of the two experimental forests.

Characteristics	Forest A	Forest B
Coordinates	44°29′06″ N 25°27′16″ E	44°28′24″ N 25°29′53″ E
Area (ha)	190	160
Dominant forest species	*Quercus cerris, Q. frainetto, Q. robur*	*Quercus cerris, Q. frainetto, Q. robur*
Secondary forest species	*Acer platanoides, A. campestre*	*Acer platanoides, A. campestre*
Regeneration mode	natural regeneration (sprout)	natural regeneration (sprout)
Age (years)	60	60
Average diameter (cm)	23	25
Average height (m)	19	20
Crown density of the stand (%)	70	90
Standing volume (m^3 ha^{-1})	150	170

After monitoring OLB biology in previous years (data not yet published), we noticed adult activity starting in mid-April, egg-laying after mid-May, and the appearance of nymphs at the beginning of June. The experimental treatments were timed for the nymph stage. Nymphs lack wings; therefore, this was determined the most suitable time, given their inability to fly.

During the experiment, two factors were taken into account: the mode of action of the insecticide (contact or systemic) and the treatment volume (low-volume (LV) at 30 L/ha and ultra-low volume (ULV) at 3 L/ha).

To avoid factors overlapping and influencing each other, the sprays were applied in the following ways:

- Forest A was sprayed with a contact insecticide (Alfametrin 10CE) over an area of 190 ha. Both LV (0.1 L/ha commercial product in 29.9 L/ha water) and ULV (0.1 L/ha commercial product in 2.9 L/ha water) treatments were applied. Alfametrin 10CE is a synthetic pyrethroid insecticide that acts on harmful insects by contact and ingestion, and it is based on the active substance alpha-cypermethrin.
- Forest B was sprayed with a systemic insecticide (APIS 200 SE) over an area of 160 ha. Both LV (0.2 L/ha commercial product in 29.8 L/ha water) and ULV (0.2 L/ha commercial product in 2.8 L/ha water) treatments were applied. APIS 200 EC is a neonicotinoid insecticide that acts on harmful insects by systemic action, penetrat-

ing rapidly into the plants' leaves and acting on all stages of insect development, and it is based on the active substance acetamiprid.

The sprays were applied on 24 June 2020, when the majority of the OLB population was in the nymph stage of the first generation, but adults from the hibernating generation were also present. The medium air temperature at the time of the experiment was 25–26 °C. The manufacturer of Alfametrin recommends applying treatment when temperatures do not exceed 23–25 °C; therefore, the temperature should have no influence on the treatment efficacity. It is also worth noting that on the treatment day and the following days no precipitation was registered in the atmosphere.

The sprays were applied using a Kamov KA-26 helicopter (Figure 2) equipped with an Aera 660 GPS device (Garmin) for loading and tracking the preset route. Orientation during the flights was made based on the flight polygons loaded in the GPS equipment, with the distance between flight lines established at 35 m. Analysis of the flight routes recorded by GPS device established that the distance between flight lines ensured adequate coverage with the solution in both Forest A (Figure 2a) and Forest B (Figure 2b). The flight speed was between 100 and 120 km/h, and the altitude above the canopy was between 10 and 15 m. To spray the LV treatment, the helicopter was fitted with a PIRNA installation, which pumped the solution through nozzles mounted under the helicopter planes and evacuated it through slits in the nozzle body, creating a spectrum of spray droplets with diameters between 150 and 400 μ. To spray the ULV treatment, the helicopter was fitted with a Micronair AU5000 installation, which used a rotating woven wire gauze to produce approximately equal spray droplets with diameters of about 100 μ.

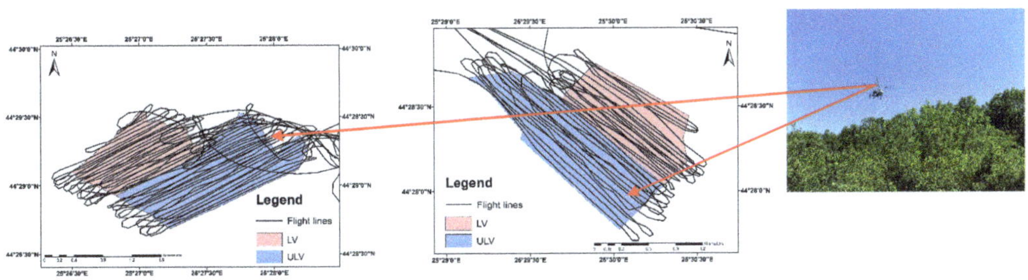

(a) (b)

Figure 2. Analysis of the flight routes recorded by the GPS device in (**a**) Forest A (contact insecticide) and (**b**) Forest B (systemic insecticide).

To establish the effectiveness of each treatment, the density of nymphs on the day of application (Moment 0) was compared with the density after approximately 15 days (Moment 1). Next, to track the re-infestations, the density of nymphs was registered every 5–9 days until the beginning of August, when the nymphs of the second generation appear (Moments 2–5). We have a total of 6 moments during which monitoring was performed, as follows:

- Moment 0 (24 June 2020): the day of spraying
- Moment 1 (9 July 2020): checking the effectiveness of control methods
- Moment 2 (16 July 2020): checking re-infestations
- Moment 3 (24 July 2020): checking re-infestations
- Moment 4 (29 July 2020): checking re-infestations
- Moment 5 (5 August 2020): checking re-infestations

The nymphs population density was determined by collecting 30 leaves from 10 test trees for a total of 300 leaves at each monitoring moment. The number of OLB nymphs were counted, thus allowing the average number of nymphs on a leaf to be calculated.

To ensure as uniform a distribution of leaves as possible, half of the leaves was collected from the lower canopy and the other half from the upper canopy of each tree.

To ensure a uniform working method, the test trees initially chosen were numbered so that the same 10 trees were used in subsequent checks.

Furthermore, because OLB nymphs and adults attack host trees by piercing the epidermis and drawing out the cellular sap material, we considered that another way to verify the effectiveness of control methods might be to measure the level of damage caused to insects in the active stage throughout the growing season. Thus, to assess the level of damage caused by *Corythucha arcuata* (Say, 1832), in the two forests, a network of observation points was evenly spaced over the area (Figure 3), in September, toward the end of the vegetation season, using a Juno SB GPS device (Trimble) with Terrasync software (Trimble), with an accuracy of five meters. Around each observation point, assessments were made on the attack intensity of the first 10 closest point trees (Figure 3). The method established by Tomescu et al. [39] was used to assess the intensity of the insect attack by visually evaluating the degree of foliage fading as a deviation from the normal color of the host oak leaves. Therefore, the intensity of the attack was evaluated by percentages of discoloration divisible by 5, ranging from 0% (not attacked) to 100% (strongly attacked). For clarity and homogeneity of the evaluations, we created a scale of discoloration (Figure 4).

The collected data regarding the attack intensity was statistically (see sub-chapter 2.2) and graphically (image analysis) interpreted on thematic maps based on the outline of the two forests. The thematic maps rendering was obtained through the ArcMap 10.5 software (ESRI) using ArcToolbox-Special Analyst Tools–Interpolation–IDW, assuming that each measured point had a local influence that diminished with distance and the measured values closest to the prediction location had more influence on the predicted value than those farther away. Subsequently, the 'renderers' raster data' function was applied for mapping, through the 'classified' method. To highlight the spatial differences between the observation points' attack by the degree of damage, we placed breaks at predefined threshold values for the percentage of discoloration taken as an average of the 10 trees in the point, as follows:

- 0–10%—very weak discoloration indicated by yellow ( )
- 11–25%—weak discoloration, indicated by pink ( )
- 26–50%—medium discoloration, indicated by green ( )
- 51–75%—strong discoloration, indicated by light blue ( )
- 76–100%—very strong discoloration, indicated by dark blue ( )

2.2. Data Analysis

To assess the effects of both applied insecticides (contact vs. systemic) and both treatment volumes (LV and ULV), and both treatment volume within each type of insecticide on the nymph population, a nonparametric test was performed. Additionally, to assess the effects of each type of insecticide on the lower canopy and upper canopy, and to compare the effects of both types of insecticide within each canopy layer on the nymph population, the same nonparametric test was applied. Before testing, Shapiro–Wilk and Levene's tests were used to check the normality of data and homoscedasticity of variances, respectively. Because no data complied with the requirements of parametric tests such as the two-way ANOVA (analysis of variance), we applied the Mann–Whitney U test. All statistical significance tests were performed using STATISTICA 12.1 (StatSoft, Inc., Tulsa, OK, USA).

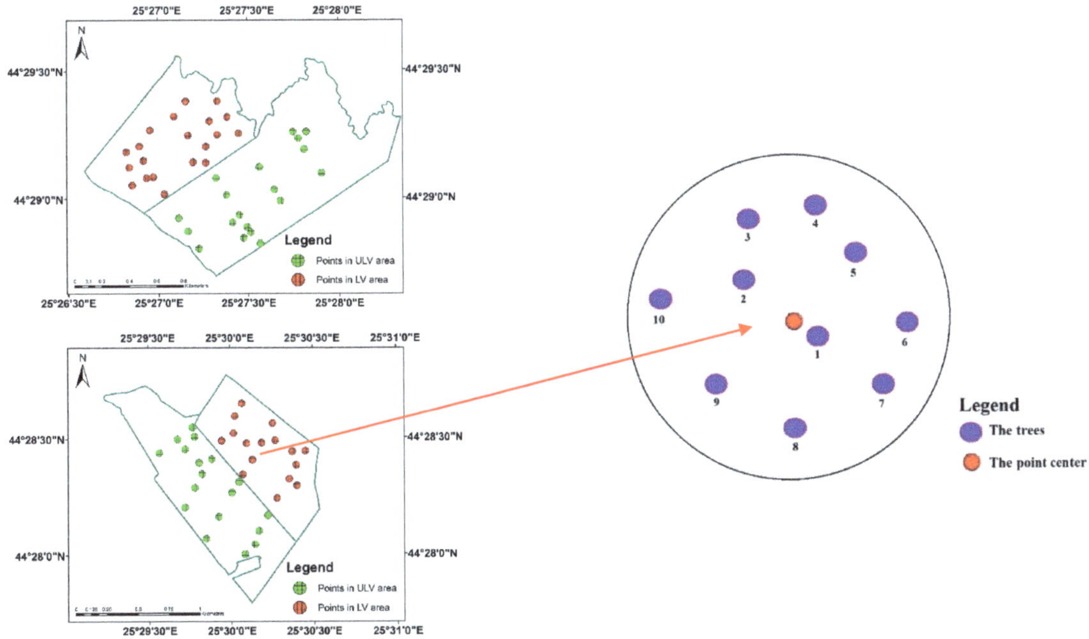

Figure 3. The network of observation points established to assess the level of damage caused by *Corythucha arcuata*.

Figure 4. Scale showing different attack intensities of *Corythucha arcuata* nymphs and adults on oak leaves: (**a**) 5%; (**b**) 25%; (**c**) 50%; (**d**) 75%; and (**e**) 100%.

3. Results

3.1. Effects of Applied Insecticide and Treatment on Nymphs' Density

At the population level (Figure 5a,b), a high density of nymphs was observed at Moment 0, before treatment. The nymph population immediately after spraying was reduced by up to 95% in both forests under both LV and ULV treatments.

Next, there was a slight re-infestation over time, starting at the end of July and being much more apparent in Forest A (Figure 5a), which was sprayed with a contact insecticide, than in Forest B (Figure 5b), which was sprayed with a systemic insecticide. In both forests, the population remained at a certain level and never rose to the initial level.

Given that at Moment 0 (24 June 2020), the average number of nymphs on a leaf was calculated between 4.26 and 6.51, and no significant differences in population density were observed among the four experimental areas in the two forests (Figure 6a), we presumed that all four areas in the experiment were homogeneously infested before the experiment. In checking the treatment effectiveness, at Moment 1 (09 July 2020), we found that the nymph population density was reduced to 0.24–0.50 nymphs per leaf, without significant differences among the four experimental areas (Figure 6b). Thus, the population was reduced by 91–96%.

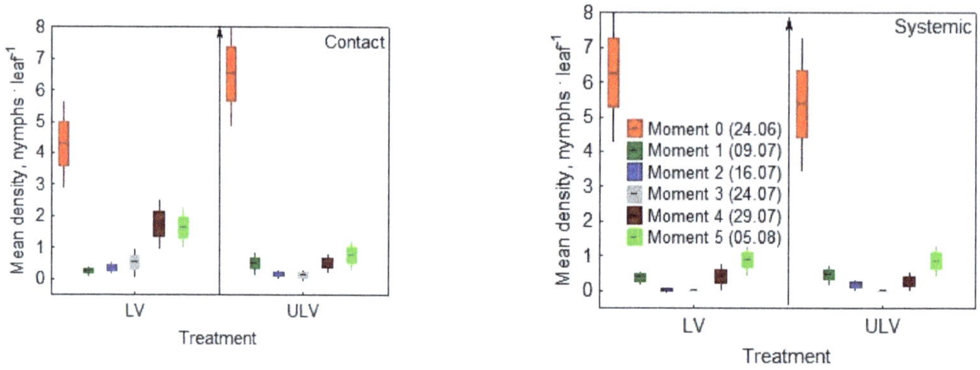

(a) (b)

Figure 5. Dynamics of nymphs over the observed period (Moment 0, before spraying, and M1–M5, five moments after spraying) separated by treatment volume (LV or ULV) and insecticide type: (**a**) contact insecticide; and (**b**) systemic insecticide.

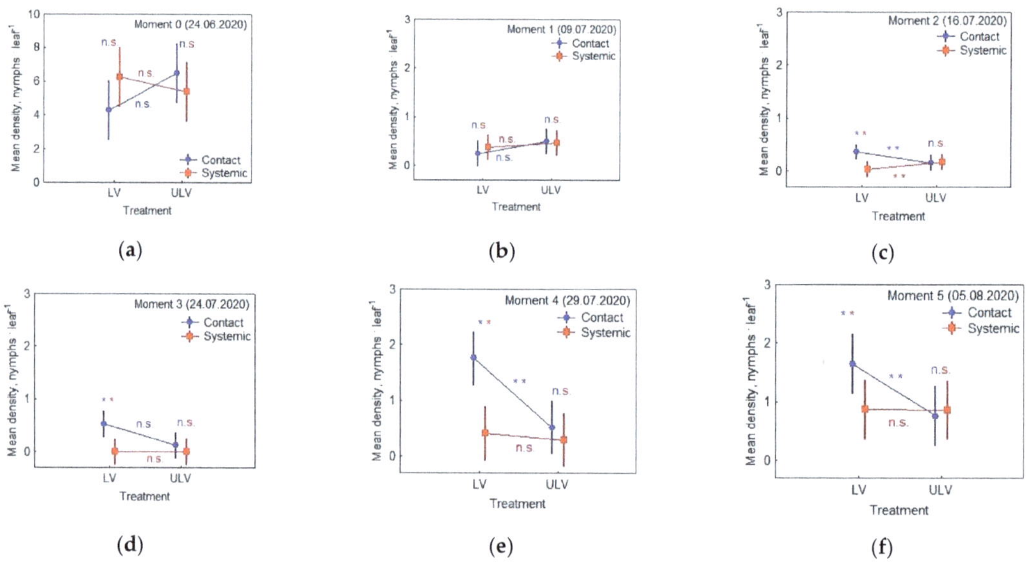

Figure 6. Effects of applied insecticide type (contact vs. systemic) and treatment volume (LV vs. ULV) on nymph density at each measurement time: (**a**) M0 (before spraying)-24 June 2020; (**b**) M1-09 July 2020; (**c**) M2-16 July 2020; (**d**) M3-24 July 2020; (**e**) M4-29 July 2020; and (**f**) M5-05 August 2020. ** = significant differences (U test, $p < 0.05$); n.s. = nonsignificant.

We began to record the re-infestation of the experimental area starting with Moment 2. At Moment 2 (Figure 6c), there was a slight increase in the density of nymphs in the Forest A–LV area, while the other three areas (Forest A-ULV, Forest B-LV and Forest B-ULV) showed a clear decrease in nymph population.

At Moment 3 (Figure 6d), while the Forest A-LV population noticeably increased, the other three areas showed decreases in nymph density to values close to 0 nymphs per leaf. At this time, there were significant differences between the population density in contact-LV areas and systemic-LV areas.

At Moments 4 and 5 (Figure 6e,f) the hypothesis made in previous moments regarding the strong re-infestation of the Forest A-LV area was confirmed, and an attempted re-infestation was also observed in Forest A-ULV but was not statistically assured. Furthermore, in Forest B–LV and Forest B-ULV, there were slight increases in the population density without statistical differences between these areas. At both moments, there were significant differences between the population density in Forest A-LV and the population density in Forest B-LV and Forest A-ULV.

To identify the main reason Forest A was re-infested, we compared the nymph population density at the upper and lower canopy after systemic and contact treatments.

The infestation was initially homogeneous at both upper and lower levels of the canopy in both forest B (Figure 7a). Furthermore, even after treatment, at Moment 1 (Figure 7b), nymph population density decreased consistently at both canopy levels in both forests.

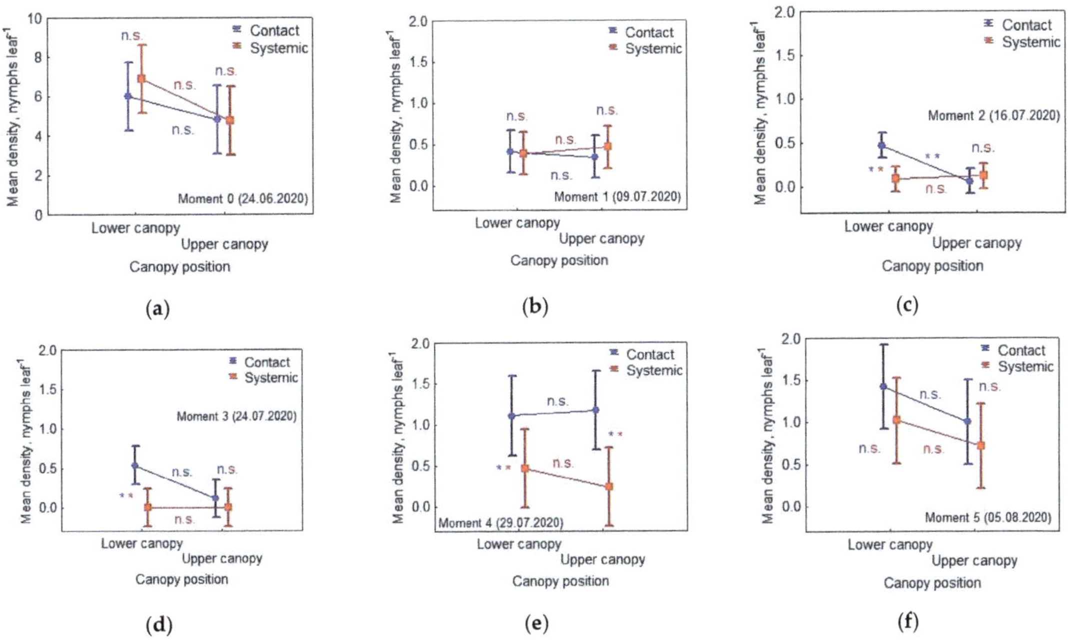

Figure 7. Effects of applied insecticide type (contact vs. systemic) on nymph density by leaf position (lower canopy vs. upper canopy) at each measurement time: (**a**) M0 (before spraying)-24 June 2020; (**b**) M1-09 July 2020; (**c**) M2-16 July 2020; (**d**) M3-24 July 2020; (**e**) M4-29 July 2020; and (**f**) M5-05 August 2020. ** = significant differences (U test, $p < 0.05$); n.s. = nonsignificant.

At Moment 2 (Figure 7c), there was a clear increase in the lower canopy population in Forest A. This increase was statistically significant both compared with the population density in the upper canopy in Forest A but also with that in the upper canopy in Forest B.

At Moment 3 (Figure 7d), the population density in Forest B insecticide remained low, with close to 0 nymphs per leaf and with no statistical differences between the two canopy levels. However, in Forest A, the population density increased even in the upper canopy, this time without any statistical differences between the two canopy areas. At the same time, the population level in the lower canopy in Forest A was statistically significantly higher than that in Forest B.

At Moment 4 (Figure 7e), the re-infestation of Forest A was apparent through homogenized population density at both canopy levels. Similarly, there was a slight increase in population density with similar values between both canopy levels, in Forest B, although the population values were statistically significantly lower than those in Forest A.

At Moment 5 (Figure 7f), we continued to observe an increase in population density in both forests, without statistical differences between the indicators, but with lower population values for the area treated with a systemic insecticide.

3.2. Attack Intensity Assessment at the End of Growing Season

The attack intensity of OLB nymphs and adults at the forest level (Figure 8) shows that Forest A experienced the highest attack intensities, at 80.9% in the ULV area and 76.5% in the LV area (no significant differences between them). Forest B showed statistically significant differences both depending on the application method and compared with Forest A, at 53.2% in the ULV area and 27.6% in the LV area.

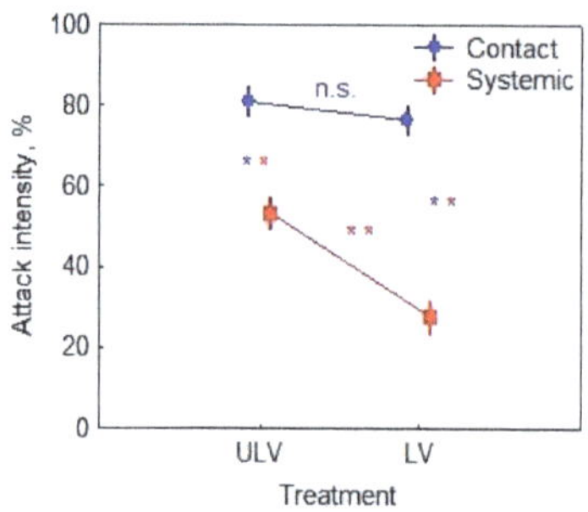

Figure 8. Attack intensity assessment at the end of the growing season (three months after treatment) by applied insecticide type and treatment volume. ** = significant differences (U test, $p < 0.05$); n.s. = nonsignificant.

The thematic maps show higher discoloration intensity in Forest A (Figure 9a) compared with Forest B (Figure 9b). Moreover, the treatment volumes resulted in higher discoloration intensity ULV areas compared with LV areas.

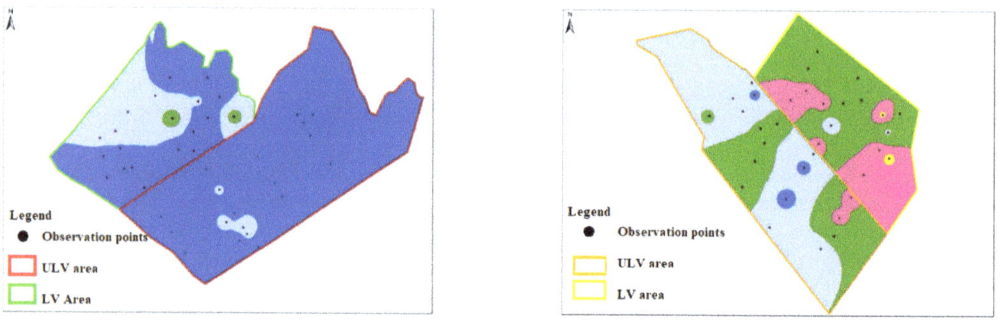

(a) (b)

Figure 9. Thematic maps showing attack intensity by insecticide type and treatment volume: (**a**) contact insecticide (Forest A); and (**b**) systemic insecticide (Forest B). very weak discoloration (0–10%); weak discoloration (11–25%); medium discoloration (26–50%); strong discoloration (51–75%); and very strong discoloration (76–100%).

4. Discussion

Considering the initial nymph population, we can conclude that the experimental treatments were 91–96% effective at Moment 1. If we look at the short term, immediately after spraying, our results at the stand level match the results at the nursery level [48]. Both insecticide types and treatment volume were highly effective. However, in the long term and regarding the dynamics of the population as a whole, we observed a re-infestation tendency at the end of July (Moment 4, 29 July 2020) in all four experimental areas, overlapping with the development of a new generation of nymphs, as observed in OLB populations in Italy [49]. Consequently, this re-infestation phenomenon in all experimental areas can be explained by the particular phenology of OLB and the hypothesis that the residual population was sufficient to allow re-infestations over time.

Forest B showed an exponential increasing in population (the population density increased sharply at Moment 4), while Forest A showed a linear increase (the population density increased steadily from Moment 2). Therefore, Forest A registered population increases immediately after spraying, which leads us to conclude that other factors affected the experiment and, as a result, it was not successful.

One of the hypotheses to explain Forest A re-infestation could be the action mode of the insecticide. The contact insecticide used may have performed best in the upper part of the canopy, where the toxic particles reached. According to this hypothesis, the re-infestation happened in the lower canopy (Figure 7), where the toxic substance would not reach. This could be studied in future research by measuring the drop dispersion pattern in the upper and lower canopy.

Another possible justification for the increase in nymph population at Moment 2 (16 July 2020) in Forest A is the known fact that contact insecticides generally persist for a very short period. There are studies [50,51] in which experimental injections of cypermethrin-based contact insecticide into the trunks of affected trees have shown longer persistence against bark beetles. The injection method is even recommended for systemic insecticides used for lace bug control [45]. Although this option is considered highly effective in the native habitats of lace bugs, we believe this procedure would be difficult to implement, given the large size of infested oak areas in Europe.

The highest nymph mortality was recorded in Forest B-LV, which received better dispersion of particles and a slow-acting insecticide. In Forest A, although mortality was high immediately after the application, the effect diminished over time and allowed for faster re-infestations and higher population densities compared with Forest B.

The analysis of the attack intensity of OLB nymphs and adults shows that both the action mode of the insecticide used (contact or systemic) and the treatment volume (ULV or LV) have a significant impact on the degree of foliage discoloration.

The registered discolorations toward the end of the growing season were more reduced after a LV application of a systemic insecticide, which offered a better spread of toxic particles inside the canopy and thus ensured better foliage protection.

The significant difference in damage recorded in the two forests after treatment can be explained by the different times of re-infestation. Forest A was re-infested at Moment 2 and Forest B only at Moment 5. During this entire time, in Forest A, both adults and nymphs OLB fed continuously.

In addition to the damage caused to the foliage, if we consider that 80% of the radial growth of oaks occurs in the first part of the growing season, until the end of July [52–54], the treatments used on Forest B offer another advantage in protecting the trees during the period of greatest growth, allowing re-infestation to begin in early August.

Therefore, although all four treatments seemed effective immediately after spraying, by the end of the experiment, we could conclude that the systemic insecticide, especially applied at LV, proved more effective than the contact insecticide, judging by the OLB damage caused by the end of the growing season. Our results were therefore in agreement with the hypothesis Paulin et al. [41] put forward, that systemic insecticides are likely to be the best treatments options because both larvae and adults are relatively concentrated on the undersides of the leaves.

However, if we consider that in Europe, 27 oak species have been proven suitable hosts for OLB [29], it is difficult for the results of this experiment to be put into practice. For example, as of autumn of 2019, it is estimated that OLB have infested over 1.7 million hectares of forest in: Croatia, Hungary, Romania, Serbia, and Russia (in Europe) [41].

Furthermore, if we take into account the recommendations of Dreistadt and Perry [45] and Shetlar [47] that, for adequate control to prevent re-infestations, the undersides of leaves must be thoroughly wet and spraying with toxic substances should be repeated, we conclude that chemical control methods are not economically justifiable. However, considering that lately, citizens have complained of discomfort caused by OLB stings or bites [55], chemical control might still be applied in isolated cases, such as frequented parks, park forests, private forests, forests of tourist interest, gardens, or isolated trees. In these cases, based on the results of this study, we recommend the use of a systemic insecticide, applied with discernment due to its persistence over time that could affect natural OLB enemies or pollinators.

It is worth noting that in North America, OLB populations are not significant risks to oak forests due to different control tactics such as cultural methods and, especially, biological methods. Only after these methods are chemical methods considered [45–47]. Based on this model of good practice, we support the statement of Paulin et al. [41] that the only viable OLB control method in the large forests recently invaded in Europe is classical biological control, but this must also be preceded by isolated experimental studies on ecological balance. Moreover, this method would be consistent with the general public's preference for biological control of invasive species over chemical control, as reported by Japelj et al. [56].

5. Conclusions

In the short term, the population density of OLB nymphs was reduced in all four experimental areas by 91–96%.

However, in Forest A, sprayed with a contact insecticide, re-infestation was observed starting 16 July 2020, 22 days after treatment. In Forest B, sprayed with a systemic insecticide, the re-infestation was observed more than a month later, starting 05 August 2020.

The analysis of attack intensity shows stronger discoloration in Forest A (sprayed with contact insecticide) compared with Forest B (sprayed with systemic insecticides). Furthermore, when using the same type of insecticide, stronger discoloration was observed in ULV application areas compared with LV application areas.

The only viable option to control OLB in large forests is classical biological control, but this must be preceded by isolated experimental studies on ecological balance.

Author Contributions: Conceptualization, F.B., C.N. and R.T.; methodology, F.B. and C.N.; software, F.B., A.B. and I.C.P.; validation F.B., C.N., R.T., A.B., D.T., D.C.S. and I.C.P.; formal analysis F.B., A.B. and I.C.P.; investigation, F.B., A.B. and D.T.; resources, F.B., C.N. and A.B.; data curation, F.B., D.T. and A.B.; writing—original draft preparation, F.B.; writing—review and editing, C.N., R.T., A.B., D.C.S. and I.C.P.; visualization, R.T. and D.C.S.; supervision, C.N.; project administration, F.B.; funding acquisition, R.T. All authors have read and agreed to the published version of the manuscript.

Funding: The data collection for this paper was carried out within the project PN 19070201 "Assessment of the risk of new species of harmful insects with potential for outbreak of deciduous forests in Romania", and the writing was made in the project PN 19070202 "Improving methods of surveillance and control of harmful insects using modern technologies".

Acknowledgments: We thank the staff of the Forest Protection Department–ROMSILVA, Giurgiu Forest District and INCDS 'Marin Drăcea' for their assistance in fieldwork. Likewise, we acknowledge that this study is part of Flavius Bălăcenoiu PhD thesis (Bioecology of the invasive alien species *Corythucha arcuata* (Say.) in Romania). We would also like to thank the two anonymous reviewers who made suggestions for improving the work.

Conflicts of Interest: The authors declare no conflict of interest. The funders had no role in the design of the study; in the collection, analyses, or interpretation of data; in the writing of the manuscript, or in the decision to publish the results.

References

1. Trumbore, S.; Brando, P.; Hartmann, H. Forest health and global change. *Science* **2015**, *349*, 814–818. [CrossRef] [PubMed]
2. Williamson, M.; Fitter, A. The varying succes of invaders. *Ecology* **1996**, *77*, 1661–1666. [CrossRef]
3. Juliano, S.A.; Lounibos, L.P. Ecology of invasive mosquitoes: Effects on resident species and on human health. *Ecol. Lett.* **2005**, *8*, 558–574. [CrossRef]
4. European Environment Agency. Invasive Alien Species: A Growing Problem for Environment and Health. 2013. Available online: https://www.eea.europa.eu/highlights/invasive-alien-species-a-growing (accessed on 26 April 2021).
5. Davis, M.A.; Grime, J.P.; Thompson, K. Fluctuating resources in plant communities: A general theory of invasibility. *J. Ecol.* **2000**, *88*, 528–534. [CrossRef]
6. Lounibos, L.P. Invasions by insect vectors of human disease. *Annu. Rev. Entomol.* **2002**, *47*, 233–266. [CrossRef] [PubMed]
7. Simberloff, D.; Martin, J.L.; Genovesi, P.; Maris, V.; Wardle, D.A.; Aronson, J.; Courchamp, F.; Galil, B.; Garcia-Berthou, E.; Pascal, M.; et al. Impacts of biological invasions: What's what and the way forward. *Trends Ecol. Evol.* **2013**, *28*, 58–66. [CrossRef] [PubMed]
8. Pimentel, D.; Lach, L.; Zuniga, R.; Morrison, D. Environmental and economic costs of nonindigenous species in the United States. *BioScience* **2000**, *50*, 53–65. [CrossRef]
9. Lovell, S.J.; Stone, S.F.; Fernandez, L. The economic impacts of aquatic invasive species: A review of the literature. *Agric. Resour. Econ. Rev.* **2006**, *35*, 195–208. [CrossRef]
10. Meyerson, L.A.; Mooney, H.A. Invasive alien species in an era of globalization. *Front. Ecol. Environ.* **2007**, *5*, 199–208. [CrossRef]
11. Vilà, M.; Basnou, C.; Pyšek, P.; Josefsson, M.; Genovesi, P.; Gollasch, S.; Nentwig, W.; Olenin, S.; Roques, A.; Roy, D.; et al. How well do we understand the impacts of alien species on ecosystem services? A pan-European, cross-taxa assessment. *Front. Ecol. Environ.* **2010**, *8*, 135–144. [CrossRef]
12. Vilà, M.; Espinar, J.L.; Hejda, M.; Hulme, P.E.; Jarošík, V.; Maron, J.L.; Pergl, J.; Schaffner, U.; Sun, Y.; Pyšek, P. Ecological impacts of invasive alien plants: A meta-analysis of their effects on species, communities and ecosystems. *Ecol. Lett.* **2011**, *14*, 702–708. [CrossRef] [PubMed]
13. Jeschke, J.M.; Keesing, F.; Ostfeld, R.S. Novel organisms: Comparing invasive species, GMOs, and emerging pathogens. *Ambio* **2013**, *42*, 541–548. [CrossRef] [PubMed]
14. Blackburn, T.M.; Essl, F.; Evans, T.; Hulme, P.E.; Jeschke, J.M.; Kühn, I.; Kumschick, S.; Marková, Z.; Mrugała, A.; Nentwig, A.; et al. A unified classification of alien species based on the magnitude of their environmental impacts. *PLoS Biol.* **2014**, *12*, e1001850. [CrossRef] [PubMed]
15. Hulme, P.E. Invasive species challenge the global response to emerging diseases. *Trends Parasitol.* **2014**, *30*, 267–270. [CrossRef] [PubMed]
16. Schindler, S.; Staska, B.; Adam, M.; Rabitsch, W.; Essl, F. Alien species and public health impacts in Europe: A literature review. *NeoBiota* **2015**, *27*, 1–23. [CrossRef]
17. Bernardinelli, I.; Zandigiacomo, P. Prima segnalazione di *Corythucha arcuata* (Say) (Heteroptera, Tingidae) in Europa. *Informatore Fitopatologico* **2000**, *50*, 47–49.
18. Forster, B.; Giacalone, I.; Moretti, M.; Dioli, P.; Wermelinger, B. Die Amerikanishe Eichennetzwaanze *Corythucha arcuata* (Say) (Heteroptera, Tingidae) hat die Südschweitz erreicht. *Mitteilungen-Schweiz. Entomol. Ges.* **2005**, *78*, 317–323.
19. Mutun, S. First report of the oak lace bug, *Corythucha arcuata* (Say, 1832) (Heteroptera: Tingidae) from Bolu, Turkey. *Isr. J. Zool.* **2003**, *49*, 323–324.
20. Dobreva, M.; Simov, N.; Georgiev, G.; Mirchev, P.; Georgieva, M. First record of *Corythucha arcuata* (Say) (Heteroptera: Tingidae) on the Balkan Peninsula. *Acta Zool. Bulg.* **2013**, *65*, 409–412.
21. Csóka, G.; Hirka, A.; Somlyai, M. A tölgy csipkéspoloska (*Corythucha arcuata* Say, 1832—Hemiptera, Tingidae) els"o észlelése Magyarországon. *Növényvédelem* **2013**, *49*, 293–296.
22. Hrašovec, B.; Posarić, D.; Lukić, I.; Pernek, M. Prvi nalaz hrastove mrežaste stjenice *Corythucha arcuata* u Hrvatskoj. *Šumar. List* **2013**, *9*, 499–503.
23. Poljakovic-Pajnik, L.; Drekic, M.; Pilipovic, A.; Nikolic, N.; Pap, P.; Vasic, V.; Markovic, M. Pojava velikih šteta od *Corythucha arcuata* (Say) (Heteroptera: Tingidae) u šumama hrasta u Vojvodini. In Proceedings of the XIII savetovanje o zaštiti bilja, Zlatibor, Serbia, 23–26 November 2015.
24. Pap, P.; Drekic, M.; Poljakovic-Pajnik, L.; Markovic, M.; Vasic, V. Monitoring zdravstvenog stanja šuma na teritoriji Vojvodine u 2015. *God. Topola* **2015**, *195–196*, 117–133.
25. Glavendekić, M. Fauna iekologija insekata koji naseljavaju invazivne i nativne ukrasne biljke. In *Ukrasne i Invazivne Biljke u Uslovima Klimatskih Promena—Uticaji i Adaptacije*; Obratov-Petkovic, D., Ed.; Univerzitet u Beogradu-Šumarski fakultet: Beograd, Serbia, 2017; pp. 240–264. [CrossRef]

26. Neimorovets, V.V.; Shchurov, V.I.; Bondarenko, A.S.; Skvortsov, M.M.; Konstantinov, F.V. First documented outbreak and new data on the distribution of *Corythucha arcuata* (Say, 1832) (Hemiptera: Tingidae) in Russia. *Acta Zool. Bulg.* **2017**, *9*, 139–142.
27. Don, I.; Don, C.D.; Sasu, L.R.; Vidrean, D.; Brad, M.L. *Insect Pests on the Trees and Shrubs from the Macea Botanical Garden*; Studia Universitatis 'Vasile Goldiş' Arad Engineering Sciences and Agrotourism Series; 2016; Volume 11, pp. 23–28. Available online: http://www.facultateadeinginerie.ro/studia/studia112/112-don.pdf (accessed on 6 May 2021).
28. Chireceanu, C.; Teodoru, A.; Chiriloaie, A. First record of oak lace bug Corythucha arcuata (Tingidae: Heteroptera) in Romania. In Proceedings of the 7th ESENIAS Workshop with Scientific Conference Networking and Regional Cooperation towards Invasive Alien Species Prevention and Management in Europe, Sofia, Bulgaria, 28–30 March 2017.
29. Csóka, G.; Hirka, A.; Mutun, S.; Glavendekic, M.; Mikó, Á.; Szőcs, L.; Paulin, P.; Eötvös, C.B.; Gáspár, C.; Csepelényi, M.; et al. Spread and potential host range of the invasive oak lace bug [*Corythucha arcuata* (Say, 1832)—Heteroptera: Tingidae] in Eurasia. *Agric. For. Entomol.* **2019**, *22*, 61–74. [CrossRef]
30. Jurc, M.; Jurc, D. The first record and the beginning the spread of Oak lace bug, *Corythucha arcuata* (Say, 1832) (Heteroptera: Tingidae), in Slovenia. *Šumar. List* **2017**, *141*, 485–488. [CrossRef]
31. Glavendekic, M.; Vukovic-Bojanovic, V. Prvi nalaz hrastove mrežaste stenice *Corythucha arcuata* (Say) (Hemiptera: Tingidae) u Bosni i Hercegovini i novi nalazi u Srbiji. In *Book of Abstracts of XI Symposium of Entomologists of Serbia, Goč, Serbia*; Glavendekic, M., Ed.; pp. 70–71. Available online: https://knlitd.pan.pl/index.php/en/news-events/357-xi-symposium-of-serbian-entomologists-17-21-09-2017-serbia (accessed on 6 May 2021).
32. Dautbašic, M.; Zahirovic, K.; Mujezinovic, O.; Margaletic, J. Prvi nalaz hrastove mrežaste stjenice (*Corythucha arcuata*) u Bosni i Hercegovini. *Šumar. List* **2018**, *142*, 179–181. [CrossRef]
33. Streito, J.C.; Balmès, V.; Aversenq, P.; Weill, P.; Chapin, E.; Clément, M.; Piednoir, F. *Corythucha arcuata* (Say, 1832) et Stephanitis lauri Rietschel, 2014, deux espèces invasives Nouvelles pour la faune de France (Hemiptera, Tingidae). *L. Entomol.* **2018**, *74*, 133–136.
34. Zúbrik, M.; Gubka, A.; Rell, S.; Kunca, A.; Vakula, J.; Galko, J.; Nikolov, C.; Leontovyc, R. First record of *Corythucha arcuata* in Slovakia. Short Communication. *Plant Protect. Sci.* **2019**, *55*, 129–133. [CrossRef]
35. Sallmannshofer, M.; Ette, M.S.; Hinterstoisser, W.; Cech, T.L.; Hoch, G. Erstnachweis der Eichennetzwanze, *Corythucha arcuata*, in Österreich. *Forsch. Aktuell* **2019**, *66*, 1–6.
36. Mutun, S.; Ceyhan, Z.; Sözen, C. Invasion by the oak lace bug, *Corythucha arcuata* (Say) (Heteroptera: Tingidae) in Turkey. *Turk. J. Zool.* **2009**, *33*, 263–268. [CrossRef]
37. Csepelényi, M.; Hirka, A.; Szénási, Á.; Mikó, Á.; Szőcs, L.; Csóka, G. Az inváziós tölgycsipkéspoloska (*Corythucha arcuata* (Say, 1832)) gyors terjeszkedése és tömeges fellépése Magyarországon. *Erdtud. Közlemények* **2017**, *7*, 127–134. [CrossRef]
38. Simov, N.; Grozeva, S.; Langourov, M.; Georgieva, M.; Mirchev, P.; Georgiev, G. Rapid expansion of the Oak lace bug *Corythucha arcuata* (Say, 1832) (Hemiptera: Tingidae) in Bulgaria. *Hist. Nat. Bulg.* **2018**, *27*, 51–55.
39. Tomescu, R.; Olenici, N.; Neţoiu, C.; Balacenoiu, F.; Buzatu, A. Invasion of the oak lace bug *Corythucha arcuata* (Say.) in Romania: A first extended reporting. *Ann. For. Res.* **2018**, *61*, 161–170. [CrossRef]
40. Bernardinelli, I. Potential host plants of *Corythucha arcuata* (Het., Tingidae) in Europe: A laboratory study. *J. Appl. Ecol.* **2006**, *130*, 480–484. [CrossRef]
41. Paulin, M.; Hirka, A.; Eötvös, C.S.; Gáspár, C.; Fürjes-Mikó, A.; Csóka, G. Known and predicted impacts of the invasive oak lace bug (*Corythucha arcuata*) in European oak ecosystems—A review. *Folia Oecol.* **2020**, *47*, 131–139. [CrossRef]
42. Connell, W.A.; Beacher, J.H. Life history and control of the oak lace bug. *Bull. Univ. Del. Agric. Exp. Stn.* **1947**, *265*, 28.
43. Wheeler, A.G.; Benjamin, J.R.; Stinner, R.; Henry, T.J. Biology and nymphal stages of *Deraeocoris nebulosus* (Hemiptera: Miridae), a predator of arthropod pests on ornamentals. Ann. *Entomol. Soc. Am.* **1975**, *68*, 1063–1068. [CrossRef]
44. Puttler, B.; Bailey, W.C.; Triapitsyn, S.V. Notes on distribution, host associations, and bionomics of Erythmelus klopomor Triapitsyn (Hymenoptera, Mymaridae), an egg parasitoid of lace bugs in Missouri, USA, with particular reference to its primary host *Corythucha arcuata* (Say) (Hemiptera, Tingidae). *J. Entomol. Acarol. Res.* **2014**, *46*, 30–34. [CrossRef]
45. Dreistadt, S.H.; Perry, E.J. Lace Bugs—Integrated Pest Management for Home Gardeners and Landscape Professionals. Pest Notes Publication. University of Calfornia, Agriculture and Natural Resources, Statewide Integrated Pest Management Program. 2014. Available online: http://ipm.ucanr.edu/PMG/PESTNOTES/pn7428.html (accessed on 1 May 2021).
46. Boggs, J. More Lace Bugs. Ohio State University Extension. Available online: https://bygl.osu.edu/node/1073 (accessed on 5 May 2021).
47. Shetlar, D.J. Lace Bugs. Ohio State University Extension. Available online: https://ohioline.osu.edu/factsheet/HYG-2150-10 (accessed on 5 May 2021).
48. Drekić, M.; Pajnik, L.P.; Pilipović, A.; Nikolić, N. *Suppression of Oak Lace Bug Corythucha Arcuata Say*; UDK: Berlin, Germany, 2019; pp. 215–223. Available online: https://www.researchgate.net/publication/350176869_SUPPRESSION_OF_OAK_LACE_BUG_Corythucha_arcuata_Say (accessed on 6 May 2021).
49. Berniardeli, I.; Zandigiacomo, P. Corythucha arcuata (Say): A new pest for European oaks. In *Methodology of Forest Insect and Disease Survey in Central Europe. Proceedings of the IUFRO WP 7.03.10 Workshop, Buşteni, Romania, 24–28 Septemer 2000*; Knizek, M., Ed.; IUFRO: Buşteni, Romania, 2000.
50. Jin, H.; Webster, G.R.B. Persistence, Penetration, and Surface Availability of Cypermethrin and Its Major Degradation Products in Elm Bark. *J. Agric. Food Chem.* **1998**, *46*, 2851–2857. [CrossRef]

51. Carrillo, D.; Crane, J.H.; Peña, J.E. Potential of Contact Insecticides to Control *Xyleborus glabratus* (Coleoptera: Curculionidae), a Vector of Laurel Wilt Disease in Avocados. *J. Econ. Entomol.* **2013**, *106*, 2286–2295. [CrossRef]
52. Szőnyi, L. Adatok néhány fafaj vastagsági növekedéséhez. *Az Erdő* **1962**, *97*, 289–300.
53. Járó, Z.; Tátraaljai, E. A fák éves növekedése. *Erd. Kut.* **1985**, *76–77*, 221–234.
54. Hirka, A. Bükk, luc és kocsánytalan tölgy éves kerületnövekedési menetének vizsgálata. *Erd. Kut.* **1991**, *82–83*, 15–23.
55. *Euphresco Project—'Corythucha arcuata (Heteroptera, Tingidae): Evaluation of the Pest Status in Europe and Development of Survey, Control and Management Strategies'*; Unpublished work; 2020; Available online: https://zenodo.org/record/4898795#.YMLV9kwRWUk (accessed on 6 May 2021).
56. Japelj, A.; Veenvliet, J.K.; Malovrh, J.; Verlič, A.; de Groot, M. Public preferences for the management of different invasive alien forest taxa. *Biol. Invasions* **2019**, *21*, 3349–3382. [CrossRef]

forests

MDPI

Article

Comparative Efficacy Trials with Two Different *Bacillus thuringiensis* Serovar *kurstaki* Strains against Gypsy Moth in Mediterranean Cork Oak Forests

Maurizio Olivieri [1], Roberto Mannu [1,*], Luca Ruiu [1], Pino A. Ruiu [2] and Andrea Lentini [1]

1 Dipartimento di Agraria, Università degli Studi di Sassari, 07100 Sassari, Italy; molivieri@uniss.it (M.O.); lucaruiu@uniss.it (L.R.); lentini@uniss.it (A.L.)

2 Servizio Ricerca per la Sughericoltura e la Selvicoltura, AGRIS Sardegna, 07029 Tempio Pausania, Italy; paruiu@agrisricerca.it

* Correspondence: rmannu@uniss.it

Abstract: The efficacy of two formulations (Foray® 76B AVIO and Rapax® AS AIR) containing different *Bacillus thuringiensis kurstaki* (*Btk*) strains (ABTS-351 and EG-2348, respectively) was evaluated against *Lymantria dispar* larval populations in cork oak forests in Sardinia (Italy), in 2018 and 2019. The experimental design involved the following treatments: (I) untreated control; (II) Foray® 76B at the dose of 2.0 L/ha; (III) Foray® 76B at the dose of 2.5 L/ha; (IV) Rapax® AS AIR at the dose of 2.0 L/ha. Aerial applications were carried out using a helicopter equipped with four electronic rotary atomizers adjusted to sprinkle 160 micron-sized drops. *Btk* efficacy was evaluated by assessing the larval density reduction 7, 14, and 21 days after the application in each experimental plot in comparison with an untreated check. In addition to field surveys, the mortality of second and third instar larval samples, randomly collected from each plot after treatment and fed with foliage from the same plot, was determined in the laboratory. All *Btk* treatments were similarly effective, and no differences in larval density reduction among *Btk* strains and doses were found in either year. Twenty-one days after application, the average larval density reduction in the field was approximately 70% in all treated plots in 2018, whereas in 2019 it reached 80% only in areas treated with Foray 76B at 2.5 L/ha. Laboratory observations showed that all *Btk*-based products were effective against gypsy moth larvae, with significant differences in mortality between untreated control and the different *Btk* treatments. Our results shed light on the possibility of alternating different *Btk* strains for resistance management purposes and of applying lower doses than labeled, in order to achieve cost savings for product shipment and distribution and to reduce the environmental impact.

Keywords: *Lymantria dispar*; biopesticides; aerial application; entomopathogens; *Bacillus thuringiensis*

1. Introduction

Cork oak (*Quercus suber* L.) is an evergreen species typical of Mediterranean pure and mixed forests or included in extensive agroforestry systems [1]. In addition to being biodiversity hotspots [2,3], cork oak forests are important economic resources in regard to cork production and their usefulness as pastureland.

A major threat to Mediterranean cork oak is the gypsy moth, *Lymantria dispar* (L.) (Lepidoptera: Erebidae), one of the main forest pest defoliators worldwide [4–6]. Gypsy moth is a univoltine species whose larvae can feed and develop on more than 300 different plant species [7–9], even though *Quercus* species are the main hosts in natural environments [10,11]. Spring larval feeding behavior, mostly targeting young oak shoots, causes a significant loss of the photosynthesizing surface, which leads to a tree growth decrease and consequent cork production reduction of up to 60% in totally defoliated trees and 40% in partially defoliated trees [12]. Moreover, the ordinary development of infested trees might be affected in the following year too, resulting in delayed bud burst and additional cork growth reduction [13].

In order to reduce gypsy moth infestations and consequent damage to trees, different control strategies have been developed [14–18]. Of these, the application of *Bacillus thuringiensis* serovar *kurstaki* (*Btk*) is considered one of the most effective methods to reduce larval population density [16,17] with few biological and practical limitations [19,20]. *Btk* application success against *L. dispar* strictly depends on several technical aspects, including timing, which should target second instar larvae [19,21], appropriate dose [14,22], and droplet size [23]. In Sardinia (Italy), according to a multi-year experiment in which various strains and formulations were applied by employing diverse aerial distribution methods against different gypsy moth generations [24,25], *Btk* was able to protect cork oak forests from this pest [25]. Based on these studies, Foray® 48B and Foray® 76B (Valent Bioscience Corporation, Libertyville, Illinois, USA) suspension concentrates (SC) of strain ABTS-351 authorized against gypsy moth were the *Btk*-based formulations providing the highest efficacy [26]. Among these two products with equal effectiveness, Foray® 76B is recommended to be used at a nearly half dose, which allows cost saving for product shipment and distribution [19].

Btk-based formulations normally used to control Lepidopteran defoliators are suspension concentrates (SC) of spores and crystals, containing insecticidal *Cry* proteins, acting by ingestion. Activated *Cry* proteins bind to specific receptors covering the plasma membrane of the insect's midgut epithelium, determining the formation of amphiphilic pores and a subsequent abnormal flux of ions and water into the epithelial cells [27]. Consequently, affected gut cells lyse [28] and infected larvae undergo paralysis and death, eventually exhibiting bacterial septicemia [29]. Although *Btk* acts specifically against Lepidoptera, susceptibility to toxins can vary depending on *Btk* strains [30,31]. For example, the tent caterpillar, *Malacosoma neustria* (L.) (Lepidoptera Lasiocampidae), which is often considered a secondary pest of cork oak forests [5,11], was more susceptible to ABTS-351 strain than *L. dispar* in application experiments conducted in cork oak forests in Sardinia [18]. Such differences were shown to be related to different degrees of *Cry* toxin affinity for midgut receptors, resulting in different amounts of amphiphilic pore formation on epithelial cells in susceptible insect species [32]. On the other hand, full insecticidal action of a *Btk*-based formulation is achieved when a sufficient number of spores and *Cry* proteins are ingested by young larvae [22], which is strictly related to the intrinsic features of the formulation and the way it is applied [33]. Therefore, the characteristics of the formulation and the insecticidal traits of the *Btk* strain it contains represent key factors to maximize gypsy moth containment efficacy in the field, enabling a longer residual effect.

Since the dose, formulation, and strain of *Btk* products are recognized as the main features determining a successful application against *L. dispar*, large-scale comparative trials with different *Btk*-based products and doses were performed in 2018 and 2019 in two cork oak forests in Sardinia. Experiments, including field and laboratory work, involved the two *Btk* strains ABTS-351 and EG-2348, containing a different assortment of *Cry* toxins.

2. Materials and Methods

2.1. Experimental Products

Two aqueous suspension concentrate (SC) *Btk* formulations were used in this study: (1) Foray® 76B AVIO (Sumitomo Chemical Agro Europe S.A.S., Machelen, Belgium), containing 20 billion international units (BIU)/L of *Btk* strain ABTS-351 expressing the insecticidal crystal toxins Cry1A(a), Cry1A(b), Cry1A(c), and Cry2A and (2) Rapax® AS AIR (Biogard, CBC Europe S.r.l., Grassobbio, Italy), containing 24 BIU/L of *Btk* strain EG-2348 and Cry1A(a), Cry1A(c), and Cry2A proteins. Foray and Rapax were authorized for aerial applications in Sardinian forests by Reg. n. 17,191 and 17,190 in 2018 and by Reg. n. 17,393 and 17,394 in 2019, issued by the Italian Ministry of Health according to art. 53, p. 1, Regulation (CE) n. 1107/2009.

2.2. Study Area and Experimental Design

Trials were carried out in 2018 and 2019 in pure cork oak forests in northern and eastern Sardinia, respectively, according to a randomized complete block design with three replicates (plots) of approximately 200 ha each, including the following treatments: (1) Foray® 76B AVIO applied at the dose of 2.0 L/ha (hereafter called Foray 2.0); (2) Foray® 76B AVIO applied at the recommended dose of 2.5 L/ha (hereafter called Foray 2.5); (3) Rapax® AS AIR applied at the dose of 2.0 L/ha (hereafter called Rapax); (4) untreated check (control). *Btk*-based formulations were sprayed at an ultralow volume using a helicopter equipped with a 12-m long bar with four rotary atomizers (model AU) regulated to sprinkle 160 micron-sized drops. Spraying equipment included a flow control system allowing for the management of insecticide distribution rates at variable helicopter speeds. *Btk* applications were made on 12 May and 11 May in 2018 and 2019, respectively, under optimal weather conditions, corresponding to an average temperature ranging from 19 to 28 °C, 30–60% relative humidity, and wind speeds below 1–2 m/s. Treatments targeted a gypsy moth larval population mainly in the second instar, being highly susceptible to the *Btk* action [34].

In both years, the level of gypsy moth infestation in the study areas, estimated by assessing the number of egg masses along a standardized 40-plant transect (economic damage threshold = 100 eggs), clearly indicated this species fell in a retrogradation phase within its multi-year cycle, as determined by analysis of data from network stations established in Sardinia since 1980 [35] and including more than 680 permanent monitoring sites [36]. Based on the close correlation between egg mass density and defoliation level in the following spring, in Sardinia, total defoliation is expected when the 100-egg threshold is reached [19]. Despite being in a retrogradation phase, the average density of *L. dispar* recorded in both 2018 and 2019 was higher than the action threshold, thus justifying the need for a phytosanitary application.

2.3. Evaluation of Btk Treatments

The effect of *Btk* applications against gypsy moth larvae was evaluated under field and laboratory conditions. In the field, gypsy moth larval population density was estimated in each plot before spraying and 7, 14, and 21 days after spraying. Larval density was assessed by counting the number of larvae on four randomly chosen shoots (approximately 30 cm in length) per tree from 10 consecutive cork oak trees selected along a linear transect (40 shoots/site/sampling date). In addition, larvae were identified according to different instars. Larval density reduction in treatment x after t days (ΔD_{xt}) was calculated as

$$\Delta D_{xt} = \frac{D_{x0} - D_{xt}}{D_{x0}} \times 100 \quad (1)$$

where D_{x0} is the initial larval density in treatment x at sampling time 0 (i.e., before spraying), and D_{xt} is the larval density t days after the *Btk* applications in treatment x.

In the laboratory, the larval mortality was evaluated on second and third instars randomly collected from each plot a few hours after each spraying. Groups of one hundred larvae from each plot were transferred into aerated plastic containers 18 cm in diameter and 25 cm in height (Kartell™ Inc., New York, NY, USA) with a perforated cup. To prevent larvae escaping from the upper holes, a portion of non-woven fabric was inserted between the cap and the top edge of the container. Throughout the entire breeding period, the larvae were fed ad libitum with fresh cork leaves from their original treated plot. Gypsy moth larvae were reared until pupation and the number of live and/or dead larvae, as well as the total number of pupae, was recorded every 2–3 days.

2.4. Statistical Analysis

All statistical analyses were performed using R statistical software version 3.10 [37]. Differences among treatments in instars composition were evaluated before *Btk* application for all years using a χ^2 test for independence ($p < 0.05$). Moreover, one-way Analysis of Variance (ANOVA) followed by Tukey post hoc test at 0.05 level of significance were performed to test for differences in larval infestation (i.e., average number of larvae counted on 40 shoots) before treatment.

The efficacy of different *Btk* treatments was evaluated by considering larval density reduction compared to initial density. In order to take into account the natural population decrease, larval reduction due to *Btk* treatments were corrected for natural mortality (C_{xt}) using the Schneider–Orelli formula [38]:

$$C_{xt} = \frac{\Delta D_{xt} - \overline{D}_{Ct}}{100 - \overline{D}_{Ct}} \times 100 \quad (2)$$

where ΔD_{xt} is the larval density reduction in treatment x after t days, and $\overline{D}_{Ct}$ is the average larval density reduction t days after the *Btk* applications in untreated control. Differences in corrected larval reduction among *Btk* treatments were tested separately within sampling times (i.e., 7, 14, and 21 days after application) using ANOVA followed by Tukey's test at significance level of 0.05.

Data obtained from laboratory assessments were used to evaluate both larval survival and density reduction under controlled conditions. Survival analysis was performed separately for each year of observation using a mixed effects Cox proportional hazard model using survival (version 3.2-11) [39], and coxme (version 2.2-16) [40] packages in R. In each model, treatments were considered as fixed factors and cage (i.e., replicate) as a random effect factor. Further post-hoc analysis was performed using multcomp (version 1.4-16) package in R [41], applying a Bonferroni correction for multiple testing. Larval density reduction corrected using Schneider–Orelli was calculated at 7, 14, and 21 days after treatment in order to evaluate the individual effect of *Btk* against gypsy moth larvae under laboratory conditions. One-way ANOVA followed by Tukey's test at 0.05 level of significance was used to test for differences among treatments.

3. Results

In both experimental years, *Btk* formulations were applied when the gypsy moth larval population was mainly in the second instar (Table 1). Statistical differences in instar distribution among treatments were found in 2018 (χ^2 = 108.3, df = 6, $p < 0.05$), and 2019 (χ^2 = 282.4, df = 6, $p < 0.05$). In 2018, plots treated with Foray 2.5 were characterized by a higher proportion of third instars than other plots, whereas plots sprayed with Rapax showed a significantly higher proportion of first instars (Table 1). In 2019, although treatments targeted a larval population mainly in the second instar, the presence of fourth instars was observed in all treated areas except those treated with Rapax (Table 1). No statistical differences in larval density among treatments were found before microbiological applications in either 2018 or 2019 (Table 2), indicating a similar level of infestation in all experimental areas.

Table 1. Occurrences of *Lymantria dispar* larval instars observed in 2018 and 2019 in untreated (control) areas and areas treated with Foray® 76B AVIO and Rapax® AS AIR 2018 in Sardinia (Italy). Pearson standardized residuals measuring the deviation from expected values are reported in brackets (+ = positive deviation; − = negative deviation).

Year	Treatments	First Instar	Second Instar	Third Instar	Fourth Instar
2018	Foray 2.0 L/ha	242	521	121	-
	Foray 2.5 L/ha	278	527 (−)	218 (+)	-
	Rapax	106 (+)	120	8 (−)	-
	Control	192	467 (+)	65 (−)	-
2019	Foray 2.0 L/ha	11 (−)	190 (−)	163 (+)	24 (+)
	Foray 2.5 L/ha	15	118	61 (+)	2
	Rapax	42	308	46 (−)	0
	Control	48	394 (+)	26 (−)	1 (−)

Table 2. Average (mean ± standard error) *Lymantria dispar* larval density occurred in 2018 and 2019 in untreated and *Btk* treated plots before insecticide applications. Analysis of Variance (ANOVA) statistics and *p*-value are reported.

Year	Foray 2.0 L/ha	Foray 2.5 L/ha	Rapax	Control	F	*p*
2018	294.7 ± 101.3	247.0 ± 19.2	78.0 ± 35.7	241.3 ± 65.9	2.20	0.17
2019	129.7 ± 61.6	65.3 ± 13.7	132 ± 12.2	156.3 ± 51.9	0.89	0.49

The analysis of larval density reduction in the field exclusively due to *Btk* treatments applied in 2018 showed a comparable percentage of reduction among treatments 7 ($F_{2,6} = 0.27$, $p = 0.77$), 14 ($F_{2,6} = 1.13$, $p = 0.39$), and 21 ($F_{2,6} = 0.12$, $p = 0.89$) days after spraying (Figure 1a). Similarly, no differences in corrected larval reduction were found in 2019 throughout the sampling period (7 days after application: $F_{2,6} = 3.31$, $p = 0.12$; 14 days after application: $F_{2,6} = 0.08$, $p = 0.93$; 21 days after application: $F_{2,6} = 0.88$, $p = 0.50$) (Figure 1b).

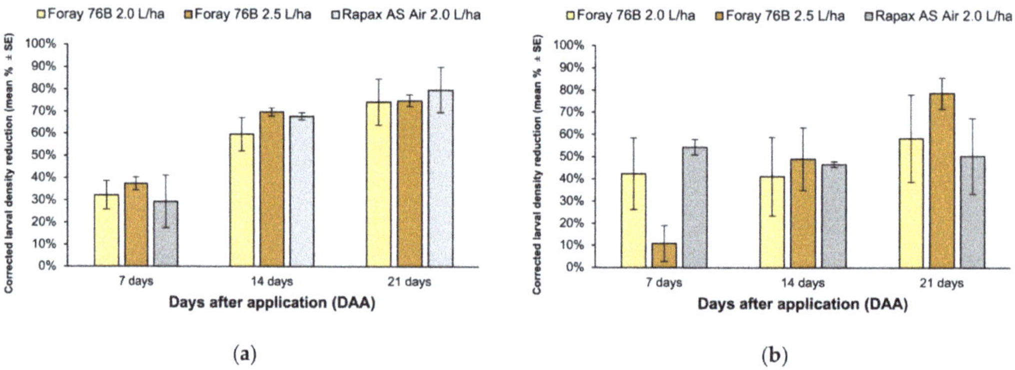

Figure 1. Percentage reduction of *Lymantria dispar* larval population caused by application of Foray 76B® (2.0 L/ha and 2.5 L/ha) and Rapax AVIO® 7, 14, and 21 days in (**a**) 2018 and (**b**) 2019 panel. Larval reductions were corrected with the Schneider–Orelli formula.

Survival of field-collected larvae maintained in the laboratory in 2018 was different among treatments ($\chi^2 = 210.03$, $p < 0.01$), with a significantly higher survival rate in untreated control than other treatments. Survival observed in larvae fed with foliage sampled from areas treated with Rapax was comparable to larvae from Foray 2.0-treated areas ($z = -2.48$, $p = 0.08$), whereas it was significantly lower in respect to larvae fed with Foray 2.5-treated foliage ($z = -3.96$, $p < 0.01$).

No difference in survival was found between larvae fed with Foray 2.0- and Foray 2.5-treated foliage ($z = 1.49$, $p = 0.81$). The survival of larvae fed with untreated foliage was approximately 80% at the end of the laboratory observations (30 days), in contrast to 24.7%, 10.0%, and 6.3% in larvae fed with foliage treated with Rapax, Foray 2.0, and Foray 2.5, respectively (Figure 2a). A similar pattern was observed in laboratory observations conducted in 2019 (Figure 2b), in which the survival of larvae fed with *Btk*-treated foliage after 30 days was slightly lower than that observed in larvae from untreated areas (Foray 2.0 = 1.3%; Foray 2.5 = 4.3%; Rapax 2.0 = 10.0%; untreated control = 68.3%). In these experiments, significant differences in survival among treatments were found ($\chi^2 = 339.92$, $p < 0.01$), with a higher survival rate of larvae fed on untreated foliage than those fed with Foray 2.0 ($z = -17.25$, $p < 0.01$), Foray 2.5 ($z = -15.66$, $p < 0.01$), and Rapax ($z = -15.76$, $p < 0.01$).

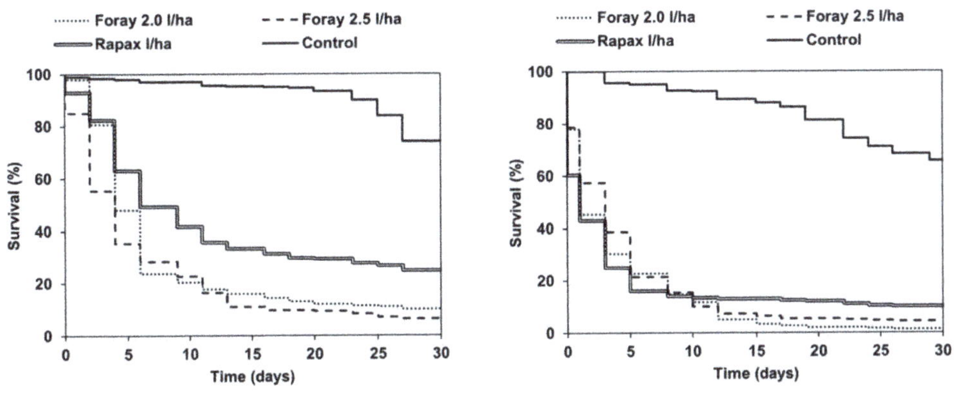

(a) (b)

Figure 2. Survival of *Lymantria dispar* larvae fed in the laboratory with untreated and *Btk*-treated foliage in (**a**) 2018 and (**b**) 2019.

The corrected larval mortality observed in the laboratory in 2018 and 2019 is illustrated in Figure 3a,b, respectively. In 2018, larval mortality attributable exclusively to *Btk* formulations was similar 7 ($F_{2,6} = 3.79$, $p = 0.09$) and 14 ($F_{2,6} = 5.24$, $p = 0.05$) days after their application in all treatments (Figure 3a). However, significant differences among *Btk*-based formulations were found 21 days after sprayings, with a lower mortality in larvae fed with Rapax than those fed with Foray at both assayed doses ($F_{2,6} = 9.69$, $p = 0.01$). In 2019, the average corrected larval mortality observed in the laboratory seven days after application was, respectively, 75.4%, 73.9%, and 59.2% for larvae fed with Foray 2.0, Rapax, and Foray 2.5. Larval density decreased over time and the reduction of reared larvae reached more than 80% in all treatments 21 days after the application. No differences in larval mortality among treatments attributable to *Btk* were found at different times of observation (7 days after application: $F_{2,6} = 0.51$, $p = 0.62$; 14 days after application: $F_{2,6} = 1.03$, $p = 0.41$; 21 days after application: $F_{2,6} = 2.39$, $p = 0.17$) (Figure 3b).

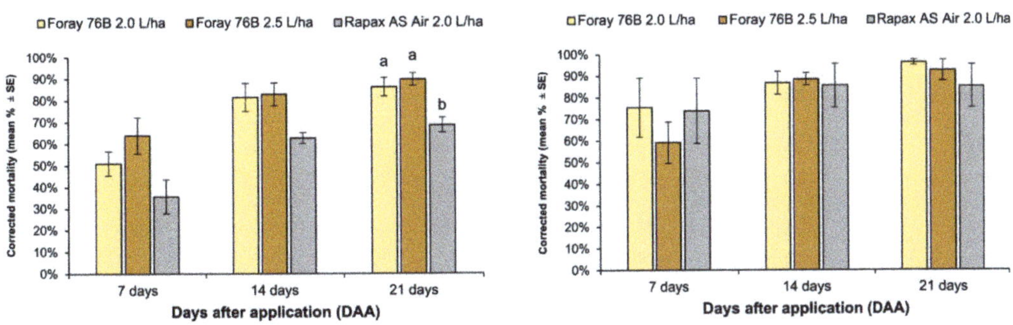

(a) (b)

Figure 3. *Lymantria dispar* mortality (%) under laboratory conditions caused by Foray 76B® (2.0 L/ha and 2.5 L/ha) and Rapax AVIO® 7, 14, and 21 days in (**a**) 2018 and (**b**) 2019 panel. Mortality was corrected with the Schneider–Orelli formula.

4. Discussion

All *Btk* applications in this study successfully contributed to the reduction in *L. dispar* larval populations in the forest. Foray® 76B AVIO, commonly used in large-scale gypsy moth management programs in Sardinia [19], caused a significant decrease in larval density when applied either at a standard (2.5 L/ha) or lower (2.0 L/ha) dose. The efficacy of Rapax® AS Air, which is characterized by a different *Btk* strain (i.e., EG-2348), was generally comparable to Foray® 76B AVIO (i.e., ABTS-351). However, the latter, as a whole, appeared more effective in controlling the *L. dispar* larval population. On the other hand, the efficacy of these *Btk* products observed in 2018 and 2019, compared to the untreated control areas, was less evident than expected due to the target population being under retrogradation. Another aspect affecting efficacy comparisons between the two experimental years is the larval age at the time of application, which is a factor correlated to *Btk* effectiveness [22]. In 2019, the larval population at the time of application was characterized by a higher larval age than that in 2018. Therefore, to achieve the same level of mortality when targeting older larvae, a higher dose would have been needed [14,22]. Nevertheless, differences in larval age are also related to other natural control factors affecting population. Older larvae which survive *Btk* treatments are exposed for a longer time to several biotic and abiotic mortality factors, which act more consistently during a gypsy moth retrogradation phase [42]. Among these, the biological control agent community, whose population is host-density dependent, expresses its highest potential during retrogradation [42–45]. The effects of the complexity of natural phenomena affecting *Btk* efficacy evaluations based on population density measurements before and after treatments and during retrogradation were mitigated in this study by including laboratory observations of larval samples collected from treated plots. This approach allowed the demonstration of clear treatment efficacy, with larvae feeding on non-treated foliage showing a significantly higher survival rate than that of larvae from treated areas in both years (Figure 2a,b).

Based on this study, both *Btk* formulations were equally effective in attaining larval density reduction in the field. The EG-2348 strain characterizing Rapax® AS Air formulation was previously assayed against *L. dispar* and *M. neustria* populations, proving to be effective when applied on individual cork oak trees [17]. However, in those experiments, *Btk* suspension was diluted with water and applied from the ground with an atomizer at a volume of 10 L/100 m^2. No recommendations for aerial application of this product were available, but our results demonstrated that the same procedures in use in Sardinia since the 1990s for aerial application of Foray® [19,24,25] can also be used effectively for Rapax®. This aspect is not secondary to achieving an appropriate size and density of suspension droplets, which is strictly dependent on the type of distribution and the physico-chemical features of the formulation [33]. These results provide important practical information for the aerial application of Foray® 76B AVIO and Rapax® AS AIR, as both are aqueous suspensions. Our results align with previous trials reporting the effectiveness of *Btk*-based SC formulations on different Lepidopteran species affecting forest and agricultural crops, under different field conditions [17,46].

Another relevant finding emerging from this work is the availability of two formulations containing *Btk* strains bearing different *Cry* genes, both resulting in a satisfactory pest control capability. This relates to the need to develop a longer-term control strategy involving the alternate or combined use of different strains, according to a resistance management approaches based on modes of action diversification. Insect resistance to *Bt* toxins was in fact reported in both laboratory and field conditions [47–49], being developed by different mechanisms, most of which are still under investigation [32]. Possible adaptation to *Bt* toxins by natural populations should therefore also be considered in the forest environment, where the selection pressure caused by the application of the same active substance for several years may facilitate the onset of resistance [50]. In addition to strains rotation, resistance management should include the maintenance of treatment-free areas as a reserve of *Bt*-sensitive populations.

The present study indicated similar efficacy of using the standard dose of 2.5 L/ha or a reduced dose of 2.0 L/ha for Foray® 76B AVIO. This dose was also adequate for Rapax® AS AIR. The consistent employment of reduced doses is recommended in order to minimize the possible environmental impact and the costs related to product shipment and distribution [19].

In conclusion, our results confirmed the high efficiency of *Btk* applications to manage *L. dispar* infestations in the Mediterranean area, which represents an effective method to protect forests from defoliation. The two formulations containing different *Btk* strains (i.e., ABTS-351, EG-2348) did not show significant differences in efficacy against gypsy moth larval population, providing a choice of alternative products to be considered in insect resistance management strategies. In addition, Foray® 76B AVIO proved to be equally effective at a lower dose than that labeled, which allows treatment cost saving or the opportunity to use locally available budgets to increase the surfaces of the forest to be protected. On the other hand, the use of different doses might be calibrated according to the typical spatial heterogeneity of gypsy moth infestations, applying lower doses in less infested areas and standard doses in areas with higher population density.

Author Contributions: Conceptualization, A.L. and P.A.R.; methodology, A.L., L.R. and R.M.; formal analysis, R.M.; investigation, M.O. and R.M.; writing—original draft preparation, L.R., M.O. and R.M. All authors have read and agreed to the published version of the manuscript.

Funding: This research was partially funded by Fondo di Ateneo per la ricerca 2019 (Lentini and Ruiu L.) and by Regione Autonoma della Sardegna, Assessorato della Difesa dell'Ambiente through a scientific–technical collaboration agreement (Lentini and Ruiu P.A.).

Institutional Review Board Statement: Not applicable.

Informed Consent Statement: Not applicable.

Data Availability Statement: The data presented in this study are available on reasonable request from the corresponding author. The data are not publicly available due to private restrictions.

Conflicts of Interest: The authors declare no conflict of interest. The funders had no role in the design of the study; in the collection, analyses, or interpretation of data; in the writing of the manuscript, or in the decision to publish the results.

References

1. Aronson, J.; Pereira, J.S.; Pausas, J.G. *Cork Oak Woodlands on the Edge: Ecology, Adaptative Management, and Restoration*; Island Press: Washington, DC, USA, 2009.
2. Verdinelli, M.; Yakhlef, S.E.B.; Cossu, C.S.; Pilia, O.; Mannu, R. Variability of ant community composition in cork oak woodlands across the Mediterranean region: Implication for forest management. *IForest* **2017**, *10*, 707–714. [CrossRef]
3. Mannu, R.; Pilia, O.; Fadda, M.L.; Verdinelli, M. Variability of beetle assemblages in Mediterranean cork oak woodlands: Does the higher taxa approach reliably characterize a specific response to grazing? *Biodivers. Conserv.* **2018**, *27*, 3599–3619. [CrossRef]
4. Elkinton, J.S.; Liebhold, A.M. Population dynamics of gypsy moth in North America. *Annu. Rev. Entomol.* **1990**, *35*, 571–596. [CrossRef]
5. Luciano, P.; Prota, R. Insect pests in Sardinian cork-oak forests. *IOBC/WPRS Bull* **1995**, *18*, 1–7.
6. Davidson, C.B.; Gottschalk, K.W.; Johnson, J.E. Tree mortality following defoliation by the European gypsy moth (*Lymantria dispar* L.) in the United States: A review. *For. Sci.* **1999**, *45*, 74–84. [CrossRef]
7. Barbosa, P.; Martinat, P.; Waldvogel, M. Development, fecundity and survival of the herbivore *Lymantria dispar* and the number of plant species in its diet. *Ecol. Entomol.* **1986**, *11*, 1–6. [CrossRef]
8. Liebhold, A.M.; Gottschalk, K.W.; Muzika, R.M.; Montgomery, M.E.; Young, R.; O'Day, K.; Kelley, B. *Suitability of North American Tree Species to the Gypsy Moth: A Summary of Field and Laboratory Tests*; General Technical Report NE-211; USDA Forest Service: Radnor, PA, USA, 1995.
9. Floris, I.; Cocco, A.; Buffa, F.; Mannu, R.; Satta, A. Insect pests of *Eucalyptus* plantations in Sardinia (Italy). *Redia* **2018**, *101*, 61–71. [CrossRef]
10. Liebhold, A.M.; Gottschalk, K.W.; Mason, D.A.; Bush, R.R. Forest susceptibility to the gypsy moth. *J. For.* **1997**, *95*, 20–24.
11. Tiberi, R.; Branco, M.; Bracalini, M.; Croci, F.; Panzavolta, T. Cork oak pests: A review of insect damage and management. *Ann. For. Sci.* **2016**, *73*, 219–232. [CrossRef]
12. Cambini, A. Valutazione Dei Danni Causati Dagli Insetti Defogliatori Alla Quercia Da Sughero. In *Proceedings of Atti del 1o Convegno Regionale del Sughero*; Stazione Sperimentale del sughero: Tempio Pausania, Italy, 1971; pp. 327–339.
13. Cambini, A. *Effetti Delle Defogliazioni Singole e Duplici Sull'accrescimento Della Quercia da Sughero Quercus Suber L. Memoria 42*; Stazione Sperimentale del Sughero: Tempio Pausania, Italy, 1975.

14. Smitley, D.R.; Davis, T.W. Aerial application of *Bacillus thuringiensis* for suppression of gypsy moth (Lepidoptera: Lymantriidae) in *Populus–Quercus* forests. *J. Econ. Entomol* **1993**, *86*, 1178–1184. [CrossRef]
15. Webb, R.E.; Peiffer, R.; Fuester, R.W.; Thorpe, K.W.; Calabrese, L.; McLaughlin, J.M. An evaluation of the residual activity of traditional, safe, and biological insecticides against the gypsy moth. *J. Arboric.* **1998**, *24*, 286–293.
16. Luciano, P.; Lentini, A. Ten years of microbiological control program against lepidopterous defoliators in Sardinian cork oak forests. *IOBC/WPRS Bull* **2012**, *76*, 175–178.
17. Ruiu, L.; Mannu, R.; Falchi, G.; Braggio, A.; Luciano, P. Evaluation of different *Bacillus thuringiensis* sv *kurstaki* formulations against *Lymantria dispar* and *Malacosoma neustria* larvae infesting *Quercus suber* trees. *Redia* **2013**, *96*, 27–31.
18. Ruiu, L.; Mannu, R.; Olivieri, M.; Lentini, A. Gypsy Moth Management with LdMNPV Baculovirus in Cork Oak Forest. *Forests* **2021**, *12*, 495. [CrossRef]
19. Lentini, A.; Mannu, R.; Cocco, A.; Ruiu, P.A.; Cerboneschi, A.; Luciano, P. Long-term monitoring and microbiological control programs against lepidopteran defoliators in Sardinian cork oak forests (Italy). *Ann. Silvic. Res.* **2020**, *45*, 21–30. [CrossRef]
20. Mannu, R.; Cocco, A.; Luciano, P.; Lentini, A. Influence of Bacillus thuringiensis application timing on population dynamics of gypsy moth in Mediterranean cork oak forests. *Pest. Manag. Sci.* **2020**, *76*, 1103–1111. [CrossRef]
21. Reardon, R.C.; Podgwaite, J.D. Summary of efficacy evaluations using aerially applied Gypchek® against gypsy moth in the USA. *J. Environ. Sci. Health* **1994**, *29*, 739–756. [CrossRef]
22. Glare, T.R.; O'Callaghan, M. Bacillus thuringiensis. In *Biology, Ecology and Safety*; John Wiley & Sons: New York, NY, USA, 2000.
23. Maczuga, S.A.; Mierzejewski, K.J. Droplet size and density effects of *Bacillus thuringiensis kurstaki* on gypsy moth (Lepidoptera: Lymantriidae) larvae. *J. Econ. Entomol.* **1995**, *88*, 1376–1379. [CrossRef]
24. Lentini, A.; Luciano, P. Bacillus thuringiensis in the management of gypsy moth (*Lymantria dispar* L.) in Sardinian cork-oak forests. *IOBC/WPRS Bull* **1995**, *18*, 104–109.
25. Lentini, A.; Cocco, A.; Luciano, P. Effects of Bacillus thuringiensis kurstaki treatments on gypsy moth population dynamics. *IOBC/WPRS Bull* **2012**, *76*, 171–174.
26. Ruiu, L.; Lentini, A.; Coinu, M.; Loi, A.; Serra, G.; Luciano, P. Comparative applications of Bacillus thuringiensis formulations against Lymantria dispar in Sardinian forests. *IOBC/WPRS Bull* **2012**, *76*, 185–190.
27. Boucias, D.G.; Pendland, J.C. *Principles of Insect Pathology*; Kluwer Academic Publisher: Boston, MA, USA, 1998.
28. Ellar, D.J.; Knowles, B.H.; Carroll, J.; Horsnell, J.; Haider, M.Z.; Ahmad, W.; Nicholls, C.N.; Armstrong, G.; Hodgman, T.C. Genetic and biochemical studies of the mechanism of action of *Bacillus thuringiensis* entomocidal-endotoxins. In *Bacterial Toxins: Zentralblatt für Bakteriologie*; International Medical Microbiology, Supplement 19; Rappuoli, R., Alouf, J., Freer, J., Fehrenbach, F., Wadstrom, T., Witholt, B., Eds.; Gustav Fischer: Stuttgart, Germany, 1990; pp. 409–506.
29. Crickmore, N. Beyond the spore–past and future developments of *Bacillus thuringiensis* as a biopesticide. *J. Appl. Microbiol.* **2006**, *101*, 616–619. [CrossRef]
30. Van Frankenhuyzen, K.; Fast, P.G. Susceptibility of Three Coniferophagous *Choristoneura* Species (Lepidoptera: Tortricidae) to *Bacillus thuringiensis* var. *kurstaki*. *J. Econ. Entomol.* **1989**, *82*, 193–196. [CrossRef]
31. Boulton, T.J.; Otvos, I.S.; Ring, R.A. Monitoring Nontarget Lepidoptera on *Ribes cereum* to Investigate Side Effects of an Operational Application of *Bacillus thuringiensis* subsp. *kurstaki*. *Environ. Entomol.* **2002**, *31*, 903–913. [CrossRef]
32. Pardo-López, L.; Soberón, M.; Bravo, A. *Bacillus thuringiensis* insecticidal three-domain *Cry* toxins: Mode of action, insect resistance and consequences for crop protection. *FEMS Microbiol. Rev.* **2013**, *37*, 3–22. [CrossRef]
33. Satinder, K.B.; Verma, M.; Tyagi, R.D.; Valéro, J.R. Recent advances in downstream processing and formulations of *Bacillus thuringiensis* based biopesticides. *Process Biochem.* **2006**, *41*, 323–342. [CrossRef]
34. Dubois, N.R. New laboratory and field developments in Bacillus thuringiensis and host susceptibility. In Proceedings of the U.S. Department of Agriculture Interagency Gypsy Moth Research Forum, Annapolis, MD, USA, 19–22 January 2003.
35. Cocco, A.; Cossu, A.Q.; Erre, P.; Nieddu, G.; Luciano, P. Spatial analysis of gypsy moth populations in Sardinia using geostatistical and climate models. *Agr. For. Entomol.* **2010**, *12*, 417–426. [CrossRef]
36. Mannu, R.; Gilioli, G.; Luciano, P. Occupancy of the territory by *Lymantria dispar* (L.) (Lepidoptera Erebidae) egg masses as a predictive index of damage. *Redia* **2017**, *100*, 157–165. [CrossRef]
37. R Development Core Team. *R: A Language and Environment for Statistical Computing*; R Foundation for Statistical Computing: Vienna, Austria, 2016.
38. Pünterner, W. Manual for field trials in plant protection. In *Agricultural Division*; Ciba-Geigy, Ltd.: Basle, Switzerland, 1981.
39. Therneau, T.M.; Grambsch, P.M. *Modeling Survival Data: Extending the Cox Model*; Springer: New York, NY, USA, 2000.
40. Therneau, T.M. Coxme: Mixed Effects Cox Models. R Package Version 2.2-5. Available online: https://CRAN.R-project.org/package=coxme (accessed on 24 April 2021).
41. Hothorn, T.; Bretz, F.; Westfall, P. Simultaneous inference in general parametric models. *Biom. J.* **2008**, *50*, 346–363. [CrossRef]
42. Alalouni, U.; Schädler, M.; Brandl, R. Natural enemies and environmental factors affecting the population dynamics of the gypsy moth. *J. Appl. Entomol.* **2013**, *137*, 721–738. [CrossRef]
43. Eichhorn, O. Experimental studies upon the parasitoid complex of the gypsy moth (*Lymantria dispar* L.) (Lep. Lymantriidae) in lower host populations in eastern Austria. *J. Appl. Entomol.* **1996**, *120*, 205–212. [CrossRef]

44. Hoch, G.; Zubrik, M.; Novotny, J.; Schopf, A. The natural enemy complex of the gypsy moth, *Lymantria dispar* (Lep. Lymantriidae) in different phases of its population dynamics in eastern Austria and Slovakia–A comparative study. *J. Appl Entomol.* **2001**, *125*, 217–227. [CrossRef]
45. Contarini, M.; Luciano, P.; Pilarska, D.; Pilarski, P.; Solter, L.; Huang, W.F.; Georgiev, G. Survey of pathogens and parasitoids in late instar Lymantria dispar larval populations in Sardinia, Italy. *Bull. Insectol.* **2013**, *66*, 51–58.
46. Ladurner, E.; Benuzzi, M.; Franceschini, S. Bacillus thuringiensis sv kurstaki strain EG 2348: Effect of formulation on efficacy against tomato leaf miner (Tuta absoluta). *IOBC/WPRS Bull* **2011**, *66*, 39–42.
47. McGaughey, W.H.; Beeman, R.W. Resistance to *Bacillus thuringiensis* in Colonies of Indianmeal Moth and Almond Moth (Lepidoptera: Pyralidae). *J. Econ. Entomol.* **1988**, *81*, 28–33. [CrossRef]
48. Ferré, J.; Van Rie, J. Biochemistry and genetics of insect resistance to *Bacillus thuringiensis*. *Annu. Rev. Entomol.* **2002**, *47*, 501–533. [CrossRef]
49. Janmaat, A.F.; Myers, J. Rapid evolution and the cost of resistance to *Bacillus thuringiensis* in greenhouse populations of cabbage loopers *Trichoplusia ni*. *Proc. Biol. Sci.* **2003**, *270*, 2263–2270. [CrossRef]
50. Hemingway, J.; Ranson, H. Insecticide resistance in insect vectors of human disease. *Ann. Rev. Entomol.* **2000**, *45*, 371–391. [CrossRef]

forests

MDPI

Article

The Difficulty of Predicting Eastern Spruce Dwarf Mistletoe in Lowland Black Spruce: Model Benchmarking in Northern Minnesota, USA

Ella R. Gray *, Matthew B. Russell and Marcella A. Windmuller-Campione

Department of Forest Resources, University of Minnesota, St. Paul, MN 55108, USA;
russellm@umn.edu (M.B.R.); mwind@umn.edu (M.A.W.-C.)
* Correspondence: grayx413@umn.edu

Abstract: Insects, fungi, and diseases play an important role in forest stand development and subsequently, forest management decisions and treatments. As these disturbance agents commonly occur within and across landscapes, modeling has often been used to inform forest planning and management decisions. However, models are rarely benchmarked, leaving questions about their utility. Here, we assessed the predictive performance of a Bayesian hierarchical model through on–the-ground sampling to explore what features of stand structure or composition may be important factors related to eastern spruce dwarf mistletoe (*Arceuthobium pusillum* Peck) presence in lowland black spruce (*Picea mariana* (Mill.) B. S. P.). Twenty-five state-owned stands included in the predictive model were sampled during the 2019 and 2020 growing seasons. Within each stand, data related to the presence of eastern spruce dwarf mistletoe, stand structure, and species composition were collected. The model accurately predicted eastern spruce dwarf mistletoe occurrence for 13 of the 25 stands. The amount of living and dead black spruce basal area differed significantly based on model prediction and observed infestation, but trees per hectare, total living basal area, diameter at breast height, stand age, and species richness were not significantly different. Our results highlight the benefits of model benchmarking to improve model interpretation as well as to inform our understanding of forest health problems across diverse stand conditions.

Keywords: forest health; lowland conifers; boreal forest; *Arceuthobium*; black spruce; eastern spruce dwarf mistletoe; modeling

1. Introduction

Insects, fungi, and diseases are important components of forest ecosystems and are frequently at the center of discussions regarding forest health. During the maximum sustained yield era of forestry, these disturbance agents were largely considered threats that needed to be "controlled" to reduce the losses to timber production and quality [1]. However, as goals and objectives have expanded beyond commercial timber production, forest health has shifted to exploring how insects, diseases, and other disturbances agents influence stand structure, composition, and development [2]. Here, we define forest health as the maintenance of resilience, defined as the recurrence and persistence of processes that lead to sustainable ecological conditions as well as timber and non-timber forest products [3].

As biotic disturbance agents interact in novel ways with changing climate conditions, non-native forest health threats continue to be introduced, and data on the subsequent impact on stand development becomes better understood [4–10], more information about the distribution, prevalence, and effects of these disturbance agents is being sought. This information, in turn, is used to inform forest management decisions that increasingly require adaptability and flexibility [11,12].

Modeling is often used to supplement data collected through forest inventories or more intensive sampling for a specific disease or insect [13]. However, in some forest communities, locations may be difficult to access or prohibitively expensive to sample [14]. Additionally, forest cover types can cover large areas, increasing the utility of modeling. The ability to model systems using previously collected forest inventory data allows researchers and natural resource managers to address many different forest health questions. Modeling has been used to predict the occurrence of forest disturbance agents [15], risk factors [8,16], rates and patterns of spread [17], and their effects on growth [1,2,18], among other objectives.

While models can provide insight into the distribution, potential impact, and other aspects of biotic disturbance agents related to forest management, they are still mathematical representations of systems, not the systems themselves [19]. As such, each model should be carefully evaluated [20]. Evaluation of a model can include, but is not limited to, examining the data used to create the model (what the original purpose was, etc.), asking if model complexity reflects the system or our understanding of the system [21], and, critically, benchmarking to assess model performance [20,22].

A system where modeling of forest health threats has occurred is in lowland black spruce (*Picea mariana* (Mill.) B. S. P.) forests of Minnesota that are infected with eastern spruce dwarf mistletoe (*Arceuthobium pusillum* Peck; hereafter ESDM) [15,17]. Lowland black spruce is an important component of the southern boreal forest, providing many ecological and economic benefits [23–26]. ESDM, the only dwarf mistletoe on *Picea* spp. in Minnesota, is a native, parasitic plant that has long been present on the landscape and is an important but not well understood component of disturbance regimes in black spruce forests [27]. In the region, black spruce is the main host of ESDM; though, at extremely high levels of infestation, it has been known to also parasitize tamarack (*Larix larcina* (Du Roi) K. Koch) [28].

Historically, black spruce forests were affected by large, stand-replacing disturbances (e.g., fire) at infrequent intervals with smaller, more frequent disturbances occurring at shorter intervals [29,30]. These fires are hypothesized to have maintained ESDM populations at a lower level than is currently observed, as ESDM requires a living host to survive and reproduce [27]. With fire largely absent from the landscape, clear-cut regeneration harvest is now the most common stand-replacing disturbance, resulting in mostly even-aged stands that often are affected by windthrow or ESDM infestations between rotations. Management of ESDM infestations occurs at the time of regeneration harvest, with best practices recommending cutting everything at least 1.5-m tall [31].

Long-range dispersal of ESDM by birds and other species such as small mammals is hypothesized to result in new infection locations [32]. When a black spruce tree is infected with ESDM, there is an initial reduction in growth accompanied by the development of witches' brooms—bushy, compact masses of branches and twigs that grow at the site of infection [27]. After 15 to 20 years, 75% of black spruce will succumb to mortality [33]. Infestations spread radially within stands and create a unique spatial distribution of ESDM at both the stand and landscape levels.

This spatial distribution of ESDM has been the focus of much of the modeling done in this system. Baker et al. [17] developed a model (DMLOSS) to predict the within-stand spread and resulting black spruce volume loss; their model estimated significant volume loss over the course of a single rotation using previously published rates of spread and mortality [33]. Moving from the stand level to the landscape, Hanks and colleagues [15], developed a model to predict the occurrence of ESDM utilizing a Bayesian hierarchical framework. However, this model has not been benchmarked to assess the model's predictive performance. Our objective is to assess the predictive performance of the model through on–the-ground sampling to explore what features of stand structure or composition may be important factors related to ESDM infection modeling at the landscape scale.

2. Materials and Methods

2.1. Study Area

Northern Minnesota, USA, falls within the Laurentian Mixed Deciduous forest and is a transition zone between northern hardwood forests to the south and east and boreal forests to the north [34]. The growing season in the region is short, ranging from 98 to 111 days [35]. The average annual temperature is between 1 °C and 4 °C; winter temperatures regularly reach −31 °C while summer temperatures above 30 °C have been recorded. Precipitation ranges between 650 and 800 mm per year, with the majority occurring during the summer months as rain [36].

The region contains mixed hardwood and conifer forests, as well as conifer swamps and bogs. Conifer bogs and swamps include peatland forests, which have deep, acidic, nutrient-poor, and poorly drained soils and are dominated by very few tree species, including black spruce and tamarack [30,35].

Black spruce peatlands are an important component of the region's forests, comprising 648,000 hectares of the 7.04 million hectares of forest in the state [37], occurring in both managed and unmanaged stands. When managed, black spruce is often found in pure stands with an even-aged distribution. However, black spruce can also occur in mixed stands with tamarack and balsam fir (*Abies balsamea* L. Mill.) with components of birch (*Betula* spp.), quaking aspen (*Populus tremuloides* Michx.), and northern white cedar (*Thuja occidentalis* L.) which are more likely to be multi-aged [35].

2.2. ESDM Model

Given that ESDM is one of the major disturbance agents of black spruce forests, there was interest in modeling the probability of its occurrence. Hanks et al. [15] developed a Bayesian hierarchical model to predict the probability of occurrence of ESDM for 25,235 state-owned stands across northeastern Minnesota, USA, using multiple data sets. These data included the Minnesota Department of Natural Resources Forest Inventory Management database (FIM; MNDNR, 2019) and an independent intensive survey for ESDM (Baker et al., unpublished manuscript). The variables used in the model include measures of species composition, stand conditions, environmental conditions, and spatial autocorrelation (Table 1). This represents the only published predictive model of ESDM occurrence in the study region. However, the predictive accuracy of the model has not been assessed.

Table 1. Variables included in the Bayesian data reconciliation model used to predict occurrence of ESDM; for full model description see Hanks et al. (2011).

Regression Coefficients for y	Regression Coefficients for Φ	Regression Coefficients for Ψ
Cover type size class	Mortality of dominant species	Cover type size class
Stand "wetness"code	Presence of tamarack	Understory size class
Stand density (1000 board-feet/acre)	Lowland black spruce cover type	Mortality of dominant species
Height of dominant species		Understory density
Mortality of dominant species		Presence of northern white cedar
Understory density		Spatial autocovariate (ac Ψ)
Presence of tamarack		
Presence of northern white cedar		
Presence of lowland black spruce		
Presence of balsam fir		
Stagnant spruce cover type		
Aspen cover type		
Jack pine cover type		
Spatial autocovariate (ac_y)		

Notes: The variable y is the presence (y_i = 1) or absence (y_1 = 0) of mistletoe in a stand, as identified in the intensive survey; Φ is the probability of the DNR survey finding mistletoe if it is present in the intensive survey; and Ψ is the probability of the DNR survey reporting mistletoe present if it is not present in the intensive survey. "DNR"is the Minnesota Department of Natural Resources.

2.3. Model Benchmarking

To assess the accuracy of the ESDM model [15], we applied the predicted probability of ESDM occurrence in black spruce stands to the MN DNR FIM database. Using ArcMap 10.5.1 (ESRI) twenty-five stands for on-the-ground sampling were selected (Figure 1). Selection criteria included stand size (≥4.04 hectares in area), age (between 60 and 100 years old at the time of the survey), dominant species (black spruce; as denoted by the DNR), surveyed since at least 1995, and within 200 m of a navigable road. This sampling frame was further reduced after selecting an equal number of stands predicted to either have ESDM present or absent. A greater than 80% predicted probability of ESDM occurrence was used to label stands where ESDM was predicted to occur. Stands with less than 20% predicted probability of occurrence were labeled as ESDM free. These percentages were chosen as conservative thresholds indicating presence or absence.

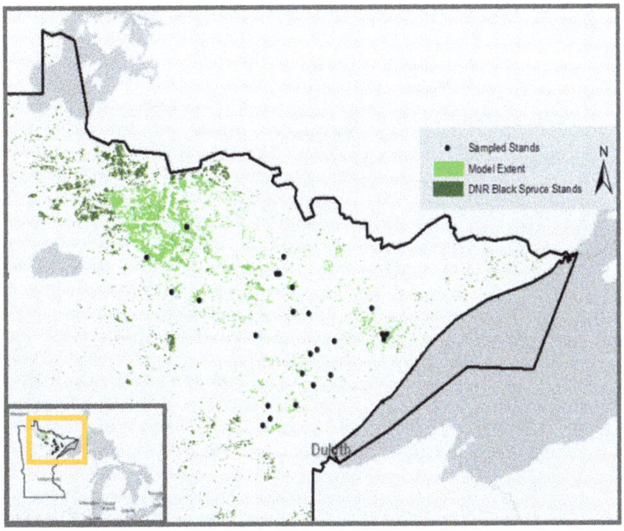

Figure 1. Study area, showing state-managed black spruce stands (green), stands including in the Hanks et al. (2011) model (light green), and stands sampled for model benchmarking (black).

Twenty-five black spruce stands were sampled during the summers of 2019 and 2020; 13 predicted to be ESDM free, and 12 predicted to have ESDM present. Within each stand, sampling occurred across three units, each comprised of six plots, consistent with previous work in this system [28]. The location of the first unit was selected by placing a GPS point within at least 30 m of the stand boundary. The second unit was 100 m away from the sixth plot of the first unit. The third unit was 100 m away from the final plot of the second unit, and all azimuths were chosen to fit within the shape of the stand (given many have highly irregular shapes). Each unit contained six 1/100th hectare overstory plots spaced 30 m × 60 m in a 2 × 3 design; in some irregularly shaped stands, the rectangular configuration could not be met, but the number of plots and spacing remained consistent.

Within each overstory plot, all trees with a diameter at breast height (DBH) of at least 3 cm were measured. For each tree, species, DBH, status (alive or dead), live crown ratio, and presence or absence of ESDM was recorded. On trees with ESDM present, the proportion of crown occupied by witches' brooms was recorded (<50%, ~50%, or >50%) in a simplified version of the dwarf mistletoe rating system [38]. A simplified version was utilized because, unlike other mistletoes, any level of ESDM will result in the eventual mortality of the black spruce. Height was measured on every fourth tree, provided the tree remained upright (leaning less than 45° from vertical) and the crown was unbroken. Height of snags was also recorded as long as these criteria were met.

2.4. Analysis

The accuracy of the model was assessed by creating a confusion matrix and calculating overall accuracy, type I error, and type II error. Additionally, stand structure characteristics were compared across predictive groups (true negative, false negative, false positive, true positive) using an analysis of variance and Tukey's honest significant difference test when significance was found. All analyses were performed in R [39].

3. Results

Of the 25 stands sampled, the model accurately predicted the presence or absence of ESDM for 13 stands—7 stands without infestation and 6 stands with infestation. We found an overall error rate of 48% and type I and II error rates of 46% (Table 2).

Table 2. Confusion matrix comparing observed occurrence of eastern spruce dwarf mistletoe (ESDM) with that predicted by Hanks et al. (2011).

Predicted	Observed		
	ESDM Absent	**ESDM Present**	**Total**
ESDM absent	7	6	13
ESDM present	6	6	12
Total	13	12	25

A comparison of stands across predicted and observed occurrence of ESDM shows groups diverging in few characteristics, while many remain similar (Table 3, Figure 2). There was no significant difference in trees per hectare across all predictive groups ($p = 0.637$, F-value = 0.593), nor was there a significant difference in total live basal area ($p = 0.389$, F-value = 1.056), diameter at breast height ($p = 0.764$, F-value = 0.386), stand age ($p = 0.603$, F-value = 0.631), or species richness ($p = 0.31$, F-value = 1.271) (Table 3).

Table 3. Stand structure characteristics of the 25 sampled stands with or without eastern spruce dwarf mistletoe (ESDM), calculated from field surveys, and classified as true negative, false negative, false positive, and true positive. Unless otherwise specified, all values reflect those of living trees only. Means followed by the same letter are not significantly different by Tukey's HSD test at a 0.05 level of significance for predictive group.

	Observed ESDM Free/Predicted ESDM Free (n = 7)					Observed ESDM/Predicted ESDM free (n = 6)				
	Mean		**SE**	**Min**	**Max**	**Mean**		**SE**	**Min**	**Max**
Trees per hectare	2218	a	91.8	1089	3117	2252	a	107.0	1238	3444
Basal area (m^2/ha)	26.3	a	0.85	22.3	34.7	25.4	a	1.26	18.6	34.9
Black spruce BA (m^2/ha)	21.6	a	0.81	17.8	31.3	19.8	a,b	0.98	14.8	24.5
Dead black spruce BA (m^2/ha)	2.3	a	0.24	0.7	3.6	3.4	a	0.34	1.3	5.0
Mean DBH (cm)	12.2	a	0.31	9.7	17.0	11.6	a	0.40	7.8	14.5
Species richness	4.4	a	0.81	3	9	4.2	a	0.87	2	7
DNR reported age	88.4	a	4.85	66	105	87.7	a	4.78	73	107
	Observed ESDM Free/Predicted ESDM (n = 6)					**Observed ESDM/Predicted ESDM (n = 6)**				
	Mean		**SE**	**Min**	**Max**	**Mean**		**SE**	**Min**	**Max**
Trees per hectare	2483	a	139.5	1244	3861	1870	a	107.6	883	3439
Basal area (m^2/ha)	22.2	a	1.0	15.1	28.3	21.9	a	1.16	14.0	29.0
Black spruce BA (m^2/ha)	17.1	b,c	0.89	13.6	25.9	15.0	c	1.08	7.1	21.8
Dead black spruce BA (m^2/ha)	1.6	a	0.18	0.6	2.4	7.4	b	0.92	2.5	18.0
Mean DBH (cm)	10.7	a	0.36	6.6	14.5	11.4	a	0.40	8.0	14.0
Species richness	4.0	a	0.93	2	7	6.2	a	1.01	2	9
DNR reported age	96.3	a	6.70	73	111	90.2	a	4.15	77	103

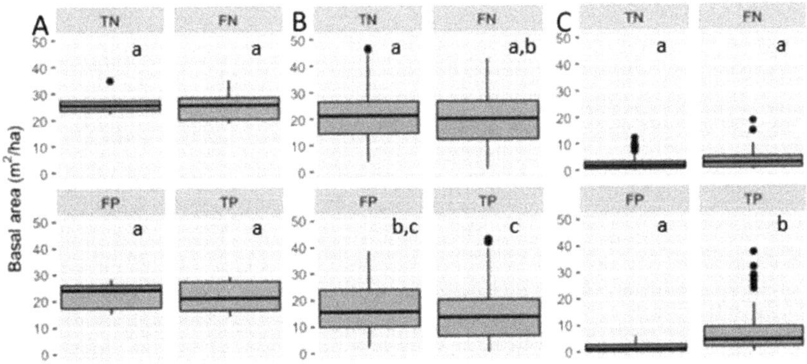

Figure 2. Living overstory basal area (**A**), living black spruce overstory basal area (**B**), and dead black spruce overstory basal area (**C**) within each of the four predictive groups (true negative (TN), false negative (FN), false positive (FP), and true positive (TP)). Panels with the same letter are not significantly different by Tukey's HSD test at a 0.05 level of significance for each predictive group.

Predictive groups diverged when looking at black spruce individually, with significant differences in both living black spruce basal area ($p = 1.97 \times 10^{-6}$, F-value = 10.09) and dead black spruce basal area ($p = 2.42 \times 10^{-15}$, F-value = 26.4) (Table 3). The amount of living black spruce basal area was not significantly different between true negative and false negative groups, between false negative and false positive groups, nor between false positive and true negative groups (Table 3). However, the true negative group had significantly more living black spruce basal area than both the false positive and true positive groups; similarly, the true positive group had significantly less living black spruce basal area than both the true negative and false negative groups (Table 3). The true positive group also had significantly more dead black spruce basal area than all other predictive groups (Table 3).

4. Discussion

The results from this analysis showed the challenges of predicting the occurrence of ESDM, as indicated by the overall accuracy of 52% (Table 2). Model performance did not differ between absence and presence of ESDM (true negative rate = 54%, true positive rate = 50%, Table 2), indicating it is not biased towards either over or under prediction.

Some of the limitations of the model developed by Hanks and colleagues [15] for ESDM are common across many models. Data used to create the Hanks model represents a single point in time, creating static predictions of ESDM occurrence. In this case, it is for the inventory year for each stand in the FIM dataset, which ranges from the early 1980s to the current year (all stands used in field verification were sampled no earlier than 1995; inventory year used in original modeling not available). Given the lag between stand inventory, model development, and field verification (up to 25 years in our analysis), some discrepancies would be expected. However, if this were the main reason for the inaccurate predictions, we would expect a lower false-positive rate and a higher false-negative rate (i.e., with the passing of time, more stands would have become infested).

With this temporal lag, it is also important to consider climate change and its potential effects on this disturbance regime as the elapsed time represents a period of shifting climate in the region. While the role of climate in the distribution and prevalence of ESDM has not been studied, the effects of climate change on black spruce are more well known. Overall, climate change is predicted to significantly negatively affect black spruce at the landscape scale [40], which would likely increase stress in black spruce and potentially the amount and rate of mortality due to ESDM infestation. However, the effects of climate change at the stand (and tree) level, may rely more on microsite conditions and water table dynamics [41] which could further increase the spatial complexity of ESDM across the state.

The patterns observed in stand structure and composition among all four predictive groups provide insight into ESDM presence. Most of the included stand structure characteristics did not differ significantly across predictive groups; the only significant differences were found in living and dead black spruce basal areas (Table 3). Live black spruce basal area shows a gradient of significant differences, with the true-negative group having more living black spruce basal area than both the false and true positive groups, the true positive group having significantly less black spruce basal area than the false and true negative groups and the false negative and positive groups having intermediate amounts (Figure 2). This gradient shows that, as stands become infested with ESDM, the amount of living black spruce declines, but that low levels of black spruce basal area do not always indicate infestation. Stands in the true positive group also had significantly more dead black spruce basal area than any other predictive group, perhaps indicating more advanced levels of infestation (and thus mortality due to ESDM infestation) (Figure 2).

The difference in patterns observed between the predicted/observed groups indicates that the Hanks model may be characterizing stand structure trends that could be, but are not necessarily, related to ESDM infestation or are only identifying stands at a specific stage of infestation. The model appears to be well suited to identifying stands with more established or more severe infestations, picking up trends of lower living black spruce basal area and higher dead black spruce basal area (Figure 2). Our results show, however, that there may simply be greater variation in black spruce stand conditions than originally hypothesized, in both the presence and absence of ESDM. Though ESDM in black spruce has been studied for decades, it has largely been from a plant pathology perspective and focused on spread and mortality rates, rather than effects on stand structure and composition [33,42]. It is only in recent years that the effects of ESDM on stand structure have started to be explored [28].

This new research has shown that stands at different levels of ESDM infestation exhibit differing stand structures. Stands with no ESDM present are predominately composed of living black spruce, while stands with low (<50% of stand infested) levels of ESDM still have low levels of black spruce mortality as well as additional species becoming established, and stands with high (>50% of stand infested) levels of ESDM had both increased black spruce mortality as well as a greater number of species present [28]. These differences indicate that predicting the occurrence of ESDM without considering the severity or extent of the infestation may not be possible, as different levels of infestation appear to have differing effects on stand structure.

Though mortality of the dominant species was included as a variable in this model, all infested stands were included as one group (ESDM presence was coded as a binary variable), which reflects opinions that any ESDM in a stand will result in an undesirable loss of timber volume [17]. A better approach may be to include multiple levels of severity to better reflect how ESDM infestation progresses or to focus on a specific stage of infestation (establishment, for example) that may have a more unique signature for a model to identify.

Our results show some of the difficulty of modeling the presence of disturbance agents that affect forest health—particularly those with strong spatial considerations. In modeling ESDM, mountain pine beetle, spruce budworm, or similar forest health threats, considering not just the presence or absence but the level of infestation could allow for more useful predictions [43,44]. Our results also indicate the value of looking beyond the main effect of the disturbance agent—mortality, in the case of ESDM—and considering broader ecosystem effects, such as species composition shifts or changes to stand development [28,45]. These considerations are important not only for predicting the occurrence of important disturbance agents but also for accurate growth and yield models that are necessary for forest management decision-making [2].

The Hanks et al. 2011 model, as well as our analysis, highlights the benefits of model benchmarking to improve model interpretation as well as the importance of understanding the data used in modeling and basing models on the ecological understanding of a system (here, how stand structure characteristics vary based on the severity of infestation). Our analysis also provides direction for future modeling work in this, and other, systems that may improve our ability to predict the occurrence or severity of disturbance agents and aid in monitoring and managing for improved forest health.

Author Contributions: Conceptualization, E.R.G., M.B.R., and M.A.W.-C.; methodology, E.R.G., M.B.R., and M.A.W.-C.; formal analysis, E.R.G.; investigation, E.R.G.; resources, M.B.R., and M.A.W.-C.; writing—original draft preparation, E.R.G.; writing—review and editing, M.B.R., and M.A.W.-C.; funding acquisition, M.B.R., and M.A.W.-C. All authors have read and agreed to the published version of the manuscript.

Funding: Funding for this project was provided by the Minnesota Invasive Terrestrial Plants and Pests Center through the Environment and Natural Resources Trust Fund as recommended by the Legislative-Citizen Commission on Minnesota Resources (LCCMR) through MITPPC Subproject #4: Dwarf Mistletoe Detection and Management in Minnesota. This research received additional support from the Minnesota Agricultural Experiment Station (projects MIN-42-101 and MIN-42-068).

Data Availability Statement: The Minnesota Department of Natural Resources Forest Inventory Management database used in the original model development is publicly available and can be accessed at https://gisdata.mn.gov/dataset/biota-dnr-forest-stand-inventory.

Acknowledgments: The authors would like to thank our collaborators on funded project—Brian Anderson, Fredrick Baker, Robert Blanchette, Charles Blinn, Christopher Edgar, Sarah Fellows, Heather Koop, Amit Pradhananga, Stephanie Snyder, Raychel Skay, and Robert Vennette. We appreciate the assistance of Anna Stockstad and Christopher Bru with fieldwork. Finally, we appreciate support from the Minnesota DNR for access to the study stands and the publicly available inventory data.

Conflicts of Interest: The authors declare no conflict of interest.

References

1. Woods, A.J.; Watts, M. The extent to which an unforeseen biotic disturbance can challenge timber expectations. *For. Ecol. Manag.* **2019**, *453*, 117558. [CrossRef]
2. Woods, A.; Coates, K.D. Are biotic disturbance agents challenging basic tenets of growth and yield and sustainable forest management? *Forestry* **2013**, *86*, 543–554. [CrossRef]
3. Forest Health US Forest Service. Available online: https://www.fs.usda.gov/science-technology/forest-health (accessed on 4 May 2021).
4. Hantula, J.; Muller, M.M.; Uusivuori, J. International plant trade associated risks: Laissez- faire or novel solutions. *Environ. Sci. Policy* **2014**, *37*, 158–160. [CrossRef]
5. Kappel, A.P.; Trotter, R.T.; Keena, M.A.; Rogan, J.; Williams, C.A. Mapping of the Asian longhorned beetle's time to maturity and risk to invasion at contiguous United States extent. *Biol. Invasions* **2017**, *19*, 1999–2013. [CrossRef]
6. Kolka, R.K.; D'Amato, A.W.; Wagenbrenner, J.W.; Slesak, R.A.; Pypker, T.G.; Youngquist, M.B.; Grinde, A.R.; Palik, B.J. Review of ecosystem level impacts of emerald ash borer on black ashwetlands: What does the future hold? *Forests* **2018**, *9*, 179. [CrossRef]
7. Baker, F.A.; Slivitsky, M.; Knowles, K. Impact of dwarf mistletoe on jack pine forests in Manitoba. *Plant Dis.* **1992**, 76. [CrossRef]
8. Windmuller-Campione, M.A. Assessing the future susceptibility of mountain pine beetle (Dendroctonus ponderosae) in the Great Lakes Region using forest composition and structural attributes. *Can. J. For. Res.* **2018**, *48*, 451–459. [CrossRef]
9. Mckee, F.R.; Aukema, B.H. Influence of temperature on the reproductive success, brood development and brood fitness of the eastern larch beetle Dendroctonus simplex LeConte. *Agric. For. Entomol.* **2015**, *17*, 102–112. [CrossRef]
10. Ward, S.F.; Aukema, B.H. Anomalous outbreaks of an invasive defoliator and native bark beetle facilitated by warm temperatures, changes in precipitation and interspecific interactions. *Ecography* **2019**, *42*, 1068–1078. [CrossRef]
11. Nagel, L.M.; Palik, B.J.; Battaglia, M.A.; D'Amato, A.W.; Guldin, J.M.; Swanston, C.W.; Janowiak, M.K.; Powers, M.P.; Joyce, L.A.; Millar, C.I.; et al. Adaptive silviculture for climate change: A national experiment in manager-scientist partnerships to apply an adaptation framework. *J. For.* **2017**, *115*, 167–178. [CrossRef]
12. Waring, K.M.; O'Hara, K.L. Silvicultural strategies in forest ecosystems affected by introduced pests. *For. Ecol. Manag.* **2005**, *209*, 27–41. [CrossRef]

13. Host, T.K.; Russell, M.B.; Windmuller-Campione, M.A.; Slesak, R.A.; Knight, J.F. Ash presence and abundance derived from composite landsat and sentinel-2 time series and lidar surface models in Minnesota, USA. *Remote Sens.* **2020**, *12*, 1341. [CrossRef]
14. Campbell, E.M.; Antos, J.A. Distribution and severity of white pine blister rust and mountain pine beetle on whitebark pine in British Columbia. *Can. J. For. Res.* **2000**, *30*, 1051–1059. [CrossRef]
15. Hanks, E.M.; Hooten, M.B.; Baker, F.A. Reconciling multiple data sources to improve accuracy of large-scale prediction of forest disease incidence. *Ecol. Appl.* **2011**, *21*, 1173–1188. [CrossRef] [PubMed]
16. Crocker, S.J.; Liknes, G.C.; McKee, F.R.; Albers, J.S.; Aukema, B.H. Stand-level factors associated with resurging mortality from eastern larch beetle (Dendroctonus simplex LeConte). *For. Ecol. Manag.* **2016**, *375*, 27–34. [CrossRef]
17. Baker, F.A.; French, D.W.; Rose, D.W. DMLOSS: A simulator of losses in dwarf mistletoe infested black spruce stands. *For. Sci.* **1982**, *28*, 590–598.
18. Patton, S.R.; Russell, M.B.; Windmuller-Campione, M.A.; Edgar, C.B. Modeled diameter growth response to intermediate treatments of planted white spruce (Picea 2 glauca) affected by eastern spruce budworm (Choristoneura fumiferana) in Minnesota, USA. *Can. J. For. Res.* **2019**, *55*, 259–267. [CrossRef]
19. Box, G.E.P.; Draper, N.R. *Empirical Model-Building and Response Surfaces*; John Wiley & Sons: New York, NY, USA, 1987.
20. Weiskittel, A.R.; Hann, D.W.; Kershaw, J.A.; Vanclay, J.K. Model evaluation and calibration. In *Forest Growth and Yield Modeling*; Wiley-Blackwell: Chichester, UK; Hoboken, NJ, USA, 2011; pp. 295–309.
21. Bell, D.M.; Schlaepfer, D.R. On the dangers of model complexity without ecological justification in species distribution modeling. *Ecol. Modell.* **2016**, *330*, 50–59. [CrossRef]
22. Kimmins, J.P.; Blanco, J.A.; Seely, B.; Welham, C.; Scoullar, K. Complexity in modelling forest ecosystems: How much is enough? *For. Ecol. Manag.* **2008**, *256*, 1646–1658. [CrossRef]
23. Brassard, B.W.; Chen, H.Y.H. Stand structural dynamics of North American boreal forests. *Crit. Rev. Plant Sci.* **2006**, *25*, 115–137. [CrossRef]
24. Seymour, R.S.; Hunter, M.L.J. New forestry in eastern spruce-fir forests: Principles and applications to Maine. *Maine Misc. Publ.* **1992**, *716*, 1–36.
25. Stockstad, A.; Gray, E.; Sebestyen, S.; Lany, N.; Kolka, R.; Windmuller-Campione, M. Analyzing trends in water table elevations at the Marcell Experimental Forest, Minnesota, U.S.A. *Am. J. Undergrad. Res.* **2021**, *17*, 19–32. [CrossRef]
26. Lavoie, M.; Paré, D.; Bergeron, Y. Impact of global change and forest management on carbon sequestration in northern forested peatlands. *Environ. Rev.* **2005**, *13*, 199–240. [CrossRef]
27. Ostry, M.E.; Nicholls, T.H. Eastern dwarf mistletoe on black spruce. *Dept. Agric. For. Serv.* **1979**, *158*, 2–7.
28. Skay, R.; Windmuller-Campione, M.A.; Russell, M.B.; Reuling, L.F. Influence of eastern spruce dwarf mistletoe on stand structure and composition in northern Minnesota. *For. Ecol. Manag.* **2021**, *481*, 118712. [CrossRef]
29. Harper, K.A.; Bergeron, Y.; Gauthier, S.; Drapeau, P. Post-fire development of canopy structure and composition in black spruce forests of Abitibi, Québec: A landscape scale study. *Silva Fenn.* **2002**, *36*, 249–263. [CrossRef]
30. Heinselman, M.L. Forest sites, bog processes, and peatland types in the Glacial Lake Agassiz Region, Minnesota. *Ecol. Monogr.* **1963**, *33*, 327–374. [CrossRef]
31. Minnesota Department of Natural Resources. *Black Spruce Timber Sale Design and Control Guidelines for Minimizing the Threat of Eastern Dwarf Mistletoe*; Minnesota Department of Natural Resources: St. Paul, MN, USA, 2019.
32. Hudler, G.; Nicholls, T.; French, D.W.; Warner, G. Dissemination of seeds of the eastern dwarf mistletoe by birds. *Can. J. For. Res.* **1974**, *4*, 409–412. [CrossRef]
33. Baker, F.A.; French, D.W. Spread of Arceuthobium pusillum and rate of infection and mortality in black spruce stands. *Plant Dis.* **1980**, *64*, 1074–1076. [CrossRef]
34. Bailey, R.G. *Description of the Ecoregions of the United States*; U.S. Department of Agriculture, Miscellaneous publication No. 1391; U.S. Department of Agriculture: Washington, DC, USA, 1980.
35. Aaseng, N.E.; Almendinger, J.C.; Rusterholz, K.; Wovcha, D.; Klein, T.R. *Field Guide to the Native Plant Communities of Minnesota: The Laurentian Mixed Forest Province*; Minnesota Department of Natural Resources: Saint Paul, MN, USA, 2003.
36. Eyre, F.H. *Forest Cover Types of the United States and Canada*, 6th ed.; Society of American Foresters: Washington, DC, USA, 1980.
37. Miles, P.D. Forest Inventory EVALIDator Web-Application, Version 1.6.0.03. Available online: http://apps.fs.fed.us/Evalidator/evalidator.jsp (accessed on 17 April 2017).
38. Hawksworth, F.G. *The 6-Class Dwarf Mistletoe Rating System*; Rocky Mountain Forest and Range Experiment Station, Forest Service, U.S. Deptartment of Agriculture: Fort Collins, CO, USA, 1977.
39. R Core Team. *R: A Language and Environment for Statistical Computing*; R Foundation for Statistical Computing: Vienna, Austria, 2020. Available online: https://www.r-project.org/ (accessed on 25 June 2020).
40. Landscape Change Research Group. *Climate Change Atlas*; Northern Research Station, U.S. Forest Service: Delaware, OH, USA, 2014.
41. Dymond, S.F.; D'Amato, A.W.; Kolka, R.K.; Bolstad, P.V.; Sebestyen, S.D.; Gill, K.; Curzon, M.T. Climatic controls on peatland black spruce growth in relation to water table variation and precipitation. *Ecohydrology* **2019**, *12*, e2137. [CrossRef]
42. Baker, F.A.; Knowles, K.R. Case Study: 36 Years of dwarf mistletoe in a regenerating black spruce stand in northern Minnesota. *North. J. Appl. For.* **2004**, *21*, 150–153. [CrossRef]

43. Aukema, B.H.; Carroll, A.L.; Zhu, J.; Raffa, K.F.; Sickley, T.A.; Taylor Aukema, S.W.; Carroll, Á.A.; Taylor, S.; Raffa, Á.K. Landscape level analysis of mountain pine beetle in British Columbia, Canada: Spatiotemporal development and spatial synchrony within the present outbreak. *Ecography (Cop.)* **2006**, *29*, 427–441. [CrossRef]
44. Boulanger, Y.; Arseneault, D. Spruce budworm outbreaks in eastern Quebec over the last 450 years. *Can. J. For. Res.* **2004**, *34*, 1035–1043. [CrossRef]
45. Diskin, M.; Rocca, M.E.; Nelson, K.N.; Aoki, C.F.; Romme, W.H. Forest developmental trajectories in mountain pine beetle disturbed forests of Rocky Mountain National Park, Colorado. *Can. J. For. Res.* **2011**, *41*, 782–792. [CrossRef]

forests

MDPI

Article

The Effects of the Management Strategies on Spruce Bark Beetles Populations (*Ips typographus* and *Pityogenes chalcographus*), in Apuseni Natural Park, Romania

Ciprian George Fora [1,*] and Adalbert Balog [2,*]

[1] Faculty of Horticulture and Forestry, Banat's University of Agricultural Sciences and Veterinary Medicine "King Michael I of Romania" from Timișoara, Calea Aradului 119, 300645 Timișoara, Romania
[2] Department of Horticulture, Faculty of Technical and Human Sciences, Sapientia Hungarian University of Transylvania, Aleea Sighișoarei 1C, 530104 Târgu Mureș/Corunca, Romania
* Correspondence: ciprian.fora@usab-tm.ro (C.G.F.); adalbert.balog@ms.sapientia.ro (A.B.)

Abstract: The population densities of *I. typographus* and *P. chalcographus* inside the Carpathian Mountains increasing mostly because of the non-synchronized and divers management strategies. The growing loss of trees from one year to another indicates assessment to determine the influence of the current management practices (or the absence of such) on bark beetle densities. A comprehensive tree-year assessment were made inside the Apuseni Natural Park, with a surface of 75,784 ha, to assess the population density of bark beetles. High abundance of both species were detected from one year to another, both in managed and unmanaged forests, the latter explained by the presence of scattered wind falling trees which represent favorable places for oviposition. General linear modelling revealed that the effect of environmental variables (forest mean age, forest density, altitude and slope) on *I. typographus* density were only significant under management systems, and only forest age has significantly negative effect on bark beetles densities. Overall, we conclude that there is little difference in population between non-intervention and intervention areas and even over a relatively short period of time (2011–2013), the population of both species expanded, and new uninfected habitats were infected both in unmanaged and managed forests, therefore a synchronized management system has to be developed. Climate variables as temperature and precipitations has more significant effects on population density than management.

Keywords: conservation areas; management systems; spruce forest; climate change; population density

1. Introduction

Bark beetles (Coleoptera: Curculionidae: Scolytinae), are the most important pests of coniferous trees [1]. Of those belonging to the genus Ips, *Ips typographus* is the most destructive species, damaging spruce forests across Eurasia [2], and causing economically-significant attacks in spruce forests [3–5]. In addition, beetles of the genus Pityogenes can also be of considerable economic importance. For example, *Pityogenes chalcographus* is one of the most abundant bark beetle species in Europe and can be detected widely in Picea (*Picea abies* (L.) H. Karst) forests, ranging from Scandinavia to the Balkans [6]. Damaged produced by this latter species was found to increase from one year to another throughout Eurasia [7], with more severe damage to Picea trees being widely reported [8,9]. Damage is frequently detected in combination with that of *I. typographus* (Linnaeus, 1758), cases when the two species occur together producing more severe losses that when found alone, as frequently described in the literature, even early papers [10]. Several studies have reported that *P. chalcographus* prefers young trees, with enhanced physiological demands and thus more sensitive to attack [11–13], with the highest density of infestation detected in trees between 25–45 years [14].

Economic thresholds (the individual number which requires already intervention to protect trees) for *I. typographus* were detected at different levels, and were reported at 8000 individuals/trap/year by Faccoli & Stergulc (2004), with a strong correlation found between individual numbers and tree numbers. Other papers have reported 10,000 individuals/trap/year [15], or sometimes even as high as 15,000 individuals/trap/year [16]. With *P. chalcographus*, economic thresholds were reported as 20,000 up to 60,000 individuals/trap/year [17]. In Romania, 3000 individuals/trap/year have been reported for *I. typographus* and 30,000 individuals/trap/year for *P. chalcographus* [18]. While high variations in economic thresholds can be detected, the effect of management, which usually follows the practices that dead trees and wood residuals are left after harvest for a longer time near healthy growing trees, are rarely considered. Previous related researches have mentioned that tree mortality was most strongly determined by standing trees killed by bark beetles in the previous year and left in the forest, followed by wind-felled trees. Precise and timely removal of infested trees is a suitable method to control bark beetle infestation [19]. Another study also reported that the application of selective control methods reduced mainly the populations of *I. typographus* (results detected by using pheromone traps), and no effect on *P. chalcographus* were detected. This is because a large volume of fresh logging residues, mainly tree tops and branches remaining in stands, offer excellent breeding conditions for *P. chalcographus*. It was also reported that the fast increase of this species population seems to be a consequence of recent human errors concerning insect control and forest management [14].

Overall, it appears that the increasing densities of these insects should be further investigated and management practices compared in other regions (i.e., Carpathian Mountains) with large protected forest areas. Also, the effect of climate change has to be considered in such assessments. Therefore in the present study, our main goal was to assess the population density of *I. typographus* and *P. chalcographus* inside the Natural Park Apuseni, Romania and to determine whether or not current management practice ('non-intervention' and 'intervention' strategies) have a marked effect on bark beetles densities.

2. Materials and Methods

2.1. Study Area

The Apuseni Natural Park (Western Carpathian Mountains, Central Europe, (46.604° N, 22.810° E), with a surface of 75,784 ha, is one of the most important protected area from Western Romanian Carpathians. Covered mainly with forests, about 75%, with a special karst heritage, the natural park of the Bihor Mountains is included in category V IUCN-Protected Landscape (Figure S1). The main trees species which compose the forests are spruce, *Picea* spp. (70%), fir, *Abies* spp. (2%) and beech, *Fagus sylvatica* (27%). Other resinous species like pine and larch and few species of deciduous trees like ash, sessile oak, sycamore, birch and elm, can be found.

2.2. Forests Status and Management

In the past, the forests of Apuseni Mountains, composed mainly by European spruce, were owned and managed, in the respect of Romanian forest protection standards, by three state forest directorates: Alba, Bihor and Cluj (common named with counties). Valuable forests of spruce or spruce mixed with beech and fir were included in conservation areas (20% from total forests surfaces) and the remaining 80% of spruce mixed with beech and fir in buffer areas in 2004, during this time the protected natural park was established. All over of these surfaces, the internal zoning of the park imposed including of the forest stands in conservation areas (unmanaged forests) and buffer areas (managed forests). In unmanaged forest, no interventions are allowed. Just in case of major disturbances created by abiotic and/or biotic factors, in unmanaged forests interventions are allowed after a long and hard procedure, which take years (elimination of dead trees as example). Parallel, in the last three decades, a complex and yet un-finalized process of restitution of the forests takes place. Either unmanaged or managed stands changed properties, from the state to the private entities or private persons. These created difficulties to apply all forest protection measures in managed forests and to respect non-intervention in unmanaged forests. In August 2006, a highly destructive storm affected the forests in both conservation and buffer areas of the park, resulting in a huge quantity of wind felled trees. In every year since then, interventions have been performed in both areas to assess natural calamities and involving delineation of the protected areas by special minimally-planted buffer zones. However, starting in 2010, no more intervention was allowed in forests belonging to the conservation areas. In this context the population densities of *I. typographus* and *P. chalcographus* inside the Apuseni Natural Park, increased, especially after wind throw from 2006 which affect more than 200,000 cubic meter of spruce trees in unmanaged and managed forests and as an effect of climate change.

2.3. Site Selection and Environmental Conditions

Selection of the research sites (Table S1), were the following: the forests are included in conservation areas (unmanaged forests), with a non-intervention strategy (i.e., no dead trees were removed from the forest or no other management activity were followed). In buffer areas (managed forests), with an intervention strategy (i.e., all dead trees investigated and removed from the forests, with pheromone traps used to assess bark beetle densities, with trap trees to decrease the bark beetles population level. All other forest protection measures applied); the forests covering a range of stand conditions; the existence of minimum three spruce trees per site colonized by the bark beetles; the forests are accessible.

In total, 48 permanent research sites have been established (Figure 1) inside the Forest Directorates/counties (Alba–site A, Bihor–site B and Cluj–site C). From these, 24 infested by *I. typographus* (16 unmanaged forests and 8 managed forests) and 24 infested by *P. chalcographus* (17 unmanaged forests and 7 managed forests) were established (Figure S1). We used site as differences (independent variables) because sites located in different counties have different administrative strategies with independent regime. Forest management reorganization is under progress in Romania and not even the property rights were clarified yet.

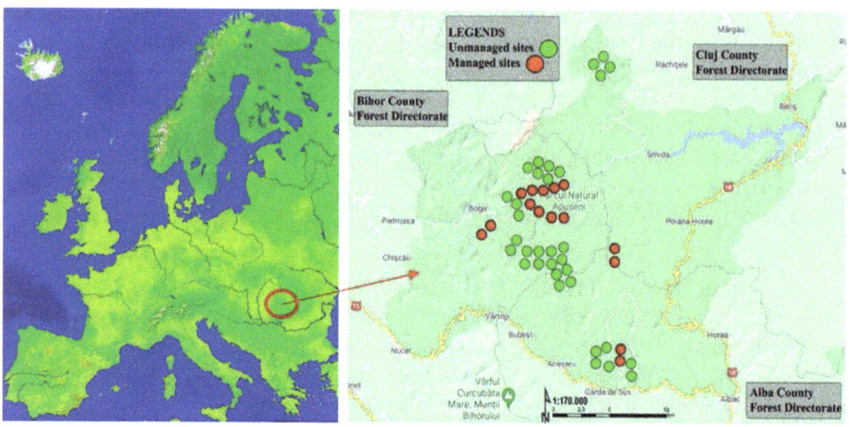

Figure 1. Map of the areas investigated in the Apuseni Natural Park (46.604° N, 22.810° E), Romania. Unmanaged sites are marked with green, managed sites with red.

Of these sites, in 80% of the research forest stands, spruce comprised between 90 to 100% of the trees, whilst 75% had a mean age between 60 to 120 years old, 28% 120 to 170 years old, and just 8% 20 to 60 years old. The altitude of the sites was between 800 to

1500 m a.s.l. The climate characterizing the whole area may be described as temperate-continental, with multiannual mean temperature of 4.6 °C and mean precipitation of 1600 mm per year. Climate conditions during the assessment period are presented in Figure 2. During the research, no significant gale conditions were registered; however, downwind trees were observed scattered around the forests, whilst dying trees caused by the bark beetles attack were also seen to be present and similarly scattered in the forest.

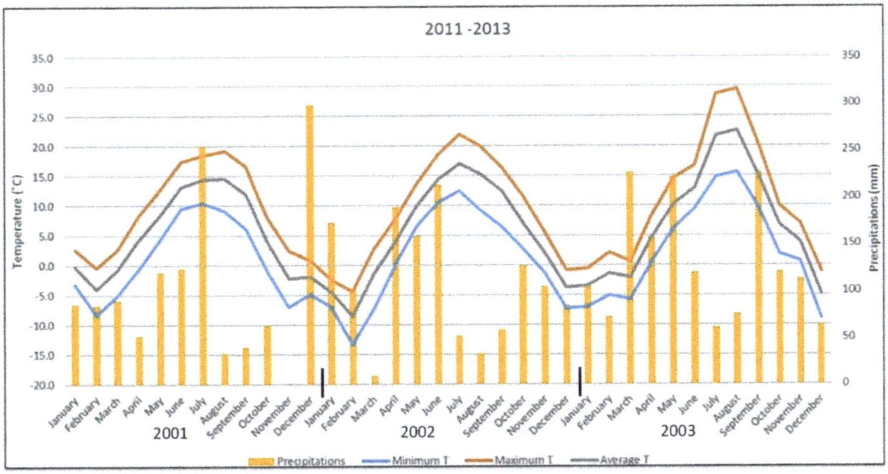

Figure 2. Temperature and precipitation values in the target areas between 2011–3013.

2.4. Data Collection

In each site, Theysohn slot traps baited with synthetic pheromones were used. Placement of the traps was performed as much as possible in sunshine, at 15–20 m distance by the spruce trees stands, and at 1.3 m above the ground. The synthetic pheromone used for *I. typographus* was S-cis verbenol 3.3%, namely (1S, 2S)-4,6,6-trimethylbicyclo (3.1.1) hept-3-en-2-ol (IT-Ecolure Tubus Mega) and for *P. chalcographus*, Chalcogran 4%, namely 2-ethyl-1,6 dioxaspiro (4,4) nonane (PC-Ecolure Tubus Mega). The pheromones were active for 22 weeks according to the manufacture's advice and cover the period of spring and summer beetle flights. The level of pheromone in the tubes was simultaneously checked with that of the beetle collection. The traps were active between the end of April until the end of September of each sampling year. Trapped beetles were collected weekly in plastic boxes, labeled according to the particular trap and date of collection, and stored in the fridge at −20 °C. Periodically, the boxes were transported from the Forest Entomology Laboratory to the University of Timisoara for analysis. Beetles from each sample were physically counted up to 500 individuals, whereafter bigger samples were measured volumetrically.

2.5. Data Analysis

Insect density values for each sampling period, location and year were averaged between the same traps. Thereafter, all data were log10 transformed to reduce the influence of high values (by adding 1 to each value and then log10 transforming it). This was followed by a normality test and homogeneity of variance. As all data were normally distributed, one-way ANOVA followed by Tukey HSD test were used to compare abundances between sites and treatments. Each year was considered separately in this analysis. Statistically significant differences were considered at a level of $p < 0.05$. Next, all data were again compared, this time as each year, site and treatment using one-way ANOVA followed by Tukey HSD test ($p < 0.05$ level). In the following analyses, heat-map matrices were constructed to compare abundances between sites, treatments and flying periods for each year and species. In this approach, matrices blue colors represents low abundances, red colours high abundances. General linear model (GLM) was used to test the effect of environmental variables (forest age, forest density, altitude and slope) on beetle density using forest mean age, forest density, altitude and slope as independent variable, and each species group as a dependent (response) variables. Lastly principal component analyses (PCoA) was used to test the effect of sites climate variables (temperature and precipitations) and treatments on species abundances, where treatments and climate variables were considered as main components and species abundances as variables. The method was replicated again using separately temperatures and precipitations as main components. All data analyses were made in PAST 4.02. Analyses were also made in PAST 4.02.

3. Results

Altogether more than 400,000 individuals of *I. typographus* and more than 600,000 individuals of *P. chalcographus* were captured and used for analyses. This means that in unmanaged forests site A the *I. typographus* abundance varied between 10.1 (early spring) and 849.7, (summer), in site B between 2.3 and 2237.3, while in site C between 2.2 and 4565.6 during the assessment period. In managed forests, this variation was between 9 and 598.2 in site A, 1.8 ad 5733.8 in site B and between 3.5 and 11,667.5 in site C.

The abundance of *P. chalcographus* wearied in unmanaged forests between 12.1 and 6172.8 in site A, between 5 and 12,695.8 in site B and between 6.8 and 15,804.5 in site C. The species variation in managed forests was 7 to 46,386.2 in site A, 3 to 7012.3 in site B and 6.3 to 19,831.7 in site C during the assessment period.

High abundance of both species were detected in all years of the investigation, both following 'unmanaged forests' (control) and 'managed forests' (treatment). While in 2011 the abundance of *I. typographus* was significantly higher in site C, its abundance increased in 2012, and high abundance in site B, under managed forests were also detected. Further increase in 2013 were observed also in site A, both in the unmanaged and managed forests (Figure 3A–C). Similar trends at *P. chalcographus* was detected, with higher abundance in 2011 on site B managed forests, in 2012 on site B unmanaged forests, and significantly higher abundances on site C managed forests in 2013 (Figure 3D–F). By comparing all the data together, significantly higher abundances on site B unmanaged forests for *I. typographus* and site C managed forests were detected; however, other differences in mean abundances can be detected (Table S2, supplementary online materials).

Heat map analyses also revealed that the abundance of *I. typographus* in 2011 was concentrated at site C, mostly during the spring flying period; this concentration, however, expanded in 2012 and in 2013, higher abundance developing also at sites B and A as well. Low abundance was observed in each year on site A, under managed forests (Figure 4A–C). A rather different trend for *P. chalcographus* was detected, but only in 2011, when higher abundances were observed during the summer flying period (Figure 4D), while the expansion of high abundance in 2012 and 2013 were also detected, although at lower density in site A under managed forests (Figure 4E,F). Additional analyses on beetles flight dynamics (Supplementary online materials, Figure S1), revealed high synchronization between the two species, but also differences between sites and years were detected.

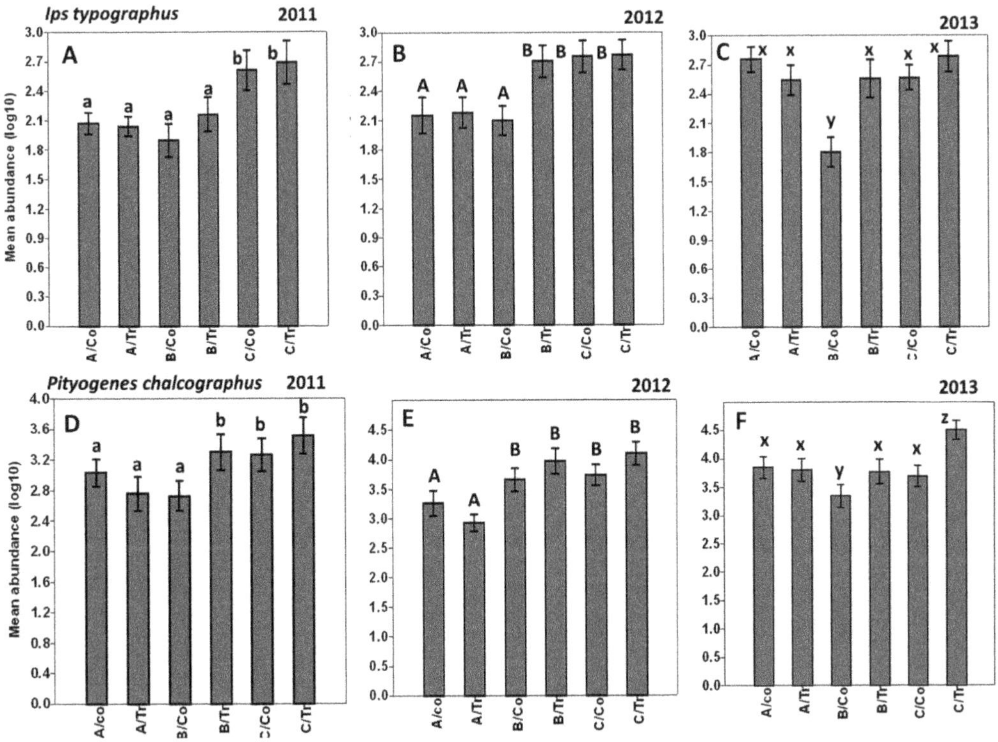

Figure 3. Mean abundances of *I. typographus* (**A–C**) and *P. chalcographus* (**D–F**), compared between sites and treatments using one-way ANOVA followed by Tukey HSD tests. Each year was considered separately. Different letters represents statistically significant differences ($p < 0.05$) according to the Tukey's multiple comparison test. Co = control (non-intervention zone), Tr = treatment (intervention zone). Differences in letters (a, A or X) are used to mark significances inside the years and species.

General linear modelling (GLM) revealed that the effect of environmental variables (forest mean age, forest density, altitude and slope) on *I. typographus* density were only significant under treatment sites, and only forest age has significantly negative effect on density in 2011 and 2012. The abundance of *P. chalcographus* was positively influenced by altitude in 2012 and negatively influenced by the forest age and altitude in 2013 in control sites. Forest altitude had also negative effect on density under intervention strategy in 2012 and 2013 (Table 1).

Principal component analyses (PCoA) revealed that the greatest variation in densities were explained primarily by sites environmental variables (temperature and precipitations) in all years and flying periods and less influenced by non-intervention or intervention strategy systems. In most of the cases, climate variables together explained up to 80%, while treatments only up to 30% (in 2011), but several times only up to 10% species densities (Figure 5). By considering separately temperature and precipitations, they explaining only up to 30–40% species densities, while no change in treatments effects were detected.

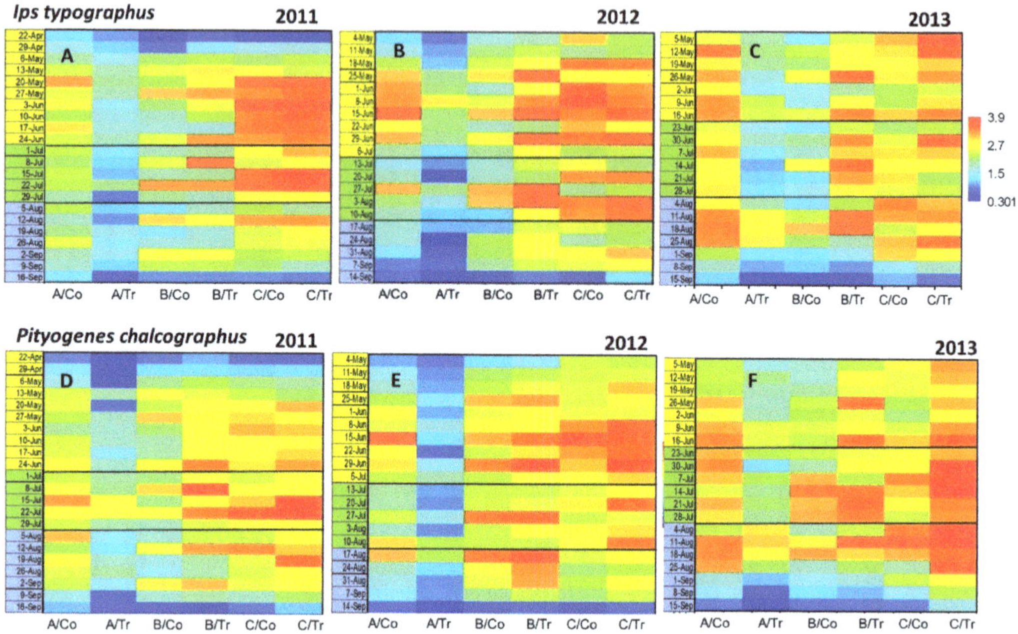

Figure 4. Heat-map matrices comparing abundances of *I. typographus* (**A–C**) and *P. chalcographus* (**D–F**), between sites, treatments and flying periods for each year and species. Blue colours represents low abundances, red colours high abundances. All data were log10 transformed and presented. Co-control (non-intervention zone), Tr-treatment (intervention zone).

Table 1. General linear model (GLM) using the effect of environmental variables on bark beetle density. Log10 transformed data were used for analyses. Arrows before value show the direction of main effects: ↑ indicates a positive relationship; ↓ a negative one. Significant values are in bold.

	I. typographus Density (Dep Variable) in Unmanaged Forests					
	2011		2012		2013	
Indep. Var.	**Std.err.**	*p*	**Std.err.**	*p*	**Std.err.**	*p*
Mean age (year)	0.135	0.062	0.282	0.111	0.185	0.384
Forest dens.	0.025	0.796	0.052	0.386	0.034	0.675
Altitude (m)	0.041	0.124	0.086	0.516	0.056	0.715
Slope	0.123	0.198	0.258	0.414	0.169	0.789
	I. typographus density (Dep variable) in managed forests					
Mean age (year)	3.557	**0.054**↓	3.557	**0.054**↓	0.619	0.669
Forest dens.	0.617	0.374	0.617	0.374	0.347	0.620
Altitude (m)	1.393	0.385	0.393	0.385	0.630	0.421
Slope	1.084	0.613	1.084	0.613	0.594	0.892
	P. chalcographus density (Dep variable) in unmanaged forests					
Mean age (year)	0.101	0.152	0.075	0.074	0.068	**0.034**↓
Forest dens.	0.046	0.409	0.033	0.103	0.030	0.110
Altitude (m)	0.033	0.270	0.024	**0.015**↑	0.022	**0.059**↓
Slope	0.116	0.786	0.085	0.730	0.078	0.871
	P. chalcographus density (Dep variable) in managed forests					
Mean age (year)	1.674	0.220	2.196	0.102	1.729	0.720
Forest dens.	0.393	0.142	0.708	0.509	0.305	0.065
Altitude (m)	4.472	**0.051**↓	4.411	**0.005**↓	5.253	0.266
Slope	1.802	0.746	2.589	0.508	1.615	0.900

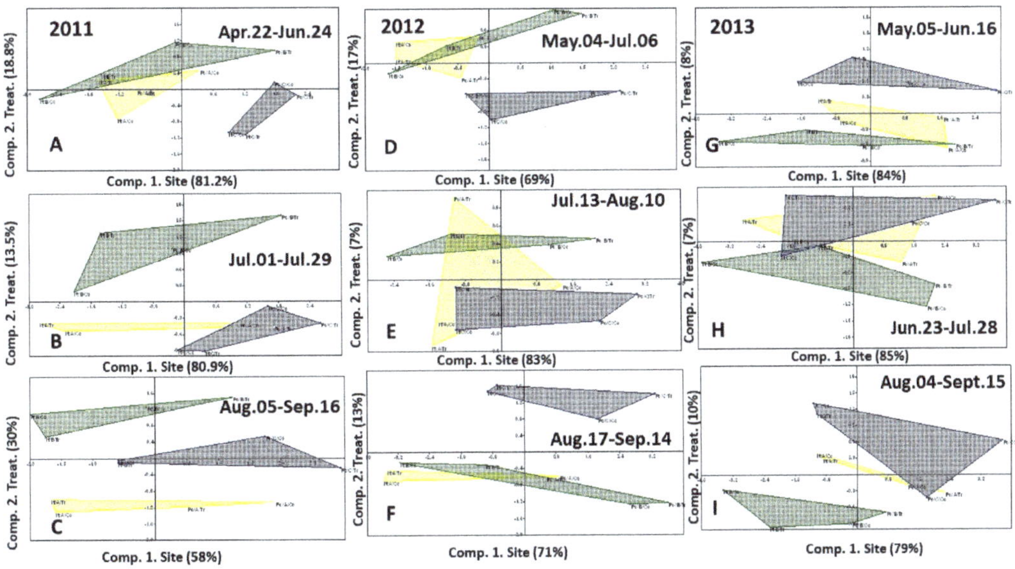

Figure 5. Principal component analyses (PCoA) to test the effect of site climate variables (temperature and precipitations) and treatments on species abundances, where climate variables and treatments were considered as the main components and species abundances as variables. Figures **A–C** represents year 2001, figures **D–F** 2002, and figures **G–I** year 2003. Yellow plains represent site **A**, green site **B**, and blue site **C**. Differentiations inside years were made to have a proper comparison between site climate variables. As example year 1–29 July 2011 was similar or very closed in mean temperature and precipitation with 13 July–10 August 2012 and with 23 June–28 July 2013.

Principal component analyses (PCoA) revealed that the greatest variation in densities were explained primarily by sites environmental variables (temperature and precipitations) in all years and flying periods and less influenced by non-intervention or intervention strategy systems. In most of the cases, climate variables together explained up to 80%, while treatments only up to 30% (in 2011), but several times only up to 10% species densities (Figure 5). By considering separately temperature and precipitations, they explaining only up to 30–40% species densities, while no change in treatments effects were detected.

4. Discussion

High abundance of both species was detected in all years, both after non-intervention and intervention strategies. Increase in the abundance of *I. typographus* in 2011 was significant in site C; its abundance increased in 2012, and high abundance in site B, under managed forests were also detected. A further increase in 2013 was observed also at site A, both in unmanaged forests and managed forests. This can be partially explained because of the effect of environmental variables (forest mean age, forest density, altitude and slope) on *I. typographus* density were only significant under treatment sites, and only forest age has significantly negative effect on density in 2011 and 2012.

P. chalcographus showed a similar trend (Figure 4). The increase in its abundance in the unmanaged forests can be explained by the presence of scattered wind felled trees, which represent favorable places for oviposition and by the existence of standing trees attacked by beetles, and in the managed forests, by late removal of the mentioned categories of trees.

The high insects' density due to the high wood biomass probably influenced tree mortality, which further increased the insects' abundances until the end of 2013. This can be clearly observed when insect abundances are presented in the form of heat maps (Figure 4A–F), but also a synchronization in flight of the two species (Figure S3).

Other research has reported another six bark beetle species inhabiting in the same area of the National Park Apuseni: *I. amitinus* (Eichhoff, 1872), *I. acuminatus* (Gyllenhal, 1827), *P. chalcographus*, *Dryocoetes autographus* (Ratzeburg, 1837), *Hylurgops palliatus* (Gyllenhal, 1813), and *H. glabratus* (Zetterstedt, 1828) [20].

The decision not to extract highly infected and dead trees by 2010 from unmanaged forests may also explain the high insect abundances. Some studies have also revealed that the reproduction of *P. chalcographus* increased significantly in wood residuals when these are collected near forest for a longer time [20–22]; however, damages to the nearest young and healthy trees through drying out were not significant [23].

Overall, general linear modelling (GLM) revealed that the effect of environmental variables (forest mean age, forest density, altitude and slope) on *I. typographus* density were only significant under the managed system, and only forest age has significantly negative effects on density in 2011 and 2012, but not in 2013. This further explains why the management strategies employed had increasing effect on insect abundance and this was followed by higher levels of captures in 2012 and 2013. A more detailed field survey revealed that within the unmanaged forests in 2011–2013, the felled tree biomass and old tree densities favoured insect reproductions and thereby density increase that influenced the nearest regions too. Principal component analyses (PCoA) revealed however that the greatest variation in densities was explained primarily by site climate variables (temperature and precipitations) in all years and less influenced by management systems. In most of the cases examined, site explained up to 80%, while management only up to 30% (in 2011), and on several occasions, only up to 10% of recorded species' densities (Figure 5). Considering separately temperature and precipitations they explained only up to 30–40% density variations of both species. This result seems to contradict our previous findings; however, if the management strategies are considered in the intervention zone (i.e., late extraction of felled wood biomass and noting of the wood residuals, which favorited mostly *P. chalcographus* developments), we conclude that site has generally a more accentuating effect than management strategies.

Earlier studies claim that the two species cause increasing tree mortality when occurring together. The effect of *P. chalcographus* alone on *Picea abies* produced 8% tree mortality, while together with *I. typographus*, damage increased significantly [24].

Effect of the multiannual change in temperature and precipitations has to be more considered in the future. Higher temperature in association with low precipitations also had a catalytic effect on both species abundance, as observed when annual flight periods are compared (Figure 4, Figure S1). While in both 2011 and 2012, three flight period per year can clearly be discerned, this is less obvious in 2013, whereupon a more continuous flight period was observed. Species started showing activity at 16.8–17 °C until a temperature of 35.6 °C was reached [25]. The effect of climate change on these species' activity are now more often detected, the result of increasing global temperatures in forests with decreasing precipitation (e.g., in the Carpathian Mountains during the last few years), effects that have had a deleterious effect on tree physiology, thereby increasing their susceptibility to these insects [26,27].

5. Conclusions

It can be concluded that even over a relatively short time period (three years), as here examined, the population of both bark beetle species studied expanded, with new uninfected habitats being infected both under unmanaged and managed regimes, mostly due to the management involved, but also as a consequence of climate change. No relevant differences in population densities between unmanaged and managed forests were detected, this underlines the question as to how forest management, even in protected areas, should be followed in light of the insidious effects of ongoing climate change. If such practices continue to be followed in the future, a mass-extinction of resinous forests, mainly composed by spruce in the Carpathian Mountains, is to be expected, primarily due to these two bark beetle species (and evidently other species occurrences which have yet to be investigated), even under the guise of so called 'management' systems, as currently practiced. New habitat conservation, wood-management protocols and legislation urgently need to be introduced and followed to prevent these predicted forest mass extinctions, which seemingly are already happening over a relatively small time period (less than a decade), as here highlighted. From the standpoint of Biological Pest Management involving natural biological control agents, i.e., predators, parasitoids and/or natural products (i.e., entomopathogenic fungal treatments with *Metarhisium* or *Beauveria*), to combat these economically serious bark beetle pests, such an approach is to be recommended, especially when wood residuals are collected near healthy trees [28].

Supplementary Materials: The following are available online at https://www.mdpi.com/1999-4907/12/6/760/s1, Figure S1: Flight dynamics in 2011, 2012 and 2013 Ips + Pityogenes, Table S1: Research sites and details, Table S2: Statistical analyses comparing insect densities between sites, treatments and years (ANOVA). Above diagonal are probability (p) values; significant values are in bold, below diagonal are F values.

Author Contributions: Conceptualization, C.G.F.; methodology, C.G.F.; software, A.B.; validation, C.G.F. and A.B.; formal analysis, C.G.F. and A.B.; investigation, C.G.F.; resources, C.G.F.; data curation, C.G.F. and A.B.; writing—original draft preparation, C.G.F. and A.B.; writing—review and editing, C.G.F. and A.B.; visualization, C.G.F. and A.B.; supervision, C.G.F. and A.B.; project administration, C.G.F.; funding acquisition, C.G.F. and A.B. All authors have read and agreed to the published version of the manuscript.

Funding: This research was funded by POSDRU/89/1.5/S/6237, co-financed by the European Social Fund, through Sectorial Operational Program for Human Resources Development 2007–2013.

Acknowledgments: We thank to the administration of Natural Park Apuseni for technical support and general permission to sample insects. This paper is published from the own funds of the Banat's University of Agricultural Sciences and Veterinary Medicine from Timisoara and Research Institute for Biosecurity and Bioengineering Timisoara.

Conflicts of Interest: The authors of this article have no financial or other conflict of interest to declare.

References

1. Grégoire, J.-C.; Evans, H.F. Damage and Control of Bawbilt Organisms an Overview. In *Bark and Wood Boring Insects in Living Trees in Europe, a Synthesis*; Lieutier, F., Day, K.R., Battisti, A., Grégoire, J.-C., Evans, H.F., Eds.; Springer: Dordrecht, The Netherlands, 2004; pp. 19–37. ISBN 978-1-4020-2241-8.
2. Christiansen, E.; Bakke, A. The Spruce Bark Beetle of Eurasia. In *Dynamics of Forest Insect Populations: Patterns, Causes, Implications*; Population Ecology; Berryman, A.A., Ed.; Springer: Boston, MA, USA, 1988; pp. 479–503. ISBN 978-1-4899-0789-9.
3. Hlásny, T.; Turčáni, M. Persisting Bark Beetle Outbreak Indicates the Unsustainability of Secondary Norway Spruce Forests: Case Study from Central Europe. *Ann. For. Sci.* **2013**, *70*, 481–491. [CrossRef]
4. Marini, L.; Lindelöw, Å.; Jönsson, A.M.; Wulff, S.; Schroeder, L.M. Population Dynamics of the Spruce Bark Beetle: A Long-Term Study. *Oikos* **2013**, *122*, 1768–1776. [CrossRef]
5. Wermelinger, B. Ecology and Management of the Spruce Bark Beetle Ips Typographus—A Review of Recent Research. *For. Ecol. Manag.* **2004**, *202*, 67–82. [CrossRef]
6. Avtzis, D.N.; Arthofer, W.; Stauffer, C.; Avtzis, N.; Wegensteiner, R. Pityogenes Chalcographus (Coleoptera, Scolytinae) at the Southernmost Borderline of Norway Spruce (Picea Abies) in Greece. *Entomol. Hell.* **2010**, *19*, 3–13. [CrossRef]
7. Pfeffer, A. Bark and Ambrosia Beetles from the Central and West Palaearctic Region (Coleoptera, Scolytidae, Platypodidae). Available online: https://www.cabi.org/isc/abstract/19951115912 (accessed on 9 March 2021).
8. Bertheau, C.; Salle, A.; Roux-Morabito, G.; Garcia, J.; Certain, G.; Lieutier, F. Preference–Performance Relationship and Influence of Plant Relatedness on Host Use by Pityogenes Chalcographus L. *Agric. For. Entomol.* **2009**, *11*, 389–396. [CrossRef]
9. Bouget, C.; Duelli, P. The Effects of Windthrow on Forest Insect Communities: A Literature Review. *Biol. Conserv.* **2004**, *118*, 281–299. [CrossRef]
10. Schwerdtfeger, F. A Contribution to the Biology of Reproduction of the Bark-Beetle, Ips Chalcographus. Available online: https://www.cabi.org/isc/abstract/19300500054 (accessed on 9 March 2021).
11. Christiansen, E.; Waring, R.H.; Berryman, A.A. Resistance of Conifers to Bark Beetle Attack: Searching for General Relationships. *For. Ecol. Manag.* **1987**, *22*, 89–106. [CrossRef]

12. Sauvard, D. General Biology of Bark Beetles. In *Bark and Wood Boring Insects in Living Trees in Europe, a Synthesis*; Lieutier, F., Day, K.R., Battisti, A., Grégoire, J.-C., Evans, H.F., Eds.; Springer: Dordrecht, The Netherlands, 2004; pp. 63–88. ISBN 978-1-4020-2241-8.
13. Zahradník, P.; Zahradníková, M. The Relationships between Pityogenes Chalcographus and Nemosoma Elongatum in Clear-Cuts with Different Types of Management. *Plant Prot. Sci.* **2019**, *56*, 30–34. [CrossRef]
14. Grodzki, W. Entomofauna of Dying Young Spruces Picea Abies (L.) Karst. in the Area after Forest Decline in the Izerskie Mountains. *Folia For. Pol. Ser. A* **2009**, *51*, 161–170. [CrossRef]
15. Weslien, J. Monitoring Ips Typographus (L.) Populations and Forecasting Damage1. *J. Appl. Entomol.* **1992**, *114*, 338–340. [CrossRef]
16. Lindelöw, Å.; Schroeder, M. Spruce Bark Beetle, Ips Typographus (L.), in Sweden: Monitoring and Risk Assessment. *J. For. Sci.* **2001**, *47*, 40–42.
17. Zúbrik, M.; Raši, R.; Vakula, J.; Varínsky, J.; Nikolov, C.; Novotný, J. Bark Beetle (Ips Typographus L., Pityogenes Chalcographus L., Col.: Scolytidae) Pheromone Traps Spatial Distribution Optimisation in Central Slovakian Mountains. *Lesn. Časopis For. J.* **2008**, *54*, 235–248.
18. Simionescu, A.; Mihalache, G.; Mihalciuc, V.; Ciornei, C.; Chira, D.; Olenici, N.; Lupu, D.; Nețoiu, C.; Vlăduleasa, A.; Iliescu, M.; et al. *Protecția Pădurilor*; Mușatinii: Suceava, Romania, 2003.
19. Mezei, P.; Blaženec, M.; Grodzki, W.; Škvarenina, J.; Jakuš, R. Influence of Different Forest Protection Strategies on Spruce Tree Mortality during a Bark Beetle Outbreak. *Ann. For. Sci.* **2017**, *74*, 65. [CrossRef]
20. Fora, C.G.; Lauer, K.F.; Moatar, M.M.; Stanciu, S. The Evaluation of the Bark Beetles Populations on Spruce Logging Debris. *Bull. Univ. Agric. Sci. Vet. Med. Cluj-Napoca. Hortic.* **2015**, *72*, 350–355. [CrossRef]
21. Kacprzyk, M. Feeding Habits of Pityogenes Chalcographus (L.) (Coleoptera: Scolytinae) on Norway Spruce (Picea Abies) L. (Karst.) Logging Residues in Wind-Damaged Stands in Southern Poland. *Int. J. Pest Manag.* **2012**, *58*, 121–130. [CrossRef]
22. Maes, D.; Titeux, N.; Hortal, J.; Anselin, A.; Decleer, K.; De Knijf, G.; Fichefet, V.; Luoto, M. Predicted Insect Diversity Declines under Climate Change in an Already Impoverished Region. *J. Insect Conserv.* **2010**, *14*, 485–498. [CrossRef]
23. Hedgren, P.O.; Weslien, J.; Schroeder, L.M. Risk of Attack by the Bark Beetle Pityogenes Chalcographus (L.) on Living Trees Close to Colonized Felled Spruce Trees. *Scand. J. For. Res.* **2003**, *18*, 39–44. [CrossRef]
24. Hedgren, P.O. The Bark Beetle Pityogenes Chalcographus (L.) (Scolytidae) in Living Trees: Reproductive Success, Tree Mortality and Interaction with Ips Typographus. *J. Appl. Entomol.* **2004**, *128*, 161–166. [CrossRef]
25. Lobinger, G. *Air Temperature as a Limiting Factor for Flight Activity of Two Species of Pine Bark Beetles, Ips Typographus L. and Pityogenes Chalcographus L. (Col., Scolytidae)*; Landesanstalt fuer Wald und Forstwirtschaft: Freising, Germany, 1994.
26. Boyer, J.S. Biochemical and Biophysical Aspects of Water Deficits and the Predisposition to Disease. *Annu. Rev. Phytopathol.* **1995**, *33*, 251–274. [CrossRef] [PubMed]
27. Kalkstein, L.S. Effects of Climatic Stress Upon Outbreaks of the Southern Pine Beetle 12. *Environ. Entomol.* **1976**, *5*, 653–658. [CrossRef]
28. Fora, C.G.; Banu, C.; Chisăliţă, I.; Moatăr, M.; Oltean, I. Parasitoids and Parasitoids and Predators of Ips Typographus (L.) in Unmanaged and Managed Spruce Forests in Natural Park Apuseni, Romania. *Not. Bot. Horti Agrobot.* **2014**, *42*. [CrossRef]

forests

MDPI

Communication

Quantification of One-Year Gypsy Moth Defoliation Extent in Wonju, Korea, Using Landsat Satellite Images

Won-IL Choi *, Eun-Sook Kim, Soon-Jin Yun, Jong-Hwan Lim and Ye-Eun Kim

Division of Forest Ecology, National Institute of Forest Science, 57 Hoegi-ro Dongdae mun-gu, Seoul 130-712, Korea; drummer12@korea.kr (E.-S.K.); yunsj86@korea.kr (S.-J.Y.); limjh@korea.kr (J.-H.L.); yaeeun0910@naver.com (Y.-E.K.)

* Correspondence: wchoi71@korea.kr; Tel.: +82-(2)-961-2604; Fax: +82-(2)-961-2629

Abstract: We quantified the extent and severity of Asian gypsy moth (*Lymantria dispar*) defoliation in Wonju, Korea, from May to early June in 2020. Landsat images were collected covering Wonju and the surrounding area in June from 2017 to 2020. Forest damage was evaluated based on differences between the Normalized Difference Moisture Index (NDMI) from images acquired in 8 June 2020 and the prior mean NDMI estimated from images in June from 2017 to 2019. The values of NDMI ranged from −1 to 1, where values closer to 1 meant higher canopy cover. The NDMI values for 7825 ha of forests were reduced by more than 0.05 compared to the mean NDMI values for the prior 3 years (2017 to 2019). The NDMI values of 1350 ha of forests were reduced by >0.125 to 0.2, and the NDMI values for another 656 ha were reduced by more than 0.2. A field survey showed that these forests were defoliated by gypsy moth and that forests with NDMI reductions of more than 0.2 were heavily defoliated by gypsy moth. A 311 ha area of Japanese larch (*Larix kaempferi*) was severely damaged by gypsy moth and the proportion of larch damaged was higher than that of other tree species. This intense damage to larch suggests that gypsy moths preferentially attack Japanese larch in Wonju. Our study shows that the use of NDMI values to detect areas defoliated by gypsy moth from satellite images is effective and can be used to measure other characteristics of gypsy moth defoliation events, such as host preferences under field conditions.

Keywords: gypsy moth; Landsat; normalized difference moisture index (NDMI); remote sensing; satellite image

1. Introduction

The quantification and mapping of forest areas damaged by outbreaks of forest pests offer insights into the causes of outbreaks and their impacts on forest stands, which is information that can be used for pest management [1]. Several methods exist for the quantification and mapping of forest insect damage, including field surveys, aerial sketching, and remote sensing [1,2]. Among these options, remote sensing using satellite images is considered to have merit due to the lower amount of labor required, the speed of data processing, and improved accuracy [1].

The quantification of area damaged by forest pests using remote sensing technologies has been extensively reviewed [3–8]. In Canada, forest areas damaged by eight major forest pests, including mountain pine beetle (*Dendroctonus ponderosae* Hopkins), forest tent caterpillar (*Malacosoma disstria* Hübner), and gypsy moth (*Lymantria dispar* (L.)), were quantified using satellite images from MODIS, Landsat, and Kompsat-3, etc. [3]. That study also showed that the Normalized Difference Moisture Index (NDMI) was useful in detecting damage by mountain pine beetle. The forest area damaged by gypsy moth outbreaks caused by the spring drought from 2015 to 2017 in southern New England, USA, was quantified and mapped using Landsat images and the index of forest canopy greenness [5]. In Pennsylvania and Maryland, USA, Landsat satellite images were used to relate the size of gypsy moth outbreaks to nitrogen concentrations in stream water in the Fifteenmile Creek watershed [9].

Gypsy moth is native to Europe and Asia [10]. In North America, gypsy moth was introduced from Europe in the 1860s, and is considered a major forest pest and a significant invasive species in North America [5]. In contrast, gypsy moth in Korea is the Asian subspecies and is only an occasional pest [11,12], although local outbreaks were observed in the 1990s and 2000s [13,14]. Until the 1990s, there was little information about the outbreak of gypsy moth in Korea, except for a small outbreak in 1200 deciduous trees in Seoul in 1959 [12]. Gypsy moth in Korea has a wide host range, including both deciduous and coniferous trees, but the host preference of the moth has not been recorded [14]. The fact that the egg masses of gypsy moth were collected in mainly *Quercus* spp. forests in the early 1990s suggest that *Quercus* spp. are potentially preferred hosts in Korea [12]. Similarly, gypsy moth in North America prefers *Quercus*, *Populus*, and *Salix* spp., and outbreaks of the moth usually occurred in these forests [10]. In Korea, gypsy moth outbreaks are more commonly observed in forests near human residences than in natural forests [14].

The aim of this study was to quantify the extent and severity of gypsy moth defoliation in and around Wonju, Korea, in 2020 using Landsat Satellite Images and NDMI, documenting the extensive outbreak of gypsy moth reported there by Jung et al. [15]. By analyzing defoliated areas, we determined the pest's field host preference and the physical characteristics of the areas defoliated by gypsy moth in Korea.

2. Materials and Methods

2.1. Study Site

The forest area examined was located in Wonju, Gangwon-do, Korea (37°21′5″ N, 127°56′43″ E). Wonju is an urban area with 350,000 residents and an area of 867.30 km^2. The climate at the survey area is continental, with an annual mean temperature of 12.3 °C and an annual precipitation of 1276 mm over the last decade (Korean Meteorological Administration, http://www.weather.go.kr (accessed on 26 April 2021)).

2.2. Satellite Image Acquisition and Analysis

To assess the area of forest damaged by gypsy moth, serial Landsat images (Path/Row 115/34) covering Wonju and the surrounding area from 2017 to 2020 were obtained from the USGS EarthExplorer homepage (https://earthexplorer.usgs.gov (accessed on 26 April 2021)). Basically, one scene of Landsat covered Wonji city and its vicinity. To estimate baseline information for each year, 20 images per year were collected. Therefore, 60 images were collected to estimate 3 prior years baseline information. The Landsat-8 images used in this study were orthorectified, radiometrically corrected, and included a cloud mask. Landsat images from 2017 to 2019 were used to produce baseline information before the event using a pixel-based time-series gap filling method [16]. Additionally, images from 2020 were used to evaluate the damage caused by the event compared to the past. Damaged forest areas were evaluated based on the difference between the Normalized Difference Moisture Index (NDMI) estimated from images acquired in 8 June 2020, and the mean NDMI estimated from images acquired in June from 2017 to 2019. NDMI is an effective index to evaluate the extent of forest damage from natural or anthropogenic disturbances [3]. The values of NDMI ranged from −1 to 1, with values closer to 1 meaning higher canopy cover [17]. The mean NDMI value from 2017 to 2019 (pre-defoliation) was calculated by the methods proposed by Kim et al. [16], such as NDMI = (NIR−SWIR1)/(NIR + SWIR1), where NIR is the pixel value from the near infrared band (0.85–0.88 μm) and SWIR1 is the pixel value from the short-wave infrared 1 band (1.57–1.65 μm) [18]. A mean NDMI map from 2017 to 2019 was produced by the time-series gap-filling method using clear sky pixels [16]. This map provided baseline information before the event.

Through the visual inspection of satellite images for Wonju in June 2020, forest areas with a red or brown color were selected, and those areas where the NDMI value had decreased by 0.2 compared to the values from 2017 to 2019 were noted. To minimize errors during the imaging process, areas where the NDMI was reduced by over 0.05 were considered to be defoliated by gypsy moth. The degree of forest defoliation was classified into three levels: (1) Severe (NDMI reduction of $\geq$0.2), (2) moderate (NDMI reduction of >0.125 to 0.2) or (3) light (NDMI reduction of 0.05 to 0.125). The forest types used for the classification of stands were based on a forest type map (http://www.forest.go.kr (accessed on 26 April 2021)). The Korean forest type map was produced with 5-year intervals using aerial photography and field observations at two spatial scales: 1:5000 and 1:25,000 [19]. In our analysis, a 1:5000 scale map was used. Forests with >75% coniferous trees or >75% deciduous trees were classified as coniferous or deciduous. Coniferous stands were further divided into Korean red pine (*Pinus densiflora* Siebold & Zucc.), Korean white pine (*P. koraiensis* Siebold & Zucc.), Japanese larch (*Larix kaempferi* (Lamb.) Carrière), and other forests. The deciduous stands were divided into oak (*Quercus* spp.) forests and other forests. Areas with >25% but <75% of either coniferous or deciduous trees were considered as mixed forests. To exclude reduction in NDMI due to the inclusion of non-forested area, forest areas were selected based on the forest type map of the Korean Forest Service (http://www.forest.go.kr (accessed on 26 April 2021)). Forests that had been heavily thinned or artificially damaged were also excluded from the analysis. For satellite image analysis and spatial information analysis, the ENVI + IDL software (Version 5.5, Harris Geospatial, Boulder, CO, USA) and ArcGIS (Version 10.4, ESRI, Redlands, CA, USA) were used.

2.3. Field Survey

To validate the results of the analysis of satellite images, a field survey was conducted at Wonju on 17 June 2020. At first, information on stands that were heavily defoliated by gypsy moth was collected in the center of Wonju, where the damage caused by gypsy moth was concentrated. Stands that are heavily defoliated look red and the density of gypsy moth larvae there was extremely high. These stands were considered as defoliated stands and were delimited based on forest address and forest type map according to the tree species defoliated. In addition to the heavily defoliated stands, damage and the presence of gypsy moth in nearby stands were also observed. The degree of damage by gypsy moth in the field was not classified at the stand level since the level of defoliation was dependent on tree species [15] and the stands were heavily fragmented (Figure 1).

2.4. Data Analysis

To measure the host preference of gypsy moth, an χ^2-test was conducted based on the area of each forest type and the area of forest damaged by gypsy moth. The degree of preference was estimated based on Manly's alpha feeding preference [20]. This was performed using the package selectapref [21] and R (http://cran.r-project.org (accessed on 26 April 2021)).

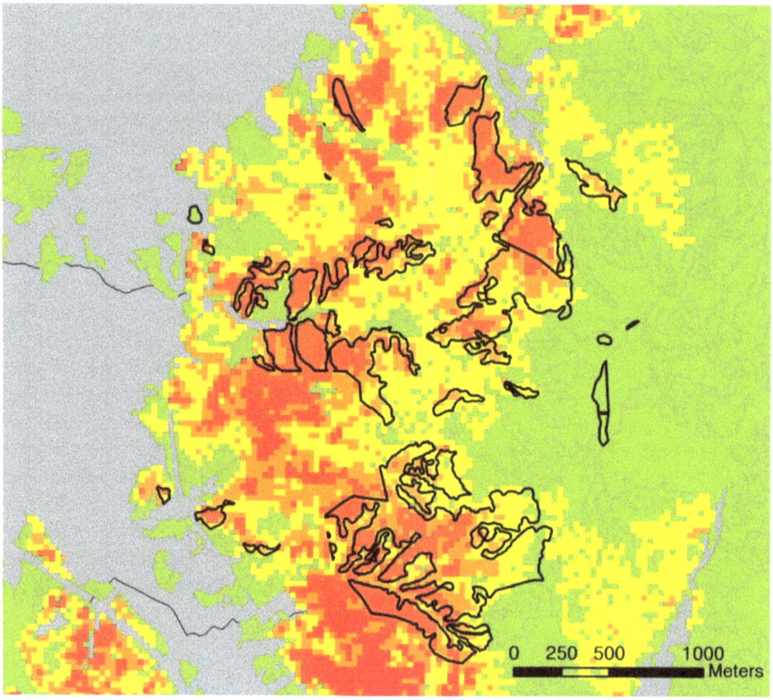

Figure 1. Comparison between field survey results and Landsat-based gypsy moth defoliation maps in the center of Wonju. Damage degree represents the mean reduction in Normalized Difference Moisture Index (NDMI) scores calculated using all the available observations in June from 2017 to 2019. The closed curves indicated the larch stands defoliated by gypsy moth observed through a field survey.

3. Results

For the total forested area (60,186 ha) in Wonju, the NDMI values of 7825 ha were reduced more than 0.05 (our criterion for at least light defoliation) when compared to the location mean NDMI values for the prior 3 years (2017 to 2019), suggesting that this area had been affected by gypsy moth defoliation (Figure 2, Table 1). The estimates of areas with moderate and heavy defoliation were 1350 and 656 ha, respectively. At least 13.0% of the forest area in Wonju was at least partially defoliated by gypsy moth.

The field survey results showed that eighteen stands were defoliated by gypsy moth and all of them were *L. kaempferi* stands (Figure 1). Among them, 93.0% of stands were judged as defoliated stands by our satellite image analysis, with severe defoliation at 37.4%, moderate defoliation at 30.6%, and light defoliation at 25.0%. In addition to the larch stands, extensive defoliation by gypsy moth was observed (Figure 2). These results suggested that the Landsat satellite image analysis using NDMI is a useful method to detect the extent and severity of defoliation caused by forest pests in Korea.

The host preferences of gypsy moth based on the satellite image analysis showed that the occurrence of gypsy moth concentrated on the specific forest type ($\chi^2 = 759.3$, df = 6, $p < 0.01$). The preference indexes of gypsy moth ranged from 36.1 to 100 according to the forest stand type and was the highest for the larch stand. The preference for *P. koraiensis* was lowest and the preference for *Quercus* spp. was 37.6.

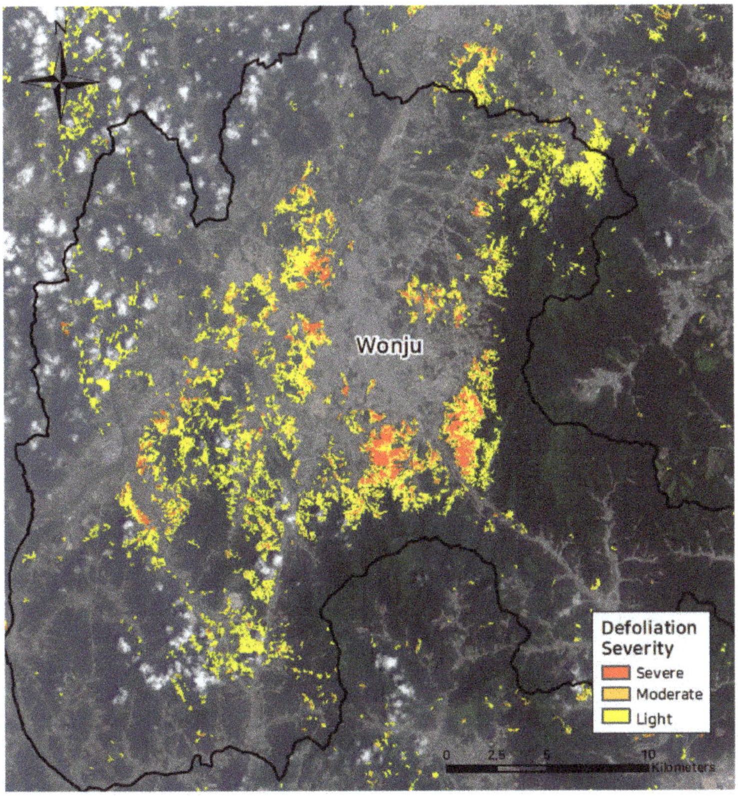

Figure 2. Landsat-based gypsy moth defoliation maps in Wonju and the surrounding area in 2020. Mapped results represent the mean reduction in Normalized Difference Moisture Index (NDMI) scores calculated using all the available observations in June from 2017 to 2019.

Table 1. Area defoliated by gypsy moth at Wonju in 2020 according to the defoliation severity and forest type. The areas were estimated based on the reduction in NDMI in 2020 compared to the prior 3 years.

Defoliation Severity	Forest Type							Total
	Pinus densiflora	*Pinus koraiensis*	*Larix kaempferi*	Other Coniferous	*Quercus* Spp.	Other Deciduous	Mixed	
Severe	27 (0.3%)	14 (0.5%)	274 (5.2%)	24 (2.2%)	82 (0.4%)	178 (1.1%)	57 (0.7%)	656 (1.1%)
Moderate	119 (1.5%)	44 (1.6%)	301 (5.7%)	47 (4.4%)	235 (1.2%)	404 (2.5%)	198 (2.6%)	1350 (2.2%)
Light	788 (10.1%)	184 (6.7%)	626 (11.8%)	108 (10.1%)	1457 (7.6%)	1573 (9.7%)	1084 (14.0%)	5820 (9.7%)
None	6843 (88.0%)	2491 (91.1%)	4105 (77.4%)	894 (83.3%)	17,511 (90.8%)	14,139 (86.8%)	6378 (82.6%)	52,361 (87.0%)
Total	7777	2734	5306	1073	19,285	16,294	7717	60,186

4. Discussion

Remote sensing, including satellite images, has been extensively used in Canada and the USA to quantify and map outbreaks of forest pests [3,5]. In Korea, remote sensing studies have been carried out mainly for oligophagous forest pests, such as pine wilt disease and Korean Oak wilt [22,23]. Our study shows that remote sensing images, with the use of an appropriate index, are also useful for detecting the damage of polyphagous forest pests. Moreover, unexpected outbreaks of new or occasional forest pests are increasing due to climate change and invasions of exotic pests. Under these circumstances, remote sensing using satellite images is useful for detecting or monitoring outbreaks of forest pests that cause widespread defoliation. For example, in Sweden, the invasive scale insect *Physokermes inopinatus* was effectively monitored using satellite images [4].

Outbreaks of gypsy moths have been extensively studied in North America since gypsy moth is one of the most serious forest pests in the region, where it has a long history of spreading and attempted control [10]. After the first occurrence of gypsy moth in the 1860s, it dispersed eastward and south. An entomophagous fungus, *Entomophaga maimaiga*, is considered one of key natural enemies currently regulating its population [15]. In southern New England, a wide outbreak of gypsy moth was observed from 2015 to 2017, following a 30-year period without significant outbreaks. The recent outbreak was due to reduced levels of larval mortality from the fungus due to a series of dry springs from 2014 and 2016 [3]. In addition, higher winter temperatures may have increased the survival of gypsy moth eggs [24].

Interestingly, the 5.2% of the stands of Japanese larch severely damaged by gypsy moth in Korea was greater than the amount damaged for other stand types (e.g., oak at 0.4%) (Table 1). Among forests defoliated by gypsy moth, the area of Japanese larch stands was the highest (22.7% of the total of the larch stand) (Table 1). The preference index was also highest for the larch forests. The results of the field survey also showed that Japanese larch was the most heavily attacked tree species. Although deciduous stands were not identified at the species level by forest type maps or satellite images, our results showed that oak forests are less heavily defoliated than other deciduous forests, suggesting that *Quercus* spp. were not the most preferred species among deciduous species (Table 1). Previous field surveys in Wonju showed that *Euonymus alatus*, *L. kaempferi*, *P. rigida*, *Populus davidiana*, *Q. serrata*, *Prunus sargentii*, and a shrub, *Rhododendron schlippenbachii*, were severely defoliated tree species, while oaks such as *Q. acutissima*, *Q. aliena*, *Q. mongolica*, and *Q. palustris* were only moderately defoliated [15]. These results contradicted with previous studies showing that *Quercus* spp. are some of the most preferred hosts for gypsy moth [10,12]. Further studies on host preferences are necessary to clarify the host preference of Korean gypsy moth.

Our analysis showed that defoliated areas were mainly concentrated near urban areas rather than more remote mountain areas. This was also observed in Chungju, Jecheon, and Hoengseong, in the vicinity of Wonju (Figure 1). This phenomenon was similar to earlier outbreaks of gypsy moth in Korea that were observed primarily in or near urban areas [14]. It has been hypothesized that the natural enemy regulation of gypsy moth may be weaker in or near urban areas compared to more remote forests in mountains. In the case of *Lymantria mathura*, a species similar to *L. dispar*, many natural enemies such as the parasitoid *Cotesia melanosceifus* (Ratzeburg) and several entomophagous pathogens stabilized the density of *L. mathura* in Korea [25]. The activities of these natural enemies probably decreased due to dryness or disturbance in or near urban forests. The second possible explanation for the spatial pattern of gypsy moth defoliation is that urban environments offer more opportunity for increase in gypsy moth populations. The lights from urban areas attract adult moths, which may then lay their eggs locally. Moreover, the winter temperatures near urban areas can be higher than those in forested areas, and this may increase the winter survival of gypsy moth eggs.

5. Conclusions

Our study showed that areas defoliated by gypsy moth may be effectively detected using satellite images and that such information may help us understand the characteristics of outbreaks of gypsy moths. Gypsy moth populations in Wonju severely defoliated

Japanese larch trees at higher rates than other tree species or groups. Additionally, gypsy moth damage was concentrated in or near urban areas. Our results showed that remote sensing methods are useful for detecting or monitoring damage caused by some forest pests, such as those that occasionally reach outbreak densities or that are newly invasive species. Moreover, the defoliation maps estimated by remote sensing techniques can be basic information for outbreak risk analysis using numerical and mathematical methods [26–30].

Author Contributions: Conceptualization, W.-I.C., E.-S.K., and J.-H.L.; methodology, E.-S.K.; validation, W.-I.C., E.-S.K., and J.-H.L.; formal analysis, W.-I.C. and E.-S.K.; data curation, E.-S.K., S.-J.Y., and Y.-E.K.; writing—original draft preparation, W.-I.C. and E.-S.K.; writing—review and editing, W.-I.C. and E.-S.K.; visualization, S.-J.Y. and Y.-E.K. All authors have read and agreed to the published version of the manuscript.

Funding: This research received no external funding.

Conflicts of Interest: The authors declare no conflict of interest. The funders had no role in the design of the study; in the collection, analysis, or interpretation of data; in the writing of the manuscript; or in the decision to publish the results.

References

1. Pasquarella, V.J.; Bradley, B.A.; Woodcock, C.E. Near-Real-Time Monitoring of Insect Defoliation Using Landsat Time Series. *Forests* **2017**, *8*, 275. [CrossRef]
2. Housman, I.W.; Chastain, R.A.; Finco, M.V. An evaluation of forest health insect and disease survey data and satellite-based remote sensing forest change detection methods: Case studies in the United States. *Remote Sens.* **2018**, *10*, 1184. [CrossRef]
3. Hall, R.J.; Castilla, G.; White, J.C.; Cooke, B.J.; Skakun, R.S. Remote sensing of forest pest damage: A review and lessons learned from a Canadian perspective. *Can. Entomol.* **2016**, *148*, S296–S356. [CrossRef]
4. Olsson, P.-O.; Jönsson, A.M.; Eklundh, L. A new invasive insect in Sweden—*Physokermes inopinatus*: Tracing forest damage with satellite based remote sensing. *For. Ecol. Manag.* **2012**, *285*, 29–37. [CrossRef]
5. Pasquarella, V.A.; Elkinton, J.S.; Bradley, B.A. Extensive gypsy moth defoliation in Southern New England characterized using Landsat satellite observations. *Biol. Invasions* **2018**, *20*, 3047–3053. [CrossRef]
6. Rahimzadeh-Bajgiran, P.; Weiskittel, A.R.; Kneeshaw, D.; MacLean, D.A. Detection of Annual spruce budworm defoliation and severity classification using Landsat imagery. *Forests* **2018**, *9*, 357. [CrossRef]
7. Spruce, J.P.; Sader, S.; Ryan, R.E.; Smoot, J.; Kuper, P.; Ross, K.; Prados, D.; Russell, J.; Gasser, G.; McKellip, R.; et al. Assessment of MODIS NDVI time series data products for detecting forest defoliation by gypsy moth outbreaks. *Remote Sens. Environ.* **2011**, *115*, 427–437. [CrossRef]
8. Zhang, J.; Huang, Y.; Pu, R.; Gonzalez-Moreno, P.; Yuan, L.; Wu, K.; Huang, W. Monitoring plant diseases and pests through remote sensing technology: A review. *Comput. Electron. Agric.* **2019**, *165*, 104943. [CrossRef]
9. Townsend, P.A.; Eshleman, K.N.; Welcker, C. Remote sensing of gypsy moth defoliation to assess variations in stream nitrogen concentrations. *Ecol. Appl.* **2004**, *14*, 504–516. [CrossRef]
10. Elkinton, J.S.; Liebhold, A.M. Population dynamics of gypsy moth in North America. *Ann. Rev. Entomol.* **1990**, *35*, 571–596. [CrossRef]
11. Lyu, D.; Lee, D.W. Occurrence of Gypsy Moth, *Lymantria dispar* around pier in Korea and comparison of insecticide response depending on instar. *Korean J. Pestic. Sci.* **2019**, *23*, 1–7. [CrossRef]
12. Pemberton, R.W.; Lee, J.H.; Reed, D.K.; Carlson, R.W.; Han, H.Y. Natural Enemies of the Asian Gypsy Moth (Lepidoptera: Lymantridae) in South Korea. *Ann. Entomol. Soc. Am.* **1993**, *86*, 423–440. [CrossRef]
13. Lee, H.-P.; Lee, H.-M. Seasonal occurrence and monitoring of gypsy moth, *Lymantria dispar* L. (Lepidoptera: Lymantriidae) by sex pheromone trap in Mt. Namsan, Seoul. *Korean J. Ecol.* **1999**, *22*, 299–303.
14. Shin, S.C.; Choi, K.S.; Choi, W.I.; Chung, Y.J.; Lee, S.G.; Kim, C.S. *An New Illustrated Book of Forest Insect Pests*; Upgo MunHwa: Seoul, Korea, 2008.
15. Jung, J.-K.; Nam, Y.W.; Kim, D.S.; Lee, S.H.; Lim, J.H.; Choi, W.I.; Kim, E.S. Tree-crown defoliation caused by outbreak of forest insect pests in Korea during 2020. *Korean J. Appl. Entomol.* **2020**, *59*, 409–410.
16. Kim, E.-S.; Lee, B.; Lim, J.-H. Forest damage detection using daily normal vegetation index based on time series Landsat Images. *Korean J. Remote Sens.* **2019**, *35*, 1133–1148.
17. Lastovicka, J.; Svec, P.; Paluba, D.; Kobliuk, N.; Svoboda, J.; Hladky, R.; Stych, P. Sentinel-2 Data in an Evaluation of the Impact of the Disturbances on Forest Vegetation. *Remote Sens.* **2020**, *12*, 1914. [CrossRef]
18. USGS. Landsat Surface Reflectance-Derived Spectral Indices-Normalized Difference Moisture Index. 2020. Available online: https://www.usgs.gov/core-science-systems/nli/landsat/normalized-difference-moisture-index (accessed on 26 April 2021).

19. Kang, Y.; Jang, E.; Im, J.; Kwon, C.; Kim, S. Developing a New Hourly Forest Fire Risk Index Based on Catboost in South Korea. *Appl. Sci.* **2020**, *10*, 8213. [CrossRef]
20. Chesson, J. Measuring preference in selective predation. *Ecology* **1978**, *59*, 211–215. [CrossRef]
21. Richardson, J. Analysis of Field and Laboratory Foraging 'Selectapref' R Package Version. 2017. Available online: https://CRAN.R-pro-ject.org/package=selectapref (accessed on 26 April 2021).
22. Cha, S.; Lee, W.-K.; Kim, M.; Lee, S.-G.; Jo, H.-W.; Choi, W.I. Characterization the spatial distribution of oak wilt disease using remote sensing data. *J. Korean For. Soc.* **2017**, *106*, 310–319.
23. Kim, S.-R.; Kim, E.-S.; Nam, Y.W.; Choi, W.I.; Kim, C.-M. Distribution characteristics analysis of pine wilt disease using time series hyperspectral aerial imagery. *Korean J. Remote Sens.* **2015**, *31*, 385–394. [CrossRef]
24. Smitley, D.; Andresen, J.; Priest, R.; Mech, R.; Mccullough, D. Winter mortality of gypsy moth (Lepidoptera: Lymantriidae) eggs in Michigan. *Environ. Entomol.* **1998**, *3*, 700–708. [CrossRef]
25. Lee, J.-H.; Lee, H.-P. Parasites and phenology of *Lymantria mathura* Moore (Lymantriidae: Lepidoptera) in Kyonggi Province, Korea. *Korean J. Entomol.* **1996**, *26*, 393–401.
26. Nam, Y.; Koh, S.-H.; Jeon, S.-J.; Youn, H.-J.; Park, Y.-S.; Choi, W.I. Hazard rating of coastal pine forests for a black pine bast scale using self-organizing map (SOM) and random forest approaches. *Ecol. Inform.* **2015**, *29*, 206–213. [CrossRef]
27. Dargahi, M.; Newson, T.; Moore, J.R. A Numerical Approach to Estimate Natural Frequency of Trees with Variable Properties. *Forests* **2020**, *11*, 915. [CrossRef]
28. Kayet, N.; Chakrabarty, A.; Pathak, K.; Sahoo, S.; Dutta, T.; Hatai, B.K. Comparative analysis of multi-criteria probabilistic FR and AHP models for forest fire risk (FFR) mapping in Melghat Tiger Reserve (MTR) forest. *J. For. Res.* **2020**, *31*, 565–579. [CrossRef]
29. He, S.; Lin, L.; Wu, Z.; Chen, Z. Application of Finite Element Analysis in Properties Test of Finger-jointed Lumber. *J. Bioresour. Bioprod.* **2020**, *5*, 124–133. [CrossRef]
30. Ibitoye, S.E.; Adegun, I.K.; Omoniyi, P.O.; Ogedengbe, T.S.; Alabi, O.O. Numerical investigation of thermo-physical properties of non-newtonian fliud in a modelled intestine. *J. Bioresour. Bioprod.* **2020**, *5*, 211–221. [CrossRef]

forests

MDPI

Article

Assessment of Machine Learning Algorithms for Modeling the Spatial Distribution of Bark Beetle Infestation

Milan Koreň [1,*], Rastislav Jakuš [2,3], Martin Zápotocký [1], Ivan Barka [4], Jaroslav Holuša [3], Renata Ďuračiová [5] and Miroslav Blaženec [2]

1 Faculty of Forestry, Technical University in Zvolen, T. G. Masaryka 24, 960 01 Zvolen, Slovakia; martin.zapotocky@tuzvo.sk
2 Institute of Forest Ecology, Slovak Academy of Sciences, Ľ. Štúra 2, 960 53 Zvolen, Slovakia; rasti.jakus@gmail.com or jakus@fld.czu.cz (R.J.); blazenec@ife.sk (M.B.)
3 Faculty of Forestry and Wood Sciences, Czech University of Life Sciences Prague, Kamýcká 129, 165 00 Prague 6, Czech Republic; holusa@fld.czu.cz
4 National Forest Centre—Forest Research Institute, T. G. Masaryka 22, 960 01 Zvolen, Slovakia; ivan.barka@nlcsk.org
5 Faculty of Civil Engineering, Slovak University of Technology in Bratislava, Radlinského 11, 810 05 Bratislava, Slovakia; renata.duraciova@stuba.sk
* Correspondence: milan.koren@tuzvo.sk; Tel.: +421-45-520-6523

Abstract: Machine learning algorithms (MLAs) are used to solve complex non-linear and high-dimensional problems. The objective of this study was to identify the MLA that generates an accurate spatial distribution model of bark beetle (*Ips typographus* L.) infestation spots. We first evaluated the performance of 2 linear (logistic regression, linear discriminant analysis), 4 non-linear (quadratic discriminant analysis, k-nearest neighbors classifier, Gaussian naive Bayes, support vector classification), and 4 decision trees-based MLAs (decision tree classifier, random forest classifier, extra trees classifier, gradient boosting classifier) for the study area (the Horní Planá region, Czech Republic) for the period 2003–2012. Each MLA was trained and tested on all subsets of the 8 explanatory variables (distance to forest damage spots from previous year, distance to spruce forest edge, potential global solar radiation, normalized difference vegetation index, spruce forest age, percentage of spruce, volume of spruce wood per hectare, stocking). The mean phi coefficient of the model generated by extra trees classifier (ETC) MLA with five explanatory variables for the period was significantly greater than that of most forest damage models generated by the other MLAs. The mean true positive rate of the best ETC-based model was 80.4%, and the mean true negative rate was 80.0%. The spatio-temporal simulations of bark beetle-infested forests based on MLAs and GIS tools will facilitate the development and testing of novel forest management strategies for preventing forest damage in general and bark beetle outbreaks in particular.

Keywords: machine learning; classifier; spatial distribution model; geographical information system; forest damage; bark beetle

1. Introduction

Wind damage to forests and subsequent spruce bark beetle (*Ips typographus* L.) outbreaks have increased in Central Europe in recent decades [1]. Bark beetle outbreaks are closely related to wind-caused forest damage, and the two factors interact to create wind–bark beetle disturbance systems [2]. There are also connections between bark beetle outbreaks and other forest disturbances related to climate change [3,4]. Rising mean annual temperatures, increases in drought duration and intensity, and shifts in growing seasons increase the risk of forest infestation by bark beetles [5]. To respond to this problem, forest managers require adequate forest management strategies based on knowledge of the forest damage spatial distribution and factors influencing the risk of wind damage and bark beetle dispersal [6]. The understanding of bark beetle dispersal processes and spatial patterns is important for the effective management of infested forests [7]. Forestry decision support systems profit from the integration of high-resolution remote sensing data, forest mapping and field inventories, advances in silviculture and forest ecology, and the use of modern statistical and machine learning approaches [8–10].

Spaceborne and aerial images are used to map the spatial and temporal distributions of forest damage caused by wind or bark beetles [11]. Damaged and infested forests are identified, and bark beetle spots are localized by classification of medium- or high-resolution images. Spatial and temporal changes in forest damage over large territories can be detected in time series of satellite images. Intra-annual satellite image series improve the detection of forest disturbance [12]. The derived multi-temporal digital maps of human and natural forest disturbances can be used to study the dynamics of forest damage.

The spatial dispersal of bark beetle infestations is driven by various environmental factors, including solar radiation, temperature, wind speed and direction, precipitation, soil moisture, distance to forest damage areas, spruce age, diameter at breast height, and stocking [13]. Bark beetle outbreaks and the spatial distribution of forest damage are influenced by the spatial structure of bark beetle populations, forest and landscape characteristics, and by factors at a wider, regional-scale [14]. Because the ecological and spatial relationships are complex, the development of a reliable spatial forest damage model (FDM) is a difficult and computationally intensive task.

Machine learning algorithms (MLAs) have been applied to solve non-linear and high-dimensional problems in forestry and ecology. A support vector machine and a random forest algorithm have recently been used to estimate the volume and basal area of eucalyptus stands from satellite images [15]. Proportions of planned end products were forecasted by Dirichlet regression and neural networks [16]. The stand volume of a rapidly growing forest plantation was estimated by random forest, support vector machine, and neural network regressions from aerial laser scanning data [17].

As noted earlier, forest damage by wind, insects, fire, and other factors is a complex spatial phenomenon that is difficult to model and predict. Advanced spatial modeling techniques are needed to develop accurate and reliable FDMs. Many MLAs are designed to handle large volumes of multi-dimensional data, including geographical data. MLAs learn from existing data and can adapt to hidden spatial patterns and unknown relationships among environmental variables. Rodrigues and de la Riva [18] modeled the risk of human-caused wildfire by random forest, boosted regression tree, and support vector machine algorithms. Mayfield et al. [19] calculated the risk of deforestation with generalized linear mixed models, Bayesian networks, neural networks, and Gaussian processes. Evolutionary and non-evolutionary MLAs were tested for predicting forest burned areas [20]. The potential of modeling forest insect dynamics by cellular automata was demonstrated by [21]. Neural network-based regression was used by Hlásny and Turčáni [22] to analyze the influence of site and stand characteristics on forest damage caused by the spruce bark beetle.

Here, we summarize the results of numerical experiments with MLAs for modeling the spatial distribution of forest damage. Our main objective was to select the MLA with the highest predictive accuracy for modeling the spatial distribution of forest damage in an open-source geographical information system (GIS).

2. Materials and Methods

2.1. Study Area

MLAs were tested for the Horní Planá region in Central Europe (Figure 1). The forests in this region are managed by the Military Forests and Farms of the Czech Republic, State Enterprise [23]. The Military Forests and Farms of the Czech Republic, State Enterprise, is a special-purpose organization that manages 19,960 ha of land in military training areas.

Forests represent 16,569 ha of this area, and water reservoirs cover 203 ha; the remaining area is represented by grassland that is used for intensive military training.

The region is characterized by hills that differ in elevation by 100–150 m. Most of the study area is located at 600 to 800 m a.s.l., and the highest point (Lysá Mt) is at 1228 m a.s.l. Part of the area belongs to the Bohemian Forest (the Šumava Mts.), and part belongs to the Šumavské podhůří Mts. The annual mean temperature ranges from 5 to 7 °C, and annual precipitation ranges from 700 to 800 mm [24].

The dominant tree species in forests is Norway spruce (*Picea abies* L. Karst) (69%). Less common are Scotch pine (*Pinus sylvestris* L.) (12%), Silver fir (*Abies alba* Mill.) (6%), and European beech (*Fagus sylvatica* L.) (5%). Among the forest stands, 23.5% are younger than 40 years, and 32.0% are older than 100 years. At higher altitudes, forests are parts of complexes that include meadows and pastures.

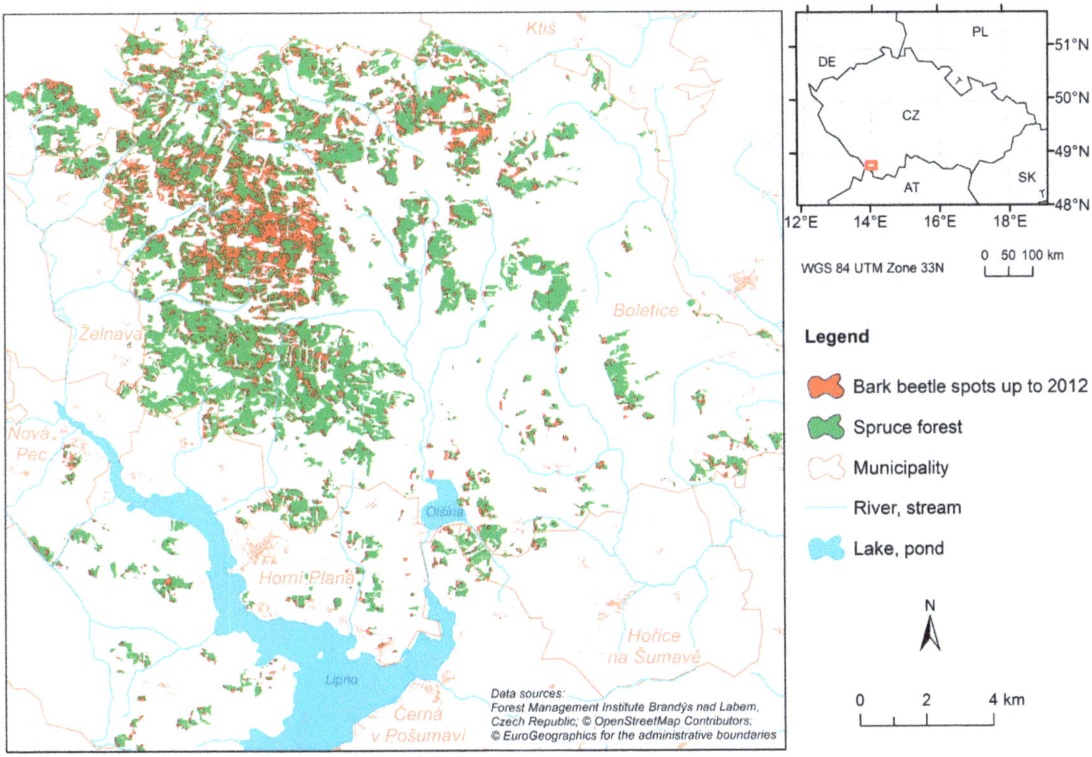

Figure 1. Location of the study area in the Czech Republic.

Spruce forests in the Horní Planá region are disturbed mainly by snow, bark beetles, and especially wind. In recent years, incidental felling caused by forest damage has represented about 50% of total felling. Bark beetle outbreaks in the region follow the trends exhibited over the whole of the Czech Republic, with peaks in the mid-1980s and mid-1990s. The last long-term outbreak of *I. typographus* began in 2003 as a result of a severe drought that occurred throughout Central Europe. This outbreak was partly extended by the winter storm "Kyrill" (January 2007), which destroyed more wood than any other factor over the last 30 years. At the beginning of 2014, the volume of wood infested by bark beetles decreased to under 0.5 $m^3 \cdot ha^{-1}$ [25]. The cyclic nature of forest damage in the study region is depicted in Figure 2. The bark beetle is a prevailing cause of forest damage in the study region.

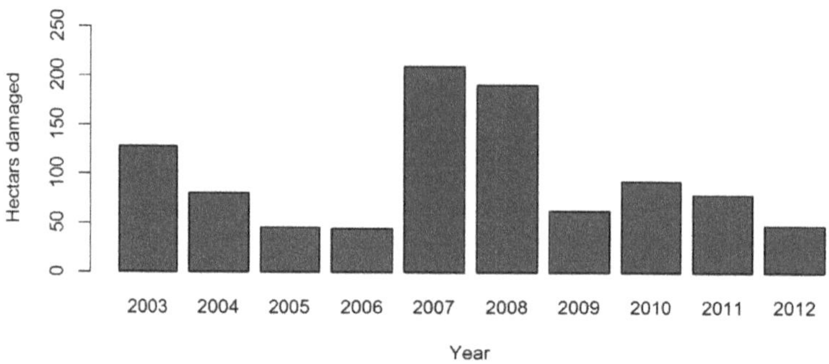

Figure 2. Size of forest damaged areas in the study region (calculated from classified LANDSAT images).

2.2. Input Data

In our study, spatial distribution models of spruce forest damage were developed for the period 2003–2012. The period was limited by years for which raster layers representing explanatory variables were available (Table 1). The spatial resolution of the raster layers was 30 m, which corresponds to the spatial resolution of LANDSAT images.

Table 1. List of the explanatory variables used in forest damage models (FDMs).

Code	Explanatory Variable
DAS	distance to forest damage spots from previous year
DFE	distance to spruce forest edge
PSR	potential global solar radiation
NDVI	normalized difference vegetation index
AGE	spruce forest age
PCT	percentage of spruce
VOL	volume of spruce wood per hectare
STO	stocking

A time series of LANDSAT images were used to identify damaged forest locations in the study area. Raster maps of forest health status were used to delineate the damaged forest locations. These forest health maps are based on LANDSAT images and have been prepared by standardized methods since 1984 by the Forest Management Institute, Brandýs nad Labem, Czech Republic [26]. Forest damage locations were not classified by forest damage factors. Only classes of strong and very strong damage of spruce forest stands [27] were considered for this study. A similar approach was used by [28,29].

Normalized difference vegetation index (NDVI) values were derived from LANDSAT scenes (Table 2). Cloud-free scenes from the growing season were preferred. Processing included manual removal of clouds and shadows and the mosaicking of two scenes based on linear regression using corresponding pixels that represented forest.

Information on spruce age (AGE), percentage of spruce in forest stands (PCT), volume of spruce per hectare (VOL), and stocking (STO) was imported from forest management plans. There are four management units within the study area, each with its own management plan. The forest management plans are prepared for 10-y periods. The age of a forest stand was specified in 5-year increments, and the percentage of spruce was based on basal area. Volume was derived from the stand mean diameter at breast height and the stand mean tree height for each species. Stocking was calculated as the relative density using yield tables. The raster layer representing spruce forest stands included only spruce stands with age >49 years and stocking >49%.

Distance to forest damage areas was calculated from the layer of actual forest damage. The actual forest damage layer was subtracted from the spruce forest stand layer to derive a layer of actual spruce forest. The raster layer of the actual spruce forest was used to calculate the distance to the spruce forest edge layer. Potential global solar radiation (PSR) was computed from a digital elevation model (DEM) by the GRASS GIS module r.sun [30] with a 1-h step.

Table 2. LANDSAT scenes used for NDVI calculation.

Date	Scene
16.07.2003	LT05_L1TP_191026_20030716_20161205_01_T1
10.08.2004	LT05_L1TP_192026_20040810_20161130_01_T1
28.07.2005	LT05_L1TP_192026_20050728_20161125_01_T1
29.08.2005	LT05_L1TP_192026_20050829_20161125_01_T1
15.07.2006	LT05_L1TP_192026_20060715_20161120_01_T1
24.07.2006	LT05_L1TP_191026_20060724_20161120_01_T1
25.06.2007	LT05_L1TP_191026_20070625_20161112_01_T1
19.07.2007	LE07_L1TP_191026_20070719_20170102_01_T1
29.06.2008	LT05_L1TP_191026_20070625_20161112_01_T1
21.08.2008	LT05_L1TP_192026_20080821_20180116_01_T1
24.08.2009	LT05_L1TP_192026_20090824_20161021_01_T1
10.07.2010	LT05_L1TP_192026_20100710_20161014_01_T1
23.08.2011	LT05_L1TP_191026_20110823_20161007_01_T1
23.07.2012	LE07_L1TP_192026_20120723_20161130_01_T1
01.08.2012	LE07_L1TP_191026_20120801_20161130_01_T1

EU-DEM 25 [31] was used as an input to compute PSR. Its original spatial resolution of 25 m is close to the resolution of LANDSAT scenes. The DEM was projected to the national spatial reference system (EPSG 5514, epsg.io/5514/) and resampled at a resolution of 30 m.

Samples representing damaged and undamaged forest were generated for each year of the period. All grid cells representing damaged forest were used as samples. An equal number of cells representing undamaged forest was randomly generated. Among the samples, 75% were used for model training, and 25% were used as controls. All FDMs were trained and validated using the same sample sets.

2.3. Computer Simulations and Data Processing

The FDM consists of a forest damage probability function and a classification function. The FDM F can be expressed in the following form:

$$F(x,y,t) = C\left(P\left(\vec{u}(x,y,t)\right)\right) \tag{1}$$

where C is a classification function; P is a forest damage probability function; $\vec{u}$ is an environment vector function; x, y are point (grid cell) coordinates; and t is a time.

The probability function P calculates the risk of forest damage at a given location (x, y) and time t. Environmental factors, e.g., distance to existing forest damage areas, drought stress, forest stand openness, and solar radiation, are described by independent variables of the forest damage probability function P. Each component $u_i(x, y, t)$ of the environment vector function $\vec{u}$ corresponds to an independent variable that varies over space and time. In GIS, the independent variable is represented by a time series of raster layers.

The open source-software GRASS GIS [32] was used for computer simulations. All suitable MLAs from the GRASS GIS add-on *r.learn.ml* were tested for modeling the spatial distribution of forest damage (Table 3). *r.learn.ml* is a front-end to the *scikit-learn* toolkit [33] for the Python programming language. A set of scripts in Python programming language was developed to automate the processing of forest damage layers, the processing of training and control samples, model training, and the analysis of FDM performance.

In the study, the spatial distribution of forest damage was modeled by linear MLAs (LR and LDA), non-linear MLAs (QDA, KNC, GNB, and SVC), and classification trees (DTC, RFC, ETC, and GBC).

Table 3. List of the tested machine learning algorithms (MLAs).

Code	Algorithm
LR	logistic regression
LDA	linear discriminant analysis
QDA	quadratic discriminant analysis
KNC	k-nearest neighbors classifier
GNB	Gaussian naive Bayes
DTC	decision tree classifier
RFC	random forest classifier
ETC	extra trees classifier
GBC	gradient boosting classifier
SVC	support vector classification

Each MLA was trained and tested on all subsets of the explanatory variables (Table 1). All combinations of explanatory variables were used as inputs of FDMs because the most suitable combinations of explanatory variables for different MLAs were unknown. The total number of FDMs tested was a product of the number of MLAs (10) × the number of combinations of exploratory variables (255). Each of the 2550 models was calculated for each year of the period 2003–2012. The probability of forest damage was calculated by each MLA. The internal classifier of each MLA was applied to identify locations (grid cells) of forest damage.

In a computer environment, a spatial forest damage model is defined by MLA m and a non-empty subset S of the explanatory variables. The forest damage spatial distribution model $fdm(m, S, y)$ was calculated for every year y of the period. The confusion matrix, true positive rate $TPR(m, S, y)$ true negative rate $TNR(m, S, y)$, and phi coefficient $\phi(m, S, y)$ were calculated for each $fdm(m, S, y)$.

The true positive rate (sensitivity) describes how many locations (grid cells) of forest damage estimated by the FDM correspond to locations of actual forest damage. Similarly, the true negative rate (specificity) is a measure of the correspondence between estimated and actual undamaged forest locations. A reliable spatial FDM is characterized by high sensitivity and high specificity.

The overall performance of the spatial FDM was measured by the arithmetic mean of the phi coefficient for the period under study Y:

$$\overline{\Phi}(m,\ S) = \frac{1}{l} \sum_{y \in Y} \Phi(m,\ S,\ y), \tag{2}$$

where l is the length of the period Y, m is an used MLA, S is a non-empty subset of the explanatory variables, and y is a year of the period. The mean true positive rate $\overline{TPR}(m,S)$ and the mean true negative rate $\overline{TNR}(m,S)$ of the spatial FDM were calculated similarly.

Data were statistically processed with the R package [34]. All statistical hypotheses were tested at a 0.05 significance level. Modules from the packages *lmPerm* and *RVAide-Memoire* were used for permutation tests of statistical hypotheses. Permutation one-way repeated measures ANOVA was used to compare the mean phi coefficients of the FDMs. Significances of mean phi coefficient differences were tested by a permutation pairwise *t*-test with Benjamini, Hochberg, and Yekutieli corrections [35].

The generated FDMs were sorted in descending order by $\overline{\phi}$ (Equation (2)). A unique ranking number r was assigned to each FDM. The ranking number 1 corresponded to the FDM with the highest mean phi coefficient for the period Y. Results were evaluated in terms of FDM performance and simplicity.

3. Results

As stated earlier, the current study evaluated the use of MLAs for modeling the spatial distribution of forest damage. A simple spatial dispersion model was used (Equation (1)). The locations of damaged forest areas were modeled with the machine learning classifiers.

The top FDM (i.e., the FDM with the highest mean phi coefficient for the period 2003–2012) was selected for each studied MLA (Table 4). As suspected, the optimal combination of input explanatory variables differed among the top FDMs.

Table 4. Top FDMs generated by machine learning algorithms (MLAs).

r [1]	nv [2]	MLA [3]	Variable [4]								$\overline{TPR}$ [5]	$\overline{TNR}$ [6]	$\overline{\phi}$ [7]
			DAS	DFE	PSR	NDVI	AGE	PCT	VOL	STO			
1	5	ETC	+		+		+	+	+		0.804	0.800	0.605
5	7	RFC	+	+	+		+	+	+	+	0.814	0.778	0.594
89	3	SVC					+	+	+		0.724	0.818	0.546
219	3	GBC					+	+	+		0.761	0.733	0.495
273	2	DTC					+		+		0.750	0.728	0.479
286	8	LDA	+	+	+	+	+	+	+	+	0.764	0.707	0.474
295	5	KNC	+	+			+	+	+		0.790	0.678	0.472
388	4	LR	+	+		+	+				0.769	0.679	0.452
421	7	QDA	+	+	+	+	+	+		+	0.815	0.618	0.444
436	5	GNB	+	+		+	+			+	0.827	0.602	0.441

[1] ranking number, [2] number of explanatory variables, [3] machine learning algorithm used to generate FDM, the MLA codes are given in Table 3, [4] codes of the variables are listed in Table 1, [5] mean true positive rate, [6] mean true negative rate, [7] mean phi coefficient, "+" indicates the presence of an explanatory variable in FDM.

The arithmetic mean and median of phi coefficients were higher for the top ETC-based model with five explanatory variables than for the other top FDMs for the period 2003–2012 (Table 4, Figures 3 and 4). The influence of MLA on the mean phi coefficient ($\overline{\phi}$) of the top FDMs was significant (one-way repeated measures ANOVA, p-value < 2.2×10−16). Results of the permutation pairwise t-test of the top FDMs' ϕ are presented in Table 5.

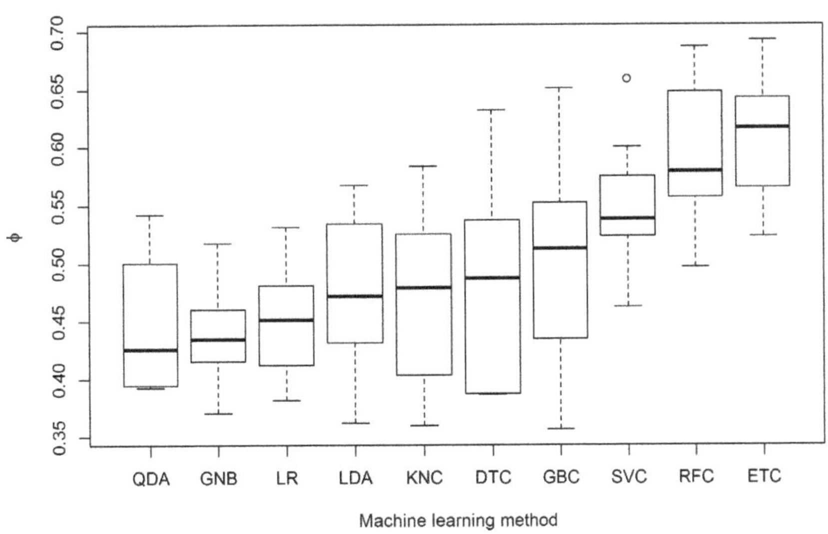

Figure 3. Boxplot of phi coefficients (ϕ) of the top FDMs generated by MLAs. Box plots show median plus upper and lower quartiles of the phi coefficient for the period 2003–2012. Minimum and maximum values are indicated by the upper and lower whiskers (1.5 × interquartile range). Circles are outliers. The MLA codes are given in Table 3.

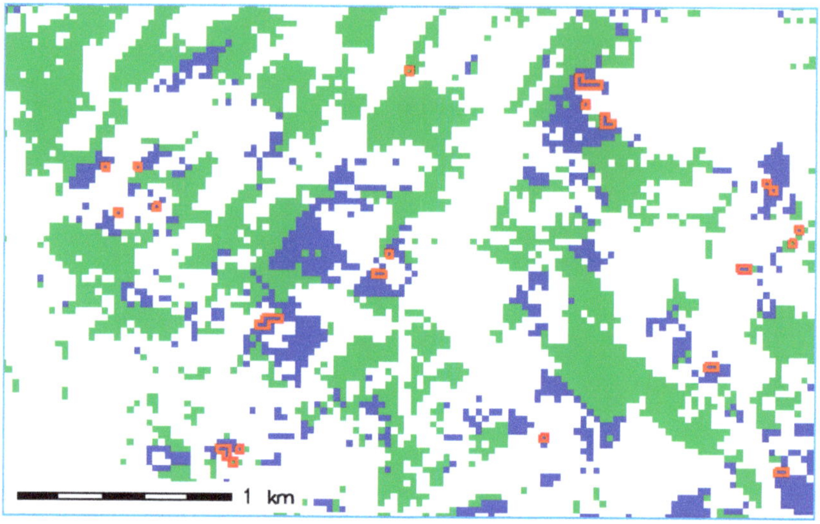

Figure 4. Preview of the bark beetle spots generated by the top ETC-based model. Green color represents unaffected spruce forest; blue color represents bark beetle infestation spots generated by the top ETC-based model; red color represents damaged forest identified on LANDSAT images. The MLA codes are given in Table 3, and the variables of the top models are listed in Table 4.

Table 5. *p*-values of the pairwise permutation *t*-test of ϕ of the top FDMs.

MLA [1]	GNB	QDA	LR	KNC	LDA	DTC	GBC	SVC	RFC
QDA	0.827	-	-	-	-	-	-	-	-
LR	0.338	0.687	-	-	-	-	-	-	-
KNC	0.338	0.406	0.348	-	-	-	-	-	-
LDA	0.020	0.154	0.054	0.906	-	-	-	-	-
DTC	0.124	0.202	0.177	0.827	0.851	-	-	-	-
GBC	0.073	0.124	0.124	0.576	0.338	0.126	-	-	-
SVC	0.013	0.011	0.013	0.034	0.010	0.010	0.010	-	-
RFC	0.011	0.010	0.011	0.011	0.010	0.013	0.011	0.020	-
ETC	0.011	0.013	0.010	0.011	0.010	0.013	0.010	0.010	0.312

[1] the MLA codes are given in Table 3.

Overall performance as indicated by the mean phi coefficient was slightly lower for the best RFC-based model than for the best ETC-based model. However, performances for these best models were not significantly different. Performance for the other top FDM models was significantly different from performance for the best ETC- and RFC-based models (Table 5). Moreover, as evident in Figure 5, most of the ETC-based FDMs out-performed the FDMs generated by the other MLAs.

The performance of the FDMs generated by SVC was highly variable (Figure 5). SVC generated many low-performance FDMs as well as a few models that performed better than the models generated by the other tested MLAs except for those generated by ETC and RFC (Table 4). The performance of the best SVC-based model was significantly different from the performance of the other best FDMs (Table 5).

The mean TPRs were highest for the SVC-, QDA-, and GNB-based models (Figure 6). These models were highly sensitive because they significantly overestimated the area of damaged forest. On the other hand, SVC also generated several FDMs with low specificities (Figure 7).

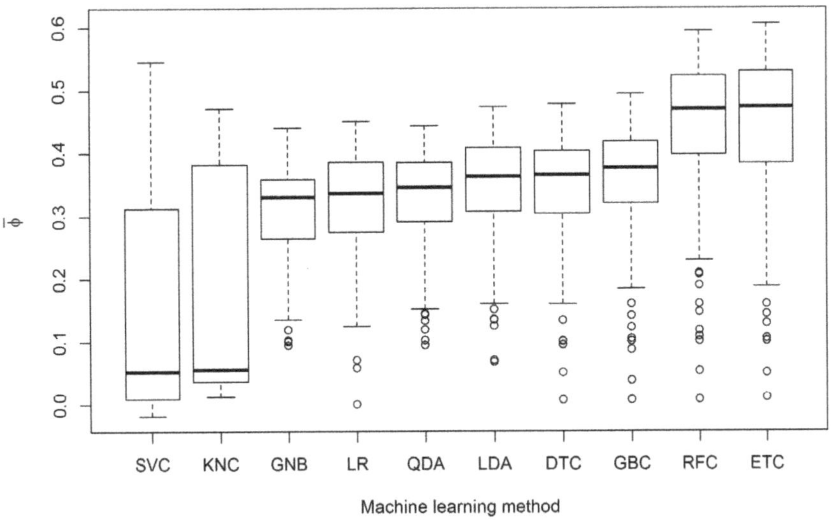

Figure 5. Box plots of mean phi coefficients ($\bar{\phi}$) of FDMs generated by MLAs. Box plots show median plus upper and lower quartiles for mean phi coefficients of all FDMs generated by a given algorithm. Minimum and maximum values are indicated by the upper and lower whiskers (1.5 × interquartile range). Circles are outliers. See Table 3 for MLA codes.

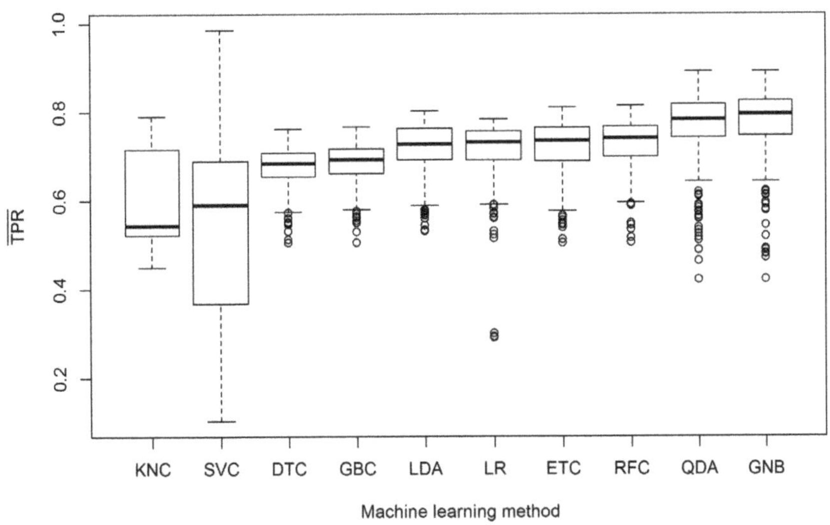

Figure 6. Box plots of mean sensitivities ($\overline{TPR}$) of FDMs generated by MLAs. Box plots show median plus upper and lower quartiles for mean sensitivities of all FDMs generated by a given algorithm. Minimum and maximum values are indicated by the upper and lower whiskers (1.5 × interquartile range). Circles are outliers. See Table 3 for MLA codes.

We then compared the performance of the FDMs when the number of explanatory variables was constant in a range from one to eight. Regardless of the number of explanatory variables, performance was always best for the models generated by the ETC, RDC, and SVC MLAs (Table 6). Performance of FDMs generated by most of the MLAs increased with the number of variables, except in the case of SVC- and KNC-based models (Appendix A).

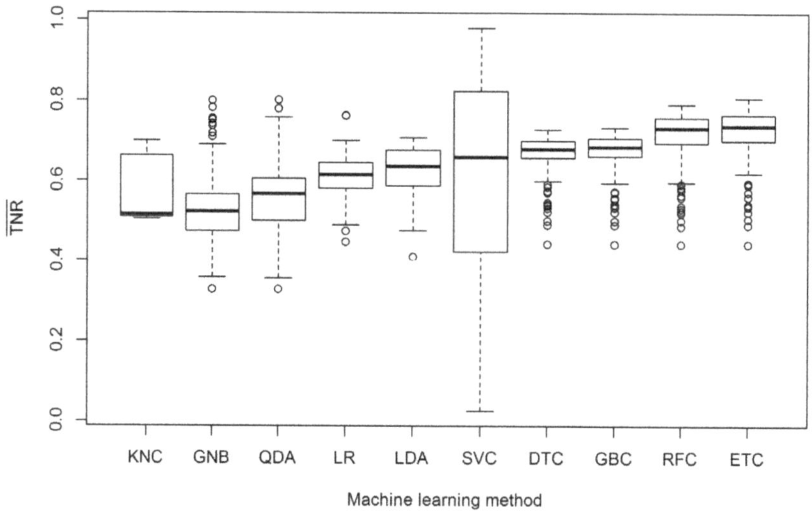

Figure 7. Box plots of mean specificities ($\overline{TNR}$) of FDMs generated by MLAs. Box plots show median plus upper and lower quartiles for mean specificities of all FDMs generated by a given algorithm. Minimum and maximum values are indicated by the upper and lower whiskers (1.5 × interquartile range). Circles are outliers. See Table 3 for MLA codes.

Table 6. Top FDMs among models with equivalent numbers of explanatory variables.

r [1]	n_V [2]	MLA [3]	Variable [4]								$\overline{TPR}$ [5]	$\overline{TNR}$ [6]	$\overline{\phi}$ [7]
			DAS	DFE	PSR	NDVI	AGE	PCT	VOL	STO			
1055	1	SVC							+		0.691	0.685	0.376
116	2	SVC					+		+		0.719	0.811	0.534
43	3	RFC	+				+		+		0.794	0.773	0.568
16	4	ETC	+		+		+		+		0.800	0.784	0.585
1	5	ETC	+		+		+	+	+		0.804	0.800	0.605
6	6	ETC	+		+	+	+	+	+		0.791	0.800	0.592
2	7	ETC	+		+	+	+	+	+	+	0.793	0.806	0.600
3	8	ETC	+	+	+	+	+	+	+	+	0.804	0.790	0.596

[1] ranking number, [2] number of explanatory variables, [3] machine learning algorithm used to generate FDM, the MLA codes are given in Table 3, [4] codes of the variables are listed in Table 1, [5] mean true positive rate, [6] mean true negative rate, [7] mean phi coefficient, "+" indicates the presence of an explanatory variable in FDM.

The Φ of ETC-based models generally increased as the number of explanatory variables increased (Figure A1). The computer simulations showed that the ETC-based model that included distance to actual forest damage, potential solar radiation, spruce forest age, percentage of spruce in forest stands, and volume per hectare had the highest mean accuracy and phi coefficient. Nearly the same performance, however, was achieved by ETC-based models that included different explanatory variables (Table A1).

4. Discussion

Given climate change and the increased emphasis on ecological and cultural services rather than on the economic services provided by forest ecosystems, new forest management strategies are needed. Forest damage by abiotic and biotic factors not only causes timber loss, but also has negative effects in terms of forest diversity, soil erosion, landslides, recreation, and landscape aesthetics. The spatial pattern of forest damage is driven by complex spatio-temporal environmental processes. Rammer and Seidl [36] have shown the effectiveness of MLAs for the spatial prediction of *I. typographus* infestations in unmanaged forests.

We found that FDMs based on traditional linear and non-linear methods generally performed less well than FDMs based on classification trees. LR and LDA are commonly used linear classification MLAs. Hernandez et al. [37] developed a spatial logistic regression model of the probability of bark beetle (*Dendroctonus frontalis* Zimmermann) attack in coniferous forests; the overall accuracy of the model was 68.7%. LR, which is a special case of the generalized linear model, models the log-odds as a linear function by minimizing the sum of the squared residuals. LDA minimizes the probability of misclassification by maximizing the separation between the classes. LR and LDA create the linear boundaries of classes in the explanatory variables' space. LDA is less sensitive than LR to correlations between explanatory variables.

QDA uses class-specific covariance matrices and separates classes by quadratic surfaces in the predictors' space. It is sensitive to collinearity between explanatory variables within a class. Because of a higher complexity of the discriminant function, QDA may perform better than linear MLAs. KNC calculates class probability as the proportion of the class in the set of k-closest neighbors from the training data. KNC is susceptible to measurement scales and local over-fitting. GNB estimates prior and conditional probabilities from the training set and then uses Bayes's rule to calculate the probability of outcome class. An important assumption of GNB is the independence of explanatory variables, which is rarely satisfied in practice.

Only the performance of the top SVC-based FDM was significantly different from the best models generated by the other tested linear and non-linear algorithms. SVC is considered one of the most flexible and effective MLAs and is widely used. It belongs to the family of kernel methods, which allow the separation of classes by non-linear boundaries.

In our study, the FDMs generated by classification trees performed better than those generated by linear and non-linear MLAs. Classification trees describe patterns in data by complex hierarchies of simple rules. The performance of a single classification tree is usually weak. The performances of DTC-based FDMs were inferior to those of the FDMs generated by MLAs based on an ensemble of classification trees.

An ensemble of classification trees usually performs better than a single classification tree, because an ensemble can detect more complex patterns in the data. The RFC MLA is an ensemble of classification trees that are trained on bootstrap samples. The randomness of the tree construction process reduces correlations between trees. Mi et al. [38] investigated Stochastic Gradient Boosting, RFC, CART (Classification and Regression Tree), and MaxEnt (Maximum Entropy) for the modeling distribution of three crane species. The RFC-generated species distribution models were found more reliable and accurate than models generated by the other algorithms under their study.

The process of GBC ensemble construction is iterative. The newly constructed classification tree of the GBC ensemble is forced to learn unexplored data. However, the boosting process was not effective in the case of FDMs. In our study, GBC-based FDMs performed less well than the RFC- and ETC-based models.

The best results in our study were achieved by FDMs generated by ETC. In the ETC ensemble, classification trees are trained on all samples. The randomly selected exploratory variable and random value are used to split the nodes of the classification trees [39]. ETC-based models were responsive to the spatial variance of environmental conditions, they accurately modeled the spatial distribution of forest damage. In addition, ETC-based models are computationally efficient.

Default settings were used for testing the MLAs in the current study. These default settings were selected by developers based on their knowledge of algorithms. Although these settings provide a good starting point for experiments with MLAs, optimization of settings may improve the performance of the tested algorithms.

Input variables for FDMs vary over space and time. Field measurements and access to historical records or remote sensing data are needed to prepare corresponding inputs for spatial distribution models. PSR, NDVI, and distance to existing forest damage areas can be calculated for past periods. Archives of satellite images can be used to calculate NDVI and to identify bark beetle infestations for past years. Thanks to the archiving of satellite images, it should be possible to identify past and current infestations and to build spatial distribution models that predict future forest damage.

We tested all combinations of the explanatory variables as input to the studied MLAs. However, only eight explanatory variables were available for our study area (Table 1). As indicated in Tables 4 and A1, some explanatory variables occurred more often than other explanatory variables in FDMs. The most common explanatory variables in the top FDMs were distance to damaged forest, potential solar radiation, forest age, and spruce volume per hectare (Table A1). These variables may carry substantial information for modeling the spatial distribution of forest damage [40–43].

5. Conclusions

In the study, we evaluated the performance of 10 MLAs and combinations of eight input variables for the spatial modeling of spruce forest damage. Our computer simulations confirmed the suitability of the ETC MLA for modeling the spatial distribution of spruce forest damage. We also found that the number of input explanatory variables could be reduced without significant spatial modeling accuracy loss. A smaller number of input explanatory variables simplifies data preparation and processing and therefore reduces financial costs.

Various MLAs are currently ready for use in spatial decision support systems. We evaluated the MLAs available in the open-source GRASS GIS. The identification of suitable MLAs and key environmental variables is an essential step in the development of GIS tools for forest damage modeling and prognosis. Our findings will facilitate the development of the open-source spatial decision system TANABBO for modeling the spatial distribution of forest damage related to *I. typographus*.

The risk of forest damage is affected by many interrelated environmental factors, forest stand parameters, and forest management practices. MLAs are effective for modeling non-linear, complex phenomena like the spatial distribution of forest damage. This study provides researchers and forest managers with an accurate method of modeling spatial forest damage. The integration of the FDM with a spatial decision support system will facilitate novel tools for managing forests.

Author Contributions: Conceptualization, M.K., R.J., and M.B.; methodology, M.K., R.J., J.H., R.Ď., and M.B.; software, M.K.; investigation, M.K., R.J., J.H., and M.B.; data curation, I.B.; writing—original draft preparation, M.K., R.J., and M.B.; writing—review and editing, M.K., R.J., M.Z., I.B., J.H., R.Ď., and M.B.; visualization, R.Ď. and M.Z.; funding acquisition, M.K., R.J., J.H., and R.Ď. All authors have read and agreed to the published version of the manuscript.

Funding: This was a cooperative study and benefited from the project Comprehensive research of mitigation and adaptation measures to diminish the negative impacts of climate changes on forest ecosystems in Slovakia (FORRES), ITMS 313011T678, Operational Programme Integrated Infrastructure (OPII) funded by the ERDF. The work was also supported by grant no. QK1920433 of the Ministry of Agriculture of the Czech Republic, by the Slovak Research and Development Agency grant no. APVV-15-0761, and grants no. VEGA 2/0176/17 and VEGA 1/0300/19 of the Scientific Grant Agency of the Ministry of Education, Science, Research, and Sport of the Slovak Republic and the Slovak Academy of Sciences.

Acknowledgments: The authors are grateful to the Military Forests and Farms of the Czech Republic, state enterprise, for its cooperation. The authors thank Bruce Jaffee (USA) for linguistic and editorial improvements.

Conflicts of Interest: The authors declare no conflict of interest. The founders had no role in the design of the study; in the collection, analyses, or interpretation of data; in the writing of the manuscript, or in the decision to publish the results.

Appendix A

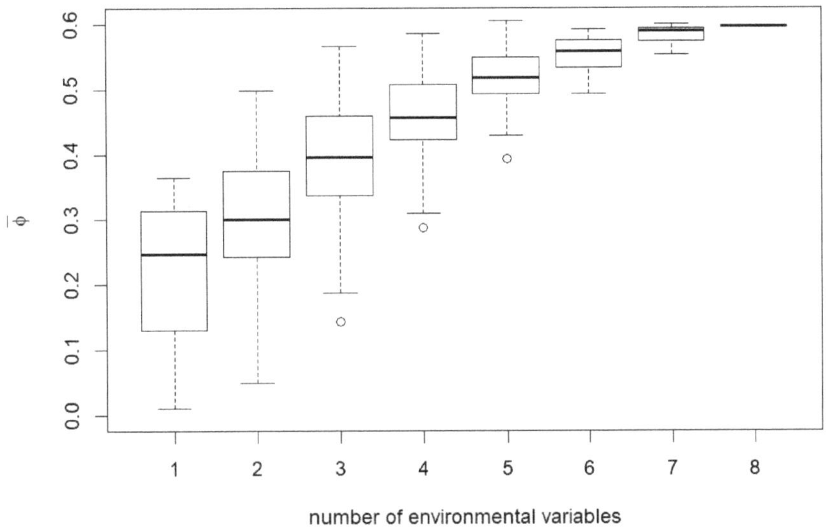

Figure A1. Box plot of mean phi coefficients ($\overline{\phi}$) of the FDMs generated by ETC as affected by the number of explanatory variables. Box plot shows median plus upper and lower quartiles for $\overline{\phi}$ of FDMs with a given number of explanatory variables. Minimum and maximum values are indicated by the upper and lower whiskers (1.5 × interquartile range). Circles are outliers.

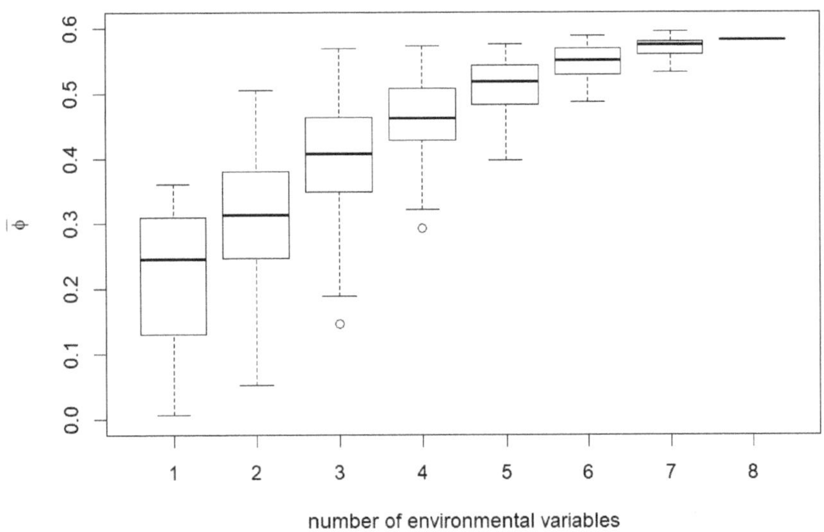

Figure A2. Box plot of mean phi coefficients ($\overline{\phi}$) of the FDMs generated by RFC as affected by the number of explanatory variables. Box plot shows median plus upper and lower quartiles for $\overline{\phi}$ of FDMs with a given number of explanatory variables. Minimum and maximum values are indicated by the upper and lower whiskers (1.5 × interquartile range). Circles are outliers.

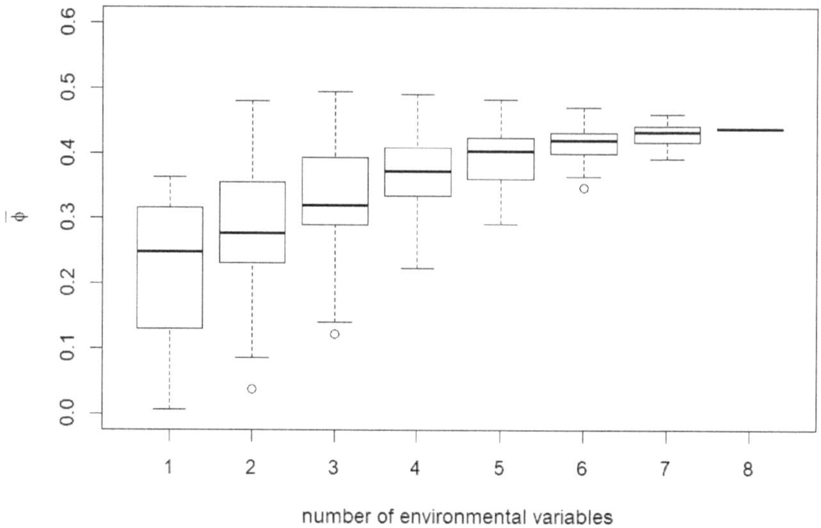

Figure A3. Box plot of mean phi coefficients ($\overline{\phi}$) of the FDMs generated by GBC as affected by the number of explanatory variables. Box plot shows median plus upper and lower quartiles for $\overline{\phi}$ of FDMs with a given number of explanatory variables. Minimum and maximum values are indicated by the upper and lower whiskers (1.5 × interquartile range). Circles are outliers.

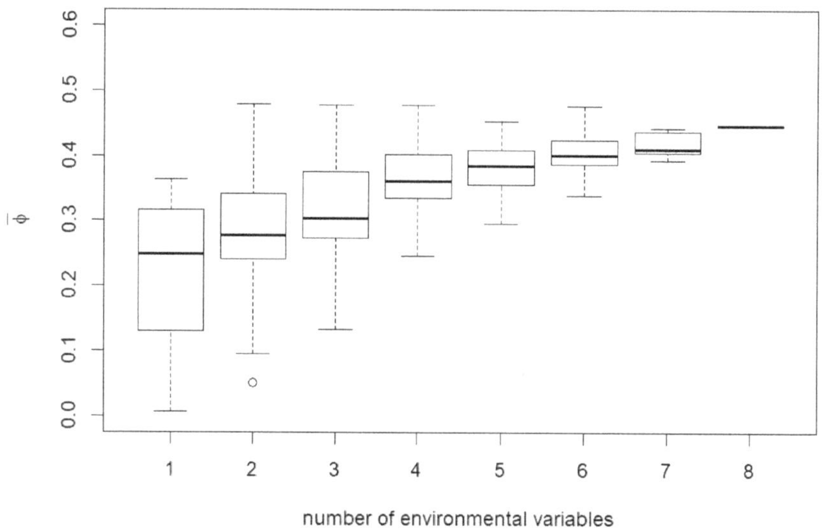

Figure A4. Box plot of mean phi coefficients ($\overline{\phi}$) of the FDMs generated by DTC as affected by the number of explanatory variables. Box plot shows median plus upper and lower quartiles for $\overline{\phi}$ of FDMs with a given number of explanatory variables. Minimum and maximum values are indicated by the upper and lower whiskers (1.5 × interquartile range). Circles are outliers.

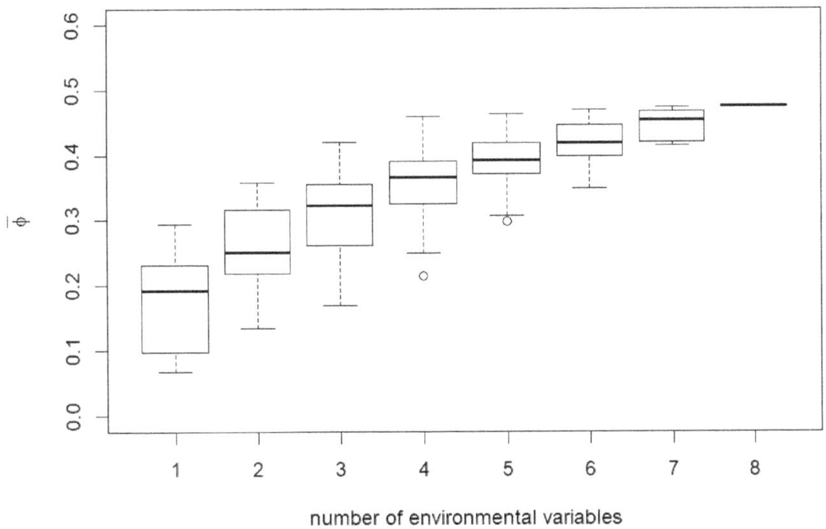

Figure A5. Box plot of mean phi coefficients ($\overline{\phi}$) of the FDMs generated by LDA as affected by the number of explanatory variables. Box plot shows median plus upper and lower quartiles for $\overline{\phi}$ of FDMs with a given number of explanatory variables. Minimum and maximum values are indicated by the upper and lower whiskers (1.5 × interquartile range). Circles are outliers.

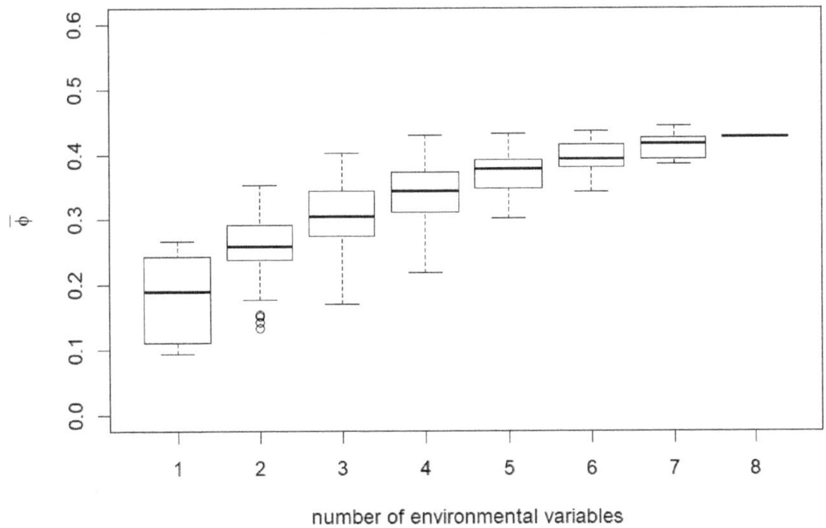

Figure A6. Box plot of mean phi coefficients ($\overline{\phi}$) of the FDMs generated by QDA as affected by the number of explanatory variables. Box plot shows median plus upper and lower quartiles for $\overline{\phi}$ of FDMs with a given number of explanatory variables. Minimum and maximum values are indicated by the upper and lower whiskers (1.5 × interquartile range). Circles are outliers.

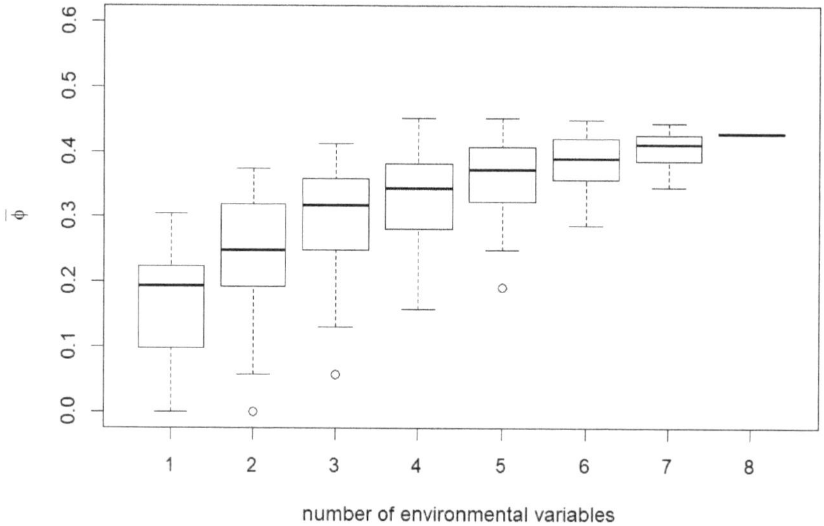

Figure A7. Box plot of mean phi coefficients ($\overline{\phi}$) of the FDMs generated by LR as affected by the number of explanatory variables. Box plot shows median plus upper and lower quartiles for $\overline{\phi}$ of FDMs with a given number of explanatory variables. Minimum and maximum values are indicated by the upper and lower whiskers (1.5 × interquartile range). Circles are outliers.

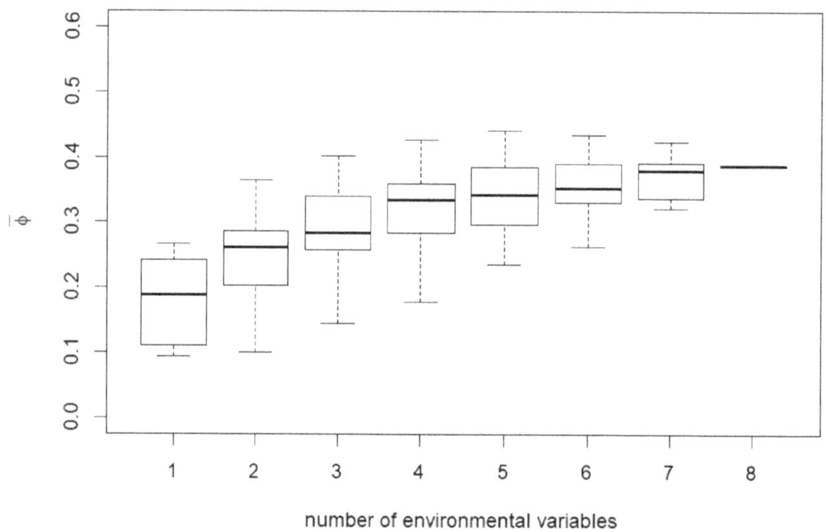

Figure A8. Box plot of mean phi coefficients ($\overline{\phi}$) of the FDMs generated by GNB as affected by the number of explanatory variables. Box plot shows median plus upper and lower quartiles for $\overline{\phi}$ of FDMs with a given number of explanatory variables. Minimum and maximum values are indicated by the upper and lower whiskers (1.5 × interquartile range). Circles are outliers.

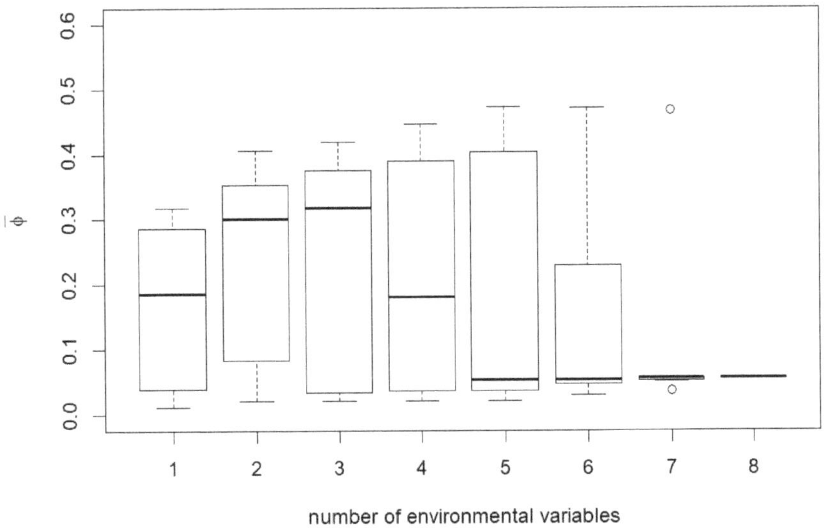

Figure A9. Box plot of mean phi coefficients ($\bar{\phi}$) of the FDMs generated by KNC as affected by the number of explanatory variables. Box plot shows median plus upper and lower quartiles for $\bar{\phi}$ of FDMs with a given number of explanatory variables. Minimum and maximum values are indicated by the upper and lower whiskers (1.5 × interquartile range). Circles are outliers.

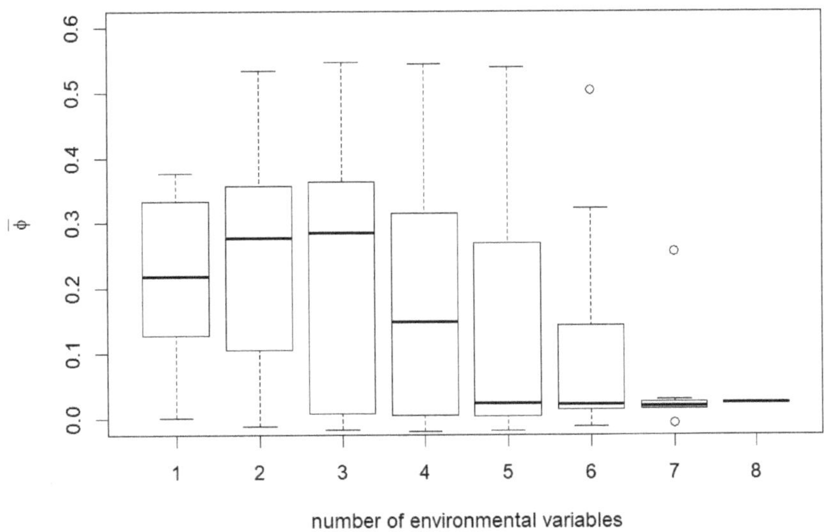

Figure A10. Box plot of mean phi coefficients ($\bar{\phi}$) of the FDMs generated by SVC as affected by the number of explanatory variables. Box plot shows median plus upper and lower quartiles for $\bar{\phi}$ of FDMs with a given number of explanatory variables. Minimum and maximum values are indicated by the upper and lower whiskers (1.5 × interquartile range). Circles are outliers.

Appendix B

Table A1. Forest damage models not significantly different from the best ETC-based model.

r [1]	nv [2]	MLA [3]	Variable [4]								$\overline{TPR}$ [5]	$\overline{TNR}$ [6]	$\overline{\phi}$ [7]	p [8]
			DAS	DFE	PSR	NDVI	AGE	PCT	VOL	STO				
1	ETC	5	+		+		+	+	+		0.8043	0.8002	0.6052	-
2	ETC	7	+		+	+	+	+	+	+	0.7927	0.8058	0.5997	0.5537
3	ETC	8	+	+	+	+	+	+	+	+	0.8038	0.7905	0.5959	0.5037
4	ETC	7	+	+	+		+	+	+	+	0.8045	0.7898	0.5957	0.4998
5	RFC	7	+	+	+		+	+	+	+	0.8140	0.7782	0.5937	0.2509
6	ETC	6	+		+	+	+	+	+		0.7910	0.8001	0.5919	0.1819
7	ETC	5	+		+		+		+	+	0.8035	0.7856	0.5902	0.1029
8	ETC	6	+	+	+		+	+	+		0.8103	0.7782	0.5898	0.2369
9	ETC	7	+	+		+	+	+	+	+	0.7989	0.7890	0.5897	0.2859
10	ETC	7	+	+	+	+	+	+		+	0.8004	0.7869	0.5896	0.1759
12	ETC	6	+		+	+		+	+	+	0.7807	0.8069	0.5888	0.1829
13	RFC	6	+	+	+		+		+	+	0.8036	0.7833	0.5882	0.0770
14	ETC	7	+	+	+	+	+	+	+		0.7980	0.7882	0.5877	0.1479
16	ETC	4	+		+		+		+		0.8004	0.7844	0.5855	0.0800
18	ETC	6	+		+	+	+		+	+	0.7813	0.8027	0.5851	0.0860
21	ETC	6	+	+	+		+		+	+	0.7983	0.7787	0.5783	0.0820
36	ETC	7	+	+	+	+		+	+	+	0.7881	0.7798	0.5693	0.0510

[1] ranking number, [2] number of explanatory variables, [3] machine learning algorithm used to generate FDM, the MLA codes are given in Table 3, [4] codes of the variables are listed in Table 1, [5] mean true positive rate, [6] mean true negative rate, [7] mean phi coefficient for the period 2003–2012, [8] *p*-value of pairwise permutation *t*-test with the best model, "+" indicates the presence of an explanatory variable in FDM.

References

1. Brázdil, R.; Stucki, P.; Szabó, P.; Řezníčková, L.; Dolák, L.; Dobrovolný, P.; Tolasz, R.; Kotyza, O.; Chromá, K.; Suchánková, S. Windstorms and Forest Disturbances in the Czech Lands: 1801–2015. *Agric. For. Meteorol.* **2018**, *250–251*, 47–63. [CrossRef]
2. Mezei, P.; Grodzki, W.; Blaženec, M.; Jakuš, R. Factors Influencing the Wind–Bark Beetles' Disturbance System in the Course of an Ips Typographus Outbreak in the Tatra Mountains. *For. Ecol. Manag.* **2014**, *312*, 67–77. [CrossRef]
3. Seidl, R.; Thom, D.; Kautz, M.; Martin-Benito, D.; Peltoniemi, M.; Vacchiano, G.; Wild, J.; Ascoli, D.; Petr, M.; Honkaniemi, J.; et al. Forest Disturbances under Climate Change. *Nat. Clim. Change* **2017**, *7*, 395–402. [CrossRef]
4. Schurman, J.S.; Trotsiuk, V.; Bače, R.; Čada, V.; Fraver, S.; Janda, P.; Kulakowski, D.; Labusova, J.; Mikoláš, M.; Nagel, T.A.; et al. Large-Scale Disturbance Legacies and the Climate Sensitivity of Primary Picea Abies Forests. *Glob. Chang. Biol.* **2018**, *24*, 2169–2181. [CrossRef] [PubMed]
5. Mezei, P.; Jakuš, R.; Pennerstorfer, J.; Havašová, M.; Škvarenina, J.; Ferenčík, J.; Slivinský, J.; Bičárová, S.; Bilčík, D.; Blaženec, M.; et al. Storms, Temperature Maxima and the Eurasian Spruce Bark Beetle Ips Typographus—An Infernal Trio in Norway Spruce Forests of the Central European High Tatra Mountains. *Agric. For. Meteorol.* **2017**, *242*, 85–95. [CrossRef]
6. Montano, V.; Bertheau, C.; Doležal, P.; Krumböck, S.; Okrouhlík, J.; Stauffer, C.; Moodley, Y. How Differential Management Strategies Affect *Ips Typographus* L. *Dispersal. For. Ecol. Manag.* **2016**, *360*, 195–204. [CrossRef]
7. Lausch, A.; Heurich, M.; Fahse, L. Spatio-Temporal Infestation Patterns of *Ips Typographus* (L.) in the Bavarian Forest National Park, Germany. *Ecol. Indic.* **2013**, *31*, 73–81. [CrossRef]
8. Økland, B.; Nikolov, C.; Krokene, P.; Vakula, J. Transition from Windfall- to Patch-Driven Outbreak Dynamics of the Spruce Bark Beetle Ips Typographus. *For. Ecol. Manag.* **2016**, *363*, 63–73. [CrossRef]
9. de Groot, M.; Ogris, N. Short-Term Forecasting of Bark Beetle Outbreaks on Two Economically Important Conifer Tree Species. *For. Ecol. Manag.* **2019**, *450*, 117495. [CrossRef]
10. Segura, M.; Ray, D.; Maroto, C. Decision Support Systems for Forest Management: A Comparative Analysis and Assessment. *Comput. Electron. Agric.* **2014**, *101*, 55–67. [CrossRef]
11. Senf, C.; Seidl, R.; Hostert, P. Remote Sensing of Forest Insect Disturbances: Current State and Future Directions. *Int. J. Appl. Earth Obs. Geoinf.* **2017**, *60*, 49–60. [CrossRef] [PubMed]
12. Oeser, J.; Pflugmacher, D.; Senf, C.; Heurich, M.; Hostert, P. Using Intra-Annual Landsat Time Series for Attributing Forest Disturbance Agents in Central Europe. *Forests* **2017**, *8*, 251. [CrossRef]
13. Thiele, J.C.; Nuske, R.S.; Ahrends, B.; Panferov, O.; Albert, M.; Staupendahl, K.; Junghans, U.; Jansen, M.; Saborowski, J. Climate Change Impact Assessment—A Simulation Experiment with Norway Spruce for a Forest District in Central Europe. *Ecol. Model.* **2017**, *346*, 30–47. [CrossRef]

14. Seidl, R.; Müller, J.; Hothorn, T.; Bässler, C.; Heurich, M.; Kautz, M. Small Beetle, Large-Scale Drivers: How Regional and Landscape Factors Affect Outbreaks of the European Spruce Bark Beetle. *J. Appl. Ecol.* **2016**, *53*, 530–540. [CrossRef]
15. dos Reis, A.A.; Carvalho, M.C.; de Mello, J.M.; Gomide, L.R.; Ferraz Filho, A.C.; Acerbi Junior, F.W. Spatial Prediction of Basal Area and Volume in Eucalyptus Stands Using Landsat TM Data: An Assessment of Prediction Methods. *N. Z. J. For. Sci.* **2018**, *48*. [CrossRef]
16. Hickey, C.; Kelly, S.; Carroll, P.; O'Connor, J. Prediction of Forestry Planned End Products Using Dirichlet Regression and Neural Networks. *For. Sci.* **2015**, *61*, 289–297. [CrossRef]
17. Görgens, E.B.; Montaghi, A.; Rodriguez, L.C.E. A Performance Comparison of Machine Learning Methods to Estimate the Fast-Growing Forest Plantation Yield Based on Laser Scanning Metrics. *Comput. Electron. Agric.* **2015**, *116*, 221–227. [CrossRef]
18. Rodrigues, M.; de la Riva, J. An Insight into Machine-Learning Algorithms to Model Human-Caused Wildfire Occurrence. *Environ. Model. Softw.* **2014**, *57*, 192–201. [CrossRef]
19. Mayfield, H.; Smith, C.; Gallagher, M.; Hockings, M. Use of Freely Available Datasets and Machine Learning Methods in Predicting Deforestation. *Environ. Model. Softw.* **2017**, *87*, 17–28. [CrossRef]
20. Castelli, M.; Vanneschi, L.; Popovič, A. Predicting Burned Areas of Forest Fires: An Artificial Intelligence Approach. *Fire Ecol.* **2015**, *11*, 106–118. [CrossRef]
21. Liang, L.; Li, X.; Huang, Y.; Qin, Y.; Huang, H. Integrating Remote Sensing, GIS and Dynamic Models for Landscape-Level Simulation of Forest Insect Disturbance. *Ecol. Model.* **2017**, *354*, 1–10. [CrossRef]
22. Hlásny, T.; Turčáni, M. Persisting Bark Beetle Outbreak Indicates the Unsustainability of Secondary Norway Spruce Forests: Case Study from Central Europe. *Ann. For. Sci.* **2013**, *70*, 481–491. [CrossRef]
23. Military Forests and Farms of the Czech Republic, State Enterprise. Available online: https://www.vls.cz/en (accessed on 25 March 2021).
24. Culek, M.; Grulich, V.; Laštůvka, Z.; Divíšek, J. *Biogeografické Regiony České Republiky*; Masarykova Univerzita: Brno, Czech Republic, 2013; ISBN 978-80-210-6693-9.
25. Forestry and Game Management Research Institute, Czech Republic. Available online: https://www.vulhm.cz/en/ (accessed on 25 March 2021).
26. Forest Management Institute (FMI), Czech Republic. Available online: http://www.uhul.cz/home (accessed on 25 March 2021).
27. Henzlik, V. Forests and Air Pollution in the Czech Republic. In *Restoration of Forests*; Gutkowski, R.M., Winnicki, T., Eds.; Springer Netherlands: Dordrecht, The Netherlands, 1997; pp. 133–149. ISBN 978-0-7923-4634-0.
28. Havašová, M.; Ferenčík, J.; Jakuš, R. Interactions between Windthrow, Bark Beetles and Forest Management in the Tatra National Parks. *For. Ecol. Manag.* **2017**, *391*, 349–361. [CrossRef]
29. Ďuračiová, R.; Muňko, M.; Barka, I.; Koreň, M.; Resnerová, K.; Holuša, J.; Blaženec, M.; Potterf, M.; Jakuš, R. A Bark Beetle Infestation Predictive Model Based on Satellite Data in the Frame of Decision Support System TANABBO. *IForest Biogeosci. For.* **2020**, *13*, 215–223. [CrossRef]
30. Hofierka, J.; Šúri, M. The Solar Radiation Model for Open Source GIS: Implementation and Applications. In Proceedings of the Open Source GIS—GRASS Users Conference 2002, Trento, Italy, 11–13 September 2002; p. 19.
31. EU-DEM v1.1—Copernicus Land Monitoring Service. Available online: https://land.copernicus.eu/imagery-in-situ/eu-dem/eu-dem-v1.1 (accessed on 25 March 2021).
32. GRASS GIS. Available online: https://grass.osgeo.org/ (accessed on 25 March 2021).
33. Scikit-Learn: Machine Learning in Python. Available online: https://scikit-learn.org (accessed on 25 March 2021).
34. *R Core Team R: A Language and Environment for Statistical Computing*; R Foundation for Statistical Computing: Vienna, Austria, 2018.
35. Benjamini, Y.; Hochberg, Y. Controlling the False Discovery Rate: A Practical and Powerful Approach to Multiple Testing. *J. R. Stat. Soc. Ser. B Methodol.* **1995**, *57*, 289–300. [CrossRef]
36. Rammer, W.; Seidl, R. Harnessing Deep Learning in Ecology: An Example Predicting Bark Beetle Outbreaks. *Front. Plant Sci.* **2019**, *10*, 1327. [CrossRef] [PubMed]
37. Hernandez, A.J.; Saborio, J.; Ramsey, R.D.; Rivera, S. Likelihood of Occurrence of Bark Beetle Attacks on Conifer Forests in Honduras under Normal and Climate Change Scenarios. *Geocarto Int.* **2012**, *27*, 581–592. [CrossRef]
38. Mi, C.; Huettmann, F.; Guo, Y.; Han, X.; Wen, L. Why Choose Random Forest to Predict Rare Species Distribution with Few Samples in Large Undersampled Areas? Three Asian Crane Species Models Provide Supporting Evidence. *PeerJ* **2017**, *5*, e2849. [CrossRef] [PubMed]
39. Geurts, P.; Ernst, D.; Wehenkel, L. Extremely Randomized Trees. *Mach. Learn.* **2006**, *63*, 3–42. [CrossRef]
40. Angst, A.; Rüegg, R.; Forster, B. Declining Bark Beetle Densities (Ips Typographus, Coleoptera: Scolytinae) from Infested Norway Spruce Stands and Possible Implications for Management. *Psyche J. Entomol.* **2012**, *2012*, 1–7. [CrossRef]
41. Kautz, M.; Schopf, R.; Ohser, J. The "Sun-Effect": Microclimatic Alterations Predispose Forest Edges to Bark Beetle Infestations. *Eur. J. For. Res.* **2013**, *132*, 453–465. [CrossRef]
42. Stadelmann, G.; Bugmann, H.; Wermelinger, B.; Meier, F.; Bigler, C. A Predictive Framework to Assess Spatio-Temporal Variability of Infestations by the European Spruce Bark Beetle. *Ecography* **2013**, *36*, 1208–1217. [CrossRef]
43. Stereńczak, K.; Mielcarek, M.; Kamińska, A.; Kraszewski, B.; Piasecka, Ż.; Miścicki, S.; Heurich, M. Influence of Selected Habitat and Stand Factors on Bark Beetle *Ips Typographus* (L.) Outbreak in the Białowieża Forest. *For. Ecol. Manag.* **2020**, *459*, 117826. [CrossRef]

forests

MDPI

Article

Comparison of *Ips cembrae* (Coleoptera: Curculionidae) Capture Methods: Small Trap Trees Caught the Most Beetles

Karolina Resnerová [1,*], Jaroslav Holuša [1], Peter Surový [2], Jiří Trombik [1] and Emanuel Kula [3]

1 Department of Forest Protection and Entomology, Czech University of Life Sciences Prague, 16500 Prague, Czech Republic; holusa@fld.czu.cz (J.H.); trombik@fld.czu.cz (J.T.)

2 Department of Forest Management, Czech University of Life Sciences Prague, 16500 Prague, Czech Republic; surovy@fld.czu.cz

3 Department of Forest Protection and Wildlife Management, Mendel University in Brno, 61300 Brno, Czech Republic; kula@mendelu.cz

* Correspondence: resnerovak@fld.czu.cz

Received: 11 October 2020; Accepted: 27 November 2020; Published: 28 November 2020

Abstract: *Ips cembrae* is the most important bark beetle pest of larches and has had several local outbreaks in recent decades in Europe. In this study, we compared the numbers of *I. cembrae* captured by pyramid-trap piles, trap trees, pheromone traps, and poisoned and baited tripods. We also studied how the properties of trap trees and trap logs (volume, sun exposure, and position relative to the ground once deployed) affected the trapping of *I. cembrae*. We found that both sexes avoided infestation at the bottom of the logs and more than 15 times the number of beetles were captured by traditional trap trees than by pheromone traps or baited and insecticide-treated tripods. The number of *I. cembrae* per trap tree did not decrease with trap volume; therefore, it is appropriate to use traps of small dimensions. Baited tripods, pyramid-trap piles, and pheromone traps could be useful for detection of the beginning of flight activity, but trap trees are the most useful for reducing *I. cembrae* numbers.

Keywords: large larch bark beetle; pheromone traps; trap trees; tripods; trap tree volume; insecticide

---

1. Introduction

European larch (*Larix decidua* Mill.) is native to the Alps and to several mountainous ranges in Central Europe. Its early introduction outside its native range, especially in the lowlands of western and northern Europe, has been problematic for ecological and phytosanitary reasons [1]. Outside of plantations, its current natural distribution is fragmented and spans about 500,000 ha. Beyond its native range, plantations of European larch cover an additional 500,000 ha [2]. In the Czech Republic, *L. decidua* covers a total area of 100,263 ha—i.e., 3.2% of the forested area in the country. Larch has been recently considered as useful for the regeneration of areas that have been cleared following bark beetle outbreaks in spruce stands [3].

Ips cembrae (Heer, 1836), one of several *Ips* species native to Europe, is an aggressive pest of larch, and local outbreaks of *I. cembrae* have been recorded in Europe [4–7]. In the years 2009–2019 in the Czech Republic, a total of 39,000 m^3 of larch was harvested, which is more than 1000 times less than the spruce harvested in the same period. This is the main reason why larch bark beetles are currently receiving less attention than spruce bark beetles by both researchers and foresters.

Ips cembrae attacks larches (*Larix* spp.) [6,8,9] of all ages without significant preference for altitude (400–2400 m) [10–12] It has no competitors for the use of the trunk and can multiply in monocultures that have been established as substitute tree stands in areas with serious air pollution, but also on individual larches within forests dominated by other tree species [13]. The beetle will occasionally attack other conifers (*Picea* spp., *Abies* spp., or *Pinus* spp.) [10,14]. In addition to reproducing in weakened trees, *I. cembrae* also reproduces in healthy trees [6,15,16] and in felled wood [6]. Predicted changes in climate are likely to increase the effects of this species [17,18], especially outside the natural range of larch where outbreaks on stressed trees may occur [19–21]. Apart from the damage caused by the adult beetles, some ophiostomatoid fungi associated with *I. cembrae* are likely to harm the attacked trees [22–24].

The life cycle of *I. cembrae* can be considered to begin when males search for and penetrate suitable host trees. One to seven (most often three) females join each male in a nuptial chamber [9], and each female then creates a maternal gallery, which first radiates in a stellate pattern and then follows the fibers of the phloem [25]. Up to about 50 eggs are laid singly along each maternal gallery. The adults often emerge from their first gallery system and start a sister brood in another tree. Depending on altitude and climate, there is one generation or two overlapping generations per year. *Ips cembrae* can benefit from climatic warming because it can complete up to two generations in a hot growing season [7,16]. Before producing their own brood, the young adults must undergo maturation feeding, either within the phloem of the tree where they developed or in 2- to 18-year-old branches [25,26].

As with all bark beetles, the most effective control of *I. cembrae* is the consistent searching for and removal of infested trees (including branches) [4,27–29]. Monitoring of *I. cembrae* includes visual inspection and capture with pheromone traps, trap trees/logs, and poisoned traps [25]. In this study, we compared these methods of capturing *I. cembrae* with the goal of both monitoring and reducing its numbers.

A review of the literature indicated that little is known about the effectiveness of methods that could be used to capture *I. cembrae*. Based on the literature for other bark beetles, we assumed that the use of pheromone lures and trap trees would have similar abilities to reduce the population densities of *I. cembrae*. We also assumed that trap trees used to capture *I. cembrae* should be relatively large as is the case with *I. typographus* (Linnaeus, 1758) spruce trap trees—i.e., the *I. cembrae* larch trap trees should have a diameter at breast height of about 30 cm [30]. We also recognize that the ability of trap trees to capture the beetles and reduce their number could be small due to trap saturation or to the limited area for establishing galleries [31]. In contrast, pheromone traps and "tripod traps" (described in the next section) have a theoretically unlimited capacity to capture bark beetles [32] and thus to reduce *I. cembrae* abundance.

When attacking larch, *I. cembrae* establishes gallery systems from the bottom to the top of the tree [13]. Consequently, *I. cembrae* is able to develop on thin trunks, thick branches, and logging residues [33]. We therefore assessed how the ability of trap trees and trap logs to capture *I. cembrae* is influenced by their diameter and volume, exposure to sunlight, and position relative to the soil. This knowledge will help forest managers optimize defense measures.

The aims of this study were to compare *I. cembrae* capture methods and to find the best ways to install and use trap trees and trap logs against this pest. To accomplish our main aims above, we propose the following specific subaims: (i) to evaluate, in experiment 1, the number of *I. cembrae* captured that were affected by exposure of deployed trap logs to sunlight and the position of trap logs relative to their location relative to the soil surface once deployed; (ii) to compare, in experiment 2, the effects of larch logs arranged in standing pyramid-trap piles vs. trap trees laid on the ground on the number of *I. cembrae* captured; (iii) to determine, in experiment 3, whether the number of imagoes of *I. cembrae* captured on trap trees is affected by the volume of trap; to determine, in experiment 4, whether the type of the baited traps affects the numbers of *I. cembrae* beetles captured and whether there will be differences in the catches of both sexes; to evaluate, in experiment 5, and select the most effective conventional *I. cembrae* capture method applicable to forestry operations.

2. Materials and Methods

2.1. Study Sites

Experiments 1 and 2 were conducted on sites with monocultures of *Larix decidua*, and experiments 3–5 were conducted on sites that were mainly monocultures of spruce, *Picea abies* (L.) H. Karst., with a 5 to 10% mixture of larch trees that were more than 60 years old. Based on the willingness of forest owners to cooperate and older data on sites with high *I. cembrae* abundance in the last 15 years [34], we selected 11 study sites and one transect (experiment 2), which were located across the entire territory of the Czech Republic and at elevations ranging from 320 to 680 m a.s.l. (Figure 1). The selection of suitable study sites for the study of *I. cembrae* populations is complicated in the Czech Republic, in part because *I. cembrae* is considered less important than spruce bark beetles. As a consequence, the records of forest managers with respect to *I. cembrae* are seldom detailed, and *I. cembrae* abundance is very often underestimated, even though the quantity of the larch timber harvested has greatly increased in recent years [35].

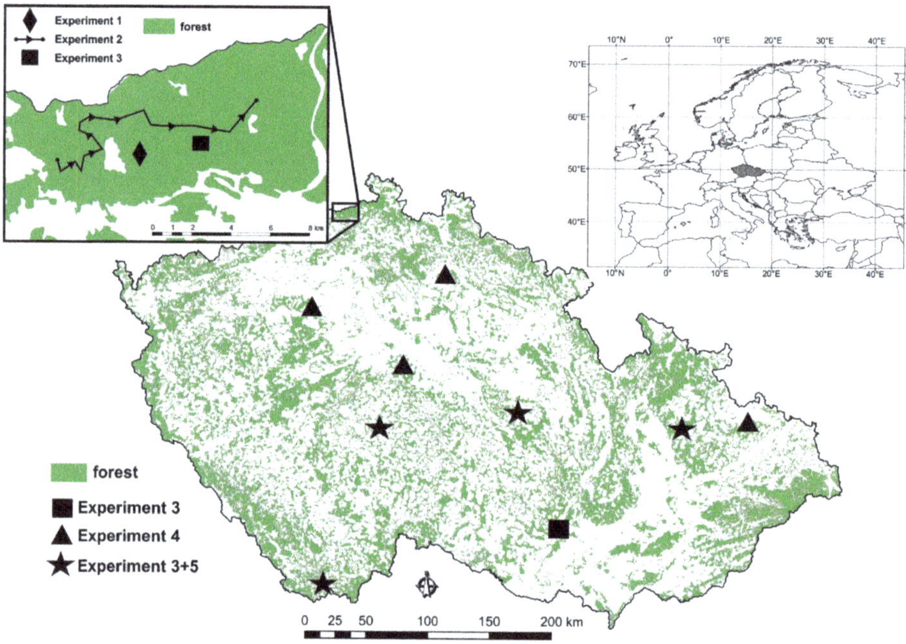

Figure 1. Study sites in the Czech Republic where experiments 1–5 were conducted from 2014 to 2019. As indicated, most experiments were conducted at more than one site, and two experiments were sometimes conducted at the same site.

2.2. Experiment 1: Influence of Sun Exposure and Position on Infestation of Trap Logs by *I. cembrae*

In February 2014, several healthy larch trees (*L. decidua*) were felled at the Sněžník study site (GPS: 50°47′43″ N, 14°05′34″ E; Figure 1) at an altitude of 600 m a.s.l. These trees were then cut into 150 cm-long logs that were leaning against a wooden railing (Figure 2). A total of 139 logs were installed in an east–west direction in an open area that was established by the harvesting of *Picea pungens* Engelm. and that was adjacent to an extensive larch monoculture. The bottoms of the logs were kept separate from the soil by a birch board. The volume, diameter in centimeters, and width of the phloem were measured for each log: averages (±SD) were 13.2 ± 10.0 dm^3 for volume, 9.9 ± 3.7 cm for diameter, and 2.3 ± 0.7 mm for phloem width.

Figure 2. Design of experiment 1: larch logs installed at the Sněžník study site in 2014. Photograph by Kula.

In May 2014, after *I. cembrae* had infested the logs and developed galleries, the numbers of entry holes and maternal galleries were determined on two sides of each log (the sunny and shaded side) and at three positions on each side (top = 30 cm below the top of the log; middle = in the middle of the log; bottom = 30 cm above the bottom of the log). The area of each bark sample was 10 × 10 cm. As a consequence, six bark samples (2 sides × 3 positions per side) were assessed on each log.

The effects of log side (sunny or shaded) on the numbers of entry holes and maternal galleries were determined with a two-sample t-test in R version 4.0.2. A GLM model (also in R version 4.0.2) with family of binomial distributions was used to compare the effects of position (top, middle, or bottom of deployed trap logs relative to the soil surface), and log diameter on the number of *I. cembrae* males (i.e., the number of nuptial chambers) and the number of *I. cembrae* maternal galleries (number of females).

2.3. Experiment 2: Numbers of I. cembrae Captured by Pyramid-Trap Piles vs. Trap Trees

Experiment 2 was conducted in 2015–2019 near the town of Sněžník (GPS 50.4849 N, 14.0717 E; 10–14 plots) at altitudes between 500 and 600 m a.s.l. (Figure 1). Pairs of traps (each pair consisting of one pyramid-trap pile and one trap tree, which are described in the following paragraphs) were located along the edge of the larch stand but at least 20 m from the stand. A total of 59 pairs of pyramid-trap piles and trap trees were deployed during the research—i.e., 13, 11, 14, 11, and 10 pairs were deployed in 2015, 2016, 2017, 2018, and 2019, respectively. Pairs of traps were always prepared and deployed in the first half of March.

Before they were cut, the trees used to make trap trees had diameters of 17.7 ± 3.7 cm at a height of 1.3 m above the soil surface. Each trap tree was a healthy larch that was cut about 0.3 m above the soil; the trap tree consisted of the top portion that was left in place on the soil surface (Figure 3). Each trap tree was cut into 1.5 m-long sections that were left in place; the upper sunlit side and the lower shaded side of each section was marked, and three positions (upper, middle, or lower) on each side were also marked with the upper section always being the thinner section.

Non-baited pyramid-trap piles were prepared to match the trap trees—i.e., the trees used were from the same site and were of similar size and age (Figure 3). The trees that were used to construct the piles were felled at the study site, and the felled portions were cut into 1.5 m-long logs; the logs were 15.6 ± 2.9 cm thick. Each pyramid-trap pile consisted of 3–8 logs (depending on the height of the tree), which were arranged as shown in Figure 3.

As indicated, the trap tree and pyramid-trap piles were arranged in pairs with 10–15 m between each member of the pair and with 100–500 m between pairs. The volume of wood was 0.1 ± 0.1 m^3 per trap tree and 0.2 ± 0.1 m^3 per pyramid-trap pile.

Figure 3. Larch trap tree (**left**) and pyramid-trap pile (**right**) and at the study site near Sněžník in 2019 (experiment 2). Photograph by Kula.

The traps were evaluated in May/June in the year of their deployment. Each trap tree was debarked at six sections (2 sides × 3 positions per side) and the number of *I. cembrae* in each section was determined. The pyramid-trap piles were evaluated in a similar manner—i.e., six sections were sampled on each of the logs in each pile (2 sides × 3 positions per side). For both kinds of traps, the number of *I. cembrae* was determined per section and was expressed as the number per dm^2.

The studied parameters were analyzed using the GLM model family of binomial distributions. Population densities on trap trees and pyramid-trap piles were compared using Wilcoxon matched pairs test. All analyses and figures were carried out in R version 4.0.2.

2.4. Experiment 3: Number of *I. cembrae* Captured as Affected by Trap Tree Wood Volume

A total of 62 trap trees were deployed at five study sites at altitudes ranging from 390 to 680 m a.s.l. during the years 2017–2019 (Table 1, Figure 1). Healthy larch trees were felled in mid-March. The trap trees were cut, debranched, and left at a distance of 20 m from the nearest trap and at least 20 m from the nearest stand. The traps trees were cut into 1.5 m-long sections.

Table 1. Study sites in years 2017–2019 in experiment 3.

Study Site	Altitude (m a.s.l.)	GPS N	GPS E	Year	Number of Trap Trees
Dolní Babákov	560	49.7985	15.8961	2017	4
Húzová	680	49.8189	17.3523	2017	3
Jiřetice u Neustupova	480	49.5917	14.7247	2017	10
Kristin Hrádek	510	50.8044	14.1301	2019	5
Rožmberk	680	48.6628	14.3776	2017	12
Soběšice	390	49.1536	16.3654	2019	28

The trees were evaluated in May/June—i.e., at the end of the *I. cembrae* infestation period at the larval stage of development. Four sections (sample areas) were designated on each trap tree according to the method of [36]. The first section (bottom) was located 0.5 to 1.0 m from the bottom of the tree; the second section (stem) was located midway between the bottom section and the beginning of the crown; the third section (middle) was located at the beginning of the crown; the fourth section (crown) was located in the center of the crown. Each trap tree was debarked, and the numbers of *I. cembrae* entry holes and maternal galleries were determined in each section. The total number of beetles on the entire surface of the trap tree was calculated based on the total numbers on the individual sections divided by the proportion of the total surface represented by the sum of the individual sections. The volume of wood in each trap tree was calculated according to the volume table of woody plants [37] using data on the height of the tree and the diameter at a height of 1.3 m. The volume of wood per trap tree ranged from 0.14 to 1.40 m^3 with an average of 0.56 ± 0.29 m^3.

The relationships between the number of captured beetles and the volume of wood per trap were assessed with a Poisson Regression model in R version 4.0.2.

2.5. Experiment 4: Numbers of *I. cembrae* Beetles Captured by Baited Tripods vs. Pheromone Traps

Five pairs of tripods and pheromone traps were installed just before the expected start of *I. cembrae* flight activity during March and early April of 2016 and 2017 at each of four study sites (Table 2, Figure 4). The study sites were at altitudes ranging from 320 to 420 m a.s.l. The pairs of traps were left in the study sites throughout the flight activity periods, which usually ended late in August. The two traps composing each pair were 20 m apart, and the pairs were located 20 m from the forest stands and from other traps. Pheromone traps of the Theyson® type were placed at a height of 1.5 m from the ground on two wooden poles; a pheromone lure was hung inside the trap on a metal wire. A Cembräwit® pheromone lure (Witasek Pflanzenschutz GmbH, Vienna, Austria) was used for both types of traps and was replaced after 8 weeks.

Table 2. Study sites for experiment 4.

Study Site	Altitude (m a.s.l.)	GPS N	GPS E	Year
Nouzov	405	50.2202	13.9445	2016
Stříbrná Skalice	420	49.9196	14.8432	2016
Žehrov	320	50.5224	15.0913	2016
Raduň	400	49.8809	17.9548	2017

Figure 4. Baited tripod (**left**) and pheromone trap (**right**) at the Nouzov study site in 2017 (experiment 4). Photograph by Resnerová.

Tripod traps were constructed from felled healthy larch trees. For each tripod trap, three logs about 2 m long and 20 cm in diameter were cut and transported to the deployment location. The tops of the three logs were firmly connected to each other with a T-shaped piece of iron. After the iron was attached to the tops of the logs with nails, the logs were arranged in a tripod shape above two partially buried logs of non-target wood (spruce). Two 12 mm-diameter pieces of iron that extended from the spruce logs (one piece of iron per spruce log) were inserted into the bottoms of two of the three logs that formed the larch tripod. A Cembräwit® pheromone lure was suspended at the top of each tripod so that it was at least partially protected from direct sunlight. The entire surface of each tripod was then treated with a mixture of Vaztak 10 SC® insecticide and a 1% solution of Scolycid® dye in water. The tripods were generally sprayed at 4- to 5-week intervals as needed, depending on the weather. Beetles that were attracted to the tripod and that died and fell were collected on a stainless-steel screen that was located beneath the tripod and that had narrow (<1 mm) openings. A second screen made of plastic and that had wide openings (15 mm) was positioned above the stainless-steel screen. The plastic screen allowed beetles to fall through but caught branches and other unwanted material and prevented animals from feeding on or otherwise removing the beetles on the stainless-steel screen.

Ips cembrae were collected every 7–10 days from the traps and tripods. In the laboratory, individuals of *I. cembrae* were identified with the aid of a stereomicroscope, and the sex was determined based on examination of genitals for at least 20 individuals per sample, datum, and site.

The total number of *I. cembrae* trapped and the proportions of males and females were compared for pheromone-baited tree traps vs. baited tripods with two-sample t-test in R version 4.0.2. We did not take into account the issue of different years though we do not expect differences.

2.6. Experiment 5: Number of I. cembrae Captured by Trap Trees, Baited Tripods, and Pheromone Traps

Experiment 5 was conducted in 2017 at four study sites at altitudes ranging from 480 to 680 m a.s.l. The experiment included 18 groups of traps with each group containing three types of traps (Table 3, Figure 1). The design of the experiment with respect to pheromone traps and baited tripods corresponded to that of experiment 4, but in addition to a pheromone trap and a tripod, each group included a trap tree.

Table 3. Study sites for experiment 5. Each group included three kinds of traps.

Study Site	Altitude (m a.s.l.)	GPS N	GPS E	Year	Number of Groups
Dolní Babákov	560	49.7985	15.8961	2017	4
Húzová	680	49.8189	17.3523	2017	3
Jiřetice u Neustupova	480	49.5917	14.7247	2017	5
Rožmberk	680	48.6628	14.3776	2017	6

The trap trees (mean wood volume 0.7 ± 0.4) were cut, debranched, and laid on the soil as described for experiment 3. Trap trees were cut and deployed in March. In May/June, the beetles captured by the trap trees were counted according to the methods described by [36] and for experiment 3. The following data were collected: the dimensions of the trap trees including the diameter at breast height (1.3 m from the base of the trunk), the number of entry holes and the number of maternal galleries on the debarked sections.

According to the phloem area of each trap tree and the number of entry holes (equivalent to the number of males) and maternal galleries (equivalent to the number of females), the total number of individuals in the entire trap tree was calculated.

Overall differences in beetle catches in the three types of traps were tested using Friedman's test (ANOVA). Poisson generalized linear mixed model (GLMM) with study site as a random effect was used to determine the relationships between the number of beetles captured and trap type. All analyses and figures were carried out in R version 4.0.2.

3. Results

3.1. Experiment 1: Influence of Sun Exposure and Position on Infestation of Trap Logs by I. cembrae

Ips cembrae penetrated logs with a diameter > 3.6 cm and a phloem width > 2 mm. The number of entry holes per dm^2 (mean ± SE) did not significantly differ between the sunny side (0.8 ± 0.6) vs. the shady side (0.8 ± 0.5). The number of maternal galleries per dm^2 was higher on the sunny side (2.9 ± 1.7) than on the shady side (2.5 ± 1.6), but the difference was statistically insignificant according to the two-sample *t*-test ($t = 1.69$; $p < 0.10$).

The GLM model indicated that the number of *I. cembrae* males (i.e., the number of nuptial chambers) and females per dm^2 was positively related to log diameter (Tables 4 and 5). Both sexes avoided forming gallery systems at the bottom position of the deployed logs (i.e., the position closest to the ground); the latter position was included in the intercept of the model (Tables 4 and 5, Figure 5).

Table 4. Results of a GLM model describing the relationship between the number of entry holes of *Ips cembrae* per dm^2 and the following variables in experiment 1: position on the log (top, middle, and bottom = included in Intercept); diameter. Corrected Akaike's Information Criterion (AIC) = 210.09. Number of Fisher Scoring iterations was 5. ***, and * indicate significance at $p < 0.001$, and 0.05, respectively.

Variable	Estimate	Std. Error	z Value	*p* Value
(Intercept)	−2.38	0.41306	−5.758	<0.001 ***
top	0.31	0.31184	1.005	0.3150
middle	0.30	0.31228	0.968	0.3330
diameter	0.30	0.03239	2.279	0.0227 *

Table 5. Results of a GLM model describing the relationship between the number of maternal galleries of *Ips cembrae* per dm^2 and the following variables in experiment 1: position on the log (top, middle, or bottom = included in Intercept); diameter. Corrected Akaike's Information Criterion (AIC) = 330.33. Number of Fisher Scoring iterations was 4. *** and ** indicate significance at $p < 0.001$ and 0.01, respectively.

Variable	Estimate	Std. Error	z Value	*p* Value
(Intercept)	−2.01	0.36960	−5.428	<0.001 ***
top	0.45	0.27417	1.657	0.09758
middle	0.14	0.28219	0.521	0.60220
diameter	0.08	0.02953	2.709	0.00675 **

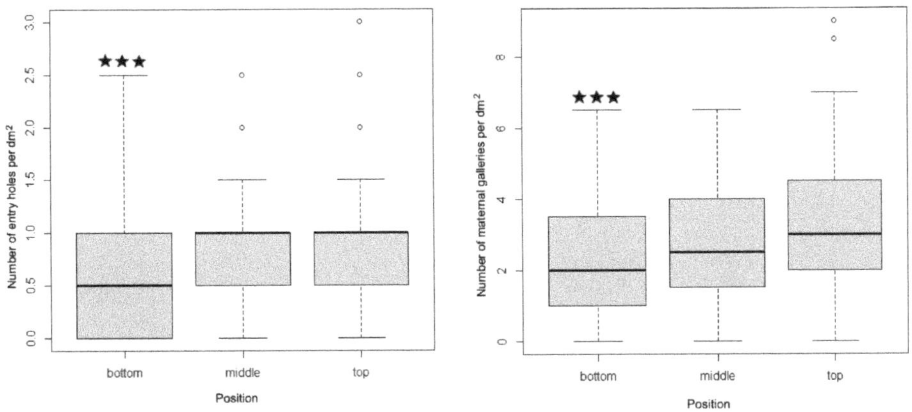

(a) (b)

Figure 5. Number of entry holes (**a**) and maternal galleries (**b**) at top, middle, and bottom positions of the logs in experiment 1. The boxes indicate medians and interquartile range, the whiskers indicate minimum and maximum values, and circles indicate outliers. *** indicate significance at $p < 0.001$.

3.2. Experiment 2: The Abundance of *I. cembrae* in Unbaited Pyramid-Trap Piles vs. Trap Trees

The number of *I. cembrae* entry holes averaged 0.6 ± 0.8 per dm^2 in trap trees and 0.4 ± 0.6 per dm^2 in pyramid-trap piles. The numbers of entry holes were not significantly affected by trap type, diameter of wood, or volume of wood according to a Wilcoxon matched-pairs signed ranks test ($z = 1.65, p > 0.05$; Figure 6) and to a GLM analysis, which also showed no effect of year.

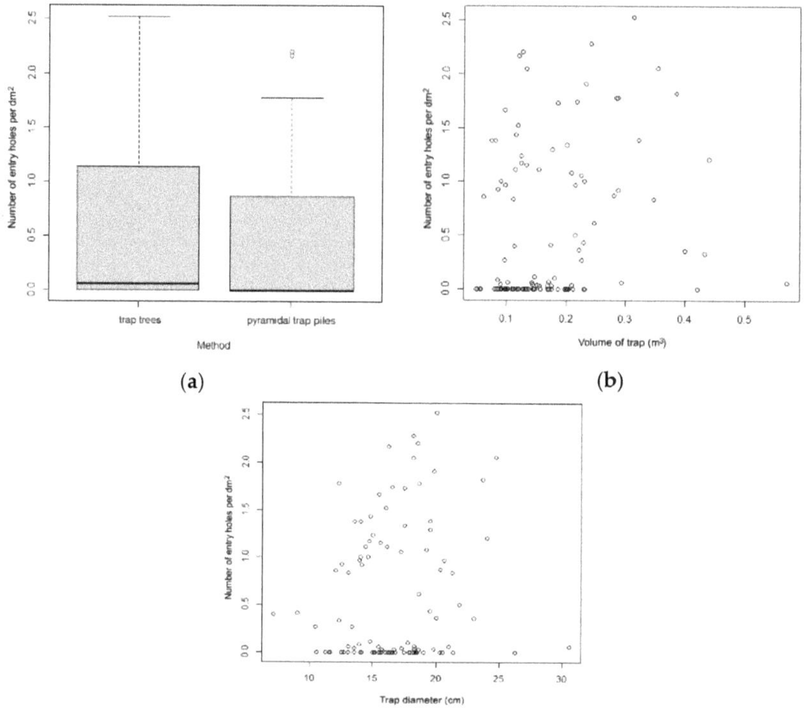

(a) (b) (c)

Figure 6. Number of entry holes as affected by trap type (trap trees vs. pyramid-trap piles) (**a**), wood volume (**b**), and wood diameter (**c**) in experiment 2. For trap type, the boxes indicate medians and interquartile range, the whiskers indicate minimum and maximum values, and circles indicate outliers.

3.3. Experiment 3: Number of *I. cembrae* Captured as Affected by Wood Volume of Trap Trees

In experiment 3, a total of 62 trap trees were analyzed at six study sites at altitudes ranging from 390 to 680 m a.s.l. A total of 257,010 *I. cembrae* were trapped, with a mean of 5039 (±5887) *I. cembrae* per trap tree.

The presented Poisson Regression model indicated that the number of *I. cembrae* per trap decreased with trap tree volume (Table 6, Figure 7).

Table 6. Results of a GLM model describing the relationship between the number of *Ips cembrae* per trap tree and the volume of wood per trap (experiment 3). Corrected Akaike's information criterion (AIC) = 312953. *** indicates significance at $p < 0.001$.

Variable	Estimate	Std. Error	z Value	*p* Value
(Intercept)	9.03	0.00398	2268.5	<0.001 ***
volume	−1.11	0.00705	−156.9	<0.001 ***

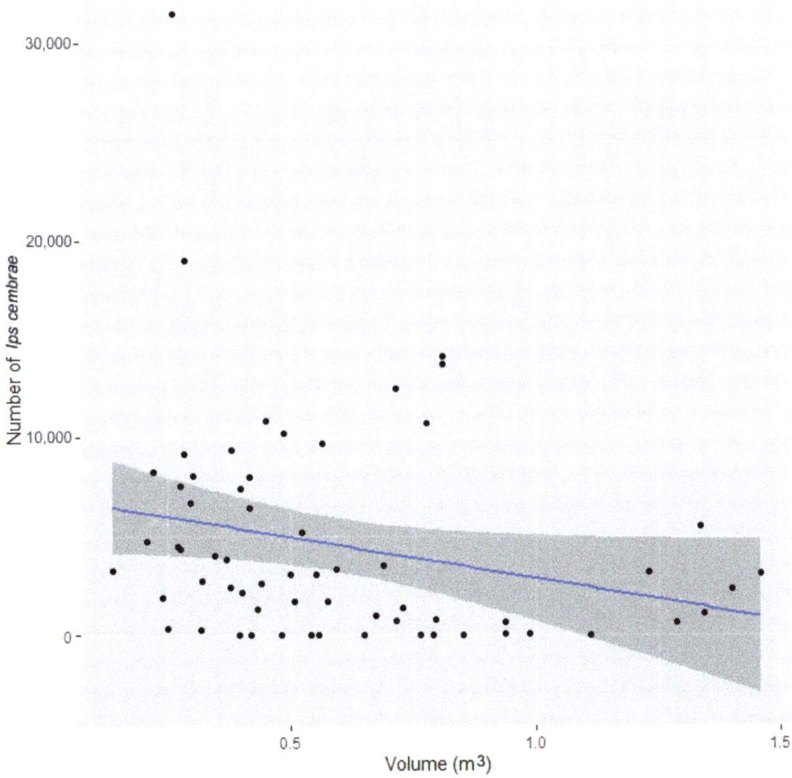

Figure 7. Relationships between the number of *I. cembrae* (points) per trap tree and wood volume per trap tree in experiment 3.

3.4. Experiment 4: Numbers of *I. cembrae* Beetles Captured by Baited Tripods vs. Pheromone Traps

A total of 13,617 *I. cembrae* were captured by baited tripods, and 10,789 were captured by pheromone traps. The sex ratio (males to females) was on average 1.5 for tripods and 0.9 for pheromone traps (Table 7). Although more individuals tended to be caught by tripods than by pheromone traps at all study sites (Table 7), the difference was not statistically significant (two-sample *t*-test: $t = 1.62$, $p > 0.05$).

Table 7. Numbers of *I. cembrae* captured by pheromone traps and tripods in experiment 4.

Capture Methods and Variables/Study Site	S. Skalice	Raduň	Žehrov	Nouzov
Tripods—total number captured	1655	3813	1200	6949
Number of beetles identified as males	490	483	606	666
Number of beetles identified as females	395	197	594	486
Mean number ± SD per trap	331 ± 269	763 ± 91	240 ± 44	1390 ± 916
Sex ratio	1.2	2.5	1.0	1.4
Pheromone traps—total number captured	340	3293	1038	6118
Number of males	144	356	353	437
Number of females	196	244	685	573
Mean ± SD per trap	68 ± 22	659 ± 135	208 ± 62	1224 ± 87
Sex ratio	0.7	1.5	0.5	0.8

Significantly more males were captured by pheromone-baited tripods than by pheromone traps (two-sample *t*-test: $t = 6.38$, $p < 0.00001$). The number of females captured did not significantly differ between the two kinds of traps (two-sample t-test: $t = -0.13$, $p > 0.05$) (Table 7, Figure 8).

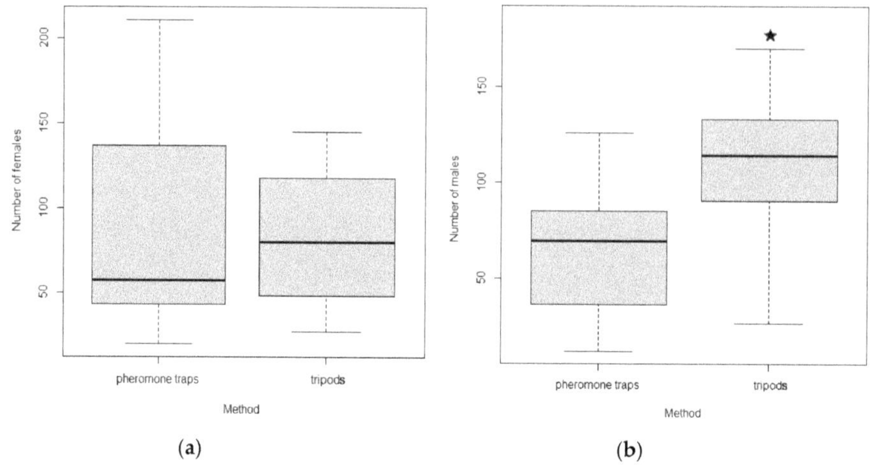

Figure 8. Number *I. cembrae* females (**a**) and males (**b**) captured in pheromone traps and tripods in experiment 4. The boxes indicate medians and interquartile range, and the whiskers indicate minimum and maximum values. * indicate significance at $p < 0.05$.

3.5. Experiment 5: Number of I. cembrae Captured by Trap Trees, Baited Tripods, and Pheromone Traps

More *I. cembrae* were captured with trap trees than with baited tripods or pheromone traps except at the Dolní Babákov study site (Table 8, Figure 9). The total number of *I. cembrae* captured was 5432 for tripods, 5676 for pheromone traps, and 91,364 for trap trees. On average, trap trees captured 16 times more individuals than pheromone traps and 17 times more individuals than tripods during *I. cembrae* flight activity (Table 8). The number of trapped beetles was significantly higher with tree traps than with pheromone traps or tripods according to Friedman's test (ANOVA) ((N = 18, df = 2) = 8.33; $p < 0.05$). The difference between tripods and pheromone traps was minimal (Table 9).

Table 8. Numbers of *I. cembrae* captured in baited tripod traps, pheromone traps, and trap trees in experiment 5.

Capture Methods and Variables/Study Site	Dolní Babákov	Húzová	Jiřetice u N.	Rožmberk
Tripods—total number captured	1881	810	1029	1712
Mean ± SD per trap	479 ± 147	270 ± 24	205 ± 137	285 ± 196
Pheromone traps—total number captured	1819	1037	2348	472
Mean ± SD per trap	455 ± 79	346 ± 154	470 ± 361	79 ± 24
Trap trees—total number captured	8	58,555	17,260	15,541
Mean ± SD per trap	2 ± 2	19,518 ± 9588	3452 ± 3121	2590 ± 1669

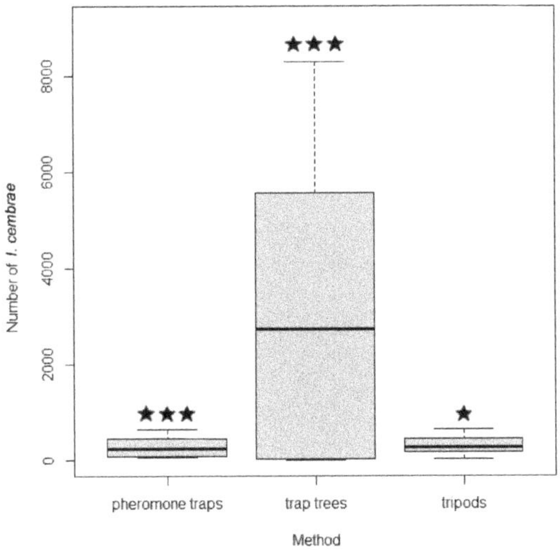

Figure 9. Numbers of *I. cembrae* captured by pheromone traps, trap trees, and tripods in experiment 5. The boxes indicate medians and interquartile range, and the whiskers indicate minimum and maximum values. *** and * indicate significance at $p < 0.001$ and 0.05, respectively.

Table 9. Results of a generalized linear mixed model (GLMM) describing the relationship between the number of *Ips cembrae* beetles per trap and trap type method and study site as random effect in experiment 5. Corrected Akaike's information criterion (AIC) = 67330. *** and * indicate significance at $p < 0.001$ and 0.05, respectively.

Variable	Estimate	Std. Error	z Value	*p* Value
(Intercept)	5.37	0.54996	9.767	<0.001 ***
trap trees	2.78	0.01368	203.126	<0.001 ***
baited tripods	−0.04	0.01898	−2.315	0.0206 *

4. Discussion

In the current research, we compared the ability of several kinds of traps to capture *I. cembrae*. For traps that use wood, we also assessed the effects of the following variables on capture of *I. cembrae*: the volume of wood in the trap, the position on the wood relative to sun exposure, and the position on the wood relative to trunk height or the soil surface. In addition to assessing several methods that use pheromone lures, we assessed traditional trap trees and trap logs without pheromone lures.

Our results indicate that larch trap trees performed better than the other methods for the capture of *I. cembrae*, apparently because the beetle responds quickly to this resource and is attracted to wood regardless of its original location along the trunk. In contrast to other bark beetles, *I. cembrae* does not have a strong preference for a certain part of the tree, although some authors have noted a preference for relatively large-diameter trees [38]. As a control measure against *I. cembrae* outbreak, it is possible to choose trap trees of low volume within a given stand (experiment 3). This is inconsistent with the hypothesis that the effectiveness of spruce trap trees could depend on the size of the trees because tree size could determine the amount of host-specific substances released [39]. Unlike trap trees deployed against spruce bark beetles [30], larch trap trees deployed to control *I. cembrae* should not be covered with branches that lack needles in March.

Our results regarding the effects of sun exposure and position (experiment 1) provide strong evidence that *I. cembrae* attacks logs with a thickness > 3 cm, which confirms a similar study [13] and that *I. cembrae* may attack young larch stands between 2 and 18 years of age [7,25,40] and branches during maturation feeding [25]. *I. cembrae* beetles probably fly at many heights above the soil, but in some cases, they apparently prefer to infest higher in the crown than in the lower sections [13]. The latter result was supported by the current results, which indicated a preference for higher positions on the traps logs. The results of experiment 1 indicate that both sexes showed no side (sunny vs. shady) preference when creating a gallery system on trap logs. One reason for the lack of preference by beetles might be that the phloem of larch trees exhibits similar defense reactions against *I. cembrae* regardless of the position along the trunk [41].

In the current research, we were interested in determining whether natural logs were more attractive to *I. cembrae* if they were positioned vertically rather than horizontally. For example, our comparison of trap trees and pyramid-trap piles indicated whether capture would be different when the same source material differed in arrangement. Trap trees should be more similar to the natural host resources used by bark beetles because they have a larger surface area than pyramid-trap pile. According to a previous study, trap logs, regardless of their arrangement, capture similar numbers of beetles whether they have or do not have a pheromone lure [42]. The latter study also found that, among trap log arrangements, crossed trap logs and pyramid-pile traps (as used in experiment 2) were the most effective at capturing *I. cembrae*. With pyramid-pile traps, *I. cembrae* produced substantially more entry holes per dm^2 in the study by [42] than in the current study. Our results (experiment 2) confirm that the addition of pheromone lures to trap logs, pyramid-trap piles, or trap trees is useless, because fresh wood is sufficiently attractive to *I. cembrae* (see also [42]). Baited and unbaited traps composed of larch logs have similar attractiveness because the traps look like a tree or its parts, or because they release attractive volatiles. The high number of beetles captured in the baited traps in experiments 4 and 5 confirm that pheromone lures attract *I. cembrae* but are more useful for monitoring *I. cembrae* in the forest (i.e., for determining the timing and peak of flight activity) than for defending against infestation [42]. A potential disadvantage of using pheromone traps for mass trapping is that they may attract beetles from long distances and thereby increase *I. cembrae* numbers in areas where the beetles' abundance was previously low, as has been reported for other bark beetle species [43]. Even though pheromone-baited traps captured large numbers of beetles, enough beetles would remain untrapped to colonize susceptible hosts, quickly replenishing the population density due to low intraspecific competition. In agreement with other researchers [44,45], we recommend using pheromone traps for detection in locations where the occurrence of *I. cembrae* has not been confirmed.

Both males and females respond to *I. cembrae* aggregation pheromones [46,47]. In the current study, the female:male sex ratio in pheromone traps fluctuated between 1 and 2, but the frequency of female and male beetles in traps was not significantly different [34]. *I. cembrae* was attracted more to tripods than to pheromone traps in experiment 4 of the current study, which differs from the results previously obtained for *I. duplicatus* (Sahlberg, 1836) and *I. typographus* [48,49]. The significantly greater number of males captured in tripods than in pheromone traps in experiment 4 was consistent with previous reports for other species of bark beetles in the genus *Ips* [48–50]. Males or females establishing gallery systems of bark beetles and ambrosia beetles are usually attracted in greater numbers to traps with visual stimuli that resemble tree trunk silhouettes than to traps baited only with host odors or pheromones [43,51–54].

Pheromone-baited tripod traps that have been sprayed with insecticide are effective against *I. cembrae* because such traps attract males, and it is males that establish the gallery systems. The possibility of underestimating the capture of *I. cembrae* on poisoned tripods because of a sublethal effect has been ruled out by a number of laboratory and field observations that used lambda-cyhalothrin [55] or alpha-cypermethrin as an insecticide [56,57], as was the case in our study.

The use of baited tripods is thus more effective than pheromone traps in monitoring *I. cembrae*, but due to the complexity of baited tripod installation, it is up to the forest managers to decide which trap to use. On the other hand, among the best defenses in addition to the timely detection and removal of bark beetle-infested trees is the use of trap trees. In studies comparing trap trees and baited tripods for monitoring *I. typographus*, trap trees were found to be more effective for all generations [58] or at least for the overwintering generation of *I. typographus* [56].

A previous study found that the area of the trapping surface of commercial traps was insufficient for optimal trapping efficiency [59]. Because a trap tree has a larger trapping surface than a commercial trap, trapping efficiency should be greater with trap trees than with commercial traps. When the population density of *I. cembrae* is low, however, the aggregation pheromones seem to be especially effective in attracting the beetles that initially fly into a site, such that baited and poisoned tripods and pheromone traps may be more effective than trap trees early in the season, as appeared to be the case at the Dolní Babákov study site in experiment 5 (Table 8).

The results of experiment 5 showed that, except at the the Dolní Babákov study site (where the population density of *I. cembrae* was probably low), trap trees captured at least 15 times more beetles than pheromone traps or baited poisoned tripods. This difference in efficacy was even greater than that previously reported for comparisons of the trapping of spruce bark beetles by spruce trap trees and artificial traps [31,60]. Similarly, standing spruce trap trees baited with pheromones and treated with insecticides caught up to 30 times more beetles than pheromone traps [61]. In contrast to our results with baited tripods vs. pheromone traps, the number of *I. cembrae* captured did not significantly differ between standing trap logs (1–3 m long) vs. pheromone traps in a previous study [32]. The use of short logs rather than complete trees seems to greatly reduce the treated bark surface and consequently the natural attractiveness. Because the amount of host volatiles decreases with the surface area of the trap trees, the surface area of trap trees greatly influences their effectiveness [32].

5. Conclusions

As is the case for control of other bark beetles, control of *I. cembrae* requires the ongoing and timely search for infested trees or logging residues and their timely elimination, or the processing of all materials suitable for bark beetle reproduction. Traditional trap trees of relatively small size are the most effective tool for *I. cembrae* capture. Unbaited tripods or pyramid-trap piles can be used, but they are less effective than trap trees. Unbaited tripods and pyramid-trap piles should be placed in sunlit locations near the forest edge and should be raised above the ground (to increase the area suitable for infestation). Baited tripods and pheromone traps are suitable for monitoring *I. cembrae* in stands, ports, or landfills with imported timber, with tripods being more efficient because they capture more males than pheromone traps. The pheromone trapping is effective and useful, especially in areas with a high proportion of larch, where early detection of *I. cembrae* flight activity is required. However, the detection of *I. cembrae* presence with pheromone traps is unnecessary if the beetle is already known to be present in a stand. Similarly, the detection of *I. cembrae* flight activity with pheromone traps is unnecessary because *I. cembrae* does not have clear peaks of flight activity—i.e., its adults fly continuously during the growing season. If *I. cembrae* is known to be present in a stand, all larch material that is infested or that is likely to be infested should be removed and processed.

Author Contributions: K.R., J.H., and E.K. provided methodology of research, collected field and laboratory data, supervised the research, and edited the manuscript; P.S. and J.T. performed the statistical analyses and the visualizations. All authors have read and agreed to the published version of the manuscript.

Funding: This research was funded by the Ministry of Agriculture of the Czech Republic, grant number QK1920433.

Acknowledgments: The authors thank Bruce Jaffee (USA) for linguistic and editorial improvements, and Jakub Zounar, Nikola Bohatá, Jiří Doskočil, Oldřich Housa, Jiří Bidmon, Radka Melecká, Richard Hurný, and Markéta Vand'urková for help with the field work.

Conflicts of Interest: The authors declare no conflict of interest.

References

1. Pâques, L.; Foffová, E.; Heinze, B.; Lelu-Walter, M.-A.; Liesebach, M.; Philippe, G. Larches (*Larix* sp.). In *Forest Tree Breeding in Europe. Managing Forest Ecosystems*; Springer: Dordrecht, The Netherlands, 2013; Volume 25.
2. Matras, J.; Pâques, L. *EUFORGEN Technical Guidelines for Genetic Conservation and Use for European Larch (Larix Decidua)*; Bioversity International: Rome, Italy, 2008.
3. Hlásny, T.; Krokene, P.; Liebhold, A.; Montagné-Huck, C.; Müller, J.; Qin, H.; Raffa, K.; Schelhaas, M.-J.; Seidl, R.; Svoboda, M.; et al. *Living with Bark Beetles: Impacts, Outlook and Management Options*; From Science to Policy 8; European Forest Institute: Joensuu, Finland, 2019; ISBN 9789525980752.
4. Grégoire, J.-C.; Evans, H.F. Damage and Control of BAWBILT Organisms—an Overview. In *Bark and Wood Boring Insects in Living Trees in Europe, a Synthesis*, 1st ed.; Lieutier, F., Day, K.R., Battisti, A., Grégoire, J.-C., Evans, H.F., Eds.; Springer: Dordrecht, The Netherlands, 2004; ISBN 978-1-4020-2240-1.
5. Krehan, T.L.; Cech, H. Larch damage in Upper Styria. An example of the complex effects of damage agents. *Forstsch. Aktuell* **2004**, *32*, 4–8.
6. Grodzki, W. *Ips cembrae* Heer. (Col.: Curculionidae, Scolytinae) in young larch stands—a new problem in Poland. *Forstsch. Aktuell* **2008**, *44*, 8–9.
7. Grodzki, W. The larch bark beetle *Ips cembrae* (Heer) (Coleoptera, Curculionidae, Scolytinae) in young and older larch stands of southern Poland. *Leś. Pr. Bad.* **2009**, *70*, 355–361. [CrossRef]
8. Pffefer, A.; Knížek, M. Coleoptera: Curculionoidea 2 (Scolytidae and Platypodidae). *Folia Fac. Sci. Nat. Univ. Masaryk. Brun. Biol.* **1996**, *94*, 601–607.
9. Grodzki, W.; Kosibowicz, M. Materiały do poznania biologii kornika modrzewiowca *Ips cembrae* (Heer) (Col., Curculionidae, Scolytinae) w warunkach południowej Polski. *Sylwan* **2009**, *153*, 587–593.
10. Postner, M. *Die Forstschädlinge Europas. II. Band. Käfer. Ips Cembrae*; Schwenke, W., Ed.; Paul Parey: Hamburg, Germany, 1974.
11. Schwerdtfeger, F. *Die Waldkrankheiten: Lehrbuch der Forstpathologie und des Forstschutzes*, 4th ed.; Paul Parey: Hamburg/Berlin, Germany, 1981.
12. Grodzki, W. On the vertical distribution of *Ips duplicatus*, *I. cembrae* and some bark- and longhorn beetles (col.: Curculionidae, Scolytinae; col.: Cerambycidae) in the Tatra national Park in Poland. *Folia For. Pol. Ser. A* **2020**, *62*, 68–77. [CrossRef]
13. Holuša, J.; Kula, E.; Wewiora, F.; Lukášová, K. Flight activity, within the trap tree abundance and overwintering of the larch bark beetle (*Ips cembrae*) in Czech Republic. *Šumar. List* **2014**, *138*, 19–27.
14. OEPP/EPPO *Ips cembrae* and *Ips subelongatus*. *EPPO Bull.* **2005**, *35*, 445–449. [CrossRef]
15. Bevan, D. *Forest Insects. A Guide to Insects Feeding on Trees in Britain. Forestry Commission Handbook No. 1*; Forestry Commission: London, UK, 1987.
16. Knížek, M. Lýkožrout modřínový. *Lesn. Pr.* **2006**, *85*, 1–4.
17. Trenberth, K.E.; Dai, A.; van der Schrier, G.; Jones, P.D.; Barichivich, J.; Briffa, K.R.; Sheffield, J. Global warming and changes in drought. *Nat. Clim. Chang.* **2014**, *4*, 17–22. [CrossRef]
18. Samaniego, L.; Thober, S.; Kumar, R.; Wanders, N.; Rakovec, O.; Pan, M.; Zink, M.; Sheffield, J.; Wood, E.F.; Marx, A. Anthropogenic warming exacerbates European soil moisture droughts. *Nat. Clim. Chang.* **2018**, *8*, 421–426. [CrossRef]
19. Luitjes, J. *Ips cembrae*, a new noxious forest insect in the Netherlands. *Ned. Bosb. Tijdschr.* **1974**, *46*, 244–246.
20. Elsner, G. Relationships between cutting time in winter and breeding success of *Ips cembrae* in larch timber. *Mitt. Dtsch. Ges. allg. angew. Entomol.* **1997**, *11*, 653–657.
21. Krehan, H.; Steyer, G. Borkenkäfer-Monitoring und Borkenkäfer-kalamität 2004. *Forstsch. Aktuell* **2005**, *33*, 12–14.
22. Stauffer, C.; Kirisits, T.; Nussbaumer, C.; Pavlín, R.; Wingfield, M.J. Phylogenetic relationships between the European and Asian eight spined larch bark beetle populations (Coleoptera, Scolytidae) inferred from DNA sequences and fungal associates. *Eur. J. Entomol.* **2001**, *98*, 99–105. [CrossRef]
23. Kirisits, T. Der Große Lärchenborkenkäfer (*Ips cembrae*) als Überträger von Bläuepilzen. *Forstsch. Aktuell* **2004**, *32*, 19–20.
24. Jankowiak, R.; Strzałka, B.; Bilański, P.; Kacprzyk, M.; Lukášová, K.; Linnakoski, R.; Matwiejczuk, S.; Misztela, M.; Rossa, R. Diversity of Ophiostomatales species associated with conifer-infesting beetles in the Western Carpathians. *Eur. J. For. Res.* **2017**, *136*, 939–956. [CrossRef]

25. Jeger, M.; Bragard, C.; Caffier, D.; Candresse, T.; Chatzivassiliou, E.; Dehnen-Schmutz, K.; Gilioli, G.; Jaques Miret, J.A.; MacLeod, A.; Navajas Navarro, M.; et al. Pest categorisation of *Ips cembrae*. *EFSA J.* **2017**, *15*, e05039. [CrossRef]
26. Schremmer, F. Beobachtungen uber den Triebfrass des achtzahnigen Larchenborkenkafers (*Ips cembrae* Heer) im Wienerwald. *Z. Angew. Entomol.* **1955**, *38*, 217–233.
27. Stadelmann, G.; Bugmann, H.; Meier, F.; Wermelinger, B.; Bigler, C. Effects of salvage logging and sanitation felling on bark beetle (*Ips typographus* L.) infestations. *For. Ecol. Manag.* **2013**, *305*, 273–281. [CrossRef]
28. Fettig, C.J.; Hilszczański, J. Chapter 14—Management Strategies for Bark Beetles in Conifer Forests. In *Bark Beetles*; Vega, F.E., Hofstetter, R.W., Eds.; Academic Press: San Diego, CA, USA, 2015; pp. 555–584, ISBN 978-0-12-417156-5.
29. Grégoire, J.-C.; Raffa, K.F.; Lindgren, B.S. Chapter 15—Economics and Politics of Bark Beetles. In *Bark Beetles*; Vega, F.E., Hofstetter, R.W., Eds.; Academic Press: San Diego, CA, USA, 2015; pp. 585–613, ISBN 978-0-12-417156-5.
30. Holuša, J.; Hlásny, T.; Modlinger, R.; Lukášová, K.; Kula, E. Felled trap trees as the traditional method for bark beetle control: Can the trapping performance be increased? *For. Ecol. Manag.* **2017**, *404*, 165–173. [CrossRef]
31. Abgrall, J.-F.; Schvester, D. Observations sur le piégeage de *Ips typographus* L. après chablis. *Rev. For. Française* **1987**, *39*, 359–377. [CrossRef]
32. Faccoli, M.; Stergulc, F. Damage reduction and performance of mass trapping devices for forest protection against the spruce bark beetle, *Ips typographus* (Coleoptera Curculionidae Scolytinae). *Ann. For. Sci.* **2008**, *65*, 309. [CrossRef]
33. Watzek, G.; Niemeyer, G. Verminderung der Borkenkäfergefähr durch Harvestertechnik und Arbeitsorganisation. *Forst Holz.* **1996**, *51*, 247–250.
34. Grucmanová, Š.; Holuša, J.; Trombik, J.; Lukášová, K. Large larch bark beetle *Ips cembrae* (Coleoptera: Curculionidae, Scolytinae) in the Czech Republic: Analysis of population development and catches in pheromone traps. *For. J.* **2014**, *60*, 143–149. [CrossRef]
35. Knížek, M.; Liška, J. *Výskyt Lesních Škodlivých Činitelů v Roce 2019 a Jejich Očekávaný Stav v Roce 2020*; Výzkumný ústav Lesního Hospodářství a Myslivosti, v. v. i.: Strnady, Czechia, 2020.
36. Grodzki, W. Some reactions of *Ips typographus* (L.) (Col.: Scolytidae) to changing breeding conditions in a forest decline area in the Sudeten Mountains, Poland. *J. Pest. Sci.* **2004**, *77*, 43–48. [CrossRef]
37. Petráš, R.; Pajtík, J. Sústava česko-slovenských objemových tabuliek drevín. *Les. čas.* **1991**, *37*, 49–56.
38. Arač, K.; Pernek, M. Occurrence and spreading of the large larch bark beetle (*Ips cembrae*) in Croatia and possiblities of monitoring by using pheromone traps. *Šumar. List* **2014**, *3–4*, 145–154.
39. Abgrall, J.-F. L'utilisation de la methode des arbres-pitges avec les pheromones de synthese dam la lutte contre le typographe. *CEMAGREF Inf. Tech.* **1987**, *67*, 1–4.
40. Hutka, D. Nowe oblicze kornika modrzewiowca. *Trybuna Leś.* **2006**, *4*, 10–11.
41. Jung, P.; Rohde, M.; Lunderstädt, J. Induzierte Resistenz im Leitgewebe der europäischen *Lärche Larix decidua* (Mill) nach Befall durch den großen Lärchenborkenkäfer *Ips cembrae* (Heer) (Col., Scolytidae). *J. Appl. Entomol.* **1994**, *117*, 427–433. [CrossRef]
42. Grodzki, W. Possibility of the reduction of *Ips cembrae* (Heer) (Coleoptera, Curculionidae, Scolytinae) populations in larch stands in southern Poland. *Sylwan* **2010**, *154*, 160–167.
43. Byers, J.A. Chemical ecology of bark beetles. *Experientia* **1989**, *45*, 271–283. [CrossRef]
44. Pavlin, R. New locations of the larch bark beetle (*Ips cembrae*) in Slovenia. *Gozd. Vestnik* **1997**, *55*, 336–342.
45. Niemeyer, H. First results with a pheromone-trap system for monitoring bark beetles in Lower Saxony and Schleswig-Holstein. *Forst Holz.* **1989**, *44*, 114–115.
46. Renwick, J.A.A.; Dickens, J.C. Control of pheromone production in the bark beetle, *Ips cembrae*. *Physiol. Entomol.* **1979**, *4*, 377–381. [CrossRef]
47. Francke, W.; Vité, J.P. Oxygenated terpenes in pheromone systems of bark beetles. *Z. Angew. Entomol.* **1983**, *96*, 146–156. [CrossRef]
48. Lubojacký, J.; Holuša, J. Comparison of spruce bark beetle (*Ips typographus*) catches between treated trap logs and pheromone traps. *Šumar. List* **2011**, *135*, 233–242.
49. Lubojacký, J.; Holuša, J. Comparison of lure-baited insecticide-treated tripod trap logs and lure-baited traps for control of *Ips duplicatus* (Coleoptera: Curculionidae). *J. Pest Sci.* **2013**, *86*, 483–489. [CrossRef]

50. Tomiczek, C. Fangtipi und Pheromonfalle: Erste Ergebnisse einer Vergleichsuntersuchung zu Fangleistung und Naturschutzaspekten. *Forstsch. Aktuell* **2009**, *48*, 6–7.
51. Lindgren, B.S.; Borden, J.H.; Chong, L.; Friskie, L.M.; Orr, D.B. Factors influencing the efficiency of pheromone-baited traps for three species of ambrosia beetles (Coleoptera: Scolytidae). *Can. Entomol.* **1983**, *115*, 303–313. [CrossRef]
52. Bombosch, S.; Engler, I.; Gossenauer, H.; Hermann, B. Über die Rolle von Pheroprax bei der Besiedlung der Fichte durch den Buchdrucker (*Ips typographus* L.)1. *Z. Angew. Entomol.* **2009**, *100*, 458–463. [CrossRef]
53. Ramisch, H. Host location by *Trypodendron domesticum* and *Trypodendron lineatum* (Coleoptera: Scolytidae). *Z. Angew. Zool.* **1986**, *73*, 159–198.
54. Chénier, J.; Philogéne, B. Field responses of certain forest Coleoptera to conifer monoterpenes and ethanol. *J. Chem. Ecol.* **1989**, *15*, 1729–1745. [CrossRef]
55. Hurling, R.; Stetter, J. Untersuchungen zur Fangleistung von Schlitzfallen und Fangholzhaufen bei der lokalen Dichteabsenkung von Buchdrucker (*Ips typographus*)-Populationen. *Gesunde Pflanz.* **2012**, *64*, 89–99. [CrossRef]
56. Lubojacký, J.; Holuša, J. Attraction of *Ips typographus* (Coleoptera: Curculionidae) beetles by lure-baited insecticide-treated tripod trap logs and trap trees. *Int. J. Pest Manag.* **2014**, *60*, 153–159. [CrossRef]
57. Lubojacký, J.; Holuša, J. Effect of insecticide-treated trap logs and lure traps for *Ips typographus* (Coleoptera: Curculionidae) management on nontarget arthropods catching in Norway spruce stands. *J. For. Sci.* **2014**, *60*, 6–11. [CrossRef]
58. Jeniš, J.; Vrba, M. Srovnání účinnosti lapáků, otrávených trojnožek a lapačů. *Lesn. pr.* **2007**, *86*, 586.
59. Vaupel, O.; Dubbel, V. Untersuchung zur Optimierung von Fangsystemen zur Verbesserung des Massenfanges beim Buchdrucker (*Ips typographus* L.). *Z. Angew. Entomol.* **1985**, *99*, 52–59. [CrossRef]
60. Drumont, A.; Gonzalez, R.; Windt, N.; Grégoire, J.-C.; de Proft, M.; Seutin, E. Semiochemicals and the integrated management of *Ips typographus* (L.) (Col., Scolytidae) in Belgium. *J. Appl. Entomol.* **1992**, *114*, 333–337. [CrossRef]
61. Raty, L.; Drumont, A.; de Windt, N.; Grégoire, J.C. Mass trapping of the spruce bark beetle *Ips typographus* L.: Traps or trap trees? *For. Ecol. Manag.* **1995**, *78*, 191–205. [CrossRef]

MDPI AG
Grosspeteranlage 5
4052 Basel
Switzerland
Tel.: +41 61 683 77 34

Forests Editorial Office
E-mail: forests@mdpi.com
www.mdpi.com/journal/forests

www.ingramcontent.com/pod-product-compliance
Lightning Source LLC
Chambersburg PA
CBHW061842210926
51850CB00006B/949